HANDBOOK OF SLEEP DISORDERS IN MEDICAL CONDITIONS

HANDBOOK OF SLEEP DISORDERS IN MEDICAL CONDITIONS

Edited by

Josée Savard
School of Psychology, Université Laval, Québec, QC, Canada
CHU de Québec-Université Laval Research Center, Québec, QC, Canada
Université Laval Cancer Research Center, Québec, QC, Canada

Marie-Christine Ouellet
School of Psychology, Université Laval, Québec, QC, Canada
Centre interdisciplinaire de recherche en réadaptation et intégration sociale, Québec, QC, Canada

Academic Press is an imprint of Elsevier
125 London Wall, London EC2Y 5AS, United Kingdom
525 B Street, Suite 1650, San Diego, CA 92101, United States
50 Hampshire Street, 5th Floor, Cambridge, MA 02139, United States
The Boulevard, Langford Lane, Kidlington, Oxford OX5 1GB, United Kingdom

Copyright © 2019 Elsevier Inc. All rights reserved.

No part of this publication may be reproduced or transmitted in any form or by any means, electronic or mechanical, including photocopying, recording, or any information storage and retrieval system, without permission in writing from the publisher. Details on how to seek permission, further information about the Publisher's permissions policies and our arrangements with organizations such as the Copyright Clearance Center and the Copyright Licensing Agency, can be found at our website: www.elsevier.com/permissions.

This book and the individual contributions contained in it are protected under copyright by the Publisher (other than as may be noted herein).

Notices

Knowledge and best practice in this field are constantly changing. As new research and experience broaden our understanding, changes in research methods, professional practices, or medical treatment may become necessary.

Practitioners and researchers must always rely on their own experience and knowledge in evaluating and using any information, methods, compounds, or experiments described herein. In using such information or methods they should be mindful of their own safety and the safety of others, including parties for whom they have a professional responsibility.

To the fullest extent of the law, neither the Publisher nor the authors, contributors, or editors, assume any liability for any injury and/or damage to persons or property as a matter of products liability, negligence or otherwise, or from any use or operation of any methods, products, instructions, or ideas contained in the material herein.

British Library Cataloguing-in-Publication Data
A catalogue record for this book is available from the British Library

Library of Congress Cataloging-in-Publication Data
A catalog record for this book is available from the Library of Congress

ISBN: 978-0-12-813014-8

For Information on all Academic Press publications
visit our website at https://www.elsevier.com/books-and-journals

Publisher: Nikki Levy
Acquisition Editor: Melanie Tucker
Editorial Project Manager: Kristi Anderson
Production Project Manager: Bharatwaj Varatharajan
Cover Designer: Christian J. Bilbow

Typeset by MPS Limited, Chennai, India

Contents

List of Contributors   ix
Preface   xiii

Section I

GENERAL ISSUES

1. Diagnostic Criteria and Assessment of Sleep Disorders

CHRISTINA JAYNE BATHGATE AND JACK D. EDINGER

Introduction   3
Current Diagnostic Classification
 Systems   5
Clinical Features and Diagnostic Considerations for
 Sleep/Wake Disorders   10
Psychological, Behavioral, and Laboratory
 Assessments to Aid Diagnosis   16
Areas for Future Research   20
Conclusion   21
References   23

2. Treatment of Insomnia

CHARLES M. MORIN, SIMON BEAULIEU-BONNEAU
AND JANET M.Y. CHEUNG

Introduction   28
Current Treatment Options   29
Summary of Outcome Evidence   38
Areas for Future Research   45
Summary and Conclusion   45
References   46

3. Treatment of Breathing-Related Disorders

ANEESA M. DAS AND BERNARDO J. SELIM

Introduction   51
Positive Airway Pressure Devices   52

Treatment of Sleep-Related Breathing Disorders by
 Disease Type   59
Areas for Future Research   70
Conclusion   70
References   71

4. Treatment of Sleep-Related Movement and Circadian Rhythm Disorders, Hypersomnolence, and Parasomnias

BRIENNE MINER, NASHEENA JIWA AND BRIAN KOO

Introduction   78
Restless Legs Syndrome and Periodic Limb
 Movement Disorder   78
Hypersomnolence   82
Circadian Rhythm Disorders   84
Parasomnias   88
Conclusion   93
References   93

Section II

SLEEP DISORDERS IN SPECIFIC MEDICAL CONDITIONS

5. Cardiovascular Disease

NANCY S. REDEKER

Introduction   100
Contributions of Sleep Deficiency to the
 Development of Cardiovascular Disease   100
Sleep Disorders and Sleep Deficiency Among
 Adults With Chronic Cardiovascular
 Disease   105
Consequences of Sleep Deficiency and Sleep
 Disorders in People With Chronic Cardiovascular
 Disease   107

vi CONTENTS

Assessment of Sleep Disturbance in People With or At-Risk for Cardiovascular Disease 108
Interventions to Improve Sleep Impairments in People With or At-Risk for Cardiovascular Disease 109
Areas for Future Research 112
Clinical Vignette 113
Conclusions 113
References 114

6. Lung Diseases

LAUREN TOBIAS AND CHRISTINE WON

Introduction 122
Respiratory Physiology During Normal Sleep 122
Sleep Disorders in Chronic Obstructive Pulmonary Disease 123
Asthma and Sleep Disorders 130
Sleep in Cystic Fibrosis 132
Sleep in Interstitial Lung Disease 134
Sleep in Musculoskeletal Disorders 136
Sleep in Chronic Respiratory Failure Due to Neuromuscular Disease 137
Obesity Hypoventilation Syndrome 139
Sleep in Pulmonary Hypertension 141
Sleep After Lung Transplantation 143
Conclusion 143
References 144

7. Obesity, Diabetes, and Metabolic Syndrome

SUNDEEP SHENOY, AZIZI SEIXAS AND MICHAEL A. GRANDNER

Introduction 154
Sleep and Overweight−Obesity 154
Sleep and Diabetes 158
Sleep and Metabolic Syndrome 161
Assessment of Sleep Disturbances 165
Treatment of Sleep Disturbances 166
Areas for Future Research 166
Conclusion 167
References 168

8. Cancer

JOSÉE SAVARD

Introduction 176
Brief Overview of the Medical Condition 176
Epidemiology of Sleep Disorders in Cancer 177

Sleep Difficulties in Special Populations 181
Etiology of Sleep Disorders in Cancer 182
Consequences of Sleep Disorders in Cancer 184
Assessment of Sleep Difficulties in Cancer 186
Treatment of Sleep Difficulties in Cancer 188
Clinical Vignette 192
Areas for Future Research 193
Conclusion 193
References 194

9. Chronic Pain

CAITLIN B. MURRAY AND TONYA M. PALERMO

Introduction 201
Sleep Deficiency in Chronic Pain Conditions 202
Sleep and Pain: Interconnection and Shared Mechanisms 205
Sleep Assessment 207
Sleep Treatment 210
Conclusions and Areas for Future Research 212
References 213

10. Traumatic Brain Injury

MARIE-CHRISTINE OUELLET, SIMON BEAULIEU-BONNEAU AND CHARLES M. MORIN

Introduction 222
Brief Overview of Traumatic Brain Injury (Prevalence, Etiology, Treatment) 223
Sleep−Wake Alterations in the Acute Phase After Traumatic Brain Injury and Impacts on Recovery 224
Sleep Architecture in the Chronic Phase After Traumatic Brain Injury 225
Pathophysiology of Sleep−Wake Disturbances After Traumatic Brain Injury 226
Potential Impacts of Sleep−Wake Disturbances on the Evolution of the Condition After Traumatic Brain Injury 228
Insomnia 231
Clinical Vignette 237
Excessive Daytime Sleepiness, Posttraumatic Hypersomnia, Sleep-Disordered Breathing, and Increased Need for Sleep 238
Circadian Rhythm Sleep−Wake Disorders 242
Areas for Future Research 243
Conclusion 243
References 244

11. Mild Cognitive Impairment and Dementia

CHENLU GAO, MICHAEL K. SCULLIN
AND DONALD L. BLIWISE

Introduction 254
Cognitive Aging 254
Sleep and Circadian Rhythm Changes With Normal
Aging, Mild Cognitive Impairment, and
Dementia 258
Sleep Disordered Breathing in Mild Cognitive
Impairment/Dementia 262
Treatment of Sleep Disorders in Mild Cognitive
Impairment and Dementia 264
Areas for Future Research 267
Conclusion 268
References 269

12. Stroke

ANNETTE STERR AND JAMES EBAJEMITO

Introduction 277
Stroke: Brief Overview 278
Sleep Disorders in Stroke 279
Sleep-Recovery Interaction Model 283
Sleep Interventions 285
Clinical Practice Considerations 286
Areas for Future Research 286
Conclusion 287
References 287

13. Human Immunodeficiency Virus/AIDS

KENNETH D. PHILLIPS, ROBIN F. HARRIS
AND LISA M. HADDAD

Introduction 294
HIV Disease 294
Sleep Disturbances and HIV Disease 294
Insomnia 296
Hypersomnia 297
Obstructive Sleep Apnea 297
Correlates of Sleep Quality in HIV/AIDS 298
Symptom Clusters in HIV/AIDS 299
Nonpharmacological Treatment for Insomnia 300
Pharmacological Treatment for Insomnia 302
Areas for Future Research 303
Conclusion 303
References 303

14. Inflammatory Arthropathies

REGINA M. TAYLOR-GJEVRE
AND JOHN A. GJEVRE

Introduction 309
Nature and Prevalence of Sleep
Disturbance/Disorders in Inflammatory
Arthritis Patients 311
Polysomnographic Studies in Inflammatory
Arthritis 316
Cytokines and Sleep 317
Biologic Therapies and Sleep in Inflammatory
Arthritis 317
Effect of CPAP on Inflammation and Pain in
Inflammatory Arthritis 318
Areas for Future Research 319
Conclusion 319
References 320
Further Reading 324

15. Chronic Fatigue Syndrome and Fibromyalgia

FUMIHARU TOGO, AKIFUMI KISHI
AND BENJAMIN H. NATEL

Introduction 326
Brief Overview of Chronic Fatigue Syndrome and
Fibromyalgia 326
Sleep Disorders and Abnormalities
in Chronic Fatigue Syndrome and
Fibromyalgia 328
Consequences of Sleep Disorders and
Abnormalities 332
Assessment 333
Treatment 333
Areas for Future Research 337
Conclusion 337
References 338

16. Multiple Sclerosis

CHRISTIAN VEAUTHIER AND FRIEDEMANN PAUL

Introduction 346
Brief Overview of Multiple Sclerosis 347
Sleep-Related Breathing Disorders 347
Insomnia 348
Restless Legs Syndrome and Periodic Limb
Movement Disorder 350
Narcolepsy and Hypersomnia 351

REM Sleep Behavior Disorder   352
Relationship of Sleep Disorders With Fatigue, Depression, Disease Course, and Disability   352
Impact and Quality of Life   353
Hypnotic Use Due to Insomnia and Its Relationship With Fatigue   353
Fatigue and Employment Status   354
Assessment of Sleep Disorders in Multiple Sclerosis   354
Treatment of Sleep Disorders in Multiple Sclerosis   355
Sleep Disorders and Fatigue in Neuromyelitis Optica Spectrum Disorders   357
Clinical Vignette   358
Areas for Future Research and Conclusion   358
References   359

17. Gastrointestinal Disorders

CLAIRE J. HAN AND MARGARET M. HEITKEMPER

Introduction   372
Common Pathophysiological Mechanisms: Sleep and Gastrointestinal Tract   372
Epidemiology and Risk Factors of Sleep Disturbances With Gastrointestinal Disorders   374
Assessment of Sleep Disorders in Gastrointestinal Disorders   379
Management and Prevention of Sleep Disturbances in Persons With Gastrointestinal Disorders   380
Clinical Vignette   384
Areas for Future Research   385
Conclusion   385
References   386

18. Sleep in Pediatric Patients

TERYN BRUNI, EMMA GILL AND DAWN DORE-STITES

Introduction   391
Pediatric Sleep Patterns   392
Consequences of Insufficient Sleep in Children and Adolescents   392
Sleep in Chronic Health Condition: Assessment and Intervention   399
Areas for Future Research   405
Conclusion   405
References   406

19. Sleep in Hospitalized Patients

MELISSA P. KNAUERT AND MARGARET A. PISANI

Introduction   411
Sleep Disruption in Hospitalized Patients   412
Environmental Factors Affecting Sleep in Hospitalized Patients   415
Preexisting Sleep Disorders and Poor Hospital Sleep   417
Risk Factors for Poor Sleep in Hospital Settings   418
Assessment of Sleep in Hospitalized Patients   421
Protocols to Improve Sleep in Hospitalized Patients   423
Medications to Treat Sleep Disruption in the Hospital   425
Clinical Vignette   428
Areas for Future Research   429
Conclusion   429
References   429

Index   439

List of Contributors

Christina Jayne Bathgate Department of Medicine, National Jewish Health, Denver, CO, United States

Simon Beaulieu-Bonneau Center for Interdisciplinary Research in Rehabilitation and Social Integration (CIRRIS), Québec, QC, Canada; School of Psychology, Université Laval, Québec, QC, Canada; CERVO Brain Research Centre, Québec, QC, Canada

Donald L. Bliwise Department of Neurology, Emory University, Atlanta, GA, United States

Teryn Bruni Department of Pediatrics, Michigan Medicine, Ann Arbor, MI, United States

Janet M.Y. Cheung School of Psychology, Université Laval, Québec, QC, Canada; School of Pharmacy, Faculty of Medicine and Health, The University of Sydney, Sydney, Australia

Aneesa M. Das Division of Pulmonary, Critical Care and Sleep Medicine, Ohio State University, Columbus, OH, United States

Dawn Dore-Stites Department of Pediatrics, Michigan Medicine, Ann Arbor, MI, United States

James Ebajemito School of Psychology, University of Surrey, Guildford, United Kingdom

Jack D. Edinger Department of Medicine, National Jewish Health, Denver, CO, United States

Chenlu Gao Department of Psychology and Neuroscience, Baylor University, Waco, TX, United States

Emma Gill University of Michigan, Ann Arbor, MI, United States

John A. Gjevre Division of Respiratory and Sleep Medicine, Department of Medicine, University of Saskatchewan, Saskatoon, SK, Canada

Michael A. Grandner Sleep and Health Research Program, Department of Psychiatry, University of Arizona, Tucson, AZ, United States

Lisa M. Haddad College of Nursing, East Tennessee State University, Johnson City, TN, United States

Claire J. Han School of Medicine, Biomedical Informatics and Medical Education, University of Washington, Seattle, WA, United States; Department of Biobehavioral Nursing and Health Informatics, School of Nursing, University of Washington, Seattle, WA, United States

Robin F. Harris College of Nursing, University of Tennessee, Knoxville, TN, United States

Margaret M. Heitkemper Department of Biobehavioral Nursing and Health Informatics, School of Nursing, University of Washington, Seattle, WA, United States; Department of Nursing, Center for Innovations in Sleep Self-Management, University of Washington, Seattle, WA, United States; Division of Gastroenterology, School of Medicine, University of Washington, Seattle, WA, United States

Nasheena Jiwa Internal Medicine, St. Mary's Hospital, Waterbury, CT, United States

Akifumi Kishi Educational Physiology Laboratory, Graduate School of Education, The University of Tokyo, Tokyo, Japan

Melissa P. Knauert Internal Medicine, Pulmonary, Critical Care, and Sleep Medicine, Yale University School of Medicine, New Haven, CT, United States

Brian Koo Neurology, Sleep Medicine Service, VA Connecticut Healthcare System, West Haven, CT, United States; Yale Center for Restless Legs Syndrome, New Haven, CT, United States; Yale University School of Medicine, New Haven, CT, United States

Brienne Miner Geriatrics and Sleep Medicine, Yale University School of Medicine, New Haven, CT, United States

Charles M. Morin School of Psychology, Université Laval, Québec, QC, Canada; CERVO Brain Research Centre, Québec, QC, Canada

Caitlin B. Murray Center for Child Health, Behavior, and Development, Seattle Children's Research Institute, Seattle, WA, United States

Benjamin H. Natelson Department of Neurology, Icahn School of Medicine at Mount Sinai, New York, NY, United States

Marie-Christine Ouellet Center for Interdisciplinary Research in Rehabilitation and Social Integration (CIRRIS), Québec, QC, Canada; School of Psychology, Université Laval, Québec, QC, Canada

Tonya M. Palermo Center for Child Health, Behavior, and Development, Seattle Children's Research Institute, Seattle, WA, United States; Department of Anesthesiology & Pain Medicine, University of Washington School of Medicine, Seattle, WA, United States

Friedemann Paul Charité—Universitätsmedizin Berlin, Corporate Member of Freie Universität Berlin, Humboldt-Universität zu Berlin, and Berlin Institute of Health, NeuroCure Clinical Research Center, Berlin, Germany

Kenneth D. Phillips College of Nursing, East Tennessee State University, Johnson City, TN, United States

Margaret A. Pisani Internal Medicine, Pulmonary, Critical Care, and Sleep Medicine, Yale University School of Medicine, New Haven, CT, United States

Nancy S. Redeker Beatrice Renfield Term Professor of Nursing, Yale University School of Nursing, West Haven, CT, United States

Josée Savard School of Psychology, Université Laval, Québec, QC, Canada; CHU de Québec-Université Laval Research Center, Québec, QC, Canada; Université Laval Cancer Research Center, Québec, QC, Canada

Michael K. Scullin Department of Psychology and Neuroscience, Baylor University, Waco, TX, United States

Azizi Seixas Department of Population Health and Psychiatry, New York University Langone Health, New York, NY, United States

Bernardo J. Selim Division of Pulmonary, Critical Care and Sleep Medicine, Mayo Clinic, Rochester, MN, United States

Sundeep Shenoy Division of Cardiovascular Medicine, University of Virginia, Charlottesville, VA, United States

Annette Sterr School of Psychology, University of Surrey, Guildford, United Kingdom

Regina M. Taylor-Gjevre Division of Rheumatology, Department of Medicine, University of Saskatchewan, Saskatoon, SK, Canada

Lauren Tobias Department of Internal Medicine, Yale University School of Medicine, New Haven, CT, United States

Fumiharu Togo Educational Physiology Laboratory, Graduate School of Education, The University of Tokyo, Tokyo, Japan

Christian Veauthier Charité—Universitätsmedizin Berlin, Corporate Member of Freie Universität Berlin, Humboldt-Universität zu Berlin, and Berlin Institute of Health, Interdisciplinary Center for Sleep Medicine, Berlin, Germany

Christine Won Yale University School of Medicine, Yale Centers for Sleep Medicine, New Haven, CT, United States

Preface

Sleep research is a very active field. As a consequence, the physiology of sleep and the nature, etiology, and multiple consequences of sleep disorders are increasingly understood. Much less research however has focused on how sleep interacts with the presence of medical conditions, particularly chronic illnesses. With the aging of the population, the prevalence of the latter is dramatically rising. In 2012, it was estimated that about half of US American adults (i.e., 117 million people) had at least one chronic disease and that one in four had two or more.[1] The Center for Disease Control and Prevention (United States) estimated that 7 of the 10 leading causes of death in 2014 were chronic diseases, with 2 explaining nearly 46% of all deaths, that is heart disease and cancer.[2] Sleep disorders are highly prevalent in chronic illnesses including cancer, diabetes, cardiovascular disease, and neurological conditions. Some sleep disorders contribute directly to the incidence of the medical disease, such as obstructive sleep apnea increasing the risk of cardiovascular disease, or insomnia increasing the risk for rheumatoid arthritis or cardiometabolic disease. In many cases, however, sleep disorders appear during the trajectory of an illness and can take various forms. For example, patients can suffer from insomnia, sleep apnea, restless legs syndrome, and hypersomnolence.

Sleep disorders may affect the course of the illness, its treatment, as well as patients' adherence to medical recommendations. For instance, insomnia increases the risk of developing infectious disease during cancer care trajectory. Furthermore, the decrease in quality of life already associated with conditions such as chronic pain or chronic fatigue can be worsened by the presence of sleep problems which add onto the burden of diseases by exacerbating the main symptoms or by increasing the risk of anxiety or depression. In patients with neurological conditions such as stroke or traumatic brain injury, sleep disorders have been shown to decrease the patient's participation in rehabilitation. In the field of HIV, patients with sleep disorders seem to be less adherent to antiretroviral therapy. Sleep disturbances of patients may also affect the sleep of their caregivers, for example, parents of children with a chronic disease.

In turn, treating sleep disorders may be beneficial in the management of the medical condition. For example, treating insomnia in persons with cardiovascular disease has been shown to reduce rates of rehospitalization and have a positive impact on biomarkers of cardiovascular risk. In multiple sclerosis (MS), treating sleep disorders can have a positive impact on MS-related fatigue. In the context of dementia, treating sleep apnea is associated with improvement in general cognition, and providing education on sleep hygiene can lead to improved sleep in both patients and their caregivers, thus optimizing the care of these patients in the long run.

Although screening, assessment, and treatment of sleep disorders in various medical populations is increasingly recognized as critical, these services are unfortunately not consistently provided to patients. The diagnosis and differential diagnosis of sleep disorders can be more complex in the presence of medical comorbidity. For example, measurement of sleep quality can be arduous during hospitalization or in patients with severe cognitive problems. Furthermore, symptoms of the illness often overlap or are confounded with those of the sleep disturbance. In terms of treatment, although the general principles of sleep medicine may often apply, an adapted approach is warranted in many instances. For example, some pharmacological agents are known to exacerbate symptoms of the medical condition, including memory issues in persons with neurological conditions, gastrointestinal symptoms in those with gastrointestinal disorders, or worsening of sleep-disordered breathing in patients with lung disease. Similarly, nonpharmacological interventions such as cognitive-behavioral therapy for insomnia or exercise often need adaptations to ensure their maximal efficacy in patients with specific medical conditions.

The association of sleep disorders with chronic medical conditions is increasingly recognized and investigated. However, the magnitude of the body of evidence and the level of knowledge on how the presence of medical disorders affects sleep is variable across conditions. As an example, the impact of obstructive sleep apnea on the incidence of cardiovascular disease has received a great deal of attention, while research on sleep disorders in other medical conditions is more in its early stages. Furthermore, the literature on sleep difficulties comorbid with a medical illness has evolved quite independently from one health condition to the other.

In this handbook, we aim to bring together and review the current state of knowledge about the nature and manifestations of sleep disorders associated with a variety of common medical conditions. Furthermore, the handbook provides insights regarding special considerations for assessment or treatment, both pharmacological and nonpharmacological, in specific medical conditions. The book is divided into two sections.

Section I provides general information on the diagnostic criteria and assessment of the sleep disorders that are most likely to be encountered in medical settings including insomnia, sleep apnea, restless legs syndrome, and hypersomnolence (Chapter 1). Subsequently, pharmacological, medical, and psychological interventions for sleep disorders in general are presented, as well as the level of evidence supporting their efficacy. More specifically, Chapter 2 covers the treatment of insomnia, Chapter 3 describes interventions for breathing-related disorders, and Chapter 4 reviews treatment options for the other common sleep disorders, that is restless legs syndrome, periodic limb movement disorder, hypersomnolence disorders, circadian rhythm sleep disorders, and parasomnias.

In Section II, readers will find more specific information for particular chronic medical conditions, for example, cardiovascular disease, lung disease, obesity, diabetes, cancer, chronic pain, traumatic brain injury, dementia, stroke, HIV, inflammatory arthropathies, chronic fatigue syndrome, MS, and gastrointestinal disorders. Two additional chapters cover sleep disorders in pediatric patients with psychiatric and medical conditions and sleep in hospitalized patients. Each chapter reviews information on epidemiological aspects, risk factors and correlates, influence of sleep issues on the course of illness and treatment, consequences for patients, and special considerations regarding evaluation and treatment.

Taking into consideration sleep in the global management of patients with medical conditions is essential. This handbook provides guidance to diagnose and treat sleep disorders in a manner that is appropriately adapted to the comorbid medical condition. By identifying the significant knowledge gaps, it is also hoped that this handbook will urge more research in these various fields.

References

1. Ward BW, Schiller JS, Goodman RA. Multiple chronic conditions among US adults: a 2012 update. *Prev Chronic Dis.* 2014;11:E62.
2. Centers for Disease Control and Prevention. *Leading causes of death and numbers of deaths, by sex, race, and Hispanic origin: United States, 1980 and 2014 (Table 19).* Health. <https://www.cdc.gov/nchs/data/hus/hus15.pdf#019>; 2015 [PDF – 13.4 MB] Accessed 04.07.17.

SECTION I

GENERAL ISSUES

CHAPTER

1

Diagnostic Criteria and Assessment of Sleep Disorders

Christina Jayne Bathgate and Jack D. Edinger

Department of Medicine, National Jewish Health, Denver, CO, United States

OUTLINE

Introduction	3	Psychological, Behavioral, and Laboratory Assessments to Aid Diagnosis	16
Current Diagnostic Classification Systems	5	Clinical History	16
		Daily Self-Monitoring	17
Clinical Features and Diagnostic Considerations for Sleep/Wake Disorders	10	Self-Report Questionnaires	17
		Actigraphy	17
Insomnia	10	Polysomnography	19
Hypersomnolence	11	Multiple Sleep Latency Test	20
Narcolepsy	12	Maintenance of Wakefulness Test	20
Breathing-Related Sleep Disorders	13	Areas for Future Research	20
Circadian Rhythm Sleep—Wake Disorders	14	Conclusion	21
Parasomnias	14	References	23
Movement-Related Sleep Disorders	16		

INTRODUCTION

Sleep is essential for survival and plays a key role in maintaining our physical and mental health. Normal human sleep can be categorized into two states, as defined by a cortical electroencephalogram (EEG) readings, typically collected during a polysomnography

Handbook of Sleep Disorders in Medical Conditions
DOI: https://doi.org/10.1016/B978-0-12-813014-8.00001-9

3

© 2019 Elsevier Inc. All rights reserved.

1. DIAGNOSTIC CRITERIA AND ASSESSMENT OF SLEEP DISORDERS

(PSG) sleep test: nonrapid eye movement (NREM, pronounced "non-REM") sleep and rapid eye movement (REM) sleep. NREM sleep produces a variably synchronous EEG that is associated with low muscle tone and minimal psychological activity[1] and is subdivided into three progressively deeper sleep stages: N1 (or "Stage 1" sleep), N2 (or "Stage 2" sleep), and N3 (or "Stage 3" sleep, commonly referred to as slow wave sleep). On the other hand, REM sleep produces a desynchronized EEG with muscular atonia and episodic bursts of REM and is not typically subdivided, except possibly for research purposes (e.g., tonic vs phasic REM).[1] Table 1.1 provides a brief outline of the four sleep stages and general characteristics often associated with each stage. Sleep begins in NREM before the first REM period occurs, typically 80–120 minutes after sleep onset. Throughout the sleep

TABLE 1.1 Sleep Stages and General Characteristics

Sleep Stage	American Academy of Sleep Medicine Terminology	Characteristics
Nonrapid eye movement Stage 1	N1	The stage between wakefulness and sleep, and the lightest stage of Nonrapid eye movement sleep. Slow eye movements are often present, and muscles may remain somewhat active. Electroencephalogram shows a transition from relatively unsynchronized beta and gamma brain waves (12–30 and 25–100 Hz, respectively) to more synchronized slower alpha waves (8–13 Hz) and theta waves (4–7 Hz)
Nonrapid eye movement Stage 2	N2	Muscle activity further decreases, and conscious awareness of the surrounding environment begins to fade. Electroencephalograms continue to show theta waves (as in N1), with the addition of sleep spindles (short bursts of brain activity, 12–14 Hz) and K-complexes (short negative high-voltage peaks, followed by a slower positive complex and then a final negative peak)
Nonrapid eye movement Stage 3, or "Slow Wave Sleep"	N3	The deepest, most restorative stage of Nonrapid eye movement sleep. Larger amount of synchronized, slow wave electroencephalogram activity (delta waves, 1–2 Hz) compared to the other stages. The brain becomes less responsive to external stimuli; this tends to be the hardest stage from which to awaken. Previously this stage was divided into two sleep stages (stages 3 and 4) but was consolidated into a singular N3 stage in the 2007 American Academy of Sleep Medicine scoring guidelines[2]
Rapid eye movement	R	Rapid side-to-side movements of closed eyes, breathing tends to be more rapids and irregular, with heart rate and blood pressure increasing to near waking levels. Virtual absence of muscle tone and skeletal muscle activity, known as atonia, is present. Electroencephalogram shows low-amplitude, mixed frequency brain waves similar to what is seen during wakefulness (e.g., alpha, beta, and theta waves); these appear as a "saw-tooth" brain wave pattern

I. GENERAL ISSUES

CURRENT DIAGNOSTIC CLASSIFICATION SYSTEMS 5

period, we continue to cycle through the NREM to REM sleep stages approximately every 90 minutes, with NREM sleep decreasing and REM sleep increasing over time. This suggests that the beginning of our total sleep period tends to contain longer periods of deeper, more restorative slow wave sleep, and as our sleep period progresses, more time will be spent in REM sleep, where dreams typically occur.

CURRENT DIAGNOSTIC CLASSIFICATION SYSTEMS

There are currently three main diagnostic classification systems used widely by clinicians to classify sleep disorders. The *International Classification of Sleep Disorders, Third Edition* (ICSD-3),[3] the Sleep Disorders section of the *Diagnostic and Statistical Manual of Mental Disorders, Fifth Edition* (DSM-5),[4] and the *International Classification of Diseases, Tenth Edition* (ICD-10).[5] A schematic crosswalk comparing the specific diagnostic labels used across these three classification systems is presented in Table 1.2.

The ICSD-3 is currently the most advanced classification of sleep disorders and is intended to be used by sleep experts. This American Academy of Sleep Medicine publication describes nearly 70 highly specific diagnoses classified into *six main clinical categories*: insomnia, sleep-related breathing disorders, central disorders of hypersomnolence, circadian rhythm sleep—wake disorders, parasomnias, and sleep-related movement disorders. It also contains a supplemental "Other" category to assist with assigning codes to conditions that do not otherwise fit into any of the above categories, and two appendices (sleep-related medical and neurological disorders, and ICD-10 coding for substance-induced sleep disorders). Pediatric conditions are fully merged within each of the clinical categories, except for the obstructive sleep apnea (OSA) section, which provides a specific delineation between adult and pediatric OSA. The ICSD-3 provides general diagnostic criteria for each condition, and if indicated, will outline variations for pediatric diagnosis and developmental issues pertaining to older adults.

The DSM-5 is a classification system intended for use by mental health and general medical clinicians who are not experts in sleep medicine; however, this most recent revision reflects an overall greater concordance with the ICSD-3 compared to previous editions. This American Psychiatric Association publication features an entire "Sleep—Wake Disorders" section, outlining *10 disorders or disorder groups*: insomnia disorder, hypersomnolence disorder, narcolepsy, breathing-related sleep disorders, circadian rhythm sleep—wake disorders, NREM sleep arousal disorders, nightmare disorder, REM sleep behavior disorder, restless legs syndrome (RLS), and substance/medication-induced sleep disorder. It also contains the additional diagnostic categories, "other specified" or "unspecified" which can be applied to insomnia disorder, hypersomnolence disorder, and a more general sleep—wake disorder to assist with assigning codes to conditions that do not otherwise fit into any of the above categories. Similar to the ICSD-3, the DSM-5 does not separate pediatric and adult diagnoses; rather, it provides general diagnostic criteria for each condition and outlines specific developmental features that may be present.

The ICD-10 is another classification system commonly used to classify sleep conditions. This World Health Organization publication is far less complex than the ICSD-3 schema

TABLE 1.2   Sleep Disorder Diagnoses by Classification System

Diagnostic Statistical Manual of Mental Disorders, Fifth Edition	International Classification of Sleep Disorders, Third Edition	International Classification of Diseases, Tenth Edition, Clinical Modification
INSOMNIA		
Insomnia disorder	Chronic insomnia disorder	G47.0 Insomnia unspecified
Specify if:	Short-term insomnia disorder	G47.01 Insomnia due to medical condition
• Episodic	Other insomnia disorder	G47.09 Other insomnia
• Persistent		F51.01 Primary insomnia
• Recurrent	*Isolated symptoms and normal variants*	F51.02 Adjustment insomnia
Other specified Insomnia disorder	• Excessive time in bed	F51.03 Paradoxical insomnia
Unspecified insomnia disorder	• Short sleeper	F51.04 Psychophysiologic insomnia
		F51.05 Insomnia due to other mental disorder
		F51.09 Other insomnia not due to a substance or known physiological condition
		Z73.810 Behavioral insomnia of childhood, sleep-onset association type
		Z73.811 Behavioral insomnia of childhood, limit setting type
		Z73.812 Behavioral insomnia of childhood, combined type
		Z73.819 Behavioral insomnia of childhood, unspecified type
HYPERSOMNOLENCE AND NARCOLEPSY		
Hypersomnolence disorder	Narcolepsy type I	G47.10 Hypersomnia unspecified
Specify if:	Narcolepsy type II	G47.11 Idiopathic hypersomnia with long sleep time
• With mental disorder	Idiopathic hypersomnia	G47.12 Idiopathic hypersomnia without long sleep time
• With medical condition	Kleine–Levin syndrome	
• With another sleep disorder	Hypersomnia due to a medical disorder	G47.13 Recurrent hypersomnia
Specify if:	Hypersomnia due to a medication or substance	G47.14 Hypersomnia due to medical condition
• Acute	Hypersomnia associated with a psychiatric disorder	G47.19 Other hypersomnia
• Subacute	Insufficient sleep syndrome	F51.1 Hypersomnia not due to a substance or known psychological condition
• Persistent		F51.11 Primary hypersomnia
Specify current severity:	*Isolated symptoms and normal variants*	F51.12 Insufficient sleep syndrome
• Mild	• Long sleeper	F51.13 Hypersomnia due to other mental disorder
• Moderate		F51.19 Other hypersomnia not due to a substance or known physiological condition
• Severe		G47.411 Narcolepsy with cataplexy
		G47.419 Narcolepsy without cataplexy, not otherwise specified
Narcolepsy		G47.421 Narcolepsy in conditions classified elsewhere with cataplexy
Specify whether:		G47.429 Narcolepsy in conditions classified elsewhere without cataplexy
• Narcolepsy without cataplexy but with hypocretin deficiency		
• Narcolepsy with cataplexy but without hypocretin deficiency		
• Autosomal dominant cerebellar ataxia, deafness, and narcolepsy		
• Autosomal dominant narcolepsy, obesity, and type 2 diabetes		
• Narcolepsy secondary to another medical condition		
• *Specify current severity:*		
• Mild		
• Moderate		
• Severe		
Other specified hypersomnolence disorder		
Unspecified hypersomnolence disorder		

(Continued)

TABLE 1.2 (Continued)

Diagnostic Statistical Manual of Mental Disorders, Fifth Edition	International Classification of Sleep Disorders, Third Edition	International Classification of Diseases, Tenth Edition, Clinical Modification
BREATHING-RELATED SLEEP DISORDERS		
Obstructive sleep apnea hypopnea *Specify current severity:* - Mild - Moderate - Severe **Central sleep apnea** *Specify whether:* - Idiopathic central sleep apnea - Cheyne–Stokes breathing - Central sleep apnea comorbid with opioid use **Sleep-related hypoventilation** *Specify whether:* - Idiopathic hypoventilation - Congenital central alveolar hypoventilation - Comorbid sleep-related hypoventilation	*Obstructive sleep apnea disorders* - **Obstructive sleep apnea, adult** - **Obstructive sleep apnea, pediatric** *Central sleep apnea syndrome* - **Central sleep apnea with Cheyne–Stokes breathing** - **Central sleep apnea due to a medical disorder without Cheyne–Stokes breathing** - **Central sleep apnea due to high-altitude periodic breathing** - **Central sleep apnea due to medication or substance** - **Primary central sleep apnea** - **Primary central sleep apnea of infancy** - **Primary central sleep apnea of prematurity** - **Treatment-emergent central sleep apnea** *Sleep-related hypoventilation disorders* - **Obesity hypoventilation syndrome** - **Congenital central alveolar hypoventilation syndrome** - **Late-onset central hypoventilation with hypothalamic dysfunction** - **Idiopathic central alveolar hypoventilation** - **Sleep-related hypoventilation due to medication or substance** - **Sleep-related hypoventilation due to medical disorder** - **Sleep-related hypoxemia disorder** *Isolated symptoms and normal variants* **Snoring** **Catathrenia**	G47.30 Sleep apnea unspecified G47.31 Primary central sleep apnea G47.32 High-altitude periodic breathing G47.33 Obstructive sleep apnea (adult) (pediatric) G47.34 Idiopathic sleep related nonobstructive alveolar hypoventilation G47.35 Congenital central alveolar hypoventilation syndrome G47.36 Sleep-related hypoventilation in conditions classified elsewhere G47.37 Central sleep apnea in conditions classified elsewhere G47.39 Other sleep apnea R06.3 Cheyne–Stokes breathing pattern R06.00 Dyspnea, unspecified R06.09 Other forms of dyspnea R06.3 Periodic breathing R06.83 Snoring R06.89 Other abnormalities of breathing R06.81 Apnea, not elsewhere specified
CIRCADIAN RHYTHM SLEEP–WAKE DISORDERS		
Delayed sleep phase type *(specify if familial or if overlapping with non-24-h sleep–wake type)* **Advanced sleep phase type** *(specify if familial)* **Irregular sleep-wake type** **Non-24-h sleep–wake type** **Shift work type** **Unspecified type**	**Delayed sleep–wake phase disorder** **Advanced sleep–wake phase disorder** **Irregular sleep–wake rhythm** **Non-24-h sleep–wake rhythm disorder** **Shift work disorder** **Jet lag disorder**	G47.20 Circadian rhythm sleep disorder, unspecified type G47.21 Delayed sleep phase type G47.22 Advanced sleep phase type G47.23 Irregular sleep wake type G47.24 Free running type G47.25 Jet lag type G47.26 Shift work type

(Continued)

TABLE 1.2 (Continued)

Diagnostic Statistical Manual of Mental Disorders, Fifth Edition	International Classification of Sleep Disorders, Third Edition	International Classification of Diseases, Tenth Edition, Clinical Modification
Specify if any of the above circadian disorders are • Episodic • Persistent • Recurrent **Other specified sleep−wake disorder** **Unspecified sleep−wake disorder**	Circadian rhythm sleep−wake disorder not otherwise specified	G47.27 Circadian rhythm sleep disorder in conditions classified elsewhere G47.29 Other circadian rhythm sleep disorder

PARASOMNIAS

Nonrapid eye movement sleep arousal disorders *Specify whether:* • Sleepwalking type *(specify if with sleep-related eating, or with sleep-related sexual behavior/sexsomnia)* • Sleep terror type **Nightmare disorder** *Specify if:* • During sleep onset • *Specify if:* • With associated nonsleep disorder • With associated other medical condition • With associated other sleep disorder *Specify if:* • Acute • Subacute • Persistent *Specify current severity:* • Mild • Moderate • Severe **Rapid eye movement sleep behavior disorder**	*NREM-related parasomnias* • **Disorders of arousal from NREM sleep** • **Confusional arousals** • **Sleepwalking** • **Sleep terrors** • **Sleep-related eating disorder** *REM-related parasomnias* • **REM sleep behavior disorder** • **Recurrent isolated sleep paralysis** • **Nightmare disorder** *Other parasomnias* • **Exploding head syndrome** • **Sleep-related hallucinations** • **Sleep enuresis** • **Parasomnia due to medical disorder** • **Parasomnia due to medication or substance** • **Parasomnia, unspecified** *Isolated symptoms and normal variants* − **Sleep talking**	G47.50 Parasomnia unspecified G47.51 Confusional arousals G47.52 REM sleep behavior disorder G47.53 Recurrent isolated sleep paralysis G47.54 Parasomnia in conditions classified elsewhere G47.59 Other parasomnia F51.3 Sleepwalking (somnambulism) F51.4 Sleep terrors (night terrors) F51.5 Nightmare disorder

SLEEP-RELATED MOVEMENT DISORDERS

Restless legs syndrome	**Restless legs syndrome** **Periodic limb movement disorder** **Sleep-related leg cramps** **Sleep-related bruxism** **Sleep-related rhythmic movement disorder** **Benign sleep myoclonus of infancy** **Propriospinal myoclonus at sleep onset** **Sleep-related movement disorder due to medical disorder** **Sleep-related movement disorder due to medication or substance** **Sleep-related movement disorder, unspecified** *Isolated symptoms and normal variants* • **Excessive fragmentary myoclonus**	G25.81 Restless legs syndrome G47.61 Periodic limb movement disorder G47.62 Sleep-related leg cramps G47.63 Sleep-related bruxism G47.69 Other sleep-related movement disorders

(Continued)

TABLE 1.2 (Continued)

Diagnostic Statistical Manual of Mental Disorders, Fifth Edition	International Classification of Sleep Disorders, Third Edition	International Classification of Diseases, Tenth Edition, Clinical Modification
	• Hypnagogic foot tremor and alternating leg muscle activation • Sleep starts (hypnic jerks)	

OTHER

Diagnostic Statistical Manual of Mental Disorders, Fifth Edition	International Classification of Sleep Disorders, Third Edition	International Classification of Diseases, Tenth Edition, Clinical Modification
Substance/Medication-induced sleep disorder _Specify whether:_ • Insomnia type • Daytime sleepiness type • Parasomnia type • Mixed type _Specify if:_ • With onset during intoxication • With onset during discontinuation/withdrawal	Other sleep disorder	G47.8 Other sleep disorders G47.9 Sleep disorder, unspecified F51.8 Other sleep disorders not due to a substance or known physiological condition F51.9 Sleep disorder not due to a substance or known physiological condition, unspecified Z72.820 Sleep deprivation Z72.821 Inadequate sleep hygiene Z73.8 Other problems related to life management difficulty _Drug induced sleep disorders_ F11.182 Opioid _abuse_ with opioid-induced sleep disorder F11.282 Opioid _dependence_ with opioid-induced sleep disorder F11.982 Opioid _use_, unspecified with opioid-induced sleep disorder F13.182 Sedative, hypnotic or anxiolytic _abuse_ with sedative, hypnotic or anxiolytic-induced sleep disorder F13.282 Sedative, hypnotic or anxiolytic _dependence_ with sedative, hypnotic, or anxiolytic-induced sleep disorder F13.982 Sedative, hypnotic, or anxiolytic _use_, unspecified with sedative, hypnotic, or anxiolytic-induced sleep disorder F14.182 Cocaine _abuse_ with cocaine-induced sleep disorder F14.282 Cocaine _dependence_ with cocaine-induced sleep disorder F14.982 Cocaine _use_, unspecified with cocaine-induced sleep disorder F15.182 Other stimulant _abuse_ with stimulant-induced sleep disorder F15.282 Other stimulant _dependence_ with stimulant-induced sleep disorder F15.982 Other stimulant _use_, unspecified with stimulant-induced sleep disorder F19.182 Other psychoactive substance _abuse_ with psychoactive substance-induced sleep disorder F19.21 Other psychoactive substance _dependence_, in remission F19.282 Other psychoactive substance _dependence_ with psychoactive substance-induced sleep disorder F19.982 Other psychoactive substance _use_, unspecified with psychoactive substance-induced sleep disorder

Note: The International Classification of Diseases, Tenth Edition, Clinical Modification "G" codes refer organic sleep disorders, "F" codes refer to nonorganic sleep disorders, "R" codes refer to symptoms, signs, and abnormal clinical and laboratory findings, not elsewhere classified, and "Z" codes refer to factors influencing health status and contact with health services.

and groups sleep disorders into global categories of *organic* and *nonorganic* origin. Organic sleep disorders, which are classified using "G" codes, focus on medically/neurologically based sleep disorders and diseases of the nervous system, such as sleep apnea or narcolepsy. Nonorganic sleep disorders, which are classified using "F" codes, focus on mental and behavioral disorders, and are organized into three types of conditions: dyssomnias, parasomnias, and sleep disorders secondary to medical and psychiatric disorders. Dyssomnias refer to a predominant disturbance in the amount, quality, or timing of sleep due to emotional causes, such as nonorganic insomnia, nonorganic hypersomnia, or nonorganic disorder of sleep—wake schedule. Parasomnias refer to abnormal episodic events occurring during sleep, such as sleepwalking, sleep terrors, teeth grinding, bed wetting, or nightmares. There are also sleep-related conditions classified using "R" and "Z" codes. "R" codes focus on symptoms, signs, and abnormal clinical and laboratory findings not elsewhere classified, such as Cheyne—Stokes breathing pattern or snoring. "Z" codes focus on factors influencing health status and contact with health services, such as sleep deprivation or inadequate sleep hygiene. Similar to the ICSD-3 and DSM-5, the ICD-10 does not separate pediatric and adult sleep diagnoses, with the exception of OSA and the behavioral insomnia of childhood diagnoses.

CLINICAL FEATURES AND DIAGNOSTIC CONSIDERATIONS FOR SLEEP/WAKE DISORDERS

Despite the existence of several classification methods used to diagnosis sleep disorders, our goal herein is to provide a brief description of each sleep disorder using the DSM-5 criteria as an overarching framework, since this classification system is intended for general medical clinicians and mental health professionals that may not necessarily be experts in sleep medicine. Diagnostic categories covered in this chapter will include insomnia, hypersomnolence, narcolepsy, breathing-related sleep disorders, circadian rhythm sleep disorders, parasomnias, and movement-related sleep disorders.

Insomnia

Insomnia disorder is characterized by difficulties falling or staying asleep and/or waking up earlier than desired with an inability to fall back asleep. The sleeping disturbance must occur at least 3 nights per week, for at least 3 months, and persist despite having an adequate opportunity for sleep. It must also cause the individual significant daytime impairment in social, occupational, educational, behavioral, or other important areas of function and cannot be better explained by another sleep—wake disorder (e.g., a breathing-related sleep disorder), the physiological effects of a substance, or a coexisting medical or mental health condition. Insomnia is considered *episodic* if the symptoms last for at least 1 month but less than 3 months, *persistent* if the symptoms last 3 months of longer, and/or *recurrent* if two (or more) episodes occur within the space of 1 year.

Insomnia that occurs in the context of another psychological or medical disorder is considered *comorbid*. Common comorbid conditions include cardiovascular diseases, metabolic disorders, autoimmune conditions, chronic pain, anxiety, and depression.[4,6,7] Among a large sample of patients with at least one physician identified chronic condition (hypertension, diabetes, congestive heart failure, myocardial infarction, or depression), 34% reported mild insomnia and 16% reported severe insomnia.[8] Insomnia was not only comorbid with these conditions at baseline but persisted despite receiving treatment for these concerns, such that at a 2-year follow-up, 59% of patients with mild insomnia at baseline and 83% with severe insomnia at baseline continued to report sleep problems.[8] This suggests that insomnia should be considered an important treatment target among clinicians, as insomnia symptoms do not always remit during treatment of a comorbid condition.

Of the various aforementioned sleep difficulties, trouble maintaining sleep is the most common, followed by trouble falling asleep; however, patients will often present with a combination of these symptoms.[4] Insomnia tends to affect women more than men[9,10] and is more common in older adults.[10] Research has shown that an average sleep-onset latency or middle-of-the-night wake time of 20 minutes maximizes the sensitivity and specificity for insomnia classification among individuals with insomnia compared to normal sleepers.[11]

Hypersomnolence

Hypersomnolence disorder (sometimes referred to as "hypersomnia") is characterized by self-reported excessive sleepiness despite a main sleep period lasting at least 7 hours. Individuals may exhibit an excessive quantity of sleep or have an increased propensity to sleep during wakefulness (e.g., recurrent periods of sleep or lapses into sleep within the same day). Individuals often have trouble being fully awake after an abrupt awakening, which is also referred to as sleep inertia or "sleep drunkenness." This sleeping disturbance must occur at least three times per week, for at least 3 months, and cause the individual significant distress or impairment in cognitive, social, occupational, or other important areas of function. Furthermore, the hypersomnolence cannot be better explained by or attributable to another sleep–wake disorder (e.g., narcolepsy), the physiological effects of a substance, or a coexisting medical or mental health condition. The DSM-5 also provides guidance on specifying whether the hypersomnolence occurs with a mental disorder, with a medical condition, and/or with another sleep disorder, the acuity and persistence of the concern, and the severity of the current concern.

Of the various aforementioned sleep–wake difficulties, approximately 80% of individuals with hypersomnolence report nonrestorative sleep or difficulties waking up in the morning, and 36%–50% report having sleep inertia.[4] Hypersomnolence tends to affect men and women equally[12] and is comorbid with a number of medical and mental health conditions, including Parkinson's disease,[13,14] Alzheimer's disease,[13] multiple system atrophy,[13] and depressive disorders.[15] Hypersomnolence has a progressive onset; symptoms tend to emerge between the ages of 15 and 25.[4] Nocturnal PSG shows that individuals with hypersomnolence have a short sleep latency, normal to prolonged sleep duration, and they tend to have a sleep efficiency (total sleep time/time in bed) >90%. When using a multiple sleep latency test (MSLT), the average sleep latency is typically less than 10 minutes, and the transition to REM sleep within 20 minutes of sleep may be present but occurs less than twice during the typical four or five MSLT napping periods.[4]

Narcolepsy

Narcolepsy is characterized by a recurrent, irrepressible need for sleep in the form of daytime naps or lapses into sleep, occurring at least 3 times per week, for at least 3 months. Distinct from hypersomnolence disorder, individuals with narcolepsy must also exhibit at least one of the following: cataplexy episodes (i.e., a sudden loss of muscle tone usually in response to strong emotion) occurring at least a few times a month, a hypocretin deficiency [measured using cerebrospinal fluid (CSF) hypocretin-1 level <1/3 of normal or ≤110 pg/mL if a Stanford reference sample is used for radioimmunoassay], and/or a short-onset REM period (SOREMP) ≤15 minutes using nocturnal PSG or a MSLT showing a mean sleep latency ≤8 minutes and two or more SOREMPs. With regard to SOREMPs, a cutoff of 15 minutes indicates that the individual is going into a REM sleep period faster than what is typical in individuals without sleep complaints.

The DSM-5 also provides guidance on specifying among five different types of narcolepsy. *Narcolepsy without cataplexy but with hypocretin deficiency* is characterized by meeting all the narcolepsy diagnostic criteria except cataplexy. *Narcolepsy with cataplexy but without hypocretin deficiency* is characterized by meeting all the narcolepsy diagnostic criteria except the individual has normal hypocretin levels. *Autosomal dominant cerebellar ataxia, deafness, and narcolepsy* is caused by mutations on exon 21 DNA cytosine-5-methyltranferase-1; onset for this subtype typically occurs between ages of 30 and 40 years and is characterized by low-to-intermediate CSF hypocretin-1 levels, deafness, cerebellar ataxia, and eventually dementia. *Autosomal dominant narcolepsy, obesity, and type 2 diabetes* are associated with a mutation in the myelin oligodendrocyte glycoprotein gene and is characterized by having low CSF hypocretin-1 levels with narcolepsy, obesity, and type 2 diabetes. *Narcolepsy secondary to another medical condition* is characterized as narcolepsy that is secondary to a traumatic or infectious medical condition (e.g., Whipple's disease, sarcoidosis) or a tumoral destruction of hypocretin neurons. The DSM-5 also provides guidance on specifying the severity of the current concern.

One commonality among the different narcolepsy subtypes is that the affected individual will experience excessive sleepiness that results in recurrent daytime naps or lapses into sleep. When cataplexy is present, it will often appear as a sudden, bilateral loss of muscle tone precipitated by emotions, such as laughter or joking, and can last as little as a few seconds or extend to minute-long episodes. During cataplexy, individuals are both awake and aware. Narcolepsy tends to affect men and women equally and is highly heritable; when an individual's first degree biological relative has narcolepsy, there is a 10- to 40-fold increased risk that the individual will also have narcolepsy.[16] Compared to individuals without narcolepsy, those who have narcolepsy (with or without cataplexy) are 3.8 times more likely to have a mental disorder, 3.7 times more likely to have nervous system complications, 3.5 times more likely to have a musculoskeletal problem, 2.7 times more likely to have a digestive illness, and 2.2 times more likely to have a genitourinary illness.[17] Cell loss of hypothalamic hypocretin (orexin)-producing cells, which causes hypocretin deficiency, tends to be autoimmune related, and approximately 90%–99% of people (99% of Caucasians) carry HLA-DQB1*06:02 allele.[4] Thus, checking for DQB1*06:02 prior to a lumbar puncture to determine CSF hypocretin-1 immunoreactivity may be useful when making a diagnosis.[4]

Breathing-Related Sleep Disorders

Commonly diagnosed breathing disorders include OSA, central sleep apnea (CSA), and sleep-related hypoventilation. These disorders are categorized by apneas (the total absence of airflow) and hypopneas (a marked reduction in airflow). To diagnose any of these disorders, a nocturnal PSG is required.

OSA is categorized by nocturnal breathing disturbances (e.g., snoring, snorting, gasping for air, or pauses in breathing), daytime sleepiness, fatigue, and experiencing poor, unrefreshing sleep despite getting an adequate sleep opportunity. Evidence from nocturnal PSG is required for diagnosis, such that an individual's apnea-hypopnea index (AHI; the number of apneas plus hypopneas per hour of sleep) is at least 5 with reported nighttime breathing disturbances and daytime disruption (e.g., sleepiness, fatigue), or the individual has an AHI ≥ 15, regardless of accompanying symptoms. OSA should not be better explained by or attributable to a coexisting medical or mental health condition. Clinicians should specify the severity of the OSA; mild is an AHI 5–14, moderate is an AHI 15–30, and severe is an AHI greater than 30. Prevalence of OSA tends to be higher among males and older adults,[18] and has been associated with a number of medical and mental health conditions, including obesity, systematic hypertension, coronary artery disease, heart failure, stroke, diabetes, increased mortality, and depression.[4]

CSA is characterized by repeated episodes of apneas and hypopneas during sleep, totaling at least 5 or more per hour, caused by variability in respiratory effort. Within the DSM-5, CSA can be classified as either *idiopathic CSA*, which refers to apneas and hypopneas caused by variability in respiratory effort without evidence of airway obstruction, *Cheyne–Stokes breathing*, which presents as a crescendo–decrescendo variation in tidal volume that results in central apneas and hypopneas that are frequently accompanied by an arousal, or *CSA comorbid with opioid use*, which is attributed to the effects of opioids on the respiratory rhythm generators in the medulla and the hypoxic and hypercapnic effects they may have on respiratory drive. Idiopathic CSA is relatively uncommon, constituting approximately 5% of patients referred to a sleep clinic,[19] but its presence may be much higher in other clinical populations, such as those with heart failure and left ventricular ejection fraction of 45%,[20] as well as among those living at higher altitudes.[21] Most studies also suggest that idiopathic CSA is most common in middle-aged to elderly adults and more frequent in men (although not all studies have found this sex difference).[3] CSA with Cheyne–Stokes breathing is generally seen in patients 60 years or older and has been reported to occur in 26%–50% of patients during the acute period following a stroke.[3] In addition, when looking at a sample of 450 men and women with congestive heart failure, risk factors for Cheyne–Stokes breathing included being 60 years or older, the presence of atrial fibrillation, male sex, and daytime hypocapnia.[22] CSA comorbid with opioid use occurs in approximately 30% of individuals receiving methadone maintenance therapy and those taking chronic opioids for nonmalignant pain.[4]

Sleep-related hypoventilation is characterized by episodes of decreased respiration associated with elevated CO_2 levels, as measured by PSG or oximetry, and is not better explained by another sleep disorder. It can be classified as *idiopathic hypoventilation*, which is not attributable to any readily identified condition, *congenital central alveolar*

14   1. DIAGNOSTIC CRITERIA AND ASSESSMENT OF SLEEP DISORDERS

hypoventilation, which is a rare disorder typically presenting in the perinatal period with shallow breathing or cyanosis and apnea during sleep, or *comorbid sleep-related hypoventilation*, which typically occurs as a consequence of another medical condition such as COPD, interstitial lung disease, muscular dystrophies, obesity, or medications (e.g., benzodiazepines, opiates).[4]

Circadian Rhythm Sleep–Wake Disorders

Circadian rhythm disorders are characterized by a misalignment or alteration between an individual's endogenous circadian rhythm and their required physically, socially, or professionally determined sleep–wake schedule. This disruption leads to insomnia and/ or sleepiness and causes significant daytime impairment in social, occupational, educational, behavioral, or other important areas of function. A circadian rhythm sleep–wake disorder is considered *episodic* if the symptoms last for at least 1 month but less than 3 months, *persistent* if the symptoms last 3 months of longer, and/or *recurrent* if two (or more) episodes occur within the space of 1 year.

Delayed sleep phase type occurs when an individual's sleep is delayed beyond the socially acceptable or conventional bedtime (the so-called night owls), causing difficulty waking up at their desired, or perhaps required, time. Prevalence in the general population is approximately 0.17%[4] and among adolescents is approximately 3.3%.[23] *Advanced sleep phase type* occurs when an individual is unable to stay awake in the early evening, goes to bed very early (e.g., between 6:00 and 8:00 p.m.) and then wakes up early in the morning (e.g., between 2:00 and 4:00 a.m.; the so-called morning lark or early birds). Prevalence is estimated to be 1% in middle-aged adults.[4] While the timing of advanced sleep phase time may be early, the sleep itself tends to be normal. *Irregular sleep–wake type* refers to a temporally disorganized rhythm, wherein the individual may take numerous naps throughout the day, have no main nighttime sleep episode, and have irregularities in sleep from day-to-day; the prevalence in the general population is unknown.[4] *Non-24-hour sleep–wake type* is a pattern of sleep–wake cycling that is not synchronized with a 24-hour environment with gradual drifts (usually in a clockwise direction) of sleep onset and wake times. While prevalence in the general population is unclear, it appears to be rare in sighted individuals and more common (approximately 50%) among blind individuals.[4] *Shift work type* occurs when an individual reports insomnia during the major sleep period and/or excessive sleepiness during the major awake period that is typically associated with a schedule of unconventional work hours. Prevalence of shift work type increases with advancement into middle-age and beyond, and is estimated to affect 5%–10% of the night worker population,[24] which is 16%–20% of the workforce.[4]

Parasomnias

Parasomnias are characterized by abnormal behavior occurring in association with sleep or sleep–wake transitions, such as sleepwalking, sleep terrors, nightmare disorder, and REM sleep behavior disorder. Other parasomnias which will not be detailed in this chapter include confusional arousals, sleep paralysis, exploding head syndrome, sleep-related

I. GENERAL ISSUES

CLINICAL FEATURES AND DIAGNOSTIC CONSIDERATIONS FOR SLEEP/WAKE DISORDERS

hallucinations, and sleep enuresis (see cited articles for more information).[25-28] All parasomnias must cause the individual significant daytime impairment in social, occupational, educational, behavioral, or other important areas of function, and cannot be better explained by another sleep–wake disorder, the physiological effects of a substance, or a coexisting medical or mental health condition.

Both sleepwalking and sleep terrors are considered NREM sleep arousal disorders, as they frequently occur during deep, slow wave sleep seen in the first third of the night. NREM sleep arousal disorders occur most often in childhood and decrease as individuals get older. *Sleepwalking* involves rising from the bed during sleep and walking about, with the individual being relatively unresponsive to people communicating with him or her, having a blank, staring face, and being awakened only with great difficulty. Sleepwalking may also present with eating or sexual behavior. Among adults, the prevalence of sleepwalking episodes is 1%–7% and is more common in males.[4] Among the children, the prevalence of at least one sleepwalking episode is 10%–30% (with 2%–3% sleepwalking often) and is more common in females.[4] Eighty percent of individuals who sleepwalk have a family history of sleepwalking or sleep terrors, and the risk for sleepwalking increases if both parents have a history of the disorder.[4] In a genetics study examining DNA of 22 family members across four generations, researchers found that sleepwalking has a genetic component with an "autosomal reduced penetrance" pattern, exhibited on a region of chromosome 20.[29] "Autosomal" means that the gene is carried on the nonsex chromosomes, and "reduced penetrance" means that sleepwalking appears to skip generations because not everyone who inherits the faulty genes has symptoms.

Sleep terrors present as abrupt arousals, usually beginning with a scream and accompanied by intense fear and autonomic arousal (e.g., rapid breathing, sweating, tachycardia, pupil dilation). Among adults, the prevalence of sleep terror episodes is 2.2%, and there are no sex-related differences.[4,30] Among children, the prevalence of sleep terrors episodes is much higher—36.9% at 18 months of age and 19.7% at 30 months of age—and tends to be more common in males than females.[4]

Nightmare disorder and REM sleep behavior disorder tend to occur during the second half of the night, when REM sleep is more prominent. In both disorders, an individual is both alert and oriented when they are awakened. *Nightmare disorder* involves dream imagery that seems real and elicits dysphoric emotions, such as anxiety or fear. Nightmares are more prevalent during childhood through young adulthood and tend to decline with age; they are also associated with female sex, increased stress, psychopathology, and dispositional traits, such as the tendency to experience events with distressing, highly reactive emotions.[31] *REM sleep behavior disorder* involves repeated episodes of arousal that tend to be accompanied by complex motor behaviors (e.g., kicking, punching) and vocalizations (e.g., loud, emotion-filled profanity). Diagnosis can be made when a PSG indicates REM sleep without atonia or if the patient has a history suggestive of REM sleep behavior disorder and an established synucleinopathy diagnosis, which refers to degenerative diseases of the central nervous system that exhibit excessive accumulation of alpha-synuclein (a brain protein) in the neurons (e.g., multiple system atrophy, Parkinson's disease, dementia with Lewy bodies). Prevalence among the general population is between 0.38% and 0.5%.[4] Tricyclic antidepressants, selective serotonin reuptake inhibitors, serotonin-norepinephrine reuptake inhibitors, and beta-blockers may result in PSG evidence characteristic of REM sleep behavior disorder; however, it is not known whether the medications result in REM sleep behavior disorder or if they are unmasking an underlying predisposition.[4]

Movement-Related Sleep Disorders

Two frequently diagnosed movement disorders are *RLS* and *periodic limb movement disorder (PLMD)*. While both conditions involve an inability to keep still, the disorders are distinctly different. With RLS, an individual will report an irresistible urge to move their legs that tends to worsen in the evening/night, during periods of rest or inactivity, and is either partially or totally relieved by movement. These symptoms occur at least 3 times per week and have persisted for at least 3 months. Predisposing factors to RLS include female sex, older age, family history of RLS, and genetic risk variants (e.g., intronic or intergenic regions of MEIS1, BTBD9, and MAP2K5 on chromosomes 2p, 6p, and 15q).[4,32] RLS is associated with pregnancy, peaking during the third trimester and often resolving itself following delivery.[4,33] RLS symptoms occur both when the individual is awake and asleep, and he or she can voluntarily respond to the uncomfortable sensations. On the other hand, PLMD is a sleep disorder in which an individual involuntarily moves his or her limbs during sleep; individuals are often unaware that these movements are occurring. PLMD movements typically occur during NREM sleep and individuals often report excessive daytime sleepiness. PLMD is diagnosed using PSG, which may reveal that these frequent, involuntary leg moves are often, but not always, accompanied by brief arousals or awakenings. Approximately 4% of adults have been diagnosed with PLMD,[32] with increasing rates among elderly females (11%).[34] Previous studies have found that approximately 80% of individuals with RLS also experience periodic leg movements in sleep (index >5 movements/hour sleep); [35] however, the reverse is not always the case. That is, those with periodic leg movements during sleep do not necessarily experience RLS.

Both RLS and PLMD must cause the individual significant daytime impairment in social, occupational, educational, behavioral, or other important areas of function, and cannot be better explained by another sleep–wake disorder, the physiological effects of a substance, or a coexisting medical or mental health condition.

PSYCHOLOGICAL, BEHAVIORAL, AND LABORATORY ASSESSMENTS TO AID DIAGNOSIS

There are a number of useful clinical tools to aid in the diagnosis and assessment of sleep–wake disorders. The most commonly used tools include gathering a thorough clinical history, daily self-monitoring of sleep–wake patterns, self-report questionnaires, actigraphy, PSG, the MSLT, and the maintenance of wakefulness test (MWT).

Clinical History

Collecting information using a structured or semistructured interview can help clinicians gather important details concerning a patient's sleep complaint, such as acuity or chronicity, course, factors alleviating or exacerbating the condition, and any previous treatment utilization. To help establish the etiology of their sleep concern(s), it is important to inquire about particular medical or mental health conditions, life events, and substance use present at the onset of the sleep problem. Clinicians might consider using the Duke Structured Interview for Sleep Disorders,[36] which was developed to assess sleep disorder symptoms according to criteria found in the ICSD and DSM. This semistructured interview is divided into four modules: insomnia disorders, sleep disorders associated with excessive daytime sleepiness and hypersomnia, circadian rhythm disorders, and parasomnias. Bed partners (or parents) are also an important source of clinical information, as they can often fill in gaps that the patient is unaware of, such as sleep apnea symptoms (breathing pauses, snoring), parasomnias (instances of sleepwalking or sleep terrors), or PLMD.

Daily Self-Monitoring

Filling out a prospective sleep diary on a daily basis can help create a more thorough clinical picture to aid in sleep disorder diagnosis and highlight a patient's sleep habits. It is recommended that clinicians gather 1–2 weeks of data since night-to-night variability of sleeping patterns is common (e.g., differences between sleep during weekdays vs weekends).[37,38] A typical sleep diary includes what time the patient got into bed, when they attempted to start sleeping, how long it took them to fall asleep, number and duration of awakenings, time of their final awakening, what time they actually got out of bed, napping behavior, use of sleep aids, and a measure of their perceived sleep quality.[39] Once the diaries are collected, a clinician can determine average sleep onset latency, wake after sleep onset (i.e., nocturnal awakenings), early morning awakenings, total wake time, total sleep time, total time in bed, and sleep efficiency [(total sleep time/total time in bed) × 100], which can help with treatment planning and decision-making to help a patient get their sleep back on track.

Self-Report Questionnaires

There are numerous questionnaires that can aid clinicians in the assessment of sleep disorder symptoms, with some being useful tools to employ during treatment to monitor change across time. Table 1.3 provides a brief outline of several commonly used self-report questionnaires.

Actigraphy

Actigraphs are small, wristwatch-like units that continuously measure movement and activity using an accelerometer. Some actigraphy recorders also have capability to measure light exposure. This noninvasive device is particularly helpful in determining sleep patterns and circadian rhythms in those with insomnia, hypersomnia, advanced sleep phase syndrome, delayed sleep phase syndrome, shift work disorder, jet lag disorder, and non-24-hours sleep/wake syndrome.[47] This useful tool can provide clinicians with several days

18   1. DIAGNOSTIC CRITERIA AND ASSESSMENT OF SLEEP DISORDERS

TABLE 1.3 Common Questionnaires Used to Assess Sleep-related Concerns

Questionnaire	Description and Use
Berlin Questionnaire[40]	A 10-item questionnaire assessing the presence of sleep apnea symptoms in three categories, including snoring severity, excessive daytime sleepiness, and history of high blood pressure or obesity. A physician or medical staff member must interpret the score (high risk obstructive sleep apnea = if 2 of 3 categories are positive; low risk obstructive sleep apnea = if 0−1 of the categories are positive)
Dysfunctional Beliefs About Sleep Questionnaire (DBAS-16)[41]	A 16-item questionnaire assessing dysfunctional beliefs about sleep. Scores are computed by taking the average of all completed 16 items. A higher score reflects greater dysfunctional beliefs about sleep. Target-specific items/beliefs rated as >5 for intervention
Epworth Sleepiness Scale (ESS)[42]	An eight-item questionnaire assessing daytime sleepiness (i.e., "the chance of dozing off") in different typical scenarios, such as watching TV or sitting and reading (score range 0−24; <10 = normal, 10−15 = possible excessive sleepiness depending on the situation, 16−24 = excessive sleepiness, consider seeking medical attention)
Insomnia Severity Index (ISI)[39]	A seven-item questionnaire used to assess insomnia severity (score range 0−28; <8 = normal, 8−13 = subthreshold insomnia, 14−21 = moderate insomnia, 22−28 = severe insomnia)
Morningness−Eveningness Questionnaire[43]	A 19-item questionnaire that provides information about circadian tendencies (score range 16−86; higher scores 59−86 = morning type, mid-range scores 42−58 = intermediate type with no stronger preference toward morning or evening, lower scores 16−41 = evening type)
Pittsburgh Sleep Quality Index (PSQI)[44]	A 19-item questionnaire that measures the quality of sleep and general sleep patterns of adults, differentiating "poor" from "good" sleep quality by measuring seven areas: subjective sleep quality, sleep latency, sleep duration, habitual sleep efficiency, sleep disturbances, use of sleeping medications, and daytime dysfunction over the last month (score range 0−21; ≤5 associated with good sleep quality; >5 associated with poor sleep quality)
Restless Legs Syndrome Scale[45]	A 10-item scale assessing the severity and impact of restless legs syndrome (score range 0−40; mild = 0−10, moderate = 11−20, severe = 21−30, very severe = 31−40)
STOP-BANG Questionnaire[46]	An eight-item questionnaire assessing the presence of sleep apnea symptoms, including snoring, tiredness, observed apneas, high blood pressure, body mass index, age over 50, neck circumference, and gender (low risk obstructive sleep apnea = yes to 0−2 questions; intermediate risk obstructive sleep apnea = yes to 3−4 questions; high risk obstructive sleep apnea = yes to 5−8 questions OR yes to 2 or more of the four STOP questions plus either male sex, BMI >35 kg/m^2, or neck circumference 17 in./43 cm in males or 16 in./41 cm in females)

(and sometimes weeks) of data and provides an acceptably accurate estimate of sleep patterns in normal, healthy adult populations and inpatients suspected of having certain sleep disorders.[47] An actigraphy recorder is typically worn on a patient's nondominant hand. The manufacturer of the actigraphy devices typically provides a corresponding program that reads the data through a sleep scoring algorithm to compute parameters such as time in bed, sleep onset latency, total wake time, total sleep time, and sleep efficiency. Since actigraphy is based on movement, there may be difficulties distinguishing wake from sleep in patients that are lying awake in bed but not moving (e.g., insomnia); however, previous research has shown that estimates of sleep and wake time in insomnia patients correlate moderately well with analogous measures derived from PSG ($r = 0.52-0.71$) and are sufficiently sensitive to detect sleep improvements resulting from behavioral insomnia therapy.[48] Increasingly clinicians are seeing more patients come in with data from wearable fitness trackers (e.g., Fitbit, Jawbone, Garmin) that claim reliable measurement of sleep/wake patterns; however, a recent systematic review found that fitness trackers overestimated total sleep time and sleep efficiency, and underestimated wake after sleep onset when comparing it to PSG metrics.[49] Therefore, clinicians should be cautious in using fitness tracker data to reliably ascertain sleep-related parameters.

Polysomnography

PSG is the gold standard for measuring sleep to confirm several sleep disorder diagnoses, such as sleep apnea, PLMD, narcolepsy (in conjunction with a MSLT), and some parasomnias, such as REM behavior disorder.[50] PSG can be used to measure a number of parameters during sleep, including heart rate, breathing rate, airflow, oxygen saturation, eye movements, snoring volume, muscle activity, body motion, positions during sleep, and brain activity. A typical PSG study occurs in a sleep center where a patient will spend the night and be continuously monitored; however, an increasing number of insurance companies allow patients to be sent home with an at-home sleep test that can generate either an AHI (the number of apneas or hypopneas recorded during the study per hour of sleep) or a respiratory disturbance index (the number of apneas, hypopneas, and other, more subtle, breathing irregularities occurring per hour of sleep), which may prompt a more detailed overnight PSG sleep study at a sleep center. PSG is also used for continuous positive airway pressure titration in patients with sleep-related breathing disorders and may be indicated for patients with sleep-related symptoms with concurrent neuromuscular or seizure-related disorders.[50] PSG is not routinely indicated for diagnosing insomnia; however, recent research has suggested that objectively defined total sleep time less than 6 hours (on average) may represent a more severe phenotype of insomnia[51,52] that responds to first-line treatments differently.[53] Therefore, if a PSG test is available for an individual with insomnia, it may provide useful information to help guide treatment, since it is the only measure that can truly distinguish between wake and sleep stages, as well as the percentage of time a person spends in different sleep stages [e.g., N1 or N2 (lighter sleep) vs N3([slow wave, deeper sleep) vs REM sleep].

Multiple Sleep Latency Test

The MSLT is the de facto standard for objectively measuring sleepiness and is used in conjunction with PSG to help diagnose patients with suspected narcolepsy or idiopathic hypersomnia.[54] It is not routinely indicated for evaluating sleepiness in medical or neurological disorders (other than narcolepsy), circadian rhythm disorders, or insomnia.[54] The MSLT typically consists of five naps separated by 2-hour breaks, with the first nap beginning 1.5–3 hours after termination of the nocturnal PSG recording.[54] The goal is to capture physiological sleep tendency (i.e., how long it takes a patient to fall asleep) and what sleep stages a patient transitions into during those naps. The naps are completed in a standardized environment aimed at minimizing external factors (e.g., temperature, light, noise, activity, etc.) that might keep a patient alert or affect their ability to fall asleep.

Maintenance of Wakefulness Test

The maintenance of MWT is intended to measure alertness during the day by assessing how well a patient is able to stay awake for a defined period of time (i.e., remaining alert in quiet times of inactivity). In contrast to the MSLT, this test focuses on a patient's ability to stay awake (rather than fall asleep). It is indicated for those whom the inability to remain awake constitutes a safety issue (e.g., pilot, truck driver), or to assess treatment response to medications in patients with narcolepsy or idiopathic hypersomnia.[54] The MWT typically consists of four 40-minute trials with 2-hour breaks in between, with the first trial beginning between 1.5 and 3 hours after the patient's usual wake-up time.[54] The wakefulness trials are completed in a standardized environment aimed at minimizing outside factors that might influence a patient's ability to fall asleep (e.g., temperature, light, noise, activity, etc.). Clinicians should rely on a combination of clinical judgment, sleep onset latency, clinical history, and compliance with treatment to determine impairment or risk for accidents.[54]

AREAS FOR FUTURE RESEARCH

Herein we have described the current and most widely used classification systems for the diagnosis of currently recognized sleep disorders. Although these systems reflect a level of sophistication in our current understanding of sleep disorders, it should be acknowledged that these systems have been devised as much on the basis of clinical experience and consensus as they have been on actual empirical evidence. This fact is due, in part, as a result of the uneven knowledge base for the currently recognized sleep disorders. For some conditions such as narcolepsy, much is known about underlying pathophysiological mechanisms, whereas for other conditions, like insomnia disorder, little such knowledge is currently available. This situation does not necessarily preclude the possibility that many sleep disorders listed in the current nosologies represent valid entities that can be reliably identified in clinical and research settings, yet it remains to be determined if this is the case. The available literature assessing the reliability and validity of sleep disorders remains rather limited and remarkably unimpressive. In fact, there have been very limited or no reliability/validity assessments of many of the sleep disorder categories listed in current nosologies. Thus, much more research is needed that is specifically designed to assess the reliability/validity of many of the sleep disorders individually and of the classification schemes taken as a whole.

A related issue is the fact that discrimination and subtyping among the various sleep disorder categories remains quite challenging. For example, it is often difficult in practice to clearly discriminate cases of hypersomnolence disorder from emerging cases of narcolepsy. Whereas there are many assays for identifying narcoleptic patient, particularly those with cataplexy, there are not definitive assays that define those with hypersomnolence disorder unrelated to narcolepsy. Similarly, there are no generally accepted assays for confirming the diagnosis of insomnia or assays that can discriminate those with mild vs severe forms of this disorder. Emerging research has suggested that objective total sleep duration may be a useful marker for separating those with severe and more treatment refractory insomnia from those with less serious and more treatable disease.[52,53,55] However, this criterion has yet to be widely accepted and integrated into diagnostic systems. Admittedly, these are just two instances that highlight the importance of further research into the development of reliable and valid assays that can confirm sleep disorder diagnoses. However, the diagnostic systems would generally benefit by such research across the range of disorders they describe.

Another area of exploration is suggested by the contrasting views of sleep disorders posed by the ICD-10 and DSM-5/ICSD-3 systems. As noted previously, the ICD-10 subdivides sleep disorders into those having an organic basis and those have nonorganic origins. In contrast, the DSM-5/ICSD-3 approach implies that specific sleep disorders can arise from a confluence of organic and nonorganic factors. Hence, the ICD-10 has a "black or white" view of sleep disorder etiology whereas the other systems appreciate varying "shades of gray" in a sleep disorder's evolution. Since it can be challenging to distinguish whether something has an organic vs nonorganic cause, many experts have recommended using the terminology "comorbid" to describe when conditions co-occur. Research to settle this controversy would be useful since such opposing views perpetuate an ongoing discordance among these diagnostic systems.

Finally, to date there has been a paucity of research to show that any of these diagnostic systems aid in and improve clinician's assignment of patients to specific treatments. This is surprising since one primary purpose of clinical diagnosis is that of aiding in treatment decision-making.[56] In this regard, simple study designs which allocate some patients to treatments randomly and others based on patients' diagnoses would be useful to validate these systems for clinical practice, yet such studies await future research efforts.

CONCLUSION

Currently available diagnostic systems for sleep disorders delineate a range of diagnoses including, but not limited to, insomnia disorder, hypersomnolence disorder, narcolepsy, sleep-disordered breathing, circadian rhythm sleep–wake disorders, parasomnias, and sleep-related movement disorders. The methods of classification

1. DIAGNOSTIC CRITERIA AND ASSESSMENT OF SLEEP DISORDERS

posed by the ICSD-3, DSM-5, and ICD-10 are overlapping but not totally concordant. This fact is due largely to differences among these systems as to whether sleep disorders should be divided into global organic and nonorganic categories, or not. Regardless of the particular system used, there are a number of helpful assessment procedures that can aid clinicians in establishing a patient's diagnosis and selecting appropriate treatment. These assessment procedures include the clinical interview, self-report instruments, and objective tools (PSG, MSLT, actigraphy, etc.). The field of sleep medicine encompasses a varied and complex range of disease states, and it is essential for the clinician who works with sleep-disordered patients to have a thorough understanding of the disorders outlined in the current sleep disorder classification schemes.

KEY PRACTICE POINTS

Consideration	Options
Diagnostic choice	Sleep disorders are categorized into multiple categories including insomnia, hypersomnolence, narcolepsy, breathing-related sleep disorders, circadian rhythm sleep wake disorders, parasomnias, and sleep related movement disorders
Diagnostic tools	A range of self-report assessment methods (clinical interview, sleep diary, sleep-specific questionnaires) can be useful for diagnosing sleep disorders. However, in some cases, additional objective methods such as PSG, MSLT, actigraphy, and MWT may be helpful and even necessary to confirm a diagnosis
Self-report assessment methods	Clinical interviews are useful for all sleep disorders, whereas sleep diary assessment is especially applicable to the assessment of insomnia and circadian rhythm sleep—wake disorders
PSG	This technique is most useful for diagnosis of narcolepsy, breathing-related sleep disorders, selected parasomnias, PLMD, and hypersomnolence disorder
MSLT	This technique is most useful for diagnosis of narcolepsy and hypersomnolence disorder
Actigraphy	This method is most useful in the diagnosis of circadian rhythm sleep—wake disorders. It may also be useful for select cases of insomnia, such as determining objective total sleep time and monitoring treatment response
MWT	Although not a classic "diagnostic test," this procedure is useful for assessing alertness in those working in industries wherein sleepiness can be dangerous (e.g., transportation industry)

PSG, Polysomnography; *PLMD*, periodic limb movement disorder; *MSLT*, multiple sleep latency test; *MWT*, maintenance of wakefulness test.

References

1. Carskadon MA, Dement WC. Monitoring and staging human sleep. In: Kryger MH, Roth T, Dement WC, eds. *Principles and Practice of Sleep Medicine*. 5th ed St. Louis, MO: Elsevier Saunders; 2011:16–26.
2. Iber C, Ancoli-Israel S, Chesson AL, Quan S. *The AASM Manual for the Scoring of Sleep and Associated Events: Rules, Terminology and Technical Specifications*. Westchester, IL: American Academy of Sleep Medicine; 2007.
3. American Academy of Sleep Medicine. *International Classification of Sleep Disorders*. 3rd ed Darien, IL: Author; 2014.
4. American Psychiatric Association. *Diagnostic and Statistical Manual of Mental Disorders*. 5th ed Washington, DC: Author; 2013.
5. World Health Organization. *The ICD-10 Classification of Mental and Behavioural Disorders: Clinical Descriptions and Diagnostic Guidelines*. Geneva: World Health Organization; 1992.
6. Taylor DJ, Mallory LJ, Lichstein KL, Durrence HH, Riedel BW, Bush AJ. Comorbidity of chronic insomnia with medical problems. *Sleep*. 2007;30(2):213–218.
7. Taylor DJ, Lichstein KL, Durrence HH, Reidel BW, Bush AJ. Epidemiology of insomnia, depression, and anxiety. *Sleep*. 2005;28(11):1457–1464.
8. Katz DA, McHorney CA. Clinical correlates of insomnia in patients with chronic illness. *Arch Intern Med*. 1998;158(10):1099–1107.
9. Zhang B, Wing YK. Sex differences in insomnia: a meta-analysis. *Sleep*. 2006;29(1):85–93.
10. Ohayon MM. Epidemiology of insomnia: what we know and what we still need to learn. *Sleep Med Rev*. 2002;6(2):97–111.
11. Lineberger MD, Carney CE, Edinger JD, Means MK. Defining insomnia: quantitative criteria for insomnia severity and frequency. *Sleep*. 2006;29(4):479–485.
12. Ohayon MM. From wakefulness to excessive sleepiness: what we know and still need to know. *Sleep Med Rev*. 2008;12(2):129–141.
13. Autret A, Lucas B, Mondon K, et al. Sleep and brain lesions: a critical review of the literature and additional new cases. *Neurophysiol Clin*. 2001;31(6):356–375.
14. Al-Qassabi A, Fereshtehnejad SM, Postuma RB. Sleep Disturbances in the prodromal stage of Parkinson disease. *Curr Treat Options Neurol*. 2017;19(6):22.
15. Kaplan KA, Harvey AG. Hypersomnia across mood disorders: a review and synthesis. *Sleep Med Rev*. 2009;13 (4):275–285.
16. Billiard M, Pasquie-Magnetto V, Heckman M, et al. Family studies in narcolepsy. *Sleep*. 1994;17(8 Suppl): S54–S59.
17. Black J, Reaven NL, Funk SE, et al. Medical comorbidity in narcolepsy: findings from the Burden of Narcolepsy Disease (BOND) study. *Sleep Med*. 2017;33:13–18.
18. Peppard PE, Young T, Barnet JH, Palta M, Hagen EW, Hla KM. Increased prevalence of sleep-disordered breathing in adults. *Am J Epidemiol*. 2013;177(9):1006–1014.
19. Malhotra A, Berry RB, White DP, eds. *Central Sleep Apnea*. Philadelphia, PA: Lippincott Williams and Wilkins; 2004. Carney PR, Berry RB, Geyer JD, eds. Central sleep disorders.
20. Javaheri S. Sleep disorders in systolic heart failure: a prospective study of 100 male patients. The final report. *Int J Cardiol*. 2006;106(1):21–28.
21. Pagel JF, Kwiatkowski C, Parnes B. The effects of altitude associated central apnea on the diagnosis and treatment of obstructive sleep apnea: comparative data from three different altitude locations in the mountain west. *J Clin Sleep Med*. 2011;7(6):610–615a.
22. Sin DD, Fitzgerald F, Parker JD, Newton G, Floras JS, Bradley TD. Risk factors for central and obstructive sleep apnea in 450 men and women with congestive heart failure. *Am J Respir Crit Care Med*. 1999;160 (4):1101–1106.
23. Sivertsen B, Pallesen S, Stormark KM, Boe T, Lundervold AJ, Hysing M. Delayed sleep phase syndrome in adolescents: prevalence and correlates in a large population based study. *BMC Public Health*. 2013;13:1163.
24. Drake CL, Roehrs T, Richardson G, Walsh JK, Roth T. Shift work sleep disorder: prevalence and consequences beyond that of symptomatic day workers. *Sleep*. 2004;27(8):1453–1462.
25. Matwiyoff G, Lee-Chiong T. Parasomnias: an overview. *Indian J Med Res*. 2010;131:333–337.

I. GENERAL ISSUES

26. Sharpless BA, Barber JP. Lifetime prevalence rates of sleep paralysis: a systematic review. *Sleep Med Rev*. 2011;15(5):311–315.
27. Ivanenko A, Relia S. Sleep-related hallucinations. In: Kothare SV, Ivanenko A, eds. *Parasomnias: Clinical Characteristics and Treatment*. New York: Springer New York; 2013:207–220.
28. Mathew JL. Evidence-based management of nocturnal enuresis: an overview of systematic reviews. *Indian Pediatr*. 2010;47(9):777–780.
29. Licis AK, Desruisseau DM, Yamada KA, Duntley SP, Gurnett CA. Novel genetic findings in an extended family pedigree with sleepwalking. *Neurology*. 2011;76(1):49–52.
30. Ohayon MM, Guilleminault C, Priest RG. Night terrors, sleepwalking, and confusional arousals in the general population: their frequency and relationship to other sleep and mental disorders. *J Clin Psychiatry*. 1999;60 (4):268–276. quiz 277.
31. Levin R, Nielsen TA. Disturbed dreaming, posttraumatic stress disorder, and affect distress: a review and neurocognitive model. *Psychol Bull*. 2007;133(3):482–528.
32. Ohayon MM, Roth T. Prevalence of restless legs syndrome and periodic limb movement disorder in the general population. *J Psychosom Res*. 2002;53(1):547–554.
33. Dunietz GL, Lisabeth LD, Shedden K, et al. Restless legs syndrome and sleep-wake disturbances in pregnancy. *J Clin Sleep Med*. 2017;13(7):863–870.
34. Hornyak M, Trenkwalder C. Restless legs syndrome and periodic limb movement disorder in the elderly. *J Psychosom Res*. 2004;56(5):543–548.
35. Montplaisir J, Boucher S, Poirier G, Lavigne G, Lapierre O, Lesperance P. Clinical, polysomnographic, and genetic characteristics of restless legs syndrome: a study of 133 patients diagnosed with new standard criteria. *Mov Disord*. 1997;12(1):61–65.
36. Edinger JD, Kirby A, Lineberger MD, Loiselle M, Wohlgemuth W, Means MK. *The Duke Structured Interview for Sleep Disorders*. Durham, NC: Duke University Medical Center; 2004.
37. Wohlgemuth WK, Edinger JD, Fins AI, Sullivan Jr. RJ. How many nights are enough? The short-term stability of sleep parameters in elderly insomniacs and normal sleepers. *Psychophysiology*. 1999;36(2):233–244.
38. Lacks P, Morin CM. Recent advances in the assessment and treatment of insomnia. *J Consult Clin Psychol*. 1992;60(4):586–594.
39. Morin CM. *Insomnia: Psychological Assessment and Management*. New York: Guilford Press; 1993.
40. Netzer NC, Stoohs RA, Netzer CM, Clark K, Strohl KP. Using the Berlin Questionnaire to identify patients at risk for the sleep apnea syndrome. *Ann Intern Med*. 1999;131(7):485–491.
41. Morin CM, Vallières A, Ivers H. Dysfunctional beliefs and attitudes about sleep (DBAS): validation of a brief version (DBAS-16). *Sleep*. 2007;30(11):1547–1554.
42. Johns MW. A new method for measuring daytime sleepiness: the Epworth sleepiness scale. *Sleep*. 1991;14 (6):540–545.
43. Horne JA, Ostberg O. A self-assessment questionnaire to determine morningness–eveningness in human circadian rhythms. *Int J Chronobiol*. 1976;4(2):97–110.
44. Buysse DJ, Reynolds III CF, Monk TH, Berman SR, Kupfer DJ. The Pittsburgh Sleep Quality Index: a new instrument for psychiatric practice and research. *Psychiatry Res*. 1989;28(2):193–213.
45. Walters AS, LeBrocq C, Dhar A, et al. Validation of the International Restless Legs Syndrome Study Group rating scale for restless legs syndrome. *Sleep Med*. 2003;4(2):121–132.
46. Chung F, Subramanyam R, Liao P, Sasaki E, Shapiro C, Sun Y. High STOP-Bang score indicates a high probability of obstructive sleep apnoea. *BJA: Br J Anaesth*. 2012;108(5):768–775.
47. Morgenthaler T, Alessi C, Friedman L, et al. Practice parameters for the use of actigraphy in the assessment of sleep and sleep disorders: an update for 2007. *Sleep*. 2007;30(4):519–529.
48. Vallieres A, Morin CM. Actigraphy in the assessment of insomnia. *Sleep*. 2003;26:902–906.
49. Evenson KR, Goto MM, Furberg RD. Systematic review of the validity and reliability of consumer-wearable activity trackers. *Int J Behav Nutr Phys Act*. 2015;12:159.
50. Kushida CA, Littner MR, Morgenthaler T, et al. Practice parameters for the indications for polysomnography and related procedures: an update for 2005. *Sleep*. 2005;28(4):499–521.
51. Bathgate CJ, Edinger JD, Wyatt JK, Krystal AD. Objective, but not subjective short sleep duration associated with increased risk for hypertension in individuals with insomnia. *Sleep*. 2016;39(5):1037–1045.

REFERENCES

52. Vgontzas AN, Fernandez-Mendoza J, Liao D, Bixler EO. Insomnia with objective short sleep duration: the most biologically severe phenotype of the disorder. *Sleep Med Rev*. 2013;17(4):241−254.
53. Bathgate CJ, Edinger JD, Krystal AD. Insomnia patients with objective short sleep duration have a blunted response to cognitive behavioral therapy for insomnia. *Sleep*. 2017;40(1):1−12.
54. Littner MR, Kushida C, Wise M, et al. Practice parameters for clinical use of the multiple sleep latency test and the maintenance of wakefulness test. *Sleep*. 2005;28(1):113−121.
55. Vgontzas AN, Fernandez-Mendoza J. Insomnia with short sleep duration: nosological, diagnostic, and treatment implications. *Sleep Med Clin*. 2013;8(3):309−322.
56. Buysse DJ, Young T, Edinger JD, Carroll J, Kotagal S. Clinicians' use of the International Classification of Sleep Disorders: results of a national survey. *Sleep*. 2003;26(1):48−51.

CHAPTER

2

Treatment of Insomnia

Charles M. Morin[1,2], Simon Beaulieu-Bonneau[3]
and Janet M.Y. Cheung[4,5]

[1]School of Psychology, Université Laval, Québec, QC, Canada [2]CERVO Brain Research Centre, Québec, QC, Canada [3]Center for Interdisciplinary Research in Rehabilitation and Social Integration (CIRRIS), Québec, QC, Canada [4]School of Pharmacy, Faculty of Medicine and Health, The University of Sydney, Sydney, Australia [5]School of Psychology, Université Laval, Québec, QC, Canada

OUTLINE

Introduction	28
Epidemiology of Insomnia	28
Diagnostic Considerations	28
Conceptual Model of Insomnia	29
Current Treatment Options	29
Cognitive-Behavioral Therapies	30
Pharmacotherapy	33
Natural Products	36
Bright Light Therapy	37
Acupuncture	37
Summary of Outcome Evidence	38

Impact of Treatment on Nighttime and Daytime Symptoms	38
Combined Cognitive-Behavioral Therapy and Medication	43
Short-Term and Long-Term Outcomes	43
Treatment Delivery Models	44
Clinical and Practical Considerations	44
Areas for Future Research	45
Summary and Conclusion	45
References	46

Handbook of Sleep Disorders in Medical Conditions
DOI: https://doi.org/10.1016/B978-0-12-813014-8.00002-0

© 2019 Elsevier Inc. All rights reserved.

INTRODUCTION

Epidemiology of Insomnia

Insomnia is a highly prevalent condition in various clinical populations. It can be the primary complaint but, more typically, is associated or comorbid with another medical or psychiatric condition, or even with another sleep disorder such as a sleep-related breathing disorder. Between 6% and 12% of the general population meets criteria for an insomnia disorder, with an additional 20%−25% who presents insomnia symptoms without necessarily meeting all criteria for a disorder.[1] Prevalence estimates are significantly higher when specific patient populations are sampled. For instance, surveys of patients with chronic pain, cancer, and depression produce rates of insomnia that are three to four times higher than for otherwise healthy control subjects. Likewise, cross-sectional studies indicate that individuals with insomnia are two to four times more likely than those without insomnia to have a comorbid cardiovascular, respiratory, gastrointestinal, neurological, or chronic pain disorder.[2–5]

Evidence from longitudinal studies shows that persistent insomnia is associated with long-term negative outcomes for mental, physical, and occupational health. For instance, individuals with insomnia are twice as likely to develop depression as those without insomnia.[6] There is also an increased risk for negative medical outcomes associated with persistent insomnia including hypertension, diabetes, and myocardial infarcts. In a sample of 4794 male workers in Japan, difficulties initiating (odds ratio (OR) 1.96) and maintaining sleep (OR 1.88) were associated with increased risk of hypertension.[7] Insomnia symptoms such as difficulties initiating or maintaining sleep, or nonrestorative sleep were associated with higher risks of new incident cases of a myocardial infarct over an 11-year follow-up period among a large population-based sample ($n = 52, 610$) of adults in Norway.[8] When combined with an objectively measured short sleep duration, insomnia is associated with heightened risks for cardiometabolic morbidity and mortality.[9] In a recent study, mortality risks were significantly higher among individuals with persistent insomnia (adjusted hazard ratios of 1.58) relative to those with intermittent insomnia or no insomnia.[10] Finally, persistent insomnia is also associated with increased absences from work. Population-based data from the HUNT study in Norway showed that persistent insomnia was associated with significant risk of sick leaves lasting more than 4 weeks, as well as for permanent disability.[11]

Diagnostic Considerations

Insomnia is characterized by both nocturnal and diurnal symptoms. It involves a predominant complaint of dissatisfaction with sleep quality or duration, accompanied by difficulties initiating sleep at bedtime, frequent or prolonged awakenings during the night, or early morning awakening with an inability to return to sleep. These difficulties occur despite having adequate opportunity for sleep and these are associated with clinically significant distress or impairments of daytime functioning involving fatigue, decreased energy, impaired cognitive functions (e.g., attention, concentration, memory), and mood disturbances. Sleep difficulties must be present at least 3 nights or more per week and last for more than 3 months to meet criteria for an insomnia disorder.[12,13]

A diagnosis of insomnia disorder is made whenever these criteria are met and the sleep difficulties are not better explained or do not occur exclusively in the context of another sleep-related, mental, or medical condition. However, because insomnia is so commonly associated with medical and psychiatric disorders, and there is often a bidirectional relationship between these conditions (i.e., pain can produce sleep disturbances but sleep disturbances can also exacerbate pain), current diagnostic nosology no longer requires insomnia to be differentiated as primary or secondary to another condition. As will be argued in this chapter, psychological and behavioral factors are generally involved in persistent insomnia and, for this reason, it is better when treatments target insomnia directly, regardless of whether it is the principal diagnosis or comorbid to another disorder.

Conceptual Model of Insomnia

According to Spielman's model,[14] the course of insomnia is modulated by three types of factors: predisposing, precipitating, and perpetuating factors. Predisposing factors include female gender, increasing age, hyperarousal, and a family or personal history of insomnia; these factors increase the vulnerability to develop insomnia. However, there must be some precipitating events that trigger the initial insomnia episode, and these typically involve a significant life event (e.g., illness, hospitalization, separation, death of a loved one). Most people who are otherwise good sleepers will usually resume a normal sleep pattern once this initial trigger subsides or the individual adjusts to its more permanent presence. For other people, perhaps those with increased vulnerability, several psychological and behavioral factors are likely to come into play and transform what was an episode of insomnia into a chronic problem. Those perpetuating factors are generally learned, modifiable habits, behaviors, attitudes, and beliefs and usually play a role in the maintenance of insomnia over time, regardless of what caused the sleep difficulties initially.[15–17] For example, a person may have developed sleep disturbances following an accident with injuries causing significant pain. Over time, the individual may spend excessive amounts of time in bed trying to rest or to nap during the day, while maintaining irregular sleep–wake schedules, along with worrying about not sleeping and its impact on daytime functioning. Even after the pain has resolved, insomnia may have taken a life of its own due to conditioning factors. Contrary to pharmacotherapy, cognitive-behavioral therapy (CBT) will specifically seek to identify and modify these perpetuating factors.

CURRENT TREATMENT OPTIONS

Current therapeutic options for insomnia include CBTs, various classes of sleep-promoting medications, and complementary and alternative therapies. Although medication is the most widely used approach for managing insomnia, CBT is now recognized in several practice guidelines as first-line therapy for persistent insomnia.[18,19] A variety of complementary and alternative therapies are increasingly used for sleep difficulties, but there is little evidence on either their risks or benefits. In the next section, we describe these different therapeutic options and summarize the evidence regarding their benefits and limitations, with a greater emphasis on CBT.

Cognitive-Behavioral Therapies

Cognitive-behavioral therapies for insomnia (CBT-I) include several behavioral and cognitive interventions, and combinations of interventions, which have been adapted and validated for persistent insomnia. In the following text, the main CBT-I interventions will be described. For additional details on the nature and implementation of CBT-I, readers can consult one of the several treatment manuals intended for clinicians.[20–22] The evidence documenting the efficacy and durability of CBT-I on several sleep parameters and some daytime variables is robust (please see the "Summary of Outcome Evidence" section). Several medical and sleep organizations have recommended CBT-I as the first-line treatment for all adults with chronic insomnia, including the American College of Physicians,[18] the European Sleep Research Society,[19] and the Australasian Sleep Association.[23]

Sleep Restriction

Individuals with insomnia typically spend excessive amounts of time in bed, for instance, by going to bed early in the evening or sleeping in on the weekend. While this may be effective as a short-term solution, it will inevitably lead to fragmented sleep and poorer sleep quality in the end. Sleep restriction, or more appropriately restriction of time in bed, consists of limiting the amount of time spent in bed during the night to reflect the actual amount of sleep time.[24] This strategy creates a mild sleep deprivation in the first few nights or weeks of application, contributing to an enhanced homeostatic sleep drive and improved sleep efficiency. The main side effect related to this mild sleep deprivation is increased daytime sleepiness, which can lower vigilance and prolong reaction time.[25] Daytime sleepiness has to be monitored throughout treatment, and to avoid a potentially harmful degree of sleepiness, time spent in bed should not be reduced to less than 5 hours per night.

The first step in applying this strategy is to set an initial sleep window, which is the nightly amount of time allowable in bed. This sleep window, which generally involves a constant bedtime and rising time across both weekdays and weekends, is based on the average nightly sleep time over the past 1–2 weeks, which ideally is derived from the patient's daily sleep diary. For example, if a person sleeps on average 6 hours per night out of 9 hours spent in bed, the initial sleep window will be 6 hours. Adding 30 minutes to this base rate is an option in order to enhance compliance among patients who might be resistant to following this recommendation. The sleep window is usually prescribed for 1 week and then readjusted on a weekly basis based on mean sleep efficiency (total sleep time per night/time spent in bed per night $\times$ 100). The sleep window is increased by 15–20 minutes (advancing bedtime or postponing rising time) for the following week if sleep efficiency exceeds 85%, kept unchanged if sleep efficiency falls between 80% and 85%, or decreased by 15–20 minutes (postponing bedtime or advancing rising time) if sleep efficiency is lower than 80%. These guidelines should be kept flexible and adapted to each patient in order to optimize compliance and success. Weekly adjustments in the sleep window are made until an optimal sleep duration is reached, while maintaining good sleep efficiency ($>90\%$); typically, this can take 4–8 weeks. Restriction of time in bed is a very effective strategy, but it can be distressing for some individuals. For this reason, it can be adapted to foster compliance. For example, sleep compression is a variant of sleep restriction in which time in bed is gradually decreased on a weekly basis rather than reduced drastically at first and then gradually increased.

Stimulus Control Therapy

Individuals with insomnia often experience feelings of anticipatory anxiety, frustration, or discouragement that build up during the day and culminate around bedtime. While bedtime and the sleeping environment can be soothing for good sleepers, it can trigger the opposite for persons with insomnia, who may become aroused (e.g., tension, cognitive, or emotional arousal) and worried. Furthermore, these individuals can develop habits that are counterproductive for nighttime sleep, such as lying in bed for prolonged periods trying to force sleep, or napping during the day to recover from a poor night's sleep. The objectives of stimulus control therapy[26] are to reassociate temporal (bedtime and nighttime) and environmental (bed and bedroom) stimuli with sleep rather than arousal, anxiety, frustration, and insomnia and to establish a regular sleep–wake schedule. The main recommendations are to (1) reserve the last 30–60 minutes before bedtime to unwind (i.e., relaxing activities, presleep routine); (2) go to bed only when sleepy (not just fatigued); (3) when unable to fall asleep or fall back asleep in 15–20 minutes, get out of bed, go to another room, and engage in a calm activity until sleepiness returns; (4) use the bed and bedroom for sleep only and avoid other sleep-incompatible activities such as watching television, using mobile devices, or planning the next day (sexual activities are the only allowed exception); (5) maintain a regular rising time in the morning, regardless of the quantity or quality of sleep obtained; and (6) avoid daytime napping, especially after 3 pm. While they may appear quite simple initially, sustained adherence to all stimulus control instructions for several weeks is required for optimal outcome. Some strategies to enhance compliance with behavioral recommendations include seeking help and support from the bed partner and planning for potential challenges (e.g., devising a list of activities to counter sleepiness in the evening, preparing nonstimulating activities to engage in when having to get out of bed in the middle of the night).

Relaxation-Based Interventions and Mindfulness

The objective of relaxation-based interventions is to reduce physiological or cognitive arousal. Most relaxation procedures produce equivalent benefits so individual preferences and the type of arousal interfering with sleep should guide the selection of a particular method. Examples of popular techniques include progressive muscle relaxation for physiological arousal and mental imagery for cognitive arousal. Regardless of the method selected, it should be practiced during the day for several days before attempting to use it at bedtime or upon nighttime awakening. Professional guidance may be necessary during the initial stage of training. Relaxation should be used with the intention to reduce arousal but not to induce sleep, and thus it should be done out of bed, especially when used in conjunction with stimulus control therapy to not contravene the prescribed instructions.

Mindfulness stems from the tradition of meditation where individuals exercise an intentional awareness to present thoughts and sensations with acceptance and no judgment.[27–29] Mindfulness-based techniques play a role in reducing stress and regulating emotional reactivity. Both mechanisms are particularly useful in attenuating the cognitive and somatic arousal associated with insomnia. Mindfulness aims to teach the individual to become more attuned to the objective experience of their insomnia and subsequently gain deeper insight into the process of sleeping.[30,31] Attenuating the emotional reactivity to the insomnia experience can play an important role in cognitive restructuring and enhance the therapeutic potential of CBT-I.[27,29]

Cognitive Therapy

The objective of cognitive therapy is to identify and modify dysfunctional cognitions and cognitive processes that contribute to perpetuating insomnia, such as rigid and unhelpful beliefs about sleep, unrealistic expectations about sleep patterns, excessive monitoring of sleep-related cues, and disproportionate worrying and anticipation about sleep loss and consequences of insomnia.[32–35] Cognitive therapy techniques are more abstract and require more introspection from patients relative to behavioral therapy components of CBT-I. Therapists may use Socratic questioning and guided discovery to challenge sleep-related beliefs, attitudes, and expectations during therapy sessions, and cognitive restructuring therapy and behavioral experiments (i.e., testing the validity and usefulness of cognitions in real-life experiences; for instance comparing strategies to generate energy vs strategies to conserve energy) as homework assignments between sessions.

Key cognitive therapy principles include (1) keep realistic expectations about sleep and daytime functioning (e.g., there are major individual differences in sleep needs and patterns; nighttime awakenings are normal, what characterizes insomnia is a difficulty to go back to sleep); (2) reconsider the etiology of insomnia (i.e., insomnia is caused by multiple factors, some of which are difficult to modify while others can be altered); (3) avoid amplifying the daytime consequences of insomnia (i.e., insomnia is not entirely responsible for fatigue or other daytime symptoms; insomnia is unpleasant but not dangerous); (4) never try to force sleep (i.e., sleep cannot be voluntarily initiated, trying so will exacerbate the problem); (5) do not give too much importance to sleep (i.e., sleep is essential, but it should not be the focus of the entire day); and (6) develop some tolerance to the impact of sleep disturbances.

Sleep Hygiene Education

The terms CBT-I and sleep hygiene education are often confounded. Sleep hygiene is the preferred term for recommendations about lifestyles and environmental factors that may either interfere or promote sleep.[36] It is not an effective intervention when used alone and is often used as a control condition in clinical trials.[37] Nonetheless, sleep hygiene education is provided routinely as it may address factors exacerbating sleep difficulties even if it is not targeting the core underlying processes. Sleep hygiene education includes the following instructions, some of which may overlap with other behavioral interventions: (1) avoid stimulants such as caffeine and nicotine several hours before bedtime; (2) avoid alcohol around bedtime; (3) exercise regularly (ideally in late afternoon or early evening); (4) keep the bedroom environment quiet, dark, and comfortable; and (5) avoid using electronic devices around bedtime and in the bedroom. Basic information about normal sleep is often provided as well, because this knowledge can help patients distinguish normal from pathological sleep, thus diminishing excessive worry and concern.

Multifaceted Cognitive-Behavioral Therapy

While each intervention described previously can be used separately (with the exception of sleep hygiene education which is often insufficient), combining behavioral, cognitive, and psychoeducational components into a multifaceted CBT-I is the preferred approach[37] because it is more likely to address the multiple factors that perpetuate insomnia and potentially optimize treatment outcome.

Indications and Rationale

CBT-I is indicated for persistent insomnia, but there is no evidence regarding its impact for acute insomnia. There is strong evidence supporting the use of CBT-I with younger and older adults, but much less and weaker evidence for children and adolescents. While chronic insomnia can be presented as a disorder on its own, it is often comorbid with another medical (e.g., chronic pain, cancer) or psychiatric disorder (e.g., depression, anxiety). CBT-I is also indicated in these comorbid cases and can be initiated as the first-line treatment or concomitantly with a treatment for the comorbid condition.[38]

There are very few contraindications to using CBT-I. Of note, however, the procedure involving restriction of time in bed should be avoided with individuals with seizures or parasomnias such as sleepwalking, and it should be used with extreme caution with people with bipolar disorder, as the mild sleep deprivation associated with the strategy could trigger a manic or hypomanic episode. Caution and careful monitoring are required when using restriction of time in bed with persons who experience excessive daytime sleepiness or for whom sleepiness could be hazardous (e.g., truck drivers, heavy machinery operators). Other behavioral strategies, such as getting out of bed when unable to sleep, should be used with caution with people who may be at risk for falls, such as frail elderly or other individuals with mobility issues, particularly if using hypnotic medications.[39]

Pharmacotherapy

Several classes of medication are used for the management of insomnia, some with specific indications for insomnia and others that are used off-label. The choice of an agent is based on clinical assessment to align the pharmacokinetic profile of the medication with the patient's presenting complaint. Other points of consideration include prior treatment responses, patient preference, cost, treatment availability, and the presence of existing comorbidities.

Until recently, benzodiazepines have been the mainstay in the pharmacological management of insomnia after superseding barbiturates in the 1970s. This was followed by the development of the nonbenzodiazepine benzodiazepine receptor agonists (BZRAs), also known as "z-drugs" (i.e., zaleplon, zolpidem, zopiclone), which offered greater selectivity at the receptor site. More recent drug developments include the melatonin receptor agonist, ramelteon, and prolonged-release melatonin, as well as the orexin receptor antagonist, suvorexant. Other agents used in the management of insomnia (often off label) because of their sedating properties include antidepressants, antipsychotics, anticonvulsants, and over-the-counter sedating antihistamines.

Benzodiazepine Receptor Agonists

Benzodiazepine (BZD) receptor agonists act on the gamma-aminobutyric acid (GABA) type A (GABA$_A$) receptor. Activation at the receptor site leads to an influx of intracellular chloride and promotes an inhibitory effect on the central nervous system.[40] The GABA$_A$ receptor comprises several subunits with the α_1 receptor subunit primarily responsible for mediating the inhibiting effects of BZRAs. BZDs have activity on other GABA$_A$ receptor subunits, contributing to additional anxiolytic, muscle relaxant, and antiepileptic properties. BZD agents with short- or intermediate half-life (e.g., temazepam) should be used for insomnia management in order to minimize next-morning residual sedation. Common adverse effects of BZRAs include next-day drowsiness, cognitive impairment, motor incoordination, ataxia, dizziness, and gastrointestinal upset. Rebound insomnia upon sudden withdrawal can occur. There is also a risk of developing tolerance and dependence.

The newer nonbenzodiazepines BZRAs, commonly referred to as z-drugs, include zaleplon, zolpidem, zopiclone and eszopiclone. z-Drugs share fundamental pharmacodynamic features with BZDs, but their structural difference confers greater specificity for the alpha subunit of GABA, which theoretically offers a more advantageous safety profile.[41] However, zolpidem has generated much public interest in Non-rapid eye movement (NREM)-related parasomnias with patients advised to avoid concomitant use with alcohol or other central nervous system depressants. Prescribing guidelines further recommends a lower starting dose of 5 mg for women to minimize adverse effects.[42] The use of z-drugs is contraindicated in patients with advanced hepatic disease, myasthenia gravis, and pulmonary insufficiency.[43]

Evidence from randomized controlled trials indicates that BZRAs are effective agents to improve sleep continuity, primarily through reductions of sleep-onset latency and the amount of time awake after sleep onset, with an increased in total sleep time. These benefits are usually very fast (first night of use), but they may also disappear with nightly and prolonged use, although some studies have reported a lack of tolerance developed over periods of 6 and 12 months with zolpidem and eszopiclone.[18,44,45] The BZRAs have significant effects on sleep architecture, with an important reduction of slow-wave and rapid eye movement (REM) sleep, and an increase in Stage 2 sleep. These effects are less marked with the z-drugs than with the traditional benzodiazepines.

Chronobiotic Agents: Melatonin Agonists and Related Compounds

Chronobiotic agents act on the melatonin receptors located in the suprachiasmatic nucleus. The MT1 receptor offsets the alertness promotion signals of the suprachiasmatic nuclei to reduce sleep latency. MT2 plays a more important role in phase-shifting to alter the timing of sleep–wake patterns.[46] Prolonged-release melatonin for insomnia has only been extensively tested in adults aged 55 years and older to address age-related decline in melatonin levels.[47] At the recommended doses of 2 mg at night, it is generally well tolerated but common adverse effects may include headache, pharyngitis, dizziness, nausea, back pain, and digestive difficulties.

Ramelteon is a melatonin receptor agonist with greater selectivity for the MT1 receptor.[40] It is currently approved by the US Food and Drug Administration (FDA) for sleep-onset insomnia at the recommended dose of 8 mg taken 30 minutes before bedtime. Common adverse effects include drowsiness, tiredness, nausea, and dizziness.[45]

Orexin Receptor Antagonist

Orexin antagonists promote sleep by preventing orexin neuropeptides from binding to the receptor site, thereby weakening the wake-promoting system. Suvorexant is currently the only FDA-approved orexin antagonist that is indicated for the treatment of sleep-onset and sleep-maintenance insomnia. It is a dual orexin receptor antagonist with activity on the OX1R and OX2R receptor subtypes. The current recommended initial starting dose is 10 mg, with a maximal FDA-approved dosage of 20 mg daily. Common adverse effects include headaches, next-day drowsiness, dizziness, dry mouth, and diarrhea.[48]

Off-Label Use of Pharmacotherapy for Insomnia

Despite the availability of drugs indicated for insomnia, concerns over the potential for dependence and the adverse effects associated with BZRAs have led to the widespread off-label use of antidepressants, antipsychotics, and anticonvulsants. The proposed sleep-promoting mechanisms and current evidence will be briefly discussed for the respective therapeutic class. However, there are ongoing concerns over the limited efficacy and safety for the off-label use of these agents.

ANTIDEPRESSANTS

Antidepressants commonly prescribed for insomnia because of their sedating effects include trazodone and the tricyclic antidepressants (e.g., amitriptyline, doxepin, and mirtazapine).[49] The sleep-promoting properties of these antidepressants are usually mediated by some level of postsynaptic inhibition at the serotonergic, adrenergic, cholinergic, or histaminergic receptor site. Among these agents, trazodone and doxepin are the most clinically relevant.

Trazodone is frequently prescribed off-label for insomnia at a dose ranging between 25 and 50 mg.[46] In one study, trazodone 50 mg improved self-reported total sleep time, but other measures such as sleep-onset latency and wake time after sleep onset did not reach clinical significance when compared to placebo.[50] Common adverse effects at low doses include residual sedation and orthostatic hypotension mediated by peripheral adrenergic blockade. Trazodone is currently not recommended for use in the management of insomnia as the risks outweigh the benefits.[45]

Doxepin is the only tricyclic antidepressant that is FDA-approved for insomnia. At low doses of up to 6 mg, doxepin exerts selective antihistaminergic activity to promote sleep. Doxepin 3 mg was found to improve patient reported wake after sleep onset, total sleep time, and sleep efficiency with minimal next-morning residual sedation in elderly patients.[51] Common adverse effects include dry mouth, dry eyes, urinary retention, and constipation. Doxepin is currently recommended, albeit with weak evidence, for use in sleep-maintenance insomnia.[45]

ANTIPSYCHOTICS

Atypical antipsychotics commonly prescribed for insomnia include quetiapine and olanzapine. The sleep-promoting effects of these atypical antipsychotics are mainly mediated by the inhibition of histaminergic and serotonergic activity. In a clinical trial, olanzapine 5 mg demonstrated improvements in total sleep time, sleep efficiency, decreased wake time, slow-wave sleep, and subjective sleep quality.[52] Common adverse effects include drowsiness, dizziness, tremor, extrapyramidal side effects, and weight gain. These agents are currently not recommended for use in the management of insomnia due to safety concerns, particularly the metabolic consequences, which have not been adequately established at these lower doses.[19,45]

ANTICONVULSANTS

Anticonvulsants with GABAergic activity such as gabapentin, tiagabine, and pregabalin are often used off-label for managing insomnia. These agents are proposed to work by addressing the deficits in slow-wave sleep associated with insomnia.[53] At a titrated dose of 1800 mg, gabapentin showed improvements in slow-wave sleep but other outcomes such as reductions in arousals and awakenings did not reach statistical significance.[54] Common adverse effects within this therapeutic class include fatigue, dizziness, ataxia, dry mouth, and weight gain. Anticonvulsants can potentially be beneficial for patients with comorbid epilepsy, but there is currently limited evidence for these agents when used solely for sleep promotion. Current practice parameter guidelines specifically recommend against the use of tiagabine for the management of insomnia due to limited evidence for efficacy and information regarding its adverse effects.[45]

Over-the-Counter Sleep Aids

Most over-the-counter sleep aids contain a first-generation sedating antihistamine (e.g., doxylamine up to 25 mg or diphenhydramine up to 50 mg) either as single-ingredient products or in combination with analgesics such as acetaminophen. Sedating antihistamines cross the blood—brain barrier and inhibit the histaminergic influence on the cortical wake-promoting system. Their relatively long half-life may confer benefits for maintenance of insomnia but also increase the potential for next-morning residual sedation.[55] Sedating antihistamines also exert antagonistic activity on muscarinic receptors, which is primarily responsible for its anticholinergic adverse effects including blurred vision, dry mouth, and constipation. The elderly or patients with comorbidities are particularly susceptible to these adverse effects. Despite the widespread use of sedating antihistamines, there is currently limited evidence for their efficacy or safety.[19,45]

Natural Products

Natural products remain a popular choice among people with insomnia because they are perceived to be safer alternatives to pharmacological agents and are widely accessible. Valerian is the most commonly used natural products; its sleep-promoting properties are thought to be mediated through glutamic acid decarboxylase or the modulation of GABA and 5-hydroxytryptophan receptors.[56] Common adverse effects include drowsiness, dizziness, and the potential for allergic reactions. There is currently insufficient evidence on the efficacy and harms of valerian with recommendation against its use for the management of insomnia.[45]

Chamomile is another widely used natural product for promoting sleep. One of its flavonoids, *apigenin*, has demonstrated anxiolytic properties in laboratory studies. However, the evidence on the efficacy and safety of chamomile as a sleep-promoting agent remains inconclusive.[57]

The amino acid L-tryptophan, a precursor to serotonin, is available in natural forms in animal and plant products (e.g., poultry, milk, and sunflower seeds). It is also available as an over-the-counter supplement or a prescription product. Animal studies suggest that the sleep-promoting properties of tryptophan could be mediated through reduced nocturnal serotonergic activity and the elevation of melatonin.[58] Tryptophan is generally well tolerated but adverse effects may include gastrointestinal symptoms, headache, light-headedness, drowsiness, dry mouth and visual blurring, muscle weakness, and sexual problems.[57] However, precaution should be exercised when other serotonergic agents (e.g., antidepressants) are used, as tryptophan may potentiate the effects of serotonin, and increase the risk of serotonin toxicity.[59] Tryptophan is currently not recommended for the management of insomnia due to limited evidence for efficacy and potential harms.[45]

Bright Light Therapy

Light is an important zeitgeber influencing sleep and wakefulness.[60] Bright light therapy can help to reentrain misaligned circadian rhythms in individuals with sleep-onset insomnia (i.e., delayed circadian rhythm) or early morning awakening insomnia (i.e., advanced circadian rhythm).[61] Critical to the success of bright light therapy is the timing of the light exposure relative to the core body temperature minimum.[62,63]

Bright light exposure in the morning is useful for sleep-onset insomnia as it helps the individual phase advance their circadian rhythm. One suggested protocol involves 1 hour of bright exposure immediately after awakening. The wake schedule and light exposure can then be advanced by 15 minutes each morning until the desired wake time is achieved. Once the target wake time is achieved, 30 minutes of bright light exposure upon awakening is recommended for a further 2 weeks.[61] For individuals with sleep-maintenance insomnia, early evening bright light exposure can phase delay the circadian rhythm.[62] Individuals can stay up for 1 hour later than their habitual bedtime with maximum ambient light exposure while avoiding morning bright light exposure and naps.[61,63] A recent systematic review concluded that bright light therapy is an effective treatment for improving circadian outcomes and insomnia symptoms.[60]

Acupuncture

Acupuncture is a procedure grounded within the principles of Traditional Chinese Medicine where meridian points are pierced with fine needles for therapeutic purposes. In insomnia, different meridian points have been used across different studies ranging from the ears, leg, arm, wrist, and scalp.[64] A number of mechanisms have been proposed for the sleep-promoting effect of acupuncture such as the downregulation of wake-promoting neurotransmitters like norepinephrine, serotonin, histamine, dopamine, and acetylcholine.[65] Another proposed mechanism is the stimulation of nocturnal melatonin secretion, which promotes sleep.[66] Despite the popularity of alternative therapies, there is

SUMMARY OF OUTCOME EVIDENCE

Table 2.1 summarizes the clinical practice recommendations from the American Academy of Sleep Medicine for CBTs and pharmacotherapies. Fig. 2.1 presents the effect sizes from three key metaanalyses of insomnia therapies, including a metaanalysis of CBT for insomnia in patients with and without comorbid disorders (A), one for pharmacotherapy for insomnia, mostly primary in nature (B), and one showing the effect sizes associated with CBT for comorbid insomnia with medical or psychiatric disorder (C). The rest of this section focuses predominantly on the efficacy of CBT-I, addressing both short and long-term efficacy, potential added benefits of combining CBT-I with medication, and concludes with practical considerations for treatment delivery and implementation.

Impact of Treatment on Nighttime and Daytime Symptoms

The empirical evidence gathered from clinical trials regarding the efficacy of CBT-I has been summarized in several metaanalyses[69,72,73] and systematic reviews.[37,74,75] For example, a recent metaanalysis of 87 randomized controlled trials observed significant sleep improvements on the following sleep diary parameters and questionnaires (from highest to lowest effect size): Insomnia Severity Index total score ($g = 0.98$), sleep efficiency ($g = 0.71$), Pittsburgh Sleep Quality Index total score ($g = 0.65$), wake after sleep onset ($g = 0.63$), sleep onset latency ($g = 0.57$), subjective sleep quality ($g = 0.40$), number of awakenings ($g = 0.29$), and total sleep time ($g = 0.16$) (see Fig. 2.1A). For comparison, the effect sizes obtained with pharmacotherapy (Fig. 2.1B) were small to moderate for sleep-onset latency ($g = -0.24$), wake after sleep onset ($g = 0.21$), sleep efficiency ($g = 0.41$), and total sleep time ($g = 0.21$), with higher effect sizes for BZRA compared with antidepressants and traditional BZDs.

In terms of absolute changes, systematic reviews suggest that sleep-onset latency and wake after sleep onset are reduced by about 50% and, by the end of treatment, attain values near or below the 30-minute cutoff often used to characterize insomnia symptoms. Improvement in total sleep time is more modest, increasing on average from 6 to 6.5 hours, but additional gains are often made at follow-ups. While sleep diary is the primary outcome measure in most clinical trials of CBT-I, there is some evidence suggesting that similar improvements, albeit of smaller magnitude, can be seen on polysomnography[76,77] and actigraphy.[78,79] Overall, it is estimated that 70%−80% of the individuals with insomnia benefit from CBT-I, with about 50% achieving full remission and no longer meeting diagnostic criteria for insomnia disorder, leaving a substantial proportion who do not benefit from CBT or who still experience residual symptoms after treatment.

While the main emphasis of most clinical trials is on documenting the changes in nighttime sleep with CBT-I, there is increasing interest and effort toward examining the impact of treatment on daytime symptoms associated with insomnia. Indeed, insomnia is

SUMMARY OF OUTCOME EVIDENCE

TABLE 2.1 Summary of Clinical Practice Recommendations for Cognitive-Behavioral Therapies and Pharmacotherapies for Insomnia

COGNITIVE-BEHAVIORAL THERAPIES

Intervention	Recommendation	Recommendation Level
Psychological and behavioral interventions	Recommended for primary insomnia	Standard
	Recommended for comorbid insomnia	Standard
Restriction of time in bed	Recommended	Guideline
Stimulus control therapy	Recommended	Standard
Relaxation-based techniques	Recommended	Standard
Cognitive therapy	Insufficient evidence	No recommendation
Sleep hygiene education	Insufficient evidence	No recommendation
Multifaceted cognitive-behavioral therapy		
With or without relaxation	Recommended	Standard
Without cognitive therapy	Recommended	Guideline

PHARMACOTHERAPIES

Intervention	Recommendation (Level of Evidence)	Strength of Recommendation	Quality of Evidence	Benefits vs Harms
Benzodiazepine receptor agonists				
Temazepam	Recommended for sleep-onset and sleep-maintenance insomnia	Weak	Moderate	Benefits > harms
Triazolam	Recommended for sleep-onset insomnia	Weak	High	Benefits ≈ harms
Eszopiclone	Recommended for sleep-onset and sleep-maintenance insomnia	Weak	Very low	Benefits > harms
Zaleplon	Recommended for sleep-onset insomnia	Weak	Low	Benefits > harms
Zolpidem	Recommended for sleep-onset and sleep-maintenance insomnia	Weak	Very low	Benefits > harms

(Continued)

40
2. TREATMENT OF INSOMNIA

TABLE 2.1 (Continued)

PHARMACOTHERAPIES

Intervention	Recommendation (Level of Evidence)	Strength of Recommendation	Quality of Evidence	Benefits vs Harms
Melatonin and melatonin agonists				
Melatonin	Not recommended for sleep-onset or sleep-maintenance insomnia	Weak	Very low	Benefits ≈ harms
Ramelteon	Recommended for sleep-onset insomnia	Weak	Very low	Benefits > harms
Orexin receptor agonists				
Surovexant	Recommended for sleep-maintenance insomnia	Weak	Low	Benefits > harms
Medications used off-label				
Doxepin	Recommended for sleep-maintenance insomnia	Weak	Low	Benefits > harms
Trazodone	Not recommended for sleep-onset or sleep-maintenance insomnia	Weak	Moderate	Harms > benefits
Tiagabine	Not recommended for sleep-onset or sleep-maintenance insomnia	Weak	Very low	Harms > benefits
Over-the-counter and natural products				
Diphenhydramine	Not recommended for sleep-onset or sleep-maintenance insomnia	Weak	Low	Benefits ≈ harms
L-Tryptophan	Not recommended for sleep-onset or sleep-maintenance insomnia	Weak	High	Harms > benefits
Valerian	Not recommended for sleep-onset or sleep-maintenance insomnia	Weak	Low	Benefits ≈ harms

Note. These recommendations, and associated levels of evidence, are based on two seminal works by the American Academy of Sleep Medicine: the Practice Parameters for the Psychological and Behavioral Therapies of Insomnia[68] (levels of evidence: standard > guideline > option) and the Clinical Practice Guideline for the Pharmacologic Treatment of Chronic Insomnia in Adults[45] (strength of recommendation; strong > weak).

I. GENERAL ISSUES

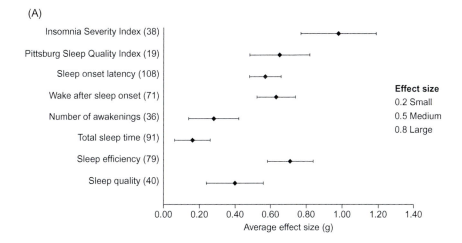

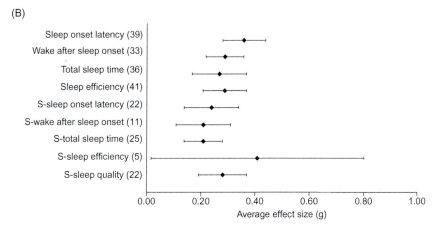

FIGURE 2.1 Effect sizes from metaanalyses of cognitive-behavioral therapy and pharmacological therapy for insomnia.
(A) Efficacy of cognitive-behavioral therapy for insomnia derived from a metaanalysis of 87 randomized controlled trial including a total of 3724 treated participants and 2579 control participants.[69] On the y axis, the numbers in parentheses refer to the number of studies for each dependent variable. The measure of effect size is Hedge's g.
(B) Efficacy of drug therapy for insomnia derived from a metaanalysis of 31 polysomnographic randomized controlled trials including a total of 3820 participants.[70] On the y axis, the numbers in parentheses refer to the number of studies for each dependent variable. The first four dependent variables are measured by polysomnography, and the last five variables are measured by sleep diary (i.e., those with "S-" before the variable name). The measure of effect size is Hedge's g.
(C) Efficacy of cognitive-behavioral therapy for comorbid insomnia derived from a metaanalysis of 23 randomized controlled trials including a total of 1379 treated participants.[71] On the y axis, the numbers in parentheses refer to the number of studies for each dependent variable. The measure of effect size is Cohen's d. CBT, Cognitive-behavioral therapy; ISI, Insomnia Severity Index; PSQI, Pittsburgh Sleep Quality Index; SOL, sleep-onset latency; WASO, wake after sleep onset; TST, total sleep time; SE, sleep efficiency.

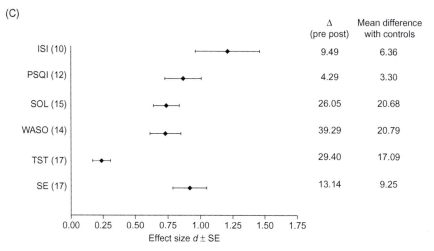

FIGURE 2.1 (Continued)

considered a 24-hour disorder and the perceived impact of daytime symptoms is often the primary reason evoked by individuals to seek treatment. Several studies found that CBT-I produces improvements in some of the main areas of daytime consequences of insomnia, including fatigue, reduced quality of life, and depression and anxiety symptoms.[80–84] However, there is no evidence of treatment-related changes in objective measures of cognitive functions such as attention, concentration, and memory.

Clinical improvements produced by CBT-I are generally similar for younger and older adults, for individuals with or without comorbidity, and for individuals using or not using sleep-promoting medications.[69] The efficacy of CBT-I in the elderly is supported by a metaanalysis reporting moderate effect sizes for sleep diary parameters (i.e., 0.52 for sleep-onset latency, 0.64 for wake after sleep onset, 0.76 for sleep quality).[73] Multifaceted CBT-I and restriction of time in bed (or sleep compression) are also recommended as validated interventions for older adults with insomnia, but not cognitive therapy, relaxation, or sleep hygiene education as standalone treatments due to insufficient evidence.[85]

With regard to comorbid insomnia, a general clinical rule is to treat both insomnia and the comorbid condition because sleep disturbances often persist if not addressed specifically when the coexisting disorder is treated, which can lead to exacerbation of or a relapse of the comorbid disorder. Data from a recent metaanalysis of 23 studies (1379 patients; see Fig. 2.1C) evaluating CBT for insomnia comorbid with medical (e.g., chronic pain, osteoarthritis, cancer) or psychiatric (e.g., posttraumatic stress disorder, major depressive disorder) disorder yielded large treatment effects (>0.8) for the Insomnia Severity Index (pre to post changes of 9.5 in total ISI score) and Pittsburgh Sleep Quality Index outcomes (change of 4.3), along with significant reductions in sleep latency (26 minutes) and wake after sleep onset (39 minutes), as well as increase in total sleep time (29 minutes) and sleep efficiency (13%). These improvements were well maintained up to 18 months after treatment completion.[71] Additional evidence regarding treatment effects for insomnia comorbid to specific medical conditions will be reviewed in the upcoming chapters. As for psychiatric comorbidity, the literature reviews indicate that CBT-I is effective for insomnia comorbid with depression, anxiety, posttraumatic stress disorder, substance abuse, and alcohol-related disorders, although the evidence is more limited for this latter class of disorder.[86,87] CBT-I can also be beneficial for improving symptoms of the comorbid condition, although with more modest results than for insomnia itself.

Combined Cognitive-Behavioral Therapy and Medication

CBT and pharmacotherapy are very different approaches for managing insomnia, notably in terms of mechanisms of action and efforts required from patients. While medication produces rapid symptomatic improvements, CBT-I takes longer to produce benefits, but these improvements are well sustained over time. Therefore, combining the two approaches could potentially lead to better outcomes, both in the short and long term. Results from the few studies comparing CBT-I (individual component or multifaceted intervention) with pharmacotherapy (temazepam, triazolam, zolpidem, or zopiclone) indicate that both modalities are effective in the short term (4−8 weeks). Sleep improvements occur rapidly after medication is initiated, but these improvements are usually not maintained once treatment is discontinued. The opposite pattern is observed with CBT-I, with a slightly delayed onset of sleep improvements, but well sustained over time.[76,77,82,88,89] When contrasting combined vs single treatment modalities, some studies report better maintenance of treatment gains with combined CBT and medication, while others report more variable outcomes with some patients showing sustained benefits and others experiencing a relapse after drug discontinuation.[76,77] Using a sequential approach (i.e., beginning with one treatment modality and switching to or adding on the other) is a promising avenue to optimize outcome.[90] Although very few studies have investigated sequential therapies yet, data seems to suggest that the most effective sequence is to use a combination of CBT-I and pharmacotherapy initially and then continue with CBT-I while discontinuing the medication.[89]

Short-Term and Long-Term Outcomes

As insomnia is often a recurrent or persistent condition, it is important to examine the efficacy of available interventions not only soon after treatment but also in the long term. As stated previously, the evidence is robust that medication produces rapid sleep improvements, often upon the very first night of use, whereas benefits derived from CBT may take a bit longer. Adding medication to CBT may produce an added value during the first 2−3 weeks of treatment, but this added benefit is often lost by the fifth or sixth week of treatment.[91] The evidence is very strong that CBT produces sustained sleep improvements over time on all sleep parameters.[37,71] Although total sleep time may improve only marginally by the end of CBT-I, mostly due to restriction of time in bed, it generally increases at follow-ups. Although most studies limit their follow-ups to 6 months after the end of treatment, a recent study reported sustained benefits on sleep diary parameters (i.e., sleep-onset latency, wake after sleep onset, sleep efficiency, and total sleep time) and the Insomnia Severity Index up to 2 years after CBT-I delivered alone or combined with medication.[92] Nevertheless, the risk for relapse remains even for individuals for whom CBT-I is effective. For this reason, long-term maintenance therapy or periodic booster sessions might help patients plan for future insomnia episodes and promote optimal sustained use of CBT-I strategies.

Treatment Delivery Models

As for most psychotherapies, CBT-I was developed initially as an individual face-to-face intervention. In order to increase access to CBT-I and minimize delivery costs, other treatment delivery models have been studied in recent years. First, group CBT-I appears to be a cost-effective alternative to individual therapy, yielding similar benefits despite reduced flexibility and personalization of treatment.[80,93] Second, self-help treatment using printed or video materials can also be effective to treat insomnia, with professional guidance through phone or email conversations yielding enhanced outcomes. [94] Third, Internet-based intervention (mostly self-help treatments with or without professional support) is a rapidly developing field with tremendous potential to limit costs and increase access, including to individuals who would not be reachable with traditional treatment models. Two recent metaanalyses, reviewing 15 and 11 clinical trials, respectively, suggest that Internet-delivered CBT-I is as effective as face-to-face therapy, with posttreatment effect sizes (Hedges' g) ranging from 0.21 to 1.09 for insomnia severity and sleep diary parameters.[95,96] These improvements were also well maintained at follow-up (ranging from 4 to 48 weeks in duration). Longer treatment duration and higher level of professional guidance were associated with greater therapeutic gains. Attrition remains an important problem with completely self-guided therapies for insomnia. Group, self-help, and Internet-based therapies can be cost effective, should be seen as complementary rather than a replacement to the direct face-to-face therapy, and could be integrated into primary care or sleep clinics.

Clinical and Practical Considerations

CBT-I is a time-limited approach, typically involving four to eight consultation sessions (usually lasting 45–60 minutes each) over a period of 4–8 weeks. The actual treatment duration may vary across patients depending on severity of symptoms, presence of comorbidity, and prior exposure to CBT-I and sleep medications. Therapy sessions are structured, sleep-focused, and involve homework assignments. There is a lack of data on the relationship between sleep improvements and compliance to homework assignments and to CBT-I recommendations in general. However, as CBT-I requires discipline and commitment for several weeks, it is essential to monitor comprehension, compliance, and reasons for noncompliance for each strategy throughout treatment. Providing reinforcement and troubleshooting early in treatment, using motivational interviewing techniques, and soliciting the collaboration of family members are ways to optimize treatment compliance.[97] While CBT-I is the first-line treatment recommended by several organizations, pharmacotherapy is generally the first option proposed to patients. Undergoing a CBT-I regimen while already taking a hypnotic medication (that may not be effective or may not be the preferred approach by the patient) is not contraindicated. In this case, the use of a structured, gradual withdrawal program, supervised by a clinician and combined with CBT-I or not, is effective in helping long-time hypnotic users discontinuing their medication.[98–100]

AREAS FOR FUTURE RESEARCH

Despite the scientific advances of recent years, there are still many challenges to ensure that treatments for insomnia are optimal and accessible for everyone. First, further research is warranted to validate optimal treatment algorithms in order to determine what should be our first-line treatment and what to do with nonresponders. Second, additional research is needed to evaluate different treatment sequences for comorbid insomnia and assess whether it is preferable to treat insomnia first, the comorbid condition first, or to treat them both concurrently. Third, research should target personalized therapies based on patient's preferences and insomnia phenotypes. Fourth, abbreviated treatment delivery models (e.g., telephone, Internet, apps) have to be further validated. Fifth, additional health-economic studies of the cost-benefits and cost-effectiveness of insomnia therapies are needed. Finally, a research should explore more efficient and judicious methods to use e-health technology to disseminate new knowledge to end users, that is, practitioners and patients.

SUMMARY AND CONCLUSION

Insomnia is the most prevalent of all sleep disorders and it is frequently comorbid with a variety of medical conditions such as pain, cancer, and neurological conditions, to name a few. It is important to treat sleep disturbances associated with medical conditions because such disturbance can exacerbate the course of the comorbid medical condition (e.g., chronic pain) and interfere with its management. Sleep medications can be helpful for acute insomnia but CBT is the first-line therapy for chronic insomnia. It is often assumed, falsely, that insomnia associated with a medical condition would not respond to treatment. However, there is increasing evidence showing that CBT-I is as effective for treating comorbid insomnia as it is for insomnia without comorbid disorders. It may be necessary to adapt some cognitive-behavioral interventions to the unique features of some medical conditions, but clinicians should be encouraged to incorporate CBT-I in the overall management of patients with insomnia and comorbid medical disorders.

KEY PRACTICE POINTS

- Insomnia is a highly prevalent condition in patients with medical or psychiatric disorders.
- Persistent insomnia is associated with negative long-term health outcomes.
- Among the different treatment options available, CBT is the treatment of choice for persistent insomnia, although there is a need to increase its accessibility in clinical practice.
- Even when insomnia is comorbid with another medical or psychiatric condition, it is essential to target directly the sleep difficulties because such difficulties often persist when treatment focuses only on the coexisting condition.
- Most CBT procedures are indicated for the management of comorbid insomnia, but it may be necessary to adapt some interventions to the unique features of comorbid disorder.
- Treatment outcomes obtained with CBT-I are slightly smaller or comparable for patients with or without comorbid medical or psychiatric conditions.

References

1. Morin CM, LeBlanc M, Belanger L, et al. Prevalence of insomnia and its treatment in Canada. *Can J Psychiatry.* 2011;56:540–548.
2. Pearson NJ, Johnson LL, Nahin RL. Insomnia, trouble sleeping, and complementary and alternative medicine: analysis of the 2002 national health interview survey data. *Arch Intern Med.* 2006;166:1775–1782.
3. Taylor DJ, Mallory LJ, Lichstein KL, et al. Comorbidity of chronic insomnia with medical problems. *Sleep.* 2007;30:213–218.
4. Roth T, Jaeger S, Jin R, et al. Sleep problems, comorbid mental disorders, and role functioning in the national comorbidity survey replication. *Biol Psychiatry.* 2006;60:1364–1371.
5. Savard J, Villa J, Ivers H, Simard S, Morin CM. Prevalence, natural course, and risk factors of insomnia comorbid with cancer over a 2-month period. *J Clin Oncol.* 2009;27:5233–5239.
6. Baglioni C, Battagliese G, Feige B, et al. Insomnia as a predictor of depression: a meta-analytic evaluation of longitudinal epidemiological studies. *J Affect Disord.* 2011;135:10–19.
7. Suka M, Yoshida K, Sugimori H. Persistent insomnia is a predictor of hypertension in Japanese male workers. *J Occup Health.* 2003;45:344–350.
8. Laugsand LE, Vatten LJ, Platou C, Janszky I. Insomnia and the risk of acute myocardial infarction: a population study. *Circulation.* 2011;124:2073–2081.
9. Fernandez-Mendoza J, Vgontzas AN. Insomnia and its impact on physical and mental health. *Curr Psychiatry Rep.* 2013;15:418.
10. Parthasarathy S, Vasquez MM, Halonen M, et al. Persistent insomnia is associated with mortality risk. *Am J Med.* 2015;128:268–275. e262.
11. Sivertsen B, Overland S, Neckelmann D, et al. The long-term effect of insomnia on work disability: the HUNT-2 historical cohort study. *Am J Epidemiol.* 2006;163:1018–1024.
12. American Psychiatric Association. Diagnostic and Statistical Manual of Mental Disorders, 5th *Edition (DSM-*5). *5th ed.* Washington, DC: American Psychiatric Publishing; 2013.
13. American Academy of Sleep Medicine. International Classification of Sleep Disorders: Diagnostic and Coding Manual, *3rd ed.* (ICSD-3). *3rd ed.* Westchester, IL: American Academy of Sleep Medicine; 2014.
14. Spielman AJ, Glovinsky PB. The varied nature of insomnia. In: Hauri P, ed. *Case Studies in Insomnia.* New York: Plenum Press; 1991:1–15.
15. Espie CA. Insomnia: conceptual issues in the development, persistence, and treatment of sleep disorder in adults. *Annu Rev Psychol.* 2002;53:215–243.

REFERENCES

16. Perlis M, Buysse DJ. Pathophysiology and models of insomnia. In: Kryger MH, Roth T, Dement WC, eds. *Principles and Practice of Sleep Medicine*. 5th ed. Philadelphia, PA: Saunders; 2011.

17. Harvey AG. A cognitive theory and therapy for chronic insomnia. *J Cogn Psychother*. 2005;19:41−59.

18. Qaseem A, Kansagara D, Forciea MA, Cooke M, Denberg TD. Management of chronic insomnia disorder in adults: a clinical practice guideline from the American College of Physicians. *Ann Intern Med*. 2016;165:125−133.

19. Riemann D, Baglioni C, Bassetti C, et al. European guideline for the diagnosis and treatment of insomnia. *J Sleep Res*. 2017;26:675−700.

20. Morin CM, Espie CA. *Insomnia: A Clinical Guide to Assessment and Treatment*. New York: Kluwer Academic/ Plenum; 2003.

21. Perlis ML, Smith MT, Jungquist C, Posner D. *Cognitive Behavioral Treatment of Insomnia: A Session-by-Session Guide*. New York: Springer-Verlag; 2005.

22. Edinger JD, Carney CE. *Overcoming Insomnia: A Cognitive-Behavioral Therapy Approach, Therapist Guide*. 2nd ed. New York: Oxford University Press, Inc; 2014.

23. Ree M, Junge M, Cunnington D. Australasian Sleep Association position statement regarding the use of psychological/behavioral treatments in the management of insomnia in adults. *Sleep Med*. 2017;36(suppl 1): S43−S47.

24. Spielman AJ, Saskin P, Thorpy MJ. Treatment of chronic insomnia by restriction of time in bed. *Sleep*. 1987;10:45−56.

25. Kyle SD, Miller CB, Rogers Z, et al. Sleep restriction therapy for insomnia is associated with reduced objective total sleep time, increased daytime somnolence, and objectively impaired vigilance: implications for the clinical management of insomnia disorder. *Sleep*. 2014;37:229−237.

26. Bootzin RR, Epstein D, Wood JM. Stimulus control instructions. In: Hauri P, ed. *Case Studies in Insomnia*. New York: Plenum Press; 1991:19−28.

27. Garland SN, Zhou ES, Gonzalez BD, Rodriguez N. The quest for mindful sleep: a critical synthesis of the impact of mindfulness-based interventions for insomnia. *Curr Sleep Med Rep*. 2016;2:142−151.

28. Lundh L-G. The role of acceptance and mindfulness in the treatment of insomnia. *J Cogn Psychother*. 2005;19:29−39.

29. Ong JC, Shapiro SL, Manber R. Combining mindfulness meditation with cognitive-behavior therapy for insomnia: a treatment-development study. *Behav Ther*. 2008;39:171−182.

30. Ong JC, Ulmer CS, Manber R. Improving sleep with mindfulness and acceptance: a metacognitive model of insomnia. *Behav Res Ther*. 2012;50:651−660.

31. Ong JC, Manber R, Segal Z, et al. A randomized controlled trial of mindfulness meditation for chronic insomnia. *Sleep*. 2014;37:1553−1563.

32. Morin CM. *Insomnia: Psychological Assessment and Management*. New York: Guilford Press; 1993.

33. Bélanger L, Savard J, Morin CM. Clinical management of insomnia using cognitive therapy. *Behav Sleep Med*. 2006;4:179−198.

34. Harvey AG. A cognitive model of insomnia. *Behav Res Ther*. 2002;40:869−893.

35. Harvey AG, Sharpley AL, Ree MJ, Stinson K, Clark DM. An open trial of cognitive therapy for chronic insomnia. *Behav Res Ther*. 2007;45:2491−2501.

36. Hauri PJ. Can we mix behavioral therapy with hypnotics when treating insomniacs? *Sleep*. 1997;20:1111−1118.

37. Morin CM, Bootzin RR, Buysse DJ, et al. Psychological and behavioral treatment of insomnia: update of the recent evidence (1998−2004). *Sleep*. 2006;29:1398−1414.

38. Smith MT, Huang MI, Manber R. Cognitive behavior therapy for chronic insomnia occurring within the context of medical and psychiatric disorders. *Clin Psychol Rev*. 2005;25:559−592.

39. Smith MT, Perlis ML. Who is a candidate for cognitive-behavioral therapy for insomnia? *Health Psychol*. 2006;25:15−19.

40. Neubauer DN. New and emerging pharmacotherapeutic approaches for insomnia. *Int Rev Psychiatry*. 2014;26:214−224.

41. Sanger DJ. The pharmacology and mechanisms of action of new generation, non-benzodiazepine hypnotic agents. *CNS Drugs*. 2004;18:9−15.

42. Krystal A, Attarian H. Sleep medications and women: a review of issues to consider for optimizing the care of women with sleep disorders. *Curr Sleep Med Rep*. 2016;2:218−222.

I. GENERAL ISSUES

48   2. TREATMENT OF INSOMNIA

43. Chaplin S, Wilson S, Nutt D. Z drugs: their properties and use in treating insomnia. *Prescriber*. 2013;24:32–33.
44. Krystal AD. A compendium of placebo-controlled trials of the risks/benefits of pharmacological treatments for insomnia: the empirical basis for U.S. clinical practice. *Sleep Med Rev*. 2009;13:265–274.
45. Sateia MJ, Buysse D, Krystal AD, Neubauer DN, Heald JL. Clinical practice guideline for the pharmacologic treatment of chronic insomnia in adults: an American Academy of Sleep Medicine clinical practice guideline. *J Clin Sleep Med*. 2017;13:307–349.
46. Roehrs T, Roth T. Insomnia pharmacotherapy. *Neurotherapeutics*. 2012;9:728–738.
47. Wade AG, Crawford G, Ford I, et al. Prolonged release melatonin in the treatment of primary insomnia: evaluation of the age cut-off for short- and long-term response. *Curr Med Res Opin*. 2011;27:87–98.
48. Bennett T, Bray D, Neville MW. Suvorexant, a dual orexin receptor antagonist for the management of insomnia. *Pharm Ther*. 2014;39:264–266.
49. Everitt H, Baldwin DS, Mayers A, Malizia AL, Wilson S. Antidepressants for insomnia. *Cochrane Database Syst Rev*. 2013;10. Available from: http://dx.doi.org/10.1002/14651858.CD010753.
50. Walsh JK, Erman M, Erwin CW, et al. Subjective hypnotic efficacy of trazodone and zolpidem in DSMIII–R primary insomnia. *Hum Psychopharmacol: Clin Exp*. 1998;13:191–198.
51. Krystal AD, Durrence HH, Scharf M, et al. Efficacy and safety of doxepin 1 mg and 3 mg in a 12-week sleep laboratory and outpatient trial of elderly subjects with chronic primary insomnia. *Sleep*. 2010;33:1553–1561.
52. Giménez S, Clos S, Romero S, et al. Effects of olanzapine, risperidone and haloperidol on sleep after a single oral morning dose in healthy volunteers. *Psychopharmacology*. 2007;190:507–516.
53. Bazil CW. Effects of antiepileptic drugs on sleep structure. *CNS Drugs*. 2003;17:719–728.
54. Foldvary-Schaefer N, De Leon Sanchez I, Karafa M, et al. Gabapentin increases slow-wave sleep in normal adults. *Epilepsia*. 2002;43:1493–1497.
55. Richardson GS, Roehrs TA, Rosenthal L, Koshorek G, Roth T. Tolerance to daytime sedative effects of H1 antihistamines. *J Clin Psychopharmacol*. 2002;22:511–515.
56. Shi Y, Dong J-W, Zhao J-H, Tang L-N, Zhang J-J. Herbal insomnia medications that target GABAergic systems: a review of the psychopharmacological evidence. *Curr Neuropharmacol*. 2014;12:289–302.
57. Yurcheshen M, Seehuus M, Pigeon W. Updates on nutraceutical sleep therapeutics and investigational research. *Evid Based Complement Alternat Med*. 2015;2015:9.
58. Esteban S, Nicolaus C, Garmundi A, et al. Effect of orally administered L-tryptophan on serotonin, melatonin, and the innate immune response in the rat. *Mol Cell Biochem*. 2004;267:39–46.
59. Fernstrom JD. Effects and side effects associated with the non-nutritional use of tryptophan by humans. *J. Nutr*. 2012;.
60. van Maanen A, Meijer AM, van der Heijden KB, Oort FJ. The effects of light therapy on sleep problems: a systematic review and meta-analysis. *Sleep Med Rev*. 2016;29:52–62.
61. Lack LC, Lovato N, Micic G. Circadian rhythms and insomnia. *Sleep Biol Rhythms*. 2017;15:3–10.
62. Kräuchi K. How is the circadian rhythm of core body temperature regulated? *Clin Auton Res*. 2002;12:147–149.
63. Lovato N, Lack L. The role of bright light therapy in managing insomnia. *Sleep Med Clin*. 2013;8:351–359.
64. Sok SR, Erlen JA, Kim KB. Effects of acupuncture therapy on insomnia. *J Adv Nurs*. 2003;44:375–384.
65. Kalavapalli R, Singareddy R. Role of acupuncture in the treatment of insomnia: a comprehensive review. *Complement Ther Clin Pract*. 2007;13:184–193.
66. Spence DW, Kayumov L, Chen A, et al. Acupuncture increases nocturnal melatonin secretion and reduces insomnia and anxiety: a preliminary report. *J Neuropsychiatry Clin Neurosci*. 2004;16:19–28.
67. Cheuk DKL, Yeung WF, Chung KF, Wong V. Acupuncture for insomnia. *Cochrane Database Syst Rev*. 2012;9. Available from: http://dx.doi.org/10.1002/14651858.CD005472.pub3.
68. Morgenthaler T, Kramer M, Alessi C, et al. Practice parameters for the psychological and behavioral treatment of insomnia: an update. An American Academy of Sleep Medicine report. *Sleep*. 2006;29:1415–1419.
69. van Straten A, Van der Swerdee T, Kleiboer A, et al. Cognitive and behavioral therapies in the treatment of insomnia: a meta-analysis. *Sleep Med Rev*. 2017;38:3–16.
70. Winkler A, Auer C, Doering BK, Rief W. Drug treatment of primary insomnia: a meta-analysis of polysomnographic randomized controlled trials. *CNS Drugs*. 2014;28:799–816.

I. GENERAL ISSUES

REFERENCES

71. Geiger-Brown JM, Rogers VE, Liu W, et al. Cognitive behavioral therapy in persons with comorbid insomnia: a meta-analysis. *Sleep Med Rev.* 2015;23:54–67.
72. Smith MT, Perlis ML, Park A, et al. Comparative meta-analysis of pharmacotherapy and behavior therapy for persistent insomnia. *Am J Psychiatry.* 2002;159:5–11.
73. Irwin MR, Cole JC, Nicassio PM. Comparative meta-analysis of behavioral interventions for insomnia and their efficacy in middle-aged adults and in older adults 55 + years of age. *Health Psychol.* 2006;25:3–14.
74. Garland SN, Johnson JA, Savard J, et al. Sleeping well with cancer: a systematic review of cognitive behavioral therapy for insomnia in cancer patients. *Neuropsychiatr Dis Treat.* 2014;10:1113–1124.
75. Trauer JM, Qian MY, Doyle JS, Rajaratnam SM, Cunnington D. Cognitive behavioral therapy for chronic insomnia: a systematic review and meta-analysis. *Ann Intern Med.* 2015;163:191–204.
76. Morin CM, Colecchi C, Stone J, Sood R, Brink D. Behavioral and pharmacological therapies for late-life insomnia: a randomized controlled trial. *JAMA.* 1999;281:991–999.
77. Wu R, Bao J, Zhang C, Deng J, Long C. Comparison of sleep condition and sleep-related psychological activity after cognitive-behavior and pharmacological therapy for chronic insomnia. *Psychother Psychosom.* 2006;75:220–228.
78. Espie CA, MacMahon KM, Kelly HL, et al. Randomized clinical effectiveness trial of nurse-administered small-group cognitive behavior therapy for persistent insomnia in general practice. *Sleep.* 2007;30:574–584.
79. Edinger JD, Wohlgemuth WK, Radtke RA, Coffman CJ, Carney CE. Dose-response effects of cognitive-behavioral insomnia therapy: a randomized clinical trial. *Sleep.* 2007;30:203–212.
80. Bastien CH, Morin CM, Ouellet MC, Blais FC, Bouchard S. Cognitive-behavioral therapy for insomnia: comparison of individual therapy, group therapy, and telephone consultations. *J Consult Clin Psychol.* 2004;72:653–659.
81. Belleville G, Cousineau H, Levrier K, St-Pierre-Delorme ME. Meta-analytic review of the impact of cognitive-behavior therapy for insomnia on concomitant anxiety. *Clin Psychol Rev.* 2011;31:638–652.
82. Jacobs GD, Pace-Schott EF, Stickgold R, Otto MW. Cognitive behavior therapy and pharmacotherapy for insomnia: a randomized controlled trial and direct comparison. *Arch Intern Med.* 2004;164:1888–1896.
83. Savard J, Simard S, Ivers H, Morin CM. Randomized study on the efficacy of cognitive-behavioral therapy for insomnia secondary to breast cancer, Part I: Sleep and psychological effects. *J Clin Oncol.* 2005;23:6083–6096.
84. Morin CM, Beaulieu-Bonneau S, Belanger L, et al. Cognitive-behavior therapy singly and combined with medication for persistent insomnia: impact on psychological and daytime functioning. *Behav Res Ther.* 2016;87:109–116.
85. McCurry SM, Logsdon RG, Teri L, Vitiello MV. Evidence-based psychological treatments for insomnia in older adults. *Psychol Aging.* 2007;22:18–27.
86. Taylor DJ, Pruiksma KE. Cognitive and behavioural therapy for insomnia (CBT-I) in psychiatric populations: a systematic review. *Int Rev Psychiatry.* 2014;26:205–213.
87. Brooks AT, Wallen GR. Sleep disturbances in individuals with alcohol-related disorders: a review of cognitive-behavioral therapy for insomnia (CBT-I) and associated non-pharmacological therapies. *Subst Abuse.* 2014;8:55–62.
88. Sivertsen B, Omvik S, Pallesen S, et al. Cognitive behavioral therapy vs zopiclone for treatment of chronic primary insomnia in older adults: a randomized controlled trial. *JAMA.* 2006;295:2851–2858.
89. Morin CM, Vallières A, Guay B, et al. Cognitive behavioral therapy, singly and combined with medication, for persistent insomnia: a randomized controlled trial. *JAMA.* 2009;301:2005–2015.
90. Morin CM. Combined therapeutics for insomnia: should our first approach be behavioral or pharmacological? *Sleep Med.* 2006;7(Suppl 1):S15–S19.
91. Morin CM, Beaulieu-Bonneau S, Ivers H, et al. Speed and trajectory of changes of insomnia symptoms during acute treatment with cognitive-behavioral therapy, singly and combined with medication. *Sleep Med.* 2014;15:701–707.
92. Beaulieu-Bonneau S, Ivers H, Guay B, Morin CM. Long-term maintenance of therapeutic gains associated with cognitive-behavioral therapy for insomnia delivered alone or combined with zolpidem. *Sleep.* 2017;40.
93. Koffel EA, Koffel JB, Gehrman PR. A meta-analysis of group cognitive behavioral therapy for insomnia. *Sleep Med Rev.* 2014;.
94. Ho FY, Chung KF, Yeung WF, et al. Self-help cognitive-behavioral therapy for insomnia: a meta-analysis of randomized controlled trials. *Sleep Med Rev.* 2014;.

I. GENERAL ISSUES

95. Seyffert M, Lagisetty P, Landgraf J, et al. Internet-delivered cognitive behavioral therapy to treat insomnia: a systematic review and meta-analysis. *PLoS One*. 2016;11:e0149139.
96. Zachariae R, Lyby MS, Ritterband LM, O'Toole MS. Efficacy of internet-delivered cognitive-behavioral therapy for insomnia—a systematic review and meta-analysis of randomized controlled trials. *Sleep Med Rev*. 2016;30:1−10.
97. Ruiter Petrov ME, Lichstein KL, Huisingh CE, Bradley LA. Predictors of adherence to a brief behavioral insomnia intervention: daily process analysis. *Behav Ther*. 2014;45:430−442.
98. Belleville G, Guay C, Guay B, Morin CM. Hypnotic taper with or without self-help treatment: a randomized clinical trial. *J Consult Clin Psychol*. 2007;75:325−335.
99. Lichstein KL, Nau SD, Wilson NM, et al. Psychological treatment of hypnotic-dependent insomnia in a primarily older adult sample. *Behav Res Ther*. 2013;51:787−796.
100. Morgan K, Gregory P, Tomeny M, David BM, Gascoigne C. Self-help treatment for insomnia symptoms associated with chronic conditions in older adults: a randomized controlled trial. *J Am Geriatr Soc*. 2012;60:1803−1810.

Treatment of Breathing-Related Disorders

Aneesa M. Das[1] and Bernardo J. Selim[2]

[1]Division of Pulmonary, Critical Care and Sleep Medicine, Ohio State University, Columbus, OH, United States [2]Division of Pulmonary, Critical Care and Sleep Medicine, Mayo Clinic, Rochester, MN, United States

OUTLINE

Introduction	51	Treatment of Central Sleep Apnea Syndromes	63
Positive Airway Pressure Devices	52	Treatment of Sleep-Related Hypoventilation Disorders	66
Modes in Positive Airway Pressure Devices	52	Areas for Future Research	70
Longitudinal Care of Patients on Positive Airway Pressure	57	Conclusion	70
Treatment of Sleep-Related Breathing Disorders by Disease Type	59	References	71
Treatment of Obstructive Sleep Apnea	59		

INTRODUCTION

The International Classification of Sleep Disorders (ICSD-2) has grouped the sleep-related breathing disorders into obstructive sleep apnea (OSA) syndromes, central sleep apnea (CSA) syndromes, and sleep-related hypoventilation/hypoxemic disorders (alveolar hypoventilation). Although these disorders have diverse pathophysiological mechanisms and they are associated with different underlying medical illnesses (e.g., obesity) or medications (e.g., opioids), they frequently share associated symptoms such as sleep fragmentation and excessive daytime sleepiness. Treatment of this diverse group of disorders shares basic principles. In general, when the underlying medical condition is amenable to effective treatment, the associated sleep-disordered breathing may improve. Therefore, reductions in opioid dose are associated with improvements in opioid-related CSA, weight loss is associated with improved ventilation in obesity hypoventilation syndrome, and pharmacological optimization of cardiac function yields improvements in Cheyne–Stokes respiration. However, at times, correction of the underlying problem is not feasible or achievable in a timely manner. In those cases, the clinician has a variety of other treatment modalities based on sleep-disordered breathing type, severity, and comorbidities. For treatment of upper airway collapse, oral appliances and noninvasive positive airway pressure (PAP) devices are among the most commonly prescribed treatments. Upper airway surgery and hypoglossal nerve stimulation are reserved for those severe cases unresponsive (or intolerant) to PAP treatment. In those sleep-disordered breathing cases associated with ventilatory failure (e.g., neuromuscular diseases, obesity hypoventilation syndrome, OSA associated with chronic obstructive pulmonary disease (COPD)), noninvasive bilevel PAP (BPAP) devices are available for supportive ventilatory treatment. This chapter will focus on the description and indication of different treatment modalities across the spectrum of sleep-disordered breathing disorders, with special emphasis on commonly applied therapies such as oral appliances and noninvasive PAP devices.

POSITIVE AIRWAY PRESSURE DEVICES

PAP devices are portable generators abled to deliver positive air pressure to a patient via a mask. PAP devices are basically composed of a blower (or turbine), a single hose (a respiratory circuit), a heated humidifier, and a mask. Based on the flow and pressure sensors located in the device, a microprocessor-based controller is constantly adjusting the turbine speed (dynamic blower) in order to reach a preset positive pressure (device output).

Based on their settings (mode), PAP devices may either deliver a single, continuous (constant) PAP (CPAP) mode at all times, or a higher inspiratory positive airway pressure (IPAP) and a lower expiratory positive airway pressure (EPAP) in the form of BPAP mode, supporting or enhancing patient's breath (tidal volumes), and, hence, ventilation (Fig. 3.1; Table 3.1).

Modes in Positive Airway Pressure Devices

CPAP Mode

In CPAP mode, a PAP device will deliver a constant (fixed) pressure during inhalation and exhalation. In sleep medicine, this mode is mainly used to pneumatically splint the upper airway and prevent collapsibility associated with OSA (Fig. 3.1; Table 3.2). Instead

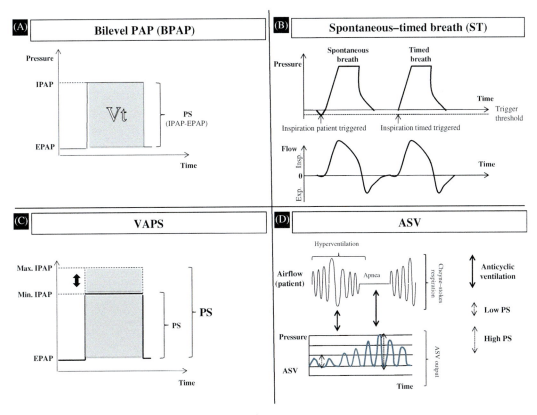

FIGURE 3.1 Positive airway pressure mode waveforms.

of a fixed pressure, autotitrating positive airway pressure (APAP) mode is set at a variable pressure and it adjusts the pressure in response to the patient's nocturnal respiratory events (e.g., obstructive apnea, hypopnea, etc.) using an internal algorithm (decision paths). The pressure is continuously monitored and adjusted in real time, based on changes in airflow.

Basic BPAP Modes: BPAP-S and BPAP-ST

BPAP mode may cycle between two pressures, a higher IPAP and a lower EPAP in response to the patient's respiratory effort (flow) or time to breathe. While the IPAP assists inspiration, the EPAP provides a pneumatic splint to maintain upper airway patency. The difference of the two pressures, called pressure support (PS) or delta pressure (ΔP), contributes to improved patient's ventilation. The larger the PS is, the larger the tidal volume (breaths' depth) generated (Fig. 3.1 and Table 3.2).

Based on the level of respiratory assistance needed by the patient, BPAP may be delivered in S (Spontaneous) mode (BPAP-S), ST (Spontaneous/Timed) mode (BPAP-ST), and T (Timed) mode:

3. TREATMENT OF BREATHING-RELATED DISORDERS

TABLE 3.1 Important Acronyms

Acronym	Definition
AHI	Apnea hypopnea index
CPAP	Continuous positive airway pressure
APAP	Autotitrating positive airway pressure
BPAP	Bilevel positive airway pressure
BPAP-S	Bilevel positive airway pressure in spontaneous mode
BPAP-ST	Bilevel positive airway pressure in spontaneous timed mode
ASV	Adaptive servo-ventilation
VAPS	Volume assured pressure support
OSA	Obstructive sleep apnea
CSA	Central sleep apnea
CSA–CSB	Central sleep apnea with Cheyne–Stokes breathing
CCHS	Congenital central alveolar hypoventilation syndrome
EPAP	Expiratory positive airway pressure
IPAP	Inspiratory positive airway pressure
HFrEF	Heart failure with reduced ejection fraction
HFpEF	Heart failure with preserved ejection fraction

BPAP-S: In this mode, the device senses when the patient is inhaling and exhaling and supplies the IPAP and EPAP accordingly.

BPAP-ST: Like in the BPAP-S, the device will support and/or augment breaths initiated by the patient, but it will also supply additional breaths should the patient breath rate fall below the preset backup rate.

Timed mode: Rarely used in sleep medicine, in this mode, a fixed breath rate and duration are applied by the device regardless of the patient's respiratory effort.

Advanced BPAP Modes

Depending on the type of software (mode), BPAP devices may interact differently with patient's breathing pattern, adjusting their pressures accordingly to propriety algorithms designed to achieve a specific respiratory assistance (e.g., ventilation support in alveolar hypoventilation, or regularization of breathing pattern in Cheyne–Stokes respiration, etc.) (Table 3.2).

VOLUME ASSURED PRESSURE SUPPORT

Volume assured pressure support (VAPS) mode proportionally self-adjusts IPAP to maintain a stable respiratory volume, hence ventilation. This mode tracks patient's

TABLE 3.2 Positive Airway Pressure Device Profiles

Positive Airway Pressure Devices				
Mode	Definition	Settings	Clinical Indications	Comments
Continuous positive airway pressure	Continuous positive airway pressure mode delivers a constant level of pressure during inspiration and expiration	• Single pressure setting, usually determined by assisted polysomnography • Range pressure setting, along which an auto-continuous positive airway pressure may self-adjust	• Obstructive sleep apnea • Obesity hypoventilation which hypercapnia improves with stabilization of upper airway resistance • Overlap syndrome (COPD + OSA)	• Continuous positive airway pressure provides a pneumatic splinting of upper airway • Continuous positive airway pressure does NOT ventilate • All breaths in this mode are spontaneous breaths

Bilevel positive airway pressure
  Spontaneous mode: The device senses patient's inhaling and exhaling flow and supplies the inspiratory positive airway pressure and expiratory positive airway pressure respectively
  Spontaneous timed mode: The device will support and/or augment breaths initiated by the patient, and supply additional breaths should the patient breath rate fall below the preset backup rate
	• Bilevel positive airway pressure mode delivers a higher inspiratory positive pressure and a lower expiratory positive airway pressure
	• Pressure support, the difference between inspiratory positive airway pressure and expiratory positive airway pressure may assist/augment tidal volume, hence ventilation
		• Inspiratory positive airway pressure
		• Expiratory positive airway pressure
		• Spontaneous mode No backup rate
		• Spontaneous timed mode: Backup respiratory rate. This ensures that the patient receives a minimum number of breaths per minute (minute ventilation)
			• Spontaneous mode: To support and/or augment breaths in patients with intact respiratory drive and respiratory muscle strength
			• Spontaneous timed mode for patients with impaired respiratory drive or neuromuscular disease with insufficient triggering of inspiratory positive airway pressure
				• Increased overall survival and quality of life in patients amyotrophic lateral sclerosis and obesity hypoventilation syndrome
				• Improved blood gases and sleep architecture
				• In CSR-HF, bilevel positive airway pressure-spontaneous mode may exacerbate the hyperventilation phase with worsening central apneas

(Continued)

TABLE 3.2 (Continued)

Positive Airway Pressure Devices				
Mode	**Definition**	**Settings**	**Clinical Indications**	**Comments**
Adaptive servo-ventilation	Adaptive servo-ventilation mode which self-adjust pressures to deliver anticyclical respiratory support to patient's breathing pattern	• Maximum and minimum inspiratory pressures • End expiratory pressure • Backup ventilatory rate and flow characteristics may be adjusted or set to default factory settings	• Treatment-emergent central sleep apnea • Opioid induced central sleep apnea *without* hypoventilation	• Contraindicated in patients with chronic, stable heart failure with reduced ejection fraction $\leq 45\%$ with predominant central sleep apneas • Not designed to provide ventilatory support in alveolar hypoventilation syndromes

Volume assured pressure support
    Average Volume Assured Pressure Support™ (Philips Respironics)
    Intelligent volume assured pressure support™ (ResMed)
	Feedback loop system which adjusts its pressures to deliver a set tidal volume or alveolar ventilation, guaranteeing a minute ventilation	• Target tidal volume or alveolar ventilation
		• End expiratory pressure
		• Minimum inspiratory pressure
		• Maximum inspiratory pressure
		• Respiratory rate
		• Flow characteristics may also be set, but default values are provided
			• In obesity hypoventilation syndrome patients who failed continuous positive airway pressure therapy
			• In obesity hypoventilation syndrome with predominant reduced respiratory drive over upper airway resistance
			• In neuromuscular disease patients
				• Different from bilevel positive airway pressure, Average Volume Assured Pressure Support adapts to disease progression
				• Average Volume Assured Pressure Support is not superior to bilevel positive airway pressure-spontaneous timed mode in obesity hypoventilation syndrome patients

spontaneous respiratory breath, proportionally adjusting IPAP, with subsequent augmentation of the patient's breathing when needed to reach a target respiratory volume set by provider. These respiratory targets could be either expiratory tidal volume or alveolar ventilation. In the case of average VAPS (AVAPS), an auto-EPAP function is also available to self-adjust to upper airway resistance (Fig. 3.1 and Table 3.2).

ADAPTIVE SERVO-VENTILATION

The adaptive servo-ventilation (ASV) device supports patient's breathing in an anticyclical fashion. When patients' ventilation decreases, the ASV mode may provide a higher respiratory support (i.e., minute ventilation or peak flow) by increasing inspiratory PS and/or respiratory rate, lowering (or none) respiratory support when the patient returns to a more stable breathing pattern. This "on-demand" support of patient's breathing allows a dampening effect of patient's drive to respiratory fluctuations, particularly useful in periodic breathing, secondary to a highly unstable respiratory drive system (Fig. 3.1 and Table 3.2).

Longitudinal Care of Patients on Positive Airway Pressure

The clinician who ordered the PAP for the patient is typically responsible for following up on the patient. Traditionally, this is either a sleep specialist, pulmonologist, or primary care provider. Nurse practitioners and physician assistants who are trained in this area can also provide longitudinal care. The first follow-up after the initial PAP set up is important in the first several weeks to assess adherence and effectiveness. Once a patient is found to be adherent to effective therapy, he/she should be seen at least annually.

While there is not a clear definition of PAP adherence, most sleep practitioners and medical insurance companies consider PAP adherence as usage of ≥ 4 hours per night for $\geq 70\%$ of nights. The greater the number of hours of PAP use during sleep, the more likely a patient is to benefit. That being said, only approximately half of the patients have long-term adherence (≥ 4 hours per night) to PAP therapy.[1] Common complaints heard as reasons for nonadherence include difficulty in tolerating pressure, mask discomfort, nasal congestion, or esthetics. Early troubleshooting of side effects with prompt intervention may improve the patient experience and thus PAP adherence. Although many strategies in mask interfaces or device comfort modalities are marketed to improve CPAP usage, there are few data to support this.[2]

Mask Fitting

Many PAP mask styles exist, although they typically fall into one of the three categories: full face, nasal, and nasal pillows or cushions. Full face masks, while larger, allow patients to breathe using both their mouth and nose. Nasal masks and nasal pillows/cushions are smaller and require that the patient breathe through their nose. All mask types can be used with the various PAP modes. While some literature supports lower pressure requirements with nasal interfaces compared to full face interfaces to control OSA, the most important factor is patient comfort and feasibility for ensured adherence.[3] A reasonable approach would be to start by offering patients nasal interfaces and moving toward full

58　　3. TREATMENT OF BREATHING-RELATED DISORDERS

face interfaces for poor tolerance or if patients are unable to keep their mouth closed during use. It is also important to note that patients with medical disorders that cause tremors or weakness may do better with a mask that slides over the head in one piece or uses magnets to fasten rather than clips for ease of use.

Extra Therapy Mode Features for Comfort

Additional features found in these devices are designed to increase comfort but do not have consistent data supporting association with improved adherence. *Pressure release during exhalation* features may offer a transient pressure relief during the expiratory phase of breathing, returning back to the prescribed pressure during inhalation. These features are generally identified by numbers, which progressively reflect increased pressure relief. The variation of pressure between inspiration and expiration using this feature is not thought to provide meaningful ventilation support.

A *ramp* feature will set the device to start at a lower positive pressure than the prescribed target setting. The pressure then gradually increases over a prespecified amount of time. Some patients may have difficulty in tolerating higher pressures when first turning on the device. Using the *ramp* feature allows them to fall asleep during the window where the pressure is still low.

Positive Airway Pressure Download Interpretation

PAP devices monitor data on PAP usage and effectiveness. Data can be downloaded using various methods, which depend upon the device manufacturer and model. While older models use secure digital cards and modems, newer devices may offer wireless transmission.

ADHERENCE MONITORING

Most PAP devices provide information on frequency and hours of use. The clinician can decide the period of time to assess adherence which is typically the past 1−3 months period.

EFFECTIVENESS MONITORING

Under optimal circumstances, apneas and hypopneas during in-laboratory polysomnography are defined by changes in airflow along with oximetry and arousals. However, since PAP devices do not measure oximetry or sleep including arousals from sleep, only airflow can be used to calculate a surrogate value for the apnea hypopnea index (AHI) known as AHI_{flow}. An AHI_{flow} of <10 events per hour is often considered sufficient control of events. However, this has to be interpreted in the context of the clinical setting for each individual patient as prospective outcomes trials have not been done to identify a clinically meaningful residual AHI_{flow}. Since the AHI_{flow} is only an estimate of true residual AHI, further testing may be required in a patient with persistent symptoms on PAP therapy.

TREATMENT OF SLEEP-RELATED BREATHING DISORDERS BY DISEASE TYPE

Treatment of Obstructive Sleep Apnea

Indications to Treat

The goals of OSA therapy are to normalize the AHI and oxygen saturation levels during sleep and to resolve symptoms associated with OSA including nonrestorative sleep, daytime sleepiness, snoring, and fragmented sleep. OSA is defined as having an AHI of 5 or greater during sleep. Treatment is indicated for all patients with moderate (AHI ≥ 15 but <30) to severe (AHI ≥ 30), independent of symptoms or comorbid disease. Treatment of patients with mild OSA (AHI of 5 or greater but <15) is somewhat controversial as most outcomes studies are associated with more significant sleep apnea. However, some patients with mild OSA may benefit from treatment if they have significant symptoms or comorbid disease. The American Academy of Sleep Medicine recommends treating mild OSA in the setting of symptoms of sleepiness or insomnia, waking with gasping or choking, hypertension, mood disorder, cognitive dysfunction, coronary artery disease, stroke, congestive heart failure (HF), atrial fibrillation, or type 2 diabetes.[4] It is also important to note which hypopnea definition was used to determine the AHI. Defining a hypopnea as a 30% decrease in nasal pressure associated with a $\geq 4\%$ oxygen desaturation may yield a much lower AHI than when defining hypopneas as a 30% decrease in nasal pressure associated with either an arousal or $\geq 3\%$ desaturation. Therefore, one may be more aggressive with treating mild OSA when using the more stringent hypopnea definition. Finally, one may take into consideration being more aggressive with treatment initiation for patients where alertness is critical including commercial drivers, pilots, or locomotive engineers.

Types of Devices

The most effective treatment for OSA remains PAP therapy which is still considered the standard of care for patients with OSA.

CONTINUOUS (CONSTANT) POSITIVE AIRWAY PRESSURE

There is robust evidence to show that CPAP can reduce the AHI, improve daytime sleepiness, and reduce blood pressure. A large metaanalysis of randomized controlled trials comparing CPAP to sham-CPAP showed a weighted mean difference in AHI reduction of -33.8 events per hour, a reduction of -2.0 scores on excessive sleepiness assessed by the Epworth Sleepiness Scale, and a reduction of -2.4 points on systolic and -1.3 points on diastolic blood pressure. There is still insufficient evidence to definitively prove that PAP therapy reduces mortality.[5] Observational studies are supportive of a mortality benefit,[6,7] but this has not been reproduced in randomized controlled trials.[5]

The most common PAP device used to treat OSA is CPAP. If CPAP is selected, then the clinician must choose between a fixed pressure or autotitrating mode of operation. In fixed pressure CPAP, the pressure remains relatively constant throughout use and respiration in order to maintain the positive transmural pressure gradient. Breaths are not augmented with additional pressure and all breaths must be initiated by the patient. The optimal required pressure is typically determined with a titration (Table 3.2). This is often done

I. GENERAL ISSUES

manually during an attended polysomnogram with either a full night study dedicated to titration or a split night with the first half being dedicated to diagnosis. While a split night study is often more efficient, it limits the time to determine the optimal pressure and may be insufficient. During the CPAP titration, the sleep technician will increase the pressure (typically by 1–2 cmH$_2$O) for observed obstructive apneas and hypopneas as well as signs of flow limitation including respiratory effort–related arousals and snoring. The optimal CPAP pressure should reduce the AHI to <5 events per hour while on supine-rapid eye movement (REM) sleep, as obstructive events are often more severe during both supine and REM sleep.[8] Of note, excessive pressure can cause central apneas to develop. Juxtacapillary stretch receptors known as J receptors are innervated by the vagus nerve and sense excessive stretching of the lung during large inspirations. The Hering–Breuer inflation reflex is then triggered, slowing the respiratory rate to prevent over-inflation of the lung.

Alternatively, an in-home titration may be done using APAP for several nights typically with a range from 5 to 20 cmH$_2$O. Then the pressure is fixed at the pressure that eliminates obstructive events for >90% or 95% of the time, depending on the device manufacturer. In-home titrations with APAP and full night attended titrations with CPAP have been compared in several randomized trials showing similar outcomes in adherence, sleepiness, and quality of life.[9–11] Most studies did not include patients with comorbid cardiorespiratory disease or other comorbid sleep disorders. Therefore, caution should be used in generalizing this data to these populations. When using an in-home titration, it is important to distinguish that the AHI$_{flow}$ reported on the APAP is only an estimate of the AHI, as oxygen desaturations and arousals are not measured.

AUTOTITRATING CONTINUOUS POSITIVE AIRWAY PRESSURE

Some patients may be prescribed APAP as an ongoing therapy rather than using the mode as a tool for titration. This may be a good option for patients where significant changes in pressure requirements are expected such as patient's undergoing weight loss surgery. APAP has not been shown to be a superior mode of treatment for OSA. Patients using APAP often do not require an initial in-laboratory CPAP titration. If using APAP long-term, the clinician should closely follow the downloaded report from the APAP. A persistently elevated AHI$_{flow}$ on the download may suggest that the pressure range needs to be narrowed if there is minimal mask leak. Alternatively, an in-laboratory titration may be required to determine both the best mode of therapy and pressure setting.

BILEVEL POSITIVE AIRWAY PRESSURE IN SPONTANEOUS MODE

BPAP-S is considered for treatment of OSA when the pressure requirement exceeds 15 cmH$_2$O or the patient develops difficulty with exhalation on the required CPAP pressure. Determining the optimal IPAP and EPAP is done with an in-laboratory titration. When switching from CPAP mode to BPAP in a titration, the EPAP is initially set at the lowest CPAP pressure where obstructive apneas were controlled, and the initial IPAP is set 4 cmH$_2$O higher.[8] For persistent obstructive apneas, both the EPAP and IPAP are increased and for obstructive hypopneas, the IPAP only is increased, thus augmenting the

TREATMENT OF SLEEP-RELATED BREATHING DISORDERS BY DISEASE TYPE

61

PS or delta pressure and subsequent mean airway pressure. The minimal recommended delta pressure is $4\,cmH_2O$ and the maximum is $10\,cmH_2O$ when treating OSA alone. However, when treating comorbid hypoventilation, larger delta pressures may be required (Fig. 3.1 and Table 3.2).

Oral Appliance Therapy

Oral appliance therapy (OAT) can be considered as an alternative treatment option for patients with mild to moderate OSA who do not tolerate or choose not to use PAP therapy. CPAP has consistently been shown to be more effective than OAT at reducing AHI and normalizing oxygen saturation in patients with OSA. However, patients often prefer OAT leading to improved adherence. The combination of both adherence and effectiveness is likely important in clinical outcomes. Both therapies have been associated with similar improvement in daily functioning, daytime sleepiness in randomized trials.[12,13] Among hypertensive patients, both therapies have shown similar reductions in blood pressure. OAT may be considered in patients with severe OSA who are intolerant to CPAP but the efficacy of therapy in this population is more variable.

There are generally two forms of OAT, mandibular advancement devices and tongue retention devices. Tongue retention devices use the force of suction to advance the tongue forward to increase the area of the posterior pharynx. It may be used in patients with minimal or no dentition. These devices do not advance the mandible itself. However, limited studies have been done on this form of therapy. Numerous mandibular advancement devices have been developed and have been studied more thoroughly. A mandibular advancement device works by advancing the mandible forward to open the airway and reduce snoring and obstructive events. Custom-made mandibular advancement devices are often made by a dentist who will help to adjust the device over time for optimal settings. Some dentition on both the upper and lower jaw is required to hold the devices in place. Once symptoms are controlled, a repeat assessment for OSA should be done with the device in place, using either a home sleep apnea test or an attended polysomnogram. Combination therapy with both a mandibular advancement device and CPAP is also available and may be appropriate for some patients but is less studied.

Behavioral Modification

Behavioral modification is indicated in conjunction with other therapies if the patient has a behaviorally modifiable risk factor. This includes weight loss and exercise, positional therapy, and avoidance of respiratory suppressants.

Weight loss and exercise should be encouraged in all patients with OSA who are overweight or obese. Weight loss in patients with OSA has been associated with improvement in AHI, blood pressure, insulin resistance, and typically daytime sleepiness.[14–16] While there is a consistent decrease in AHI proportional to the amount of weight loss, it is often insufficient for full resolution of OSA. Laparoscopic adjustable banding has not shown a more significant reduction in AHI than conventional weight loss despite greater reductions in weight.[17] However, this data may not be able to be generalized to other forms of bariatric surgery that affect absorption rather than just being restrictive. All patients with significant weight loss should be encouraged to maintain their reduced weight as increases in weight are associated with worsening of AHI.[18]

I. GENERAL ISSUES

Aerobic exercise training independent of weight loss may be associated with reductions in AHI and daytime sleepiness and can also be considered as adjunctive therapy.[19,20]

Some patients have a significant increase in severity of their OSA when in the supine position compared to nonsupine positions. Avoiding sleeping in the supine position may be encouraged in these patients. Devices ranging from sewing a tennis ball into the back of one's shirt to devices worn at the neck or chest that provide a subtle vibration that prevents patients from sleeping in a supine position can be used. A metaanalysis of prospective cohort and randomized controlled trials has shown that vibration devices produce a 54% reduction in AHI and an 84% reduction in time slept sleeping in the supine position in the short term.[21] However, long-term compliance with this therapy is less clear.

Respiratory suppressants such as alcohol, opiates, barbiturates, benzodiazepines, or antihistamines can worsen OSA. Acute alcohol ingestion is associated with increased frequency of obstructive events as well as more significant oxygen desaturations in the first hour of sleep.[22] Patients with OSA should avoid these substances at bed time if possible to reduce the risk of exacerbating their OSA.

Other Therapies

Several other therapies have been developed and are available to treat OSA. These therapies are typically considered if a patient is intolerant to CPAP and OAT is declined or ineffective.

Numerous surgeries have been described to treat OSA including uvulopalatopharyngoplasty, hyoid suspension, and maxillomandibular advancement among others. No surgeries have been found to be more effective than CPAP in adults and surgery should not be considered primary therapy in this population. In children, tonsillectomy may be considered as primary treatment. The most common site of pharyngeal collapse is the palate, so uvulopalatopharyngoplasty is one of the most commonly performed surgical procedures.[23] Treatment of nasal obstruction with either nasal steroids or surgery is sometimes considered in patients as adjunctive therapy to CPAP and can improve tolerance or reduce required pressures in patients with chronic nasal congestion.[24] Very few randomized trials have been done comparing surgery to a control. A metaanalysis pooled data from 18 small studies showed treatment success (defined as 50% reduction in AHI and/or ≤ 20) as 55%.[25] When using a definition of success of AHI $\leq$ 10, the success rate decreased to 31.5%. Surgical modification of the airway has varying effectiveness, but may be more appropriate for patients with significant correctable obstructive lesions such as tonsillar hypertrophy or micrognathia.

Neuromodulation is a growing area in the treatment of several types of sleep-disordered breathing. Hypoglossal nerve stimulation has been shown to treat OSA. A device is surgically implanted and stimulates the hypoglossal nerve to cause subsequent protrusion of the tongue in synchrony with inspiration due to a sensor placed in the intercostal muscles. The following are criteria for hypoglossal nerve stimulation therapy: age $\geq$ 22 years, AHI 20—65 events per hour, predominantly obstructive events, CPAP failure or intolerance, no complete concentric velopharyngeal collapse on drug-induced sleep endoscopy, and BMI $\leq$ 32 kg/m^2. A multicenter cohort design study showed a decreased AHI at 12 months by 68%, from a mean of 29.3 events per hour to 9.0 events per hour.[26]

Other less commonly used therapies for OSA include oral pressure therapy (OPT) and nasal EPAP devices. OPT consists of a mouthpiece that is connected to tubing and a device that applies a negative pressure inside the mouth that stabilizes the tongue and draws the soft palate forward, thus decreasing airway collapsibility. Unfortunately, this device has limited effectiveness. When treatment success was defined as $\geq 50\%$ residual AHI ≤ 10, the success rate of OPT treatment varied between 25% and 37%.[27] Nasal EPAP is a disposable valve that attaches to the nostrils with self-sticking adhesive. During inhalation, the valve opens, allowing air to enter the nose. During exhalation, the valve closes forcing air to pass through two small air holes, thus increasing nasal expiratory pressure. A metaanalysis of 18 studies shows a relative reduction in AHI by 53.2% with nasal EPAP use.[28] However, limited studies exist reporting success rates that include a maximal AHI cutoff.

Approach to Patient Care

CPAP therapy should be offered as first-line treatment for patients with moderate-to-severe OSA or symptomatic mild OSA. Some patients with mild OSA may choose to use OAT as first-line therapy. Behavioral modification, especially weight loss, should be implemented as adjunctive therapy when clinically appropriate. Due to high rates of CPAP nonadherence, close clinical follow-up is indicated. Alternative therapies including OAT should be offered to those who are persistently intolerant to CPAP.

Treatment of Central Sleep Apnea Syndromes

Based on the ICSD-3 classification, CSA syndromes are characterized by a group of sleep-related breathing disorders associated with decreased or absent respiratory efforts, combined with symptoms not different from OSA, such as excessive daytime sleepiness and/or sleep fragmentation.[29]

Central Sleep Apnea With Cheyne–Stokes Breathing

Although CSA–Cheyne–Stokes Breathing (CSB) is more prevalent in HF patients with either systolic dysfunction, known as HF with reduced ejection fraction (HFrEF) or diastolic left ventricular dysfunction, known as HF with preserved ejection fraction (HFpEF), it can also be encountered in other medical disorders such as acute ischemic stroke and renal failure.[30–32]

INDICATIONS TO TREAT

In HF patients, CSA–CSB is recognized as a marker of disease severity and impaired prognosis (e.g., increased mortality and hospitalization). Optimization of guideline-directed therapy to maximize cardiac function (e.g., cardiovascular pharmacotherapy) should be considered the first-line intervention in the management of CSA–CSB. Although considered a therapeutic option, the role of PAP continues to be controversial in this population.[33,34]

TYPE OF DEVICES

If PAP therapy is considered in HF patients with CSA−CSB, careful selection of assistive devices (e.g., CPAP, BPAP-ST, and ASV) should be based on patient's symptomatology, acuity, and cardiovascular status (e.g., left ventricular ejection fraction).

1. CPAP: Contradictory information exists regarding survival (transplant-free survival) with the use of CPAP in HFrEF patients with CSA−CSB.[35] In some studies, CPAP therapy has shown to improve hemodynamics (e.g., left ventricular ejection fraction), and a trend toward decreasing combined mortality−cardiac transplantation rate in only those HFs with CSA−CSB who comply with therapy, achieving a clinical significant suppression of the respiratory disturbance (reduction of AHI to <15 events per hour).[36,37]
2. BPAP-ST: There is consistent evidence that BPAP-ST is superior than CPAP in suppressing respiratory disturbances and equally effective to ASV in improving left ventricular ejection fraction in patients with HFrEF and CSA.[38,39]
3. ASV: Since the publication of the SERVE-HF trial, ASV use is currently contraindicated in those HF patients with a left ventricular ejection fraction of ≤45% and moderate−severe CSA due to increased risk of all-cause and cardiovascular mortality.[40] Based on the current parameter guidelines, ASV still remains an option to use in those HF patients with a left ventricular ejection fraction >45%, HFpEF, as well as in those CSA−CSB associated with acute stroke or renal failure.[41−43]
4. Phrenic nerve stimulation: Although in early research stages, studies using unilateral transvenous phrenic nerve stimulation to treat CSA in HF have shown a significant reduction of CSA and improvement of key polysomnographic parameters (oxygen desaturation index, REM sleep percentage, and sleep efficiency), sleepiness symptoms, and quality of life over 12 months of follow-up.[44,45]

APPROACH TO PATIENT CARE

If underlying HF is present, cardiac optimization with guideline-directed therapy should be prioritized before PAP therapy is contemplated. Based on the current data, it remains controversial—the decision to treat CSA with PAP therapy. However, a CPAP trial might be considered in those symptomatic HF patients with persistent CSA. If CPAP therapy fails to suppress CSA and improve symptoms, a trial of ASV could be implemented in those with HFrEF (EF > 45%), as well as in those CSA patients with HFpEF, acute stroke, or renal failure. Despite lack of consistent recommendation on the use of oxygen for the treatment of CSA in HF patients, oxygen may still be an effective therapy in reducing CSA events either alone or as a supplement to PAP therapy[46,47] (Table 3.2).

Central Sleep Apnea Due to Opioid Use

INDICATIONS TO TREAT

Patients with drug-induced sleep-related breathing disorders (e.g., opioids) often report significant excessive daytime sleepiness, usually attributed to the sedative effect of opioids. However, opioids may induce and maintain sleep-related breathing disorders such as mixed apnea (OSA and CSA), ataxic breathing as well as alveolar hypoventilation with sustained hypoxemia.[48,49]

TYPE OF DEVICES

Noninvasive ventilation studies assessing the effectiveness of PAP therapy in chronic opioid users have shown conflicting results.

1. CPAP: CPAP may reduce the number of respiratory events, generally by controlling obstructive respiratory events; however, centrally mediated sleep-disordered breathing may not only persist but also increase during the implementation of CPAP in this population.[50,51]
2. ASV: Conflicting data exist about the role of ASV in this group of patients. One study that did not show ASV efficacy likely related to suboptimal titration of the device.[50] In two other studies with optimal titrations, ASV was effective in the majority of cases.[51,52]
3. BPAP-ST: If CPAP and ASV are ineffective or if nocturnal hypercapnia develops, BPAP-ST should be considered for ventilation and respiratory support during CSA events.

APPROACH TO PATIENT CARE

As the effect of opioids on central apneas is dose dependent, discontinuation or reduction of the dose of opioid as tolerated is considered the first-line therapy to this problem.[49] However, if this approach is not feasible or effective, an initial trial of CPAP is recommended, followed by a trial of ASV, if CPAP is proven to be ineffective. If ASV fails to control respiratory events or if nocturnal hypercapnia develops, BPAP-ST should be considered (Table 3.2).

Primary or Idiopathic Central Sleep Apnea

INDICATIONS TO TREAT

Idiopathic CSA is a rare disease of unknown epidemiology, pathophysiology, and outcome. It is associated with sleep fragmentation, nonrestorative sleep, and excessive daytime sleepiness.

TYPE OF DEVICES

There are limited data on the use of assistive devices in the form of CPAP, BPAP with a backup respiratory rate (BPAP-ST), or ASV.

APPROACH TO PATIENT CARE

There are very limited data on idiopathic CSA therapy. There are no systematic studies but only small case series to support pharmacological intervention (e.g., acetazolamide, zolpidem, or triazolam), oxygen supplementation, or PAP therapy (CPAP, BPAP-ST, or ASV) in idiopathic CSA patients. If a treatment trial is considered, it should be performed only in symptomatic patients under very close supervision.

Current CSA practice parameters by the American Academy of Sleep Medicine and the European Respiratory Society recommend a trial of PAP therapy for the treatment of idiopathic CSA until further studies better define optimal therapeutic strategies. Symptomatic idiopathic CSA patients may be initiated to CPAP under polysomnographic guidance, followed by ASV or BPAP-ST, if CPAP fails to control respiratory events[53,54] (Table 3.2).

Treatment-Emergent Central Sleep Apnea (Previously Known as Complex Sleep Apnea Syndrome)

INDICATIONS TO TREAT

Treatment-emergent CSA is defined as emergence of CSA during CPAP therapy after elimination of predominantly obstructive events. However, treatment-emergent CSA is a dynamic phenomenon during which CSA may not persist over time in some patients.[55–57] Therefore, two therapeutic phenotypes can be identified, those patients with "transient CSA" which resolve with continued CPAP use, and those with "treatment-persistent" emergent CSA who will benefit from ASV.[55] Because of the difficulties in identifying these two phenotypes at PAP initiation during polysomnography, agreement on initial device mode therapy (CPAP vs ASV) continues to be somewhat controversial.[56]

TYPE OF DEVICES

1. CPAP: Even though the disorder most often becomes apparent during titration of CPAP, many studies and guidelines advocate an extended trial of CPAP. Retrospective studies suggest that the majority of patients will stabilize their breathing patterns on CPAP, although at a cost of significant dropout rates based on poor CPAP tolerance.[55,56]
2. BPAP-ST: Limited data discourage the use of BPAP-S in treatment-emergent CSA due to worsening of respiratory events.[58] On the other hand, the BPAP-ST may offer an alternative to CPAP or ASV in controlling CSA, although residual sleep fragmentation may still be present.[59]
3. ASV: A small phenotypic group of treatment-emergent CSA patients with persistently elevated CSA events may benefit from therapies focused on eliminating CSA. A prospective trial has shown that over 90% of those treated with ASV show both initial and long long-term control of CSA.[59] In comparison to CPAP and BPAP-ST, ASV treatment in patients with treatment-emergent CSA has resulted in a significantly greater decrease of the respiratory events in all positions and stages of sleep, lower sleep fragmentation, higher percentage of deep sleep (REM sleep) as well as improvement of daytime alertness.[59,60]

APPROACH TO PATIENT CARE

Unfortunately, there are no well-defined clinical or polysomnographic criteria to identify in advance the phenotype of CPAP nonresponders who may benefit from ASV therapy. Given the current limited data, patients with treatment-emergent CSA may be either initially treated with CPAP for up to 3 months before clinical evaluation and assessment for residual CSA, either using the device download or with a repeat titration. If feasibility of clinical follow-up is in question, an initial trial with ASV may be indicated (Table 3.2).

Treatment of Sleep-Related Hypoventilation Disorders

This group of disorders is defined by an elevated nocturnal $PaCO_2$ level (>45 mmHg), which may extend into the daytime.[29] The main respiratory abnormality may reside at any point along the brainstem respiratory control center [e.g., congenital central alveolar hypoventilation syndrome (CCHS), opioids' side effects], throughout the respiratory motor output unit (e.g., neuromuscular diseases), up to the thoracic–pulmonary unit (e.g., thoracic cage disorders, obstructive lung diseases such as COPD). For practical purposes, causes of nocturnal alveolar ventilation can be divided in two different groups, those patients with predominant ventilatory drive impairment (deficits in the autonomic/metabolic respiratory control system) in contraposition to those with abnormal pulmonary mechanics with increased mechanical load to breathe.

Obesity Hypoventilation Syndrome

Obesity hypoventilation syndrome is defined by the ICSD-3 classification manual as those obese patients (BMI $> 30 \, \text{kg/m}^2$) with extension of hypercapnia during daytime ($PaCO_2 > 45 \, \text{mmHg}$) when other alternative explanations for hypoventilation (e.g., COPD, use of opioids, or known CCHS) have been ruled out.[29] For therapeutic purposes, two obesity hypoventilation syndrome phenotypes can be identified— those obesity hypoventilation syndrome patients with predominant severe OSA (severity of OSA $>$ time spent in sustained oxygen desaturation), and those with milder OSA but larger sustained nocturnal hypoxemia secondary to decreased respiratory drive (severity of OSA $<$ time spent in sustained oxygen desaturation), sometimes referred to as "true Pickwickian syndrome."

INDICATIONS TO TREAT

Obesity hypoventilation syndrome commonly presents either in the intensive care unit as an acute-on-chronic hypercapnic respiratory failure, or to the sleep specialist for evaluation of suspected OSA. In the outpatient clinic, patients with obesity hypoventilation syndrome may complain of sleep fragmentation, nonrestorative sleep, with emphasis on early morning headaches, confusion, and daytime sleepiness. Left untreated, obesity hypoventilation syndrome is associated with serious cardiac and metabolic comorbidities (e.g., pulmonary hypertension) as well as increased mortality.[61,62]

TYPE OF DEVICES

1. CPAP: CPAP is generally considered the initial (first line) therapy in obesity hypoventilation syndrome patients with compensated hypercapnic respiratory failure and predominant OSA component.[63,64]
2. BPAP-S and BPAP-ST: If persistent sleep hypoventilation or sustained desaturations remain after elimination of upper airway obstruction by CPAP, a trial of BPAP-S is indicated with close follow-up of response to central events.[64,65] However, in those obesity hypoventilation syndrome patients with predominant lack of respiratory drive despite BPAP-S therapy (severity of OSA $<$ time spent in sustained oxygen desaturation), a trial of BPAP-ST or VAPS is recommended.[64] In comparison to CPAP, BPAP devices have shown to have a larger improvement of quality of life, spirometry, and 6-minute walk distance.[66]
3. VAPS: In comparison to BPAP, VAPS has shown to reach a larger improvement of $PaCO_2$ levels, and equal improvement in nocturnal oxygenation, sleep quality, and health-related quality of life.[66–69]

APPROACH TO PATIENT CARE

Although weight loss is an important long-term goal in the management of obesity hypoventilation syndrome, PAP devices are considered first-line treatment modalities until clinical significant weight loss is achieved. CPAP may be effective in many patients, especially those associated with severe OSA.[70] A trial of CPAP is indicated before BPAP-S or BPAP-ST, or AVAPS therapy is considered.[66] In patients with persistent nocturnal hypercapnia while on CPAP, or in those with predominant lack of respiratory drive ("true Pickwickian syndrome"), BPAP-ST and AVAPS are equality effective as long as both modes are set to reach the same amount of ventilatory support. If oxygen supplementation is added to obesity hypoventilation syndrome therapy, it should only be applied as an adjunct to BPAP treatment. If noninvasive mechanical ventilation fails, tracheostomy for mechanical ventilation may be required (Table 3.2).

Congenital Central Alveolar Hypoventilation Syndrome

CCHS is a rare genetic form of hypoventilation due to failure of automatic central control of breathing associated to a mutation of the PHOX2B gene (autosomal-dominant mode of inheritance with variable penetrance).[29] Although generally identified in children, this syndrome may present phenotypically later in life (adulthood) with respiratory failure following anesthesia or in associated with a minor respiratory illness.[71]

INDICATIONS TO TREAT

The spectrum of sleep-related breathing disorders in CCHS patients may range in severity from hypoventilation during non-REM sleep with adequate daytime ventilation, to complete apnea during sleep, and severe hypoventilation during wakefulness. If untreated, CCHS is associated with mental retardation, seizures, and core pulmonale with subsequent increased mortality. Patients have a better life expectancy by preventing the clinical consequences of hypoventilation with PAP-mediated ventilatory support.[54,71]

TYPE OF DEVICES

Given the current limited data (case series) of PAP therapy in CCHS, current recommendations are based on practice guidelines from expert opinions.

1. CPAP: Because CPAP is not a ventilatory device, there is no role of CPAP in the management of alveolar hypoventilation in CCHS patients.
2. BPAP-ST: BPAP-ST is considered an optional therapy in those older children and adults with the milder CCHS phenotype.[72–74]
3. Diaphragm pacing: Although limited publications are available, thoracoscopic placement of phrenic nerve electrodes for diaphragmatic pacing is a safe and effective treatment to restore alveolar ventilation and reverse PH in adult CCHS patients.[75,76]

APPROACH TO PATIENT CARE

In children with early manifestation of CCHS, portable positive pressure ventilator via tracheostomy is recommended in the first several years of life. Noninvasive ventilation with BPAP devices is a consideration in older children and adults with the milder CCHS

TREATMENT OF SLEEP-RELATED BREATHING DISORDERS BY DISEASE TYPE 69

phenotype. BPAP should not be used with a tracheostomy, only ventilators with alarm systems and backup battery are recommended.[71]

Sleep-Related Hypoventilation Due to a Medical Disorder, Neuromuscular Diseases, or Chest Wall Disorders

Neuromuscular diseases may be the cause of hypoventilation as a consequence of respiratory muscle insufficiency and/or dysfunction. Although most extensively described in amyotrophic lateral sclerosis (ALS), nocturnal alveolar hypoventilation due to neuromuscular disease can also be observed in Duchenne muscular dystrophy, myotonic dystrophy, and acid maltase deficiency among others.

INDICATIONS TO TREAT

In ALS patients, PAP therapy has not only shown to improve gas exchange and subjective sleep quality but also quality of life as well as survival, particularly in those compliant with PAP therapy ($\geq$4 hours) and without severe bulbar dysfunction.[77–80] A PAP trial should be cautiously considered only in those patients with severe bulbar impairment or with severe cognitive problems, in whom sleep-related symptoms may be related to respiratory impairment. However, low compliance with PAP therapy and a high risk of aspiration may limit the role of PAP therapy in this particular subgroup of patients.[81]

The decision to start PAP therapy in ALS patients is based on the combination of symptoms (orthopnea, sleep fragmentation, morning headaches, excessive daytime sleepiness), objective evidence of respiratory muscle insufficiency and/or dysfunction (pulmonary function tests), and the rate of disease progression. Current guidelines support initiation of PAP therapy by fulfilling any of these four objectives criteria: (1) forced vital capacity <50% of predicted (upright or supine), (2) maximal inspiratory pressure $< -60\,cmH_2O$ or Sniff Nasal Pressure $<40\,cmH_2O$ (upright or supine), (3) $PaCO_2$ >45 mmHg, or (4) overnight nocturnal desaturation <88% for 5 minutes.[81]

TYPE OF DEVICES

BPAP with backup rate (BPAP-ST or VAPS) and home-based ventilation are commonly used, frequently requiring relatively low PAP inspiratory pressures with infrequent need of adjustment in time.[82]

1. CPAP: There is no role of CPAP in the management of alveolar hypoventilation in neuromuscular disease patients.
2. BPAP-ST: BPAP-ST with relatively low inspiratory pressure is considered the first-line therapy in those neuromuscular disease patients in respiratory failure.[81]
3. VAPS: Compared to BPAP, VAPS has shown to deliver a lower median IPAP, and to improve compliance by reaching equal improvement in nocturnal gas exchange (pH, $PaCO_2$, nocturnal oxygenation), sleep quality, and respiratory muscle strength.[83]

APPROACH TO PATIENT CARE

A multidisciplinary team should coordinate and provide ongoing management and treatment for a patient with neuromuscular disease, including regular respiratory

I. GENERAL ISSUES

assessment and implementation of noninvasive ventilation. By the time PAP therapy is considered, the patient and family should be trained on using noninvasive ventilation and ventilator interfaces, as well as on emergency procedures and nighttime assistance, if the patient is unable to use the equipment independently or if the equipment fails. Initiation of therapy should be directed by personnel with extensive experience in titration of PAP therapy in neuromuscular disease patients. As neuromuscular diseases progress, nocturnal PAP support may extend during daytime, provided by home mechanical ventilators with internal battery and portability, which can be used with either mask ventilation, mouthpiece/sip ventilation, or via tracheostomy (Table 3.2).

AREAS FOR FUTURE RESEARCH

In the last two decades, medical therapies for sleep-disordered breathing have dramatically changed after the introduction of new materials in oral appliances, new generation of noninvasive PAP devices with self-adjusting capabilities, and the development of novel therapies in the field of neuromodulation (e.g., hypoglossal nerve stimulation). However, a growing gap exists between the development of new generation of medical devices and the clinical knowledge needed to support their use. As financial constraints push to move the diagnosis and treatment of breathing disorders out of the sleep laboratory into patient's home, the clinical physician should consider carefully to match the right medical device to the right clinical indication. Unfortunately, there are only a limited number of well-defined prospective clinical studies. Therefore, large randomized controlled trials with long-term follow-up are needed to better understand the proper role of some of these new technologies in the management of complex sleep disorders, including respiratory failure.

CONCLUSION

Sleep-disordered breathing is often complex and may have varied pathophysiology in a given patient. Various forms of sleep-disordered breathing including OSA, CSA, and obesity hypoventilation are not mutually exclusive and can coexist in a patient. Therapy, therefore, should be tailored to the individual. Adherence to PAP therapy can be affected by poor tolerance and early intervention of difficulties should be employed. As coexisting medical disorders change with patient's age, sleep-disordered breathing can change. Recurrent symptoms or changes in effectiveness of therapy should prompt reevaluation of therapy. The treatment of sleep-disordered breathing requires a longitudinal relationship between the clinician and the patient.

KEY PRACTICE POINTS

- Therapy must be optimized for the individual patient and tailored to their pathophysiology.
- Early troubleshooting patient intolerance and prompt intervention can improve adherence.
- Longitudinal follow-up is important to monitor effectiveness and adherence to therapy.
- A patient's disorder may be dynamic over time and require reevaluation of therapy.
- While PAP is the mainstay of therapy for sleep-disordered breathing, new treatment options are continuously emerging and may be considered.

References

1. Sawyer AM, Gooneratne NS, Marcus CL, Ofer D, Richards KC, Weaver TE. A systematic review of CPAP adherence across age groups: clinical and empiric insights for developing CPAP adherence interventions. *Sleep Med Rev*. 2011;15(6):343–356.
2. Weaver TE, Grunstein RR. Adherence to continuous positive airway pressure therapy: the challenge to effective treatment. *Proc Am Thorac Soc*. 2008;5(2):173–178.
3. Andrade RGS, Viana FM, Nascimento JA, et al. Nasal versus oronasal CPAP for obstructive sleep apnea treatment: a meta-analysis. *Chest*. 2018;153(3):665–674.
4. Duchna HW. Sleep-related breathing disorders—a second edition of the International Classification of Sleep Disorders (ICSD-2) of the American Academy of Sleep Medicine (AASM). *Pneumologie*. 2006;60(9):568–575.
5. Jonas DE, Amick HR, Feltner C, et al. Screening for obstructive sleep apnea in adults: evidence report and systematic review for the US Preventive Services Task Force. *JAMA*. 2017;317(4):415–433.
6. Campos-Rodriguez F, Martinez-Garcia MA, de la Cruz-Moron I, Almeida-Gonzalez C, Catalan-Serra P, Montserrat JM. Cardiovascular mortality in women with obstructive sleep apnea with or without continuous positive airway pressure treatment: a cohort study. *Ann Intern Med*. 2012;156(2):115–122.
7. Campos-Rodriguez F, Pena-Grinan N, Reyes-Nunez N, et al. Mortality in obstructive sleep apnea-hypopnea patients treated with positive airway pressure. *Chest*. 2005;128(2):624–633.
8. Kushida CA, Chediak A, Berry RB, et al. Clinical guidelines for the manual titration of positive airway pressure in patients with obstructive sleep apnea. *J Clin Sleep Med*. 2008;4(2):157–171.
9. Cross MD, Vennelle M, Engleman HM, et al. Comparison of CPAP titration at home or the sleep laboratory in the sleep apnea hypopnea syndrome. *Sleep*. 2006;29(11):1451–1455.
10. Masa JF, Jimenez A, Duran J, et al. Alternative methods of titrating continuous positive airway pressure: a large multicenter study. *Am J Respir Crit Care Med*. 2004;170(11):1218–1224.
11. Skomro RP, Gjevre J, Reid J, et al. Outcomes of home-based diagnosis and treatment of obstructive sleep apnea. *Chest*. 2010;138(2):257–263.
12. Phillips CL, Grunstein RR, Darendeliler MA, et al. Health outcomes of continuous positive airway pressure versus oral appliance treatment for obstructive sleep apnea: a randomized controlled trial. *Am J Respir Crit Care Med*. 2013;187(8):879–887.
13. Giles TL, Lasserson TJ, Smith BJ, White J, Wright J, Cates CJ. Continuous positive airways pressure for obstructive sleep apnoea in adults. *Cochrane Database Syst Rev*. 2006;(1)CD001106.
14. Araghi MH, Chen YF, Jagielski A, et al. Effectiveness of lifestyle interventions on obstructive sleep apnea (OSA): systematic review and meta-analysis. *Sleep*. 2013;36(10):1553–1562. 1562A-1562E.
15. Mitchell LJ, Davidson ZE, Bonham M, O'Driscoll DM, Hamilton GS, Truby H. Weight loss from lifestyle interventions and severity of sleep apnoea: a systematic review and meta-analysis. *Sleep Med*. 2014;15 (10):1173–1183.
16. Chirinos JA, Gurubhagavatula I, Teff K, et al. CPAP, weight loss, or both for obstructive sleep apnea. *N Engl J Med*. 2014;370(24):2265–2275.
17. Dixon JB, Schachter LM, O'Brien PE, et al. Surgical vs conventional therapy for weight loss treatment of obstructive sleep apnea: a randomized controlled trial. *JAMA*. 2012;308(11):1142–1149.

18. Peppard PE, Young T, Palta M, Dempsey J, Skatrud J. Longitudinal study of moderate weight change and sleep-disordered breathing. *JAMA*. 2000;284(23):3015–3021.
19. Iftikhar IH, Bittencourt L, Youngstedt SD, et al. Comparative efficacy of CPAP, MADs, exercise-training, and dietary weight loss for sleep apnea: a network meta-analysis. *Sleep Med*. 2017;30:7–14.
20. Iftikhar IH, Kline CE, Youngstedt SD. Effects of exercise training on sleep apnea: a meta-analysis. *Lung*. 2014;192(1):175–184.
21. Ravesloot MJL, White D, Heinzer R, Oksenberg A, Pepin JL. Efficacy of the new generation of devices for positional therapy for patients with positional obstructive sleep apnea: a systematic review of the literature and meta-analysis. *J Clin Sleep Med*. 2017;13(6):813–824.
22. Issa FG, Sullivan CE. Alcohol, snoring and sleep apnea. *J Neurol Neurosurg Psychiatry*. 1982;45(4):353–359.
23. Vroegop AV, Vanderveken OM, Boudewyns AN, et al. Drug-induced sleep endoscopy in sleep-disordered breathing: report on 1,249 cases. *Laryngoscope*. 2014;124(3):797–802.
24. Mickelson SA. Nasal surgery for obstructive sleep apnea syndrome. *Otolaryngol Clin North Am*. 2016;49(6):1373–1381.
25. Elshaug AG, Moss JR, Southcott AM, Hiller JE. Redefining success in airway surgery for obstructive sleep apnea: a meta analysis and synthesis of the evidence. *Sleep*. 2007;30(4):461–467.
26. Strollo Jr PJ, Soose RJ, Maurer JT, et al. Upper-airway stimulation for obstructive sleep apnea. *N Engl J Med*. 2014;370(2):139–149.
27. Nigam G, Pathak C, Riaz M. Effectiveness of oral pressure therapy in obstructive sleep apnea: a systematic analysis. *Sleep Breath*. 2016;20(2):663–671.
28. Riaz M, Certal V, Nigam G, et al. Nasal expiratory positive airway pressure devices (provent) for OSA: a systematic review and meta-analysis. *Sleep Disord*. 2015;2015:734798.
29. American Academy of Sleep Medicine. *International Classification of Sleep Disorders*. 3rd ed. Westchester, IL: American Academy of Sleep Medicine; 2014.
30. Yumino D, Wang H, Floras JS, et al. Prevalence and physiological predictors of sleep apnea in patients with heart failure and systolic dysfunction. *J Card Fail*. 2009;15(4):279–285. Available from: https://doi.org/10.1016/j.cardfail.2008.11.015. Epub2009 Jan 1021.
31. Sekizuka H, Osada N, Miyake F. Sleep disordered breathing in heart failure patients with reduced versus preserved ejection fraction. *Heart Lung Circ*. 2013;22(2):104–109. Available from: https://doi.org/10.1016/j.hlc.2012.08.006. Epub 2012 Oct 1026.
32. Johnson KG, Johnson DC. Frequency of sleep apnea in stroke and TIA patients: a meta-analysis. *J Clin Sleep Med*. 2010;6(2):131–137.
33. Khayat R, Abraham W, Patt B, et al. Central sleep apnea is a predictor of cardiac readmission in hospitalized patients with systolic heart failure. *J Card Fail*. 2012;18(7):534–540. Available from: https://doi.org/10.1016/j.cardfail.2012.05.003.
34. Khayat R, Jarjoura D, Porter K, et al. Sleep disordered breathing and post-discharge mortality in patients with acute heart failure. *Eur Heart J*. 2015;36(23):1463–1469. Available from: https://doi.org/10.1093/eurheartj/ehu522. Epub2015 Jan 1429.
35. Bradley TD, Logan AG, Kimoff RJ, et al. Continuous positive airway pressure for central sleep apnea and heart failure. *N Engl J Med*. 2005;353(19):2025–2033.
36. Kasai T, Kasagi S, Maeno K, et al. Adaptive servo-ventilation in cardiac function and neurohormonal status in patients with heart failure and central sleep apnea nonresponsive to continuous positive airway pressure. *JACC Heart Fail*. 2013;1(1):58–63. Available from: https://doi.org/10.1016/j.jchf.2012.11.002. Epub2013 Feb 1014.
37. Arzt M, Floras JS, Logan AG, et al. Suppression of central sleep apnea by continuous positive airway pressure and transplant-free survival in heart failure: a post hoc analysis of the Canadian Continuous Positive Airway Pressure for Patients with Central Sleep Apnea and Heart Failure Trial (CANPAP). *Circulation*. 2007;115(25):3173–3180. Epub 2007 Jun 3111.
38. Fietze I, Blau A, Glos M, Theres H, Baumann G, Penzel T. Bi-level positive pressure ventilation and adaptive servo ventilation in patients with heart failure and Cheyne–Stokes respiration. *Sleep Med*. 2008;9(6):652–659. Epub 2007 Nov 2019.
39. Teschler H, Dohring J, Wang YM, Berthon-Jones M. Adaptive pressure support servo-ventilation: a novel treatment for Cheyne–Stokes respiration in heart failure. *Am J Respir Crit Care Med*. 2001;164(4):614–619.

REFERENCES

40. Cowie MR, Woehrle H, Wegscheider K, et al. Adaptive servo-ventilation for central sleep apnea in systolic heart failure. *N Engl J Med*. 2015;373(12):1095–1105. Available from: https://doi.org/10.1056/NEJMoa1506459. Epub 1502015 Sep 1506451.
41. Aurora RN, Bista SR, Casey KR, et al. Updated adaptive servo-ventilation recommendations for the 2012 AASM guideline: "The treatment of central sleep apnea syndromes in adults: practice parameters with an evidence-based literature review and meta-analyses". *J Clin Sleep Med*. 2016;12(5):757–761. Available from: https://doi.org/10.5664/jcsm.5812.
42. O'Connor CM, Whellan DJ, Fiuzat M, et al. Cardiovascular outcomes with minute ventilation-targeted adaptive servo-ventilation therapy in heart failure: the CAT-HF Trial. *J Am Coll Cardiol*. 2017;69(12):1577–1587. Available from: https://doi.org/10.1016/j.jacc.2017.01.041.
43. Yoshihisa A, Suzuki S, Yamauchi H, et al. Beneficial effects of positive airway pressure therapy for sleep-disordered breathing in heart failure patients with preserved left ventricular ejection fraction. *Clin Cardiol*. 2015;38(7):413–421. Available from: https://doi.org/10.1002/clc.22412. Epub 22015 May 22412.
44. Abraham WT, Jagielski D, Oldenburg O, et al. Phrenic nerve stimulation for the treatment of central sleep apnea. *JACC Heart Fail*. 2015;3(5):360–369. Available from: https://doi.org/10.1016/j.jchf.2014.12.013. Epub2015 Mar 1011.
45. Jagielski D, Ponikowski P, Augostini R, Kolodziej A, Khayat R, Abraham WT. Transvenous stimulation of the phrenic nerve for the treatment of central sleep apnoea: 12 months' experience with the remedē® System. *Eur J Heart Fail*. 2016;18(11):1386–1393. Available from: https://doi.org/10.1002/ejhf.593. Epub2016 Jul 1384.
46. Franklin KA, Eriksson P, Sahlin C, Lundgren R. Reversal of central sleep apnea with oxygen. *Chest*. 1997;111(1):163–169.
47. Krachman SL, Nugent T, Crocetti J, D'Alonzo GE, Chatila W. Effects of oxygen therapy on left ventricular function in patients with Cheyne–Stokes respiration and congestive heart failure. *J Clin Sleep Med*. 2005;1(3):271–276.
48. Wang D, Teichtahl H, Drummer O, et al. Central sleep apnea in stable methadone maintenance treatment patients. *Chest*. 2005;128(3):1348–1356.
49. Walker JM, Farney RJ, Rhondeau SM, et al. Chronic opioid use is a risk factor for the development of central sleep apnea and ataxic breathing. *J Clin Sleep Med*. 2007;3(5):455–461.
50. Farney RJ, Walker JM, Boyle KM, Cloward TV, Shilling KC. Adaptive servoventilation (ASV) in patients with sleep disordered breathing associated with chronic opioid medications for non-malignant pain. *J Clin Sleep Med*. 2008;4(4):311–319.
51. Ramar K, Ramar P, Morgenthaler TI. Adaptive servoventilation in patients with central or complex sleep apnea related to chronic opioid use and congestive heart failure. *J Clin Sleep Med*. 2012;8(5):569–576. Available from: https://doi.org/10.5664/jcsm.2160.
52. Javaheri S, Malik A, Smith J, Chung E. Adaptive pressure support servoventilation: a novel treatment for sleep apnea associated with use of opioids. *J Clin Sleep Med*. 2008;4(4):305–310.
53. Aurora RN, Chowdhuri S, Ramar K, et al. The treatment of central sleep apnea syndromes in adults: practice parameters with an evidence-based literature review and meta-analyses. *Sleep*. 2012;35(1):17–40. Available from: https://doi.org/10.5665/sleep.1580.
54. Randerath W, Verbraecken J, Andreas S, et al. Definition, discrimination, diagnosis and treatment of central breathing disturbances during sleep. *Eur Respir J*. 2017;49(1):1600959. Available from: https://doi.org/10.1183/13993003.00959-2016. Print13992017 Jan.
55. Cassel W, Canisius S, Becker HF, et al. A prospective polysomnographic study on the evolution of complex sleep apnoea. *Eur Respir J*. 2011;38(2):329–337. Available from: https://doi.org/10.1183/09031936.00162009. Epub 09032011 Apr 09031934.
56. Liu D, Armitstead J, Benjafield A, et al. Trajectories of emergent central sleep apnea during CPAP therapy. *Chest*. 2017;152(4):751–760.
57. Kuzniar TJ, Pusalavidyasagar S, Gay PC, Morgenthaler TI. Natural course of complex sleep apnea—a retrospective study. *Sleep Breath*. 2008;12(2):135–139.
58. Allam JS, Olson EJ, Gay PC, Morgenthaler TI. Efficacy of adaptive servoventilation in treatment of complex and central sleep apnea syndromes. *Chest*. 2007;132(6):1839–1846.
59. Morgenthaler TI, Kuzniar TJ, Wolfe LF, Willes L, McLain 3rd WC, Goldberg R. The complex sleep apnea resolution study: a prospective randomized controlled trial of continuous positive airway pressure versus adaptive servoventilation therapy. *Sleep*. 2014;37(5):927–934. Available from: https://doi.org/10.5665/sleep.3662.

I. GENERAL ISSUES

60. Dellweg D, Kerl J, Hoehn E, Wenzel M, Koehler D. Randomized controlled trial of noninvasive positive pressure ventilation (NPPV) versus servoventilation in patients with CPAP-induced central sleep apnea (complex sleep apnea). *Sleep*. 2013;36(8):1163−1171. Available from: https://doi.org/10.5665/sleep.2878.
61. Nowbar S, Burkart KM, Gonzales R, et al. Obesity-associated hypoventilation in hospitalized patients: prevalence, effects, and outcome. *Am J Med*. 2004;116(1):1−7.
62. Almeneessier AS, Nashwan SZ, Al-Shamiri MQ, Pandi-Perumal SR, BaHammam AS. The prevalence of pulmonary hypertension in patients with obesity hypoventilation syndrome: a prospective observational study. *J Thorac Dis*. 2017;9(3):779−788. Available from: https://doi.org/10.21037/jtd.2017.03.21.
63. Mokhlesi B, Tulaimat A, Evans AT, et al. Impact of adherence with positive airway pressure therapy on hypercapnia in obstructive sleep apnea. *J Clin Sleep Med*. 2006;2(1):57−62.
64. Piper AJ, Wang D, Yee BJ, Barnes DJ, Grunstein RR. Randomised trial of CPAP vs bilevel support in the treatment of obesity hypoventilation syndrome without severe nocturnal desaturation. *Thorax*. 2008;63(5):395−401. Available from: https://doi.org/10.1136/thx.2007.081315. Epub 082008 Jan 081318.
65. Contal O, Adler D, Borel JC, et al. Impact of different backup respiratory rates on the efficacy of noninvasive positive pressure ventilation in obesity hypoventilation syndrome: a randomized trial. *Chest*. 2013;143(1):37−46. Available from: https://doi.org/10.1378/chest.11-2848.
66. Masa JF, Corral J, Alonso ML, et al. Efficacy of different treatment alternatives for obesity hypoventilation syndrome. Pickwick study. *Am J Respir Crit Care Med*. 2015;192(1):86−95. Available from: https://doi.org/10.1136/thoraxjnl-2016-208501.
67. Storre JH, Seuthe B, Fiechter R, et al. Average volume-assured pressure support in obesity hypoventilation: a randomized crossover trial. *Chest*. 2006;130(3):815−821.
68. Janssens JP, Metzger M, Sforza E. Impact of volume targeting on efficacy of bi-level non-invasive ventilation and sleep in obesity-hypoventilation. *Respir Med*. 2009;103(2):165−172. Available from: https://doi.org/10.1016/j.rmed.2008.1003.1013. Epub 2008 Jun 1024.
69. Masa JF, Corral J, Caballero C, et al. Non-invasive ventilation in obesity hypoventilation syndrome without severe obstructive sleep apnoea. *Thorax*. 2016;71(10):899−906. Available from: https://doi.org/10.1136/thoraxjnl-2016-208501. Epub 202016 Jul 208512.
70. Howard ME, Piper AJ, Stevens B, et al. A randomised controlled trial of CPAP versus non-invasive ventilation for initial treatment of obesity hypoventilation syndrome. *Thorax*. 2017;72(5):437−444. Available from: https://doi.org/10.1136/thoraxjnl-2016-208559. Epub 202016 Nov 208515.
71. Weese-Mayer DE, Berry-Kravis EM, Ceccherini I, Keens TG, Loghmanee DA, Trang H. An official ATS clinical policy statement: congenital central hypoventilation syndrome: genetic basis, diagnosis, and management. *Am J Respir Crit Care Med*. 2010;181(6):626−644. Available from: https://doi.org/10.1164/rccm.200807-1069ST.
72. Kerbl R, Litscher H, Grubbauer HM, et al. Congenital central hypoventilation syndrome (Ondine's curse syndrome) in two siblings: delayed diagnosis and successful noninvasive treatment. *Eur J Pediatr*. 1996;155(11):977−980.
73. Tibballs J, Henning RD. Noninvasive ventilatory strategies in the management of a newborn infant and three children with congenital central hypoventilation syndrome. *Pediatr Pulmonol*. 2003;36(6):544−548.
74. Villa MP, Dotta A, Castello D, et al. Bi-level positive airway pressure (BiPAP) ventilation in an infant with central hypoventilation syndrome. *Pediatr Pulmonol*. 1997;24(1):66−69.
75. Nicholson KJ, Nosanov LB, Bowen KA, et al. Thoracoscopic placement of phrenic nerve pacers for diaphragm pacing in congenital central hypoventilation syndrome. *J Pediatr Surg*. 2015;50(1):78−81. Available from: https://doi.org/10.1016/j.jpedsurg.2014.10.002. Epub 2014 Dec 1023.
76. Morelot-Panzini C, Gonzalez-Bermejo J, Straus C, Similowski T. Reversal of pulmonary hypertension after diaphragm pacing in an adult patient with congenital central hypoventilation syndrome. *Int J Artif Organs*. 2013;36(6):434−438. Available from: https://doi.org/10.5301/ijao.5000197. Epub 5002013 May 5000198.
77. Bourke SC, Tomlinson M, Williams TL, Bullock RE, Shaw PJ, Gibson GJ. Effects of non-invasive ventilation on survival and quality of life in patients with amyotrophic lateral sclerosis: a randomised controlled trial. *Lancet Neurol*. 2006;5(2):140−147.
78. Lo Coco D, Marchese S, Pesco MC, La Bella V, Piccoli F, Lo Coco A. Noninvasive positive-pressure ventilation in ALS: predictors of tolerance and survival. *Neurology*. 2006;67(5):761−765. Epub 2006 Aug 2009.
79. Berlowitz DJ, Howard ME, Fiore Jr JF, et al. Identifying who will benefit from non-invasive ventilation in amyotrophic lateral sclerosis/motor neurone disease in a clinical cohort. *J Neurol Neurosurg Psychiatry*. 2016;87(3):280−286. Available from: https://doi.org/10.1136/jnnp-2014-310055. Epub 312015 Apr 310059.

REFERENCES

80. Annane D, Orlikowski D, Chevret S. Nocturnal mechanical ventilation for chronic hypoventilation in patients with neuromuscular and chest wall disorders. *Cochrane Database Syst Rev.* 2014;(12)CD001941. Available from: https://doi.org/10.1002/14651858.CD001941.pub3.
81. Miller RG, Jackson CE, Kasarskis EJ, et al. Practice parameter update: the care of the patient with amyotrophic lateral sclerosis: drug, nutritional, and respiratory therapies (an evidence-based review): report of the Quality Standards Subcommittee of the American Academy of Neurology. *Neurology.* 2009;73(15):1218−1226. Available from: https://doi.org/10.1212/WNL.0b013e3181bc01a4.
82. Gruis KL, Brown DL, Lisabeth LD, Zebarah VA, Chervin RD, Feldman EL. Longitudinal assessment of noninvasive positive pressure ventilation adjustments in ALS patients. *J Neurol Sci.* 2006;247(1):59−63. Epub 2006 Apr 2024.
83. Nicholson TT, Smith SB, Siddique T, et al. Respiratory pattern and tidal volumes differ for pressure support and volume-assured pressure support in amyotrophic lateral sclerosis. *Ann Am Thorac Soc.* 2017;14 (7):1139−1146. Available from: https://doi.org/10.1513/AnnalsATS.201605-346OC.

CHAPTER

4

Treatment of Sleep-Related Movement and Circadian Rhythm Disorders, Hypersomnolence, and Parasomnias

Brienne Miner[1], Nasheena Jiwa[2] and Brian Koo[3,4,5]

[1]Geriatrics and Sleep Medicine, Yale University School of Medicine, New Haven, CT, United States [2]Internal Medicine, St. Mary's Hospital, Waterbury, CT, United States [3]Neurology, Sleep Medicine Service, VA Connecticut Healthcare System, West Haven, CT, United States [4]Yale Center for Restless Legs Syndrome, New Haven, CT, United States [5]Yale University School of Medicine, New Haven, CT, United States

OUTLINE

Introduction	78	Delayed Sleep–Wake Phase Disorder	84
		Advanced Sleep–Wake Phase Disorder	85
Restless Legs Syndrome and Periodic Limb Movement Disorder	78	Non-24-Hour Sleep–Wake Disorder	86
		Irregular Sleep–Wake Rhythm Disorder	87
Definition and Prevalence	78		
Exacerbating Factors	78	Parasomnias	88
Treatment	79	Nonrapid Eye Movement Parasomnias	88
		Rapid Eye Movement Parasomnias	90
Hypersomnolence	82		
Definition	82	Conclusion	93
Treatment	82	References	93
Circadian Rhythm Disorders	84		

Handbook of Sleep Disorders in Medical Conditions
DOI: https://doi.org/10.1016/B978-0-12-813014-8.00004-4

© 2019 Elsevier Inc. All rights reserved.

INTRODUCTION

This chapter provides a brief introduction and discusses treatment of the following disorders: restless legs syndrome (RLS), periodic limb movement disorder, hypersomnolence disorders, circadian rhythm sleep disorders, and parasomnias. When evaluating these disorders, it is important to elicit and ameliorate any exacerbating factors. If symptoms persist, many of these disorders will respond to a combination of behavioral modifications and/or psychotherapeutic modalities. Finally, there are a range of pharmacologic options available, which are employed for persistent or severe symptoms. These options will be discussed in more detail in the following sections.

RESTLESS LEGS SYNDROME AND PERIODIC LIMB MOVEMENT DISORDER

Definition and Prevalence

RLS is a neurologic disorder that is characterized by an unrelenting urge to move, that limits sitting to minutes at a time during the day and prevents sleep at night. Although all sufferers of RLS get some degree of symptom relief when they move, the relief for those with severe RLS is fleeting, and the uncomfortable urge to move may drive these persons to walk for hours at a time, preventing sleep when it is most desired. Needless to say, treatment in persons with severe RLS is essential to restore proper sleep and allow for a normal life. Any degree of RLS occurs in 5%–10% of persons in North America and Europe.[1,2] Clinically significant RLS described as moderately or severely disturbing and occurring at least twice weekly has a prevalence of 2.7% in the general population.[1,3] In a general sense, the treatment of RLS consists of the identification of any modifiable aggravating factors, which might include things in the diet, medication, medical, or social history, then an assessment for and correction of iron deficiency, and finally, if needed, behavioral or pharmacologic treatment.

Exacerbating Factors

RLS symptoms are exacerbated in severity and frequency by a number of different factors, which may include dietary, habitual, medical, or medication elements. For this reason, when evaluating a person with RLS, it is critical to determine if any of these factors may be in play. Although there are no clear dietary factors that aggravate RLS, many persons anecdotally report that meals or snacks rich in carbohydrates aggravate their RLS symptoms. It is interesting that gluten sensitivity and celiac disease have recently been associated with RLS.[4] A diet poor in iron leading to iron deficiency may also cause RLS. Alcohol is well known to exacerbate RLS; often patients are able to make this association, but this relationship also has been found in population studies.[5,6] In certain patients, caffeine may worsen RLS, but this association is not supported by epidemiologic studies.[6] Other health habits that may aggravate RLS include smoking, exercise close to bedtime, and not getting enough sleep.

RESTLESS LEGS SYNDROME AND PERIODIC LIMB MOVEMENT DISORDER

Medications, both prescribed and over-the-counter, can worsen RLS. Among over-the-counter medications, antihistamines like diphenhydramine should be avoided as much as possible in patients with RLS. Decongestant medications, such as pseudoephedrine, and antinausea medications, such as metoclopramide and prochlorperazine, also worsen RLS. The latter two medications have dopamine-blocking properties, like medications that are used to treat psychosis. Hence, it is not surprising that the antipsychotic medications cause and worsen RLS, including haloperidol, risperidol, quetiapine, clozaril, and trilafon to name a few. The antidepressant medications also worsen RLS, presumably through a serotonergic mechanism. These include the selective serotonin reuptake inhibitors (SSRIs), the tricyclic antidepressants (TCAs), and also the serotonin-norepinephrine reuptake inhibitors. Mirtazapine deserves special attention as this medicine worsens or causes RLS in up to 28% of those taking it.[7] Other medications that may be associated with worsening of RLS include the amphetamine and nonamphetamine stimulants, such as methylphenidate and dextroamphetamine, and melatonin.[8]

Once these factors have been explored, it is important to determine the iron status of the patient with RLS. First by history, is there heavy menstrual bleeding in women? Or does the patient frequently donate blood? Then serum ferritin can be checked. A serum ferritin of less than 75 μg/L should be treated with oral ferrous sulfate 325 mg along with vitamin C 500 mg before meals three times daily. If oral iron is not tolerated and/or if serum ferritin is very low (less than 25−50 μg/L), intravenous iron can be used. Intravenous iron preparations that can be used safely include low molecular weight iron dextran 500−1000 mg given in one single infusion or ferric carboxymaltose 1500 mg given in two infusions 7 days apart in time. It should be noted that anaphylaxis is rare with these iron preparations and much of the fear surrounding intravenous iron was caused by high rates of anaphylaxis caused by high molecular weight iron dextran which has since been taken off the US market.

Treatment

Once all exacerbating factors have been explored and addressed, treatment can be considered. Sleep schedule and sleep hygiene should be optimized. Even though RLS can prevent sleep, it is important to allow enough time for adequate sleep, at least 8 hours per night. In addition, the bed and wake times should be kept relatively constant. One should avoid excessive screen time in the evening. Television should not be watched in bed. The bedroom should be dark and quiet as the person attempts to go to sleep. For some persons, RLS symptoms force the sufferer to walk at night. This can be done, but more strenuous activity like running or jogging should be avoided at night.

Behavioral therapies and medication treatments can be used. Exercise is beneficial in the treatment of RLS and can consist of leg stretching or aerobic exercise.[9,10] Some have suggested that exercise especially of the legs improves circulation and decreases micro-ischemia in the legs that may be associated with RLS.[11] Devices that massage or counter stimulate the leg can be helpful in treating RLS. Pneumatic compression, vibrating devices, and near-infrared light have been shown to be effective in treating RLS.[12−14]

Of course, many patients will require medical therapy to treat RLS. Once the mainstay of medical treatment, dopamine medications have more recently fallen out of favor in the treatment of RLS. This is mainly because these medicines are associated with unwanted side effects of compulsive behavior, such as pathological gambling, shopping, or sexual behavior. Perhaps more significantly, the dopamine drugs when used at high doses and for long periods of time are associated with the development of augmentation which is a paradoxical worsening of RLS.[15] This worsening can consist of spread of symptoms to previously uninvolved body parts (e.g., arms), generalization of RLS to the daytime, a decrease in the amount of relief with movement, or a general increased intensity of the RLS symptoms.[16] Because of this, many have begun to use an alpha-2 δ ligand as first line, such as gabapentin, pregabalin, or gabapentin enacarbil.[17–19] Often, gabapentin or pregabalin can be used once nightly 1 hour before the onset of symptoms at doses between 300 and 900 mg for gabapentin and 50–150 mg for pregabalin. When symptoms occur earlier in the evening and persist until sleep, a second evening time dose of these medicines can be added 1 hour prior to usual symptom onset. In elderly patients or in those with renal impairment, it is prudent to start with doses on the order of 100 or 200 mg for gabapentin and 25 mg for pregabalin. Gabapentin enacarbil may be helpful in patients that have symptoms throughout the night that wake them up or that have daytime symptoms. Side effects from these medicines include depression, weight gain, leg edema, grogginess, and unsteadiness.

Dopaminergic medications are exquisitely efficacious in treating RLS, at least in the short term. Dopaminergic medicines that are effective in treating RLS include carbidopa/levodopa, pramipexole, ropinirole, and rotigotine. Carbidopa/levodopa is rarely used as augmentation occurs in up to 60% of persons with RLS taking this medicine.[20] Ropinirole and pramipexole are very commonly used medications for RLS. Ropinirole is dosed starting at 0.25 mg nightly, then 0.5 mg, and then 1.0 mg going up every 3 nights if symptoms persist, and there are no side effects. Pramipexole dosing is similar, but the amounts are divided by 2, so that starting and subsequent doses are 0.125, 0.25, and 0.5 mg. It is important not to exceed the recommended maximum doses of either medicine; for ropinirole, this is 4.0 mg, but we do not like to go above 2.5 mg and for pramipexole this is 0.75 mg. At doses above these maximum doses, augmentation may become more common. One needs to be cautious about increasing the dose of these medicines month after month or year after year, as this is an indication of augmentation and the longer duration of treatment with dopamine medicines, the more likely one is to develop augmentation.[21] Rotigotine is given in a patch form and is a long-acting 24-hour formulation that is best used for persons with severe day and evening symptoms. It can also be used in persons that are on a shorter acting dopamine agonist (ropinirole or pramipexole) that are having augmentation and must go off of the offending shorter acting dopamine drug.

Opiate medications are also very effective in treating RLS. Their use is often reserved for patients with severe RLS and in those who have had augmentation after being on a dopaminergic medication. RLS is best treated with long-acting opiate medications, like methadone, oxycontin, and buprenorphine. Side effects such as depression, constipation, nausea, and dizziness may limit their use. Although longitudinal studies are needed, there appears to be little risk of addiction and abuse, especially since the dose of opiate, once the effective dose is found, does not usually need to be increased.[22]

If augmentation has occurred, it is important to start a nondopaminergic agent such as an alpha-2 δ ligand or opiate and, at the same time, taper the offending dopamine drug. It is important to warn the patient that, for a period of 2–6 weeks, RLS symptoms can worsen. If dopamine agent doses are high and especially if there is comorbid depression, the drug should be tapered over 2–4 weeks. Iron and ferritin should be optimized, and caffeine and alcohol should be minimized. The alpha-2 δ ligand or opiate medicine then should be titrated to the desired effect. It is possible that the patient will need more than one agent to control RLS symptoms. Anecdotally and in these authors' experience, marijuana may also be a helpful adjunct for patients.

Periodic limb movements during sleep (PLMS) can occur in the setting of RLS or also in the absence of RLS. When PLMS occurs in the absence of RLS and when disturbed sleep and/or daytime symptoms such as fatigue or sleepiness are present, a periodic limb movement disorder (PLMD) may be present. It should be noted that special attention needs to be given to other potential causes of fatigue such as insufficient sleep, medications, medical disorders, or other sleep disorders. When these other entities are ruled out as a cause for fatigue, only then should a diagnosis of PLMD be considered. In the authors' opinion, PLMD is overdiagnosed. In the vast majority of cases, where PLMS are seen on a sleep study, no treatment is needed. In some circumstances, when an exhaustive investigation for causes of fatigue or sleepiness has been unrevealing, medical treatment for PLMS can be attempted. Often medicines that treat RLS also treat PLMS. Opiates, dopamine agents, and alpha-2 δ ligands are all effective in treating PLMS. It may be desirable to treat PLMS when the movements disrupt sleep to the point of causing fatigue or sleepiness during the day, or when they disturb a bed partner. Treatments for these movements include dopamine agonists or alpha-2 δ ligands. Benzodiazepines may also be used (Table 4.1).

TABLE 4.1 Treatment and Management of Restless Leg Syndrome and Periodic Limb Movement Disorder

Behavioral Management	Treatment of Iron Deficiency	Pharmacological Management		

1. Exercise
2. Optimize sleep hygiene
3. Identify and reduce exacerbating factors:
  a. Alcohol
  b. Caffeine
  c. Carbohydrates
  d. Smoking
4. Medications to avoid:
  a. Antidepressants
  b. Antihistamines
  c. Dopamine antagonists
  d. Melatonin
  e. Pseudoephedrine
  f. Stimulants
	1. If ferritin <75 µg/L:
	  a. Ferrous sulfate 325 mg + vitamin C 500 mg
	2. If oral iron not tolerated or ferritin <25 − 50 µg/L:
	  a. Iron dextran 500–1000 mg IV
	  b. Ferric carboxymaltose 1500 mg in two infusions
		Alpha-2 delta ligand:
		• Gabapentin: 300–900 mg
		• Pregabalin: 50–150 mg
		• Gabapentin Enacarbil: 600 mg
			Dopamine agonists:
			• Ropinirole: 0.25 mg, increase by 1 mg every 3 nights (maximum of 4 mg)
			• Pramipexole: 0.125 mg, 0.25 mg, and 0.5 mg (maximum 0.75 mg)
			• Rotigotine: 1 mg patch (max 3 mg patch)
				Opiates:
				• Methadone: 5–40 mg
				• Oxycontin: 10–80 mg
				• Buprenorphine: 0.5–6 mg

HYPERSOMNOLENCE

Definition

Hypersomnolence is a patient-reported symptom that is common in many sleep disorders as well as in other medical and psychiatric conditions. When evaluating this symptom, it is important to differentiate hypersomnolence from fatigue. Whereas fatigue may be described as generalized weakness, reduced capacity to maintain performance, or impaired mental capacity,[23] hypersomnolence, or excessive daytime sleepiness (EDS), refers specifically to an inability to stay awake or alert, with periods of irrepressible need for sleep or unintended lapses into sleep.[24] Hypersomnolence is present in 4%—6% of the general population, and in 15%—30% of people with sleep disorders.[25]

Treatment

The initial step in the treatment of hypersomnolence is to consider the differential diagnosis for this symptom. The differential includes insufficient sleep time, substance use or medication effects, hypersomnolence secondary to medical or psychiatric conditions, which includes primary sleep disorders like sleep-disordered breathing, sleep-related movement disorders, or circadian rhythm disorders, and finally the central disorders of hypersomnolence.[25,26] It is important to perform a thorough evaluation when encountering hypersomnolence, as an alerting medication may not be the most appropriate intervention, particularly if there is evidence of a primary sleep disorder.

Since multiple sections within this text are devoted to the treatment of primary sleep disorders and sleep disorders in different medical conditions, the focus for the remainder of this section will be on the treatment of central disorders of hypersomnolence, including narcolepsy with and without cataplexy, idiopathic hypersomnia (IH), and recurrent hypersomnia (Kleine—Levin syndrome). The practitioner and patient should keep in mind the goals for treatment in central disorders of hypersomnolence. Many patients may require a combination of medications. Even with multiple medications, a normal level of alertness may not be achieved, and the goals of treatment may be to alleviate daytime sleepiness and restore function.

Narcolepsy and IH, while etiologically distinct, are similar with respect to presentation and, also, treatment. Behavioral treatments for these disorders include scheduled naps, which may be more refreshing and more helpful in narcolepsy, as well as a regular bedtime schedule that ensures an adequate amount of sleep. A priority in the management of both of these disorders is to control EDS through the use of alerting medications. Narcolepsy is the more rigorously studied of these disorders, and medications for narcolepsy are often extended to off-label use in IH. According to the American Academy of Sleep Medicine (AASM) guidelines, modafinil 200 mg is the standard therapy for EDS in narcolepsy.[26] Doses up to 400 mg may be used, and there is evidence that patients may benefit from split dosing in the morning and early afternoon to improve evening sleepiness.[26] Recent evidence has shown efficacy of modafinil in IH as well.[27–29] A newer alternative to modafinil is armodafinil, which is dosed once daily at 150—250 mg.[26,30] Side effects include headache, nausea, and dry mouth, which can be reduced by progressively increasing the dosage. There are rare instances of serious allergic and skin reactions, including Stevens–Johnson syndrome. Women starting modafinil or armodafinil need to be counseled that either of these medications may decrease the effectiveness of oral contraceptives.[31] It should be noted that the AASM guidelines also recommend modafinil as a treatment for EDS in Parkinson's disease, myotonic dystrophy, and multiple sclerosis.[26]

Stimulant medications of the amphetamine class are an alternative treatment for EDS in narcolepsy and IH. Options include methylphenidate, dextroamphetamine, and methamphetamine.[26,31] Methylphenidate is also an option for the treatment of EDS due to myotonic dystrophy.[26] These are available in immediate release formulations that are dosed two to three times per day, or in extended release formulations that are generally dosed once a day. The amphetamine stimulants may not be the preferred option due to their side effect profile. There is also potential for tolerance and abuse. In addition, at doses above 60 mg/day, there is increased risk for cardiovascular and psychiatric side effects, including increased blood pressure and heart rate, palpitations, anxiety, insomnia, paranoia, and psychosis.[31]

Sodium oxybate is recommended by the AASM as standard therapy for EDS, cataplexy and disrupted sleep due to narcolepsy.[26] This medication must be given in divided doses due to a short half-life, with the first dose at bedtime and the second dose 2–3 hours later. The starting dose is 4.5 g in two divided doses, with the total dose increased over several months to between 6 and 9 g depending on the patient's response. In general, lower doses may be effective for cataplexy, while higher doses are needed to combat EDS.[31] Side effects include nausea, confusion, dizziness, enuresis, and sleepwalking. This medication should be used with caution in patients with heart failure due to high salt content. Due to central nervous system (CNS) depressive effects, use of sodium oxybate cannot be combined with use of other sedatives, narcotics, or alcohol. It is also a drug with the potential for abuse, as it is historically known as the "date rape drug." Thus, prescription of this medication is tightly controlled. Sodium oxybate is available only through a restricted distribution program which uses a certified central pharmacy, and patients and prescribers must enroll in the program.

Additional options for the treatment of cataplexy include TCAs, SSRIs, serotonin-norepinephrine reuptake inhibitors, like venlafaxine and reboxetine. Reboxetine is not available in the United States or Canada.[26] These medications exert their effect through rapid eye movement (REM) sleep suppression.[31] TCAs, SSRIs, and venlafaxine may also treat hypnagogic hallucinations and sleep paralysis.[26]

There are a handful of other medications that are less commonly used for EDS and cataplexy. These include ritanserin (not available in North America), and selegiline, which is not preferred due to the potential for drug–drug and drug–diet interactions.[26] There is emerging data suggesting that a certain percentage of persons with narcolepsy without cataplexy and idiopathic hypersomnolence may benefit from treatments that negatively modulate the gamma-aminobutyric acid (GABA) receptor. These patients may have excessive amounts of an endogenous positive allosteric modulator of $GABA_A$ receptors in their cerebrospinal fluid. As negative modulators of the $GABA_A$ receptor, the medicines clarithromycin and flumazenil may help these individuals feel less sleepy during the day. Clarithromycin can prolong the QT interval and has potential for causing arrhythmia, including torsades de pointes. Flumazenil comes only in intravenous formulation and needs be compounded into a topical cream or into a lozenge for ingestion.[32–34]

Kleine–Levin syndrome is a rare disorder that causes hypersomnolence as well as cognitive and behavioral disturbances, including confusion, irritability, compulsive eating, and hypersexuality.[35] Recommendations for treatment are less clear since there are few studies of different treatments in this group. Lithium is recommended for reduction of episode duration as well as reduction in undesirable behaviors.[26] Other options include amantadine, lamotrigine, and valproic acid.[26,36]

CIRCADIAN RHYTHM DISORDERS

Delayed Sleep–Wake Phase Disorder

Definition

Delayed sleep–wake phase disorder (DSWPD) is a disorder of circadian rhythm where there is a delayed onset of sleep and wake times, usually by more than 2 hours relative to more conventional sleep–wake times. This is accompanied by a complaint of inability to fall asleep or awaken at desired or required times and usually with dysfunction during the day.[24] DSWPD is the most common of the circadian rhythm disorders. It is estimated to affect 7%–16% of adolescents and 5%–10% of patients referred to sleep clinics.[24,37] The peak age for DSWPD is adolescence, and a familial pattern may be present in up to 40% of patients.[38]

Treatment

Treatment of DSWPD is focused on realigning the sleep–wake cycle with the desired or required conventional schedule.[39] In addition to good sleep hygiene and avoiding exposure to bright light in the evening, coexisting medical or psychiatric conditions such as anxiety, depression, or insomnia, which contribute to disturbance of the sleep–wake cycle, should be identified and treated. Several methods are used to advance the sleep–wake cycle, which include chronotherapy, bright light therapy, and the use of pharmacologic treatments such as melatonin. Combining different therapies may improve treatment success.[40]

Chronotherapy for treatment of DSWPD requires progressive and incremental phase delay until the preferred sleep time is attained. In other words, the bedtime is progressively delayed by several hours on successive days. As an example, an individual can delay his or her sleep times by 3 hours every 2 days over a 5–6-day period until the preferred sleep time is attained. This method is successful only with strict adherence to the prescribed regimen, which requires a well-structured home environment where light exposure can be controlled and sleep–wake times adhered to.[39] Light exposure at the wrong circadian time may disrupt therapeutic effectiveness. The person must also be free of societal constraints. The rigor required for chronotherapy in DSWPD may make it impractical.[40]

Light therapy is used in DSWPD to move the major sleep episode to an earlier time (i.e., phase advance). In order to advance an individual's sleep–wake cycle, bright light therapy early in the morning should be implemented. Light intensity should be between 2500 and 10,000 lux.[40] The timing of light therapy is critical. In order to advance sleep phase, light exposure should occur after the core body temperature nadir, which generally occurs 2–3 hours before spontaneous wake time.[39] To avoid exposure before the core body temperature nadir, light should be administered at or soon after the spontaneous wake time. In addition, light exposure in the early evening and especially at night should be avoided. Two to three hours of bright light exposure in the morning can successfully phase advance circadian rhythms by up to 1.4 hours per day of bright light exposure.[40] This strategy is effective in motivated patients when accompanied by behavioral education. Those who are unable to do 2–3 hours of light exposure will still advance their rhythm, but at a slower rate. Once light therapy begins, the individual should attempt to administer the light 30 minutes to 1 hour earlier each day and should also go to bed 30 minutes to 1 hour earlier each night. A limitation with this method is that individuals may have difficulty awakening in time for the implementation of bright light therapy due to their delayed phase.[40]

The main pharmacological treatment for DSWPD is melatonin. As with light therapy, exogenous melatonin can shift the circadian rhythm based on the timing of its administration. Phase advancement is achieved when melatonin is administered in the early afternoon or evening, ideally 3–6 hours before dim light melatonin onset, which generally occurs about 2 hours before bedtime and 14 hours after wake in normal phase individuals. For example, if a patient's bedtime is 2:00 a.m. then melatonin should be administered at 8:00 p.m.[39] Avoidance of bright light exposure in the evening and night must be combined with melatonin.

Another intervention that can be used to advance sleep phase in DSWPD is the use of glasses that block blue wavelength light. Blue light contains the specific wavelength of light that suppresses the secretion of melatonin from the pineal gland. By wearing glasses that block this wavelength (orange tinted), blue light entrance to the retina is limited and melatonin is allowed to rise naturally. These glasses should only be worn in the evening for the duration of time that there is exposure to bright light to prevent suppression of melatonin, usually starting approximately 3 hours before habitual bedtime.[39]

Advanced Sleep–Wake Phase Disorder

Definition

Advanced sleep–wake phase disorder (ASWPD) is an advance or early timing in the phase of the major sleep episode. The timing of sleep in this disorder leads to a complaint of difficulty staying awake until a desired conventional bedtime and/or an inability to remain asleep until the desired wake time.[24] Sleepiness in the late afternoon and early evening may impede participation in activities in the evening.[41] ASWPD is commonly seen in aging populations and may be associated with depression.[39] This is a rare disorder, thought to be present in 1% of the general population, with higher prevalence among older adults and persons with neurodevelopmental disorders.[38]

Treatment

Treatment options include chronotherapy, light therapy, and melatonin. However, relatively few studies of treatment for ASWPD exist and dosing and timing of the different therapies is less well identified than for DSWPD.[42]

Chronotherapy in ASWPD is a method that aims to progressively advance the individual's sleep time. In other words, the individual goes to bed earlier each day until the desired bedtime is achieved.[43] While there are no specific guidelines on how this should be done, it may be reasonable to have the individual advance the bedtime 1 hour earlier each night. Due to the potential for relapse, the addition of bright light therapy during the early evening is also recommended.[40] As with chronotherapy for DSWPD, strict adherence, a structured environment, and control of light exposure are needed, which may make this treatment option impractical.

The efficacy of light therapy has not been well defined in the treatment of ASWPD with respect to treatment duration, timing, and light intensity. In elderly individuals unable to maintain sleep, bright light from 7:00 to 9:00 p.m. may cause a phase delay in the major sleep episode and reduce awakenings.[40] Commonly reported side effects of light therapy include eyestrain, nausea, and agitation.[39] Patients should also avoid bright light early in the morning, since this leads to further advancement of the sleep—wake cycle. Bright light should be avoided until 5—6 hours after core body temperature nadir, or 2—3 hours after awakening.

Melatonin may be administered in a low dose in the morning upon awakening to produce a phase delay of the sleep—wake cycle.[39] Given the hypnotic effects of melatonin, particularly with higher doses, morning administration may lead to drowsiness and limit efficacy.[40] Adherence to this treatment may be higher than with chronotherapy or light therapy. Given the limited treatment options for ASWPD, it may be advisable to combine multiple treatment modalities.[42]

Non-24-Hour Sleep—Wake Disorder

Definition

Non-24-hour sleep—wake disorder (N24SWD) is defined by an endogenous circadian rhythm that cannot be entrained to light, thereby resulting in a rhythm that will free-run and has no stable phase relation to the 24-hour light—dark cycle.[24] It is characterized by progressive delay in sleep onset and offset, with the sleep period drifting 1—2 hours each day. Patients may complain of insomnia, daytime sleepiness, or no symptoms, depending on whether their circadian phase is aligned with a normal sleep—wake period. It is most often observed in blind individuals and is seen rarely in sighted persons.[42] The pathophysiology of N24SWD in sighted individuals is unknown, but it is hypothesized that they may not respond as well to the phase-resetting effects of light.[40] It is estimated that up to 25% of sighted individuals with N24SWD have a related psychiatric diagnosis.[42]

Treatment

Treatment options in N24SWD differ by whether the individual is blind or not. For the former group, melatonin is the treatment of choice. Doses of 0.5–10 mg are given 1 hour before the desired bedtime.[39] Physiologic doses (approximately 0.3 mg) may be more effective than doses above 2 mg.[42] Tasimelteon, a melatonin agonist, is a newer option for treatment of N24SWD and a reasonable alternative to melatonin. It is administered as a 20-mg dose before bedtime.[40] The efficacy of other melatonin agonists such as ramelteon and agomelatine on N24SWD is currently unknown.[40] There is some evidence that blind individuals may respond to light therapy even in the absence of conscious light perception. However, the success of this therapy likely depends on the ability to transmit photic input to the circadian pacemaker, which is unlikely to occur in an individual with retinal blindness.[41] For sighted individuals with N24SWD, treatment options include morning light exposure and melatonin administration at bedtime. These individuals have not been extensively studied, and there are no specific recommendations with respect to dosing of these treatments, though 3 mg is the most common dosage used for melatonin.[42]

Irregular Sleep–Wake Rhythm Disorder

Definition

Irregular sleep–wake rhythm disorder (ISWRD) is seen in individuals with multiple irregular sleep and wake episodes throughout a 24-hour period due to the absence of a well-defined circadian sleep–wake cycle.[24] Individuals may have insomnia symptoms during the scheduled sleep period and/or EDS. Those affected by ISWRD are usually institutionalized older adults with dementia or those with moderate-to-severe cognitive defects.[42,44] There is a paucity of data on the overall prevalence of this disorder, but its prevalence increases with age, neurologic disease, mental disability, and dementia.[38]

Treatment

The treatment goal for ISWRD is to consolidate sleep and wake periods through the use of therapies that structure sleep–wake schedules and reinforce circadian time cues.[42] Treatment options include bright light exposure, melatonin, and mixed modality therapies. Mixed modality therapies combine the use of light, physical activity, and behavioral elements (e.g., sleep hygiene and social activities) to reduce nighttime awakenings and increase rest–activity rhythms (i.e., increasing activity in the presence of light and resting in the absence of light).[40,42] The AASM recommends bright light exposure during the day and mixed modality therapies for all patients with ISWRD, while melatonin is recommended as an option only for younger patients with mental delay.

PARASOMNIAS

Nonrapid Eye Movement Parasomnias

Nonrapid eye movement (NREM) parasomnias include confusional arousals, somnambulism (sleepwalking), sleep terrors, and sleep-related eating disorders (SREDs). While common in childhood (over 30% of 18-month-old babies may experience night terrors),[38] NREM parasomnias are much less prevalent after the age of five, with prevalence estimates in adults ranging from 2% to 4% depending on the NREM parasomnia.[45] Risk factors for these disorders include family history or states that propagate slow wave sleep, such as use of CNS depressants (alcohol, sedative hypnotics, or antihistamines), fever, circadian rhythm sleep disorders, or recovery from sleep deprivation.[46] Given that these parasomnias resolve with age, management is rarely indicated. If required, treatment is generally conservative with an emphasis on behavioral modification.

Confusional Arousals

Confusional arousals are brief, partial arousals from sleep that are associated with disorientation, mental confusion, or inappropriate behavior. They may go unnoticed in the absence of a bed partner and are not associated with autonomic arousal.[47] Treatment options include avoiding sleep deprivation, preventing irregular sleep–wake patterns, avoiding exposure to CNS depressants, and addressing the management of any coexisting sleep disorder.[48] For example, confusional arousals may be precipitated by sleep-disordered breathing. In the absence of an underlying sleep disorder, parents of children with confusional arousals should be reassured that this is a benign condition and educated not to interrupt or confront the episode, since this may lead to agitation or injury.[46] Pharmacologic treatment is indicated in refractory cases. Options include TCAs, such as imipramine and clomipramine, and clonazepam.[48]

Somnambulism

Somnambulism (sleepwalking) is characterized by incomplete arousal from sleep leading to episodic typical behaviors, including sitting up in bed, walking around the room, or entering other rooms. Somnambulism can also progress to inappropriate or dangerous behaviors such as urinating, unlocking doors, or driving.[47] The cornerstones of treatment for this disorder include ensuring a safe sleep environment and behavioral modifications. Pharmacologic therapies may be considered in refractory cases or when there is concern for the individual's safety.

The initial goal for the treatment of somnambulism is to create a safe sleep environment. This includes the removal of sharp furniture or locking doors and using door alarms to alert family members if elopement from the room or house is a concern.[48] Measures should be taken to prevent injuries from tripping over objects, falling down stairs, or walking into mirrors or windows.[46]

Scheduled brief awakenings before an episode of somnambulism have been beneficial for the treatment. The success of this strategy suggests that awakening a person during presumed slow wave sleep will alter their sleeping pattern and, in turn, decrease the disruption in slow wave sleep.[48] Nightly awakenings for a 1-month duration have been shown to improve somnambulism with continued benefit at 6 months.[49] Other behavioral modifications for the treatment of somnambulism include avoiding sleep deprivation and minimizing arousals, for example, by decreasing noise or external stimuli. Precipitating medications, such as sedatives (e.g., nonbenzodiazepine receptor agonists), norepinephrine reuptake inhibitors, and atypical antipsychotics, should be avoided. In adults, psychotherapy for somnambulism thought to be related to a traumatic event and hypnosis offers an alternative to pharmacotherapy.[48,50]

Pharmacotherapy for somnambulism may be indicated if the episodes result in violent behaviors or risk of injury to the individual, bed partner, or family.[48] It may also be considered if somnambulism interferes with a patient's functional status or results in psychological distress due to sleep disruption or EDS.[46] Options include benzodiazepines, TCAs, SSRIs, or melatonin.[48]

Sleep Terrors

Sleep terrors occur during slow wave sleep and are characterized by arousal from sleep with expressions of intense fear, extreme panic, or confusion.[46] The episode is accompanied by a surge of sympathetic activity including tachycardia, tachypnea, and diaphoresis.[48] If rare, then treatment may not be indicated. However, if episodes are interfering with a patient's functional capacity, then behavioral and pharmacological treatment may be considered.

Initial management of sleep terrors includes ensuring a safe sleep environment to prevent injury or harm.[48] Safety considerations are similar to those listed for somnambulism, including locking windows and doors of the patient's sleeping area and ensuring that the patient is on the ground floor. Sharp furniture near the sleeping area should be removed, and doors alarmed to alert family members should the patient leave the room.

Behavioral modifications should be targeted toward reducing the incidence of arousals and reducing the risk factors that increase the susceptibility for arousals, such as caffeinated beverages and irregular sleep–wake cycles.[48] In addition, sleep terrors may be prevented by eliminating precipitants, such as apneic events, periodic limb movements, pain, urge to urinate, or environmental noise.[46] If the episodes occur at a consistent time, then brief awakenings 15–20 minutes prior to this time may be beneficial. Other behavioral management options include relaxation therapy, hypnosis, as well as psychotherapy and stress management techniques for patients with anxiety or other psychiatric disorders.[48]

Pharmacological therapy may be considered in severe or refractory cases of sleep terrors. Options include benzodiazepines, TCAs, SSRIs, and trazodone 25–50 mg.[48] One trial suggested reduction in sleep terrors by administering hydroxytryptophan at 2 mg/kg.[51]

Sleep-Related Eating Disorder

SRED is characterized by arousals or awakenings from NREM sleep with subsequent nocturnal sleepwalking and compulsive eating in a state of impaired consciousness. SRED can be associated with exposure to agents such as zolpidem or olanzapine, and can be precipitated by arousals in the setting of sleep-related breathing disorders, PLMD, or restless leg syndrome.[46]

Behavioral strategies to mitigate SRED are similar to those previously described for the other NREM parasomnias. If the patient is predisposed to food preparation during the

4. TREATMENT OF SLEEP-RELATED MOVEMENT

TABLE 4.2   Behavior and Pharmacological Treatment of Nonrapid Eye Movement Parasomnias

	Behavior Therapy	Pharmacological Treatment
Confusional arousal	Avoid sleep deprivation Regulate sleep–wake cycle Avoid central nervous system depressants	Imipramine 50–100 mg Clomipramine 25–100 mg Clonazepam 0.25–1 mg
Somnambulism	Avoid nonbenzodiazepine agonists Avoid norepinephrine reuptake inhibitors Avoid atypical antipsychotics Hypnosis Psychotherapy	Clonazepam 0.25–1 mg Triazolam 0.25 mg Imipramine 50–100 mg Melatonin 1–10 mg Paroxetine 20–40 mg
Sleep terror	Avoid caffeinated beverages Regulate sleep–wake cycle Scheduled awakenings 15–20 minutes prior to event Hypnosis Psychotherapy	Clonazepam 0.25–1 mg Imipramine 50–300 mg Clomipramine 25–100 mg Paroxetine 20–40 mg Hydroxytryptophan 2 mg/kg
Sleep-related eating disorder	Avoid psychotropic medications	Paroxetine 20–30 mg Topiramate 100–300 mg Clonazepam 0.25–2 mg Pramipexole 0.12–1.5 mg

SRED spell, further measures should be taken to secure the kitchen, including securing sharp utensils or removing stove knobs.[48] Precipitants for these episodes, including certain drugs or primary sleep disorders, should be removed. Pharmacologic options include benzodiazepines, dopamine agonists, SSRIs, and topiramate.[52,53] Pramipexole should be considered for patients with SRED in the setting of RLS, and clonazepam for those with SRED and sleepwalking (Table 4.2).[53]

Rapid Eye Movement Parasomnias

REM sleep-related parasomnias include REM sleep behavior disorder (RBD), nightmare disorder, and recurrent isolated sleep paralysis (RISP). The general goal of pharmacological intervention in these disorders is to prevent arousals or to suppress REM sleep, thereby decreasing the frequency of these episodes.

Rapid Eye Movement Behavior Disorder

RBD is characterized by loss of normal skeletal muscle atonia during REM sleep that results in dream enactment, which may be violent in nature.[54] There is a strong association of RBD with neurodegenerative disorders, including Parkinson's disease, multiple system atrophy, and Lewy body dementia. Acute RBD may be induced by the administration of psychiatric medications, such as TCAs, SSRIs, cholinergic agents, and monoamine oxidase inhibitors, or by withdrawal from alcohol, benzodiazepines, and barbiturates.[46] The prevalence of this disorder in the general population is estimated to be 0.38%–2.1%.[55]

I. GENERAL ISSUES

Management is focused on decreasing the frequency and severity of the abnormal behaviors and preventing injury. As with other parasomnias, patient safety and the safety of bed partners and other household members are important. Safety parameters include removing sharp and edged objects from the sleep area and keeping the bed close to the ground.[54]

The two main pharmacological therapies for RBD are melatonin and clonazepam. These two treatments are similarly effective with respect to reducing dream enactment but rarely cease the behaviors. Clonazepam and melatonin can be used exclusively or in combination. Melatonin can be used in doses of 3−15 mg, while clonazepam can be used in doses of 0.25−2.0 mg. Other drugs that have been known to be efficacious in reducing RBD symptoms are cholinesterase inhibitors, N-methyl-D-aspartate (NMDA) receptor antagonists (memantine), melatonin receptor agonists (agomelatine and ramelteon), pramipexole, zonisamide, zopiclone, and an herbal supplement known as Yi-Gan San.[54]

Nightmare Disorder

Nightmare disorder is a REM parasomnia that is characterized by vivid images that are complex in nature, and the associated emotions from these nightmares continue to manifest during wakefulness. Patients with post-traumatic stress disorder (PTSD) may have an associated nightmare disorder.[56] Frequent nightmares resulting in distress for the patient have been reported in 2%−8% of the general population, with higher prevalence rates in patients with psychiatric diagnoses who have experienced traumatic events.[38]

Image rehearsal therapy (IRT) is a type of nightmare focused cognitive behavioral therapy that is well established and effective.[57] It involves several components that implement systematic desensitization and progressive deep muscle relaxation training.[56] In IRT, patients learn that the nightmares are a sleep disorder rather than the result of an associated traumatic event, and they explore the provocative stimuli that lead to a disturbing dream. Ultimately, they are led to create a detailed, nonfrightening ending for a recurrent nightmare, which is reinforced by continually writing down and rehearsing the alternate ending.[58]

There are several pharmacologic options for the treatment of PTSD-associated nightmares. The alpha-adrenergic antagonist prazosin is the treatment with the best evidence of benefit.[56] Prazosin should be started at 1 mg and increased to 2 mg then 4 mg in increments of 3−4 nights. Dosage can be increased to effect but may be limited by orthostatic hypotension. Clonidine, an alpha-2 adrenergic receptor agonist, is another option. Clonidine suppresses sympathetic drive and may also decrease REM sleep and increase NREM sleep when taken in low doses.[57] Other treatments that have been used include trazodone, atypical antipsychotics, topiramate, low-dose cortisol, gabapentin, cyproheptadine, and TCAs.[56]

Recurrent Isolated Sleep Paralysis

Sleep paralysis occurs when REM sleep-related atonia persists into wakefulness. RISP is a disorder that is characterized by the inability to move or speak at the onset of sleep and on awakening. RISP is associated with hypnagogic and hypnopompic hallucinations and is considered among the "narcoleptic triad" along with cataplexy and daytime sleepiness. The severity of RISP depends on the frequency and intensity of these episodes, which can vary from a single lifetime episode to nightly or even multiple episodes each night.[54,56] The lifetime prevalence of RISP has been reported to be around 6%.[59]

Treatment may be indicated if sleep paralysis causes clinically significant impairment of function, anxiety, or difficulty initiating sleep.[54] Treatment decisions are not well-outlined due to the lack of randomized controlled trials, and management options are based on studies of narcolepsy and small case studies. Both psychotherapeutic and pharmacologic options are considered in the treatment for RISP.[60]

Psychotherapeutic options for RISP include psychoeducation and reassurance, sleep hygiene and insomnia treatment, and cognitive behavioral therapy. Psychoeducation mainly consists of reassurance and education about RISP, as well as addressing the associated feelings that a patient may have about the disorder. Sleep hygiene and insomnia treatment prevent insufficient sleep, which can exacerbate RISP. Cognitive behavioral therapy for RISP comprises relaxation techniques and disruption techniques during RISP episodes. Meditation and relaxation are also implemented for patients to cope with the associated distressful emotions that can be elicited by RISP episodes. TCAs, SSRIs, and sodium oxybate are pharmacological options, based on the ability of these drugs to suppress REM sleep (Table 4.3).[60]

TABLE 4.3 Behavior and Pharmacological Treatment of Rapid Eye Movement Parasomnias

	Behavior Therapy	Pharmacological Treatment
Rapid eye movement behavior disorder	–	Melatonin 3–15 mg Clonazepam 0.25–2 mg Donepezil 5–23 mg Memantine 10–20 mg Ramelteon 8 mg Agomelatine 25–50 mg Pramipexole 0.125–1.5 mg Zopiclone 1 mg Zonisamide 100–400 mg
Nightmare disorder	Image rehearsal therapy	Prazosin 1–12 mg Clonidine 0.1–0.2 mg Trazodone 25–600 mg Topiramate 50 mg Gabapentin 300–600 mg Cyproheptadine 4–12 mg
Recurrent isolated sleep paralysis	Psychoeducation Sleep hygiene and insomnia treatment Cognitive behavioral therapy	Clomipramine 25–50 mg Imipramine 25–150 mg Protriptyline 10–40 mg Desmethylimipramine 25–150 mg Sodium oxybate 6–9 g Fluoxetine 40–80 mg

CONCLUSION

In conclusion, treatment of RLS/PLMD, hypersomnolence, circadian disorders, and parasomnias relies on the use of a wide range of modalities, and pharmacotherapy is often the last resort. For many of these disorders, careful history taking may expose underlying causes (e.g., insufficient sleep in hypersomnolence; sleep-disordered breathing in parasomnias) or identify exacerbating factors (e.g., iron deficiency in the case of RLS; SSRI use in the case of RLS or RBD). Behavioral management techniques, both general (education and reassurance, psychotherapy or cognitive behavioral therapy) and specific (IRT for nightmare disorder), may be employed. Pharmacotherapy may be the mainstay of treatment (e.g., melatonin in DSWPD or ASWPDs; clonazepam or melatonin for RBD) or may be needed only in refractory cases or when safety concerns arise.

References

1. Allen RP, Walters AS, Montplaisir J, et al. Restless legs syndrome prevalence and impact: REST general population study. *Arch Intern Med*. 2005;165(11):1286–1292.
2. Koo BB. Restless leg syndrome across the globe: epidemiology of the restless legs syndrome/Willis–Ekbom disease. *Sleep Med Clin*. 2015;10(3):189–205. xi.
3. Allen RP, Stillman P, Myers AJ. Physician-diagnosed restless legs syndrome in a large sample of primary medical care patients in western Europe: prevalence and characteristics. *Sleep Med*. 2010;11(1):31–37.
4. Moccia M, Pellecchia MT, Erro R, et al. Restless legs syndrome is a common feature of adult celiac disease. *Mov Disord*. 2010;25(7):877–881.
5. Didriksen M, Rigas AS, Allen RP, et al. Prevalence of restless legs syndrome and associated factors in an otherwise healthy population: results from the Danish Blood Donor Study. *Sleep Med*. 2017;36:55–61.
6. Batool-Anwar S, Li Y, De Vito K, Malhotra A, Winkelman J, Gao X. Lifestyle factors and risk of restless legs syndrome: prospective cohort study. *J Clin Sleep Med*. 2015;12(2):187–194.
7. Rottach KG, Schaner BM, Kirch MH, et al. Restless legs syndrome as side effect of second generation antidepressants. *J Psychiatr Res*. 2008;43(1):70–75.
8. Whittom S, Dumont M, Petit D, et al. Effects of melatonin and bright light administration on motor and sensory symptoms of RLS. *Sleep Med*. 2010;11(4):351–355.
9. Dinkins EM, Stevens-Lapsley J. Management of symptoms of restless legs syndrome with use of a traction straight leg raise: a preliminary case series. *Man Ther*. 2013;18(4):299–302.
10. Aukerman MM, Aukerman D, Bayard M, Tudiver F, Thorp L, Bailey B. Exercise and restless legs syndrome: a randomized controlled trial. *J Am Board Fam Med: JABFM*. 2006;19(5):487–493.
11. Anderson KN, Di Maria C, Allen J. Novel assessment of microvascular changes in idiopathic restless legs syndrome (Willis–Ekbom disease). *J Sleep Res*. 2013;22(3):315–321.
12. Eliasson AH, Lettieri CJ. Sequential compression devices for treatment of restless legs syndrome. *Medicine*. 2007;86(6):317–323.
13. Mitchell UH. Medical devices for restless legs syndrome—clinical utility of the Relaxis pad. *Ther Clin Risk Manage*. 2015;11:1789–1794.
14. Mitchell UH, Myrer JW, Johnson AW, Hilton SC. Restless legs syndrome and near-infrared light: an alternative treatment option. *Physiother Theory Pract*. 2010;27(5):345–351.
15. Allen RP, Earley CJ. Augmentation of the restless legs syndrome with carbidopa/levodopa. *Sleep*. 1996;19 (3):205–213.
16. Garcia-Borreguero D, Kohnen R, Silber MH, et al. The long-term treatment of restless legs syndrome/Willis–Ekbom disease: evidence-based guidelines and clinical consensus best practice guidance: a report from the International Restless Legs Syndrome Study Group. *Sleep Med*. 2013;14(7):675–684.

17. Saletu M, Anderer P, Saletu-Zyhlarz GM, et al. Comparative placebo-controlled polysomnographic and psychometric studies on the acute effects of gabapentin versus ropinirole in restless legs syndrome. *J Neural Transm (Vienna, Austria: 1996)*. 2010;117(4):463–473.
18. Griffin E, Brown JN. Pregabalin for the treatment of restless legs syndrome. *Ann Pharmacother*. 2016;50(7):586–591.
19. Lee DO, Buchfuhrer MJ, Garcia-Borreguero D, et al. Efficacy of gabapentin enacarbil in adult patients with severe primary restless legs syndrome. *Sleep Med*. 2016;19:50–56.
20. Hogl B, Garcia-Borreguero D, Kohnen R, et al. Progressive development of augmentation during long-term treatment with levodopa in restless legs syndrome: results of a prospective multi-center study. *J Neurol*. 2009;257(2):230–237.
21. Allen RP, Ondo WG, Ball E, et al. Restless legs syndrome (RLS) augmentation associated with dopamine agonist and levodopa usage in a community sample. *Sleep Med*. 2011;12(5):431–439.
22. Ondo WG. Methadone for refractory restless legs syndrome. *Mov Disord*. 2005;20(3):345–348.
23. Markowitz AJ, Rabow MW. Palliative management of fatigue at the close of life: "it feels like my body is just worn out". *JAMA*. 2007;298(2):217.
24. American Academy of Sleep Medicine. *International Classification of Sleep Disorders: Diagnostic and Coding Manual*. 3rd ed. Darien, IL: American Academy of Sleep Medicine; 2013.
25. Dauvilliers Y, Buguet A. Hypersomnia. *Dialogues Clin Neurosci*. 2005;7(4):347–356.
26. Morgenthaler TI, Kapur VK, Brown T, et al. Practice parameters for the treatment of narcolepsy and other hypersomnias of central origin. *Sleep*. 2007;30(12):1705–1711.
27. Lavault S, Dauvilliers Y, Drouot X, et al. Benefit and risk of modafinil in idiopathic hypersomnia vs. narcolepsy with cataplexy. *Sleep Med*. 2011;12(6):550–556.
28. Philip P, Chaufton C, Taillard J, et al. Modafinil improves real driving performance in patients with hypersomnia: a randomized double-blind placebo-controlled crossover clinical trial. *Sleep*. 2014;37(3):483–487.
29. Mayer G, Benes H, Young P, Bitterlich M, Rodenbeck A. Modafinil in the treatment of idiopathic hypersomnia without long sleep time—a randomized, double-blind, placebo-controlled study. *J Sleep Res*. 2015;24(1):74–81.
30. Harsh JR, Hayduk R, Rosenberg R, et al. The efficacy and safety of armodafinil as treatment for adults with excessive sleepiness associated with narcolepsy. *Curr Med Res Opin*. 2006;22(4):761–774.
31. Cao MT, Guilleminault C. Narcolepsy: diagnosis and management. In: Kryger MH, Roth T, Dement WC, eds. *Principles and Practice of Sleep Medicine*. 6th ed Philadelphia, PA: Elsevier; 2017:873–882.
32. Trotti LM, Saini P, Bliwise DL, Freeman AA, Jenkins A, Rye DB. Clarithromycin in gamma-aminobutyric acid-related hypersomnolence: a randomized, crossover trial. *Ann Neurol*. 2015;78(3):454–465.
33. Trotti LM, Saini P, Koola C, LaBarbera V, Bliwise DL, Rye DB. Flumazenil for the treatment of refractory hypersomnolence: clinical experience with 153 patients. *J Clin Sleep Med*. 2016;12(10):1389–1394.
34. Khan Z, Trotti LM. Central disorders of hypersomnolence: focus on the narcolepsies and idiopathic hypersomnia. *Chest*. 2015;148(1):262–273.
35. Arnulf I, Zeitzer JM, File J, Farber N, Mignot E. Kleine—Levin syndrome: a systematic review of 186 cases in the literature. *Brain*. 2005;128(Pt 12):2763–2776.
36. Arnulf I, Rico TJ, Mignot E. Diagnosis, disease course, and management of patients with Kleine—Levin syndrome. *Lancet Neurol*. 2012;11(10):918–928.
37. Kanathur N, Harrington J, Lee-Chiong Jr. T. Circadian rhythm sleep disorders. *Clin Chest Med*. 2010;31(2):319–325.
38. Amara AW, Maddox MH. Epidemiology of sleep medicine. In: Kryger MH, Roth T, Dement WC, eds. *Principles and Practice of Sleep Medicine*. Philadelphia, PA: Elsevier; 2017:627–637.
39. Lu BS, Kwon J, Zee PC. Circadian disorders. In: Badr MS, ed. *Essentials of Sleep Medicine: An Approach for Clinical Pulmonology*. New York: Humana Press; 2016:1659–1668.
40. Abbot S, Reid KJ, Zee PC. Circadian disorders of the sleep—wake cycle. In: Kryger MH, Roth T, Dement WC, eds. *Principles and Practice of Sleep Medicine*. 6th ed Philadelphia, PA: Elsevier; 2017:413–423.
41. Gulyani S, Salas RE, Gamaldo CE. Sleep medicine pharmacotherapeutics overview: today, tomorrow, and the future (Part 1: insomnia and circadian rhythm disorders). *Chest*. 2012;142(6):1659–1668.
42. Morgenthaler TI, Lee-Chiong T, Alessi C, et al. Practice parameters for the clinical evaluation and treatment of circadian rhythm sleep disorders. An American Academy of Sleep Medicine report. *Sleep*. 2007;30(11):1445–1459.

REFERENCES

43. Auger RR, Burgess HJ, Emens JS, Deriy LV, Thomas SM, Sharkey KM. Clinical practice guideline for the treatment of intrinsic circadian rhythm sleep—wake disorders: advanced sleep—wake phase disorder (ASWPD), delayed sleep—wake phase disorder (DSWPD), non-24-hour sleep—wake rhythm disorder (N24SWD), and irregular sleep—wake rhythm disorder (ISWRD). An update for 2015: An American Academy of Sleep Medicine Clinical Practice Guideline. *J Clin Sleep Med*. 2015;11(10):1199—1236.
44. Pillar G, Shahar E, Peled N, Ravid S, Lavie P, Etzioni A. Melatonin improves sleep—wake patterns in psychomotor retarded children. *Pediatr Neurol*. 2000;23(3):225—228.
45. Ohayon MM, Guilleminault C, Priest RG. Night terrors, sleepwalking, and confusional arousals in the general population: their frequency and relationship to other sleep and mental disorders. *J Clin Psychiatry*. 1999;60 (4):268—276. quiz 277.
46. Markov D, Jaffe F, Doghramji K. Update on parasomnias: a review for psychiatric practice. *Psychiatry (Edgmont)*. 2006;3(7):69—76.
47. Vaughn BV, D'Cruz OF. Cardinal manifestations of sleep disorders. In: Kryger MH, Roth T, Dement WC, eds. *Principles and Practice of Sleep Medicine*. 6th ed Philadelphia, PA: Elsevier; 2017:576—586.
48. Avidan AY. Non-rapid eye movement parasomnias: clinical spectrum, diagnostic features, and management. In: Kryger MH, Roth T, Dement WC, eds. *Principles and Practice of Sleep Medicine*. 6th ed Philadelphia, PA: Elsevier; 2017:981—992.
49. Frank NC, Spirito A, Stark L, Owens-Stively J. The use of scheduled awakenings to eliminate childhood sleepwalking. *J Pediatr Psychol*. 1997;22(3):345—353.
50. Guilleminault C, Kirisoglu C, Bao G, Arias V, Chan A, Li KK. Adult chronic sleepwalking and its treatment based on polysomnography. *Brain*. 2005;128(5):1062—1069.
51. Bruni O, Ferri R, Miano S, Verrillo E. L-5-Hydroxytryptophan treatment of sleep terrors in children. *Eur J Pediatr*. 2004;163(7):402—407.
52. Winkelman JW. Efficacy and tolerability of open-label topiramate in the treatment of sleep-related eating disorder: a retrospective case series. *J Clin Psychiatry*. 2006;67(11):1729—1734.
53. Chiaro G, Caletti MT, Provini F. Treatment of sleep-related eating disorder. *Curr Treat Options Neurol*. 2015;17 (8):361.
54. Silber St. MH, Louis EK, Boeve BF. Rapid eye movement sleep parasomnias. In: Kryger MH, Roth T, Dement WC, eds. *Principles and Practice of Sleep Medicine*. 6th ed Philadelphia, PA: Elsevier; 2017:993—1001.
55. Ohayon MM. Epidemiology of insomnia: what we know and what we still need to learn. *Sleep Med Rev*. 2002;6(2):97—111.
56. Arnulf I. Nightmares and dream disturbances. In: Kryger MH, Roth T, Dement WC, eds. *Principles and Practice of Sleep Medicine*. 6th ed Philadelphia, PA: Elsevier; 2017:1002—1010.
57. Aurora RN, Zak RS, Auerbach SH, et al. Best practice guide for the treatment of nightmare disorder in adults. *J Clin Sleep Med*. 2010;6(4):389—401.
58. Krakow B, Zadra A. Clinical management of chronic nightmares: imagery rehearsal therapy. *Behav Sleep Med*. 2006;4(1):45—70.
59. Ohayon MM, Zulley J, Guilleminault C, Smirne S. Prevalence and pathologic associations of sleep paralysis in the general population. *Neurology*. 1999;52(6):1194—1200.
60. Sharpless BA. A clinician's guide to recurrent isolated sleep paralysis. *Neuropsychiatr Dis Treat*. 2016;12:1761—1767.

I. GENERAL ISSUES

SECTION II

SLEEP DISORDERS IN SPECIFIC MEDICAL CONDITIONS

CHAPTER

5

Cardiovascular Disease

Nancy S. Redeker

Beatrice Renfield Term Professor of Nursing, Yale University School of Nursing,
West Haven, CT, United States

OUTLINE

Introduction	100	
Contributions of Sleep Deficiency to the Development of Cardiovascular Disease	100	
Sleep Duration	100	
Insomnia and Sleep Quality	102	
Obstructive Sleep Apnea	104	
Restless Legs Syndrome	104	
Comorbid Sleep Disorders	105	
Sleep Disorders and Sleep Deficiency Among Adults With Chronic Cardiovascular Disease	105	
Consequences of Sleep Deficiency and Sleep Disorders in People With Chronic Cardiovascular Disease	107	
Assessment of Sleep Disturbance in People With or At-Risk for Cardiovascular Disease	108	

Interventions to Improve Sleep
Impairments in People With
or At-Risk for Cardiovascular Disease 109
    Population-Based Approaches 109
    *Clinical Approaches Targeting
    Sleep to Reduce Cardiovascular Disease
    Risk and Outcomes in People With
    Chronic Cardiovascular Disease* 110
    *Treatment of Sleep-Disordered
    Breathing* 111

Areas for Future Research 112

Clinical Vignette 113

Conclusions 113

References 114

Handbook of Sleep Disorders in Medical Conditions
DOI: https://doi.org/10.1016/B978-0-12-813014-8.00005-6

© 2019 Elsevier Inc. All rights reserved.

INTRODUCTION

Cardiovascular disease (CVD) leads to approximately 32% of annual deaths worldwide, and coronary heart disease (CHD) and stroke are the primary contributors.[1] The worldwide costs of CVD treatment are currently about $863 billion and expected to rise to $1044 billion by the year 2030.[2] CVD, including CHD, stroke, heart failure (HF), hypertension, and dysrhythmias, also contributes to excessive morbidity, poor quality of life, and functional performance. Several known risk factors (e.g., diet, lack of physical activity, and obesity) contribute to the development and exacerbation of these conditions, but recent epidemiological, experiwmental, and clinical evidence suggests that sleep deficiency, including shorter or longer than normative sleep, poor sleep quality, fragmentation, and timing, and specific sleep disorders [e.g., sleep apnea, insomnia, and restless legs syndrome (RLS)] may also contribute to the development of these conditions. For people who already have chronic CVD, sleep deficiency may contribute to exacerbation of disease and increased risk of death, health-care utilization, and symptom burden and decrements in functional performance and quality of life.

At least 70 million US-Americans[3] and many others throughout the world experience sleep deficiency, a multifactorial phenomenon influenced by behavioral, environmental, and health factors, as well as intrinsic sleep disorders. In this chapter, we address the extent to which sleep deficiency/sleep disorders contribute to CVD, the nature and consequences of sleep deficiency/sleep disorders across the trajectory of chronic CVD conditions, and the role of population health and clinical interventions focused on addressing the cardiovascular and related outcomes of sleep deficiency. Given the multidimensional nature of sleep deficiency (e.g., duration, perceived sleep quality/insomnia symptoms, and specific sleep disorders), the variable contributions of these attributes to CVD, and the need for specific sleep interventions that focus on these attributes, we address each of these characteristics of sleep in relationship to CVD in the following narrative.

CONTRIBUTIONS OF SLEEP DEFICIENCY TO THE DEVELOPMENT OF CARDIOVASCULAR DISEASE

Sleep Duration

Experts recommend that adults receive 7 or more hours of sleep each night.[4] However, approximately one-third of adults in western countries do not meet this goal[5] and may, therefore, be at high risk for negative cardiovascular consequences including hypertension, CHD, dysrhythmias, HF, stroke,[6] and excessive CVD-related mortality.[7,8]

There is growing evidence that short sleep duration and, in some cases, prolonged sleep duration[9] may contribute to CVD. A metaanalysis of prospective cohort studies of adults with at least 1-year follow-up revealed associations between short sleep duration and risk of hypertension, with a closer association with objectively measured [RR = 1.24 (1.04−1.49)], compared to self-reported blood pressure [RR = 1.11 (0.79−1.57)].[10] Another review revealed a 17% increased risk of hypertension for adults with short sleep,[7] and short sleep duration modified the relationship between hypertension and all-cause

CONTRIBUTIONS OF SLEEP DEFICIENCY TO THE DEVELOPMENT OF CARDIOVASCULAR DISEASE

mortality.[11] Studies that experimentally shortened sleep documented acute changes in nocturnal systolic and diastolic blood pressure, as well as heart rate,[12,13] although data are sometimes conflicting.

Cross-sectional studies suggest that long sleep duration (9 hours or longer) may also contribute to hypertension, but researchers did not find consistent associations in longitudinal studies.[14] However, a recent systematic review revealed no association between long sleep duration and hypertension,[9] despite its associations with CVD mortality.

Both short and long sleep were associated with CHD[7,15] and incident cardiac events in women who had CHD at baseline, after controlling for relevant covariates in the Nurses' Health Study. The highest relative risks were for those with 5 or fewer hours of sleep and those with 9 or more hours of sleep,[16] while a metaanalysis revealed a 23% increased risk of CHD with short sleep duration among studies with at least 1-year follow-up. A systematic review of studies that had at least 3-year follow-up revealed associations between both short and long sleep duration, incident CHD, and death from CHD, while short sleep duration was associated with particularly high risk of incident HF[17] and atrial fibrillation.[18]

Together, these studies suggest that short sleep duration increases the risk of CVD, primarily hypertension, and CHD. Although there is consistent evidence of associations between either short or long sleep duration or a U-shaped relationship with selected CVD conditions, these associations may be stronger in subsets of the population. For example, the relationship between sleep duration and hypertension seems to be higher among young and middle-aged adults than in older adults,[7,19] and there are likely sex[20] and race-related differences in this relationship. For example, black people with resistant hypertension showed a twofold increased rate of short sleep duration,[21] compared to whites. Further research is needed to determine the groups who incur the highest risk of CVD due to long or short sleep duration to assist in targeting those who may be at highest risk.

Multiple mechanisms may explain the associations between sleep and CVD.[20,22] Short sleep duration may reflect underlying autonomic dysfunction associated with hypertension,[11] as suggested by increasing urinary excretion of norepinephrine in response to experimentally induced acute sleep deprivation.[13] Short sleep also contributes to inflammatory, immunological, and metabolic alterations (i.e., elevated glucose and insulin levels),[23] oxidative stress, endothelial dysfunction, and abnormal levels of leptin/ghrelin (known risk factors for CVD) in experimental and observational studies.[22] For example, extreme sleep duration, including both long and short sleep, was associated with elevated levels of high-sensitivity C-reactive protein (hsCRP), a measure of inflammation.[24] Short sleep and decreased slow wave sleep were correlated with increased Framingham Risk Scores, computed as the composite of measures of cholesterol, blood pressure, smoking status, age, and sex,[25] and short sleep was associated with ventricular wall function and thickness—biomarkers that might predict HF—in a population-based study.[26]

From a circadian perspective, repeated exposure to short sleep leads to blunting of the normal 10% decrease in blood pressure (i.e., blood pressure dipping) that occurs during the sleep period at night.[27] Given the contribution of short sleep duration to both hypertension and diabetes, it is also possible that the pathway from sleep duration to CHD, HF, and stroke may partially be through its contributions to diabetes and hypertension.

II. SLEEP DISORDERS IN SPECIFIC MEDICAL CONDITIONS

102   5. CARDIOVASCULAR DISEASE

Behavioral factors also contribute to sleep duration and may confound apparent relationships between sleep duration and CVD risk if not considered in relevant analyses. Volitional (recreation, drug or alcohol use, smoking cigarettes) and nonvolitional activities (caregiving, work), low socioeconomic status, and being unmarried contribute to short sleep[28] Many of these may also contribute to CVD.[7] Depression is a well-documented contributor to the development of some forms of CVD, such as HF and CHD, and should be considered as a potential influence, especially given variations in sleep duration and sleep-related symptoms associated with depressive disorders. For example, higher levels of psychological distress moderated the effect of sleep duration on stroke risk and other forms of CVD.[15]

Researchers and clinicians lack a complete understanding of the mechanisms through which long sleep duration may contribute to CVD. Some biological and behavioral factors are associated with long sleep duration[29]—some of which may explain its association with CVD. For example, long sleep is associated with an altered immune function, inflammation,[30] photoperiodic abnormalities, underlying disease processes, depression, poor function, deteriorating health status, and low socioeconomic status.[29] Prolonged sleep duration may also contribute to increased sleep fragmentation, insomnia, sleep apnea, and circadian rhythm variations that may influence CVD outcomes.

Insomnia and Sleep Quality

Chronic insomnia, a disorder of initiating and maintaining sleep and/or awakening too early in the morning accompanied by daytime dysfunction, occurs in about 10% of the population and 30% of the primary care population. Because insomnia symptoms are components of descriptions of poor sleep quality, we include studies that use either term in this discussion, with the caveat that the descriptions are often not consistent (and this is a limitation to synthesizing the studies). Likewise, insomnia is sometimes, but not always, associated with short sleep duration, but short sleep in insomnia occurs with adequate sleep opportunity, while acute or chronic short sleep duration occurs with inadequate volitional or nonvolitional sleep opportunity.

Insomnia is associated with elevated physical and psychological arousal, as well as elevated nocturnal blood pressure and blunted nocturnal blood pressure dipping in insomnia patients who are normotensive during the day[31]—a finding that suggests the importance of 24-hour blood pressure monitoring. Difficulty initiating and maintaining sleep were associated with incident hypertension in Japanese male workers,[32] and increased frequency of insomnia symptoms was associated with use of antihypertensive medications.[33] Insomnia accompanied by short sleep duration confers even greater risk for hypertension than insomnia alone,[10,34-36] as well as greater variability in 24-hour blood pressure measured at home.[37]

There were correlations between insomnia and self-reported CHD in Chinese adults over the age of 60 in a cross-sectional study,[38] myocardial infarction on 11-year follow-up in the Nord-Trondelag Health (HUNT) study,[39] and with cardiovascular events, particularly among women who had both insomnia and prolonged sleep duration.[40] In a Swedish cohort of over 41,000 participants, the combination of short sleep plus insomnia symptoms, but not short sleep or insomnia separately, conferred increased risk for

II. SLEEP DISORDERS IN SPECIFIC MEDICAL CONDITIONS

CONTRIBUTIONS OF SLEEP DEFICIENCY TO THE DEVELOPMENT OF CARDIOVASCULAR DISEASE

cardiovascular events.[41] In another cohort study, the onset of disturbed sleep, rather than short or long sleep, predicted CVD [hazard ratio (HR) = 1.22, 95% confidence interval (CI): 1.04–1.44] and dyslipidemia (HR = 1.17, 95% CI: 1.07–1.29) in fully adjusted models.[42] In contrast, the combination of poor sleep quality and long, but not short sleep duration, doubled the risk of CHD in the Women's Health Initiative.[43] Difference in these findings may be due to variations in sleep measurement methods, CVD outcomes, and variability in the populations studied.

A single symptom of insomnia predicted HF [HR = 0.95 (95% CI: 0.55–1.62)], while the HR for incident HF increased for three symptoms [HR = 5.25 (95% CI: 2.25–12.22)] in analyses that controlled for clinical/demographic factors over a mean 11-year follow-up (P for trend = .0010), while insomnia also had a dose-dependent relationship with HF mortality.[44] Insomnia symptoms predicted HF in overweight, but not ideal weight men in a prospective cohort study,[45] and habitual insomnia was associated with increased cardiovascular events in people with HF.[46] On the other hand, insomnia was not linked with echocardiographic indicators of HF.[47]

Although the methods and results of studies of the biological mechanisms that explain the relationships between insomnia and CVD are heterogeneous, there is quite consistent evidence that chronic insomnia is related to inflammation and immunological dysfunction,[31,34,48,49] autonomic dysfunction,[48] and alterations in hypothalamic pituitary axis (HPA) axis functioning[50,51] that may explain its apparent contributions to CVD. Some investigators demonstrated associations between insomnia and alterations in inflammatory cytokines, such as interleukin (IL)-1, IL-6, tumor necrosis factor-alpha, and others,[49,52–54] as well as hsCRP. However, in another study, hsCRP did not link insomnia and CVD.[55] Considerable evidence supports the idea that insomnia is associated with alterations in diurnal patterns and levels of cortisol,[56–59] although research methods vary considerably. Better sleep quality was moderately positively correlated with a higher ratio of daytime to nocturnal urinary free cortisol ($r = 0.39$, $P = .04$) among patients with chronic HF,[60] while chronic insomnia was associated with activation of the renin–angiotensin–aldosterone system[46] that regulates blood pressure and fluid balance among hospitalized patients with acute HF.

Evidence of associations between chronic insomnia and sympathetic activity include elevated norepinephrine[61–63] and decreased HR variability, although authors of a recent systematic review concluded that HR variability was not consistently associated with insomnia, and the effects of insomnia treatment on HR variability could not be confirmed.[64] Taken together with evidence that insomnia with short sleep duration was associated with elevated cholesterol levels,[65] these mechanisms may explain associations between insomnia and endothelial dysfunction, as indicated by low levels of brachial artery flow-mediated dilation.[66]

An often-used explanatory model for understanding and managing insomnia includes a focus on predisposing, precipitating, and perpetuating factors for insomnia,[67] some of which may also contribute to the development of CVD. For example, anxiety, depression, and physical and psychosocial stress are not only closely associated with insomnia but also contribute to the development of CVD. In an analysis of the National Health and Nutrition Examination Survey (NHANES) data, insomnia and short sleep duration mediated the relationship between depression and hypertension.[68] As with sleep duration, there is a need to consider these potential confounding factors when addressing the role of

104

5. CARDIOVASCULAR DISEASE

insomnia about CVD. Past research which has often not addressed these multivariate influences is limited by variations in methods and measures.

Obstructive Sleep Apnea

Obstructive sleep apnea (OSA) occurs in 3%–7% of the adult male population and 2%–5% of women, when the criterion of excessive daytime sleepiness is used.[69] Prevalence rates vary among subgroups of the population and seem to be higher among people with chronic CVD and are significantly higher in some groups of racial minorities.[70] OSA is a well-documented risk factor for CVD, including hypertension and resistant hypertension,[71] dys-rhythmias,[72–74] CHD, and stroke,[75] as well as death from cardiovascular causes. OSA was associated with CVD in cross-sectional data obtained from the Sleep Heart Health Study, with dose-dependent effects depending on the severity of sleep apnea, while the strongest effects were on HF and stroke rather than on CHD.[76] The epidemiological, clinical, and experimental literature on the relationships between OSA and CVD are well documented.[77–79]

Mechanisms that explain the relationships between OSA and CVD are multifactorial. OSA-related intermittent hypoxia and sleep fragmentation contribute to increased central sympathetic outflow and activation of the hypothalamic–pituitary axis and renin–angiotensin–aldosterone systems. OSA is also associated with impairments in baroreflex function, decreased variability in heart rate and blood pressure, increases in chemoreflex sensitivity, and metabolic alterations such as glucose intolerance and reduced insulin sensitivity. These factors may contribute to oxidative stress, vascular inflammation, impaired endothelial function, and aortic stiffness[79,80] that, in turn, may lead to adverse cardiovascular consequences. Obesity is a primary risk factor for OSA, and diabetes is also often associated with this sleep disorder. Given that obesity and diabetes are also risk factors for CVD, they may partially explain the increased risk of CVD associated with OSA, although rigorously designed studies usually statistically controlled for these potentially confounding variables.

OSA is also associated with excessive daytime sleepiness, fatigue, depressed mood, and some decrements in executive function,[81,82] although the severity of these decrements may be mild.[82] These daytime effects may contribute to poor self-management behavior, including insufficient adherence to positive airway treatment, but may also contribute to CVD risk. For example, OSA was associated with nonadherence to antihypertensive medications.[83] Excessive daytime sleepiness and decrements in executive dysfunction contributed to medication nonadherence among patients with HF.[84,85] Given the importance of self-management, including treatment adherence to CVD prevention and disease management, the behavioral effects of OSA, as well as other sleep disorders that contribute to excessive daytime sleepiness and cognitive decrements, should be considered in research and treatment decisions.

Restless Legs Syndrome

RLS is a condition associated with uncomfortable sensations in the legs that may cause difficulty falling asleep, but periodic limb movements during sleep (PLMS) are common in people with this condition and lead to considerable sleep fragmentation. In a cross-sectional study of more than 18,000 European adults, the prevalence of RLS was 5.5%, with higher prevalence associated with advancing age and female sex, while data from the US National Sleep Foundation 2005 survey indicated that the prevalence of RLS was 8% in men and 11% in women,[86] with somewhat lower prevalence in the Sleep Heart Health Study.[87] RLS often cooccurs with sleep apnea and with insomnia symptoms.[86]

Considerable cross-sectional and longitudinal evidence suggests that RLS is associated with various manifestations of CVD, although data are sometimes conflicting and limited by differences in methods of eliciting RLS symptoms.[80] For example, while the severity of RLS was associated with higher risk of CHD in one study,[87] it was not associated with hypertension or other forms of CVD in a case–control study.[88]

Although the reasons for the possible associations between RLS and CVD are not known, PLMS are associated with sleep fragmentation and autonomic dysfunction. Altered heart rate variability precedes PLM events,[89,90] and PLMS are linked with acute increases in blood pressure and sympathetic surges,[71] as well as highly variable blood pressure. Gottlieb et al.[80] suggest that, aside from the intermittent hypoxia that characterized OSA, mechanistic explanations for the associations between RLS with PLMS may be similar to those of OSA. Because RLS is more common among people who have iron deficiency anemia, a possible factor for CVD, it is possible that anemia is a shared risk factor for both CVD and RLS.

Comorbid Sleep Disorders

Sleep disorders, such as OSA, RLS, and insomnia, often cooccur[86] and together contribute to short sleep duration and poor sleep quality, including reports of difficulty initiating sleep and maintaining sleep, objective sleep fragmentation, daytime sleepiness, and functional deficits. A report compiled from NHANES data revealed that sleep disorders individually were not associated with hypertension, but the combination of a sleep disorder with short sleep duration and/or disturbed sleep quality led to doubling the risk of hypertension.[91] These findings underscore the importance of considering the comorbidity of sleep disorders, potential common mechanisms that explain the relationships between sleep disorders, sleep deficiency, and CVD, as exemplified in a recent review,[80] and the need to research the synergistic relationships among these conditions. From a clinical perspective, it is important to fully explore possible contributors to general signs of sleep deficiency, such as short sleep or reports of poor sleep quality.

SLEEP DISORDERS AND SLEEP DEFICIENCY AMONG ADULTS WITH CHRONIC CARDIOVASCULAR DISEASE

Sleep deficiency is multifactorial among people who have chronic CVD and is more common in these groups than in the general population. For example, insomnia symptoms and poor sleep quality occur in about 44% of people living with CHD[92,93] and in about 50% of patients with chronic HF[94] who have poorer sleep quality and more sleep fragmentation than people without HF.[95] The prevalence rate of OSA among people living with CHD is approximately 30% but is somewhat higher in people with HF,[96] among whom a significant number also have Cheyne–Stokes breathing/central sleep apnea (CSA), often accompanied by OSA. Among myocardial infarction survivors, 65% had OSA;[97] and increased prevalence of OSA was associated with non-ST elevation myocardial infarction compared with ST-elevation myocardial infarction,[98] while PLMS occurred in about 20% of patients with stable HF.[99]

Patients with chronic CVD often suffer from multiple comorbid sleep disorders. For example, while about half of patients with stable HF had sleep disordered breathing, half also reported insomnia symptoms that were not explained by sleep-disordered breathing.[94] PLMS and sleep-disordered breathing may also cooccur in people with HF.[99] Because each of these conditions may be associated with difficulty maintaining sleep, it is important to disentangle the specific causative factors. Abnormalities in circadian rhythmicity that result from disease processes, environmental, or other factors may also contribute to CVD.[100]

The stability or acuity of cardiac conditions should be considered when interpreting research findings and during clinical assessment of sleep among these patients. The severity of sleep disturbance and the manifestations of specific sleep disorders may increase during exacerbations. For example, worsening sleep symptoms were prodromal to myocardial infarction among women[101] and sleep quality, measured by self-report and wrist actigraphy, worsened dramatically from the preoperative period to the immediate postoperative period after cardiac surgery.[102] Cheyne–Stokes breathing/CSA is often more severe during acute exacerbation of HF due to rostral fluid shifts.[103] Although there were improvements from the acute care period,[102,104] sleep deficiency continued over the course of recovery after coronary bypass surgery,[104] and 40% of people who received coronary angioplasty reported difficulty maintaining sleep at 1 year after the procedure.[105] These findings suggest the importance of understanding the trajectory of the CVD condition, including stages of stability and acuity when interpreting data on the prevalence and nature of sleep deficiency, but these changes should likely also influence the nature and timing of intervention focused on sleep.

As in the general population, factors that contribute to sleep deficiency in people with chronic CVD are multifactorial and likely include sociodemographic characteristics, disease state and pathophysiology, medications and other treatments, as well as environmental, behavioral, and emotional/stress-related influences. Given the significant role of sleep deficiency and intrinsic sleep disorders to the development of CVD, the high prevalence of sleep disturbances among this population is likely partially explained by premorbid and persistent chronic sleep deficiency and sleep disorders, as well as perpetuating thoughts and behaviors that contribute to the maintenance of sleep disturbance. For example, prehospitalization sleep disturbances explained a significant proportion of the variance in sleep during hospitalization in medically treated CHD patients[106] and patients undergoing coronary artery bypass surgery.[102] Patients with chronic HF who reported insomnia identified perpetuating factors that were quite similar to other insomnia patients (i.e., worries, negative sleep hygiene behaviors) rather than factors specific to HF,[107] a finding consistent with the idea that precipitating factors may be disease specific, but perpetuating factors are usually common across groups with various comorbid chronic conditions.

CONSEQUENCES OF SLEEP DEFICIENCY AND SLEEP DISORDERS IN PEOPLE WITH CHRONIC CARDIOVASCULAR DISEASE

Given the pathophysiological implications of sleep deficiency and sleep disorders for the development of CVD, it seems logical that the presence of these conditions may worsen the pathophysiology of the disease, CVD-related morbidity, and mortality, and daytime symptoms, cognitive function, functional performance, and quality of life. These outcomes are often already impaired by chronic CVD, but sleep impairments may contribute additional risk.

Despite the growing evidence of the importance of sleep duration to CVD risk, little is known about the specific effects of short or long sleep duration among people with chronic CVD. However, insomnia, present in about 50% of patients with stable HF,[94] was associated with alterations in the renin−angiotensin−aldosterone system, increased cardiac events, and lower functional capacity,[46] as well as decreased self-reported physical function, objective functional performance, excessive daytime sleepiness, fatigue, and depression.[94] Poor sleep quality and low sleep efficiency were associated with both worse physical function and emotional well-being after cardiac surgery in a longitudinal observational study.[108] However, the causal relationships among sleep and these outcomes are not known.

Data on the consequences of sleep disordered breathing for people with chronic CVD are conflicting. OSA contributes to death from CVD and ischemic strokes,[109,110] but the severity of OSA was not associated with worsened cardiac ischemia or arrhythmias in one study of people with CHD.[111] On the other hand, OSA predicted atrial fibrillation after cardiac surgery.[112,113] Moderate-to-severe OSA predicted death, compared to none−mild OSA among patients with HF who had reduced left ventricular ejection fraction, a marker of HF severity.[114] CSA was also associated with higher levels of brain and atrial natriuretic peptides,[115] markers of HF exacerbation, and predicted hospital readmission in people with HF.[116] However, neither OSA nor CSA were associated with excessive daytime sleepiness, fatigue, depression, or functional performance among people with stable HF.[96] Taken together, these findings suggest the importance of sleep disordered breathing for specific CVD outcomes.

Although PLMS were common in HF, there was no difference in HF patients with or without PLMS in excessive daytime sleepiness or severity of HF.[117] On the other hand, PLMS were higher among patients with decompensated HF and were associated with increased risk of death and rehospitalization.[118]

In sum, the results of studies of adults with chronic CVD suggest the importance of sleep impairments to clinical and quality-of-life outcomes. However, the majority of the literature is cross-sectional, and studies are difficult to compare due to variations in measures of sleep deficiency and the selected outcome measures, as well as other concerns.

There is a need for experimental studies and randomized controlled trials to improve understanding of the causal directions of observed associations and relevant biobehavioral mechanisms.

ASSESSMENT OF SLEEP DISTURBANCE IN PEOPLE WITH OR AT-RISK FOR CARDIOVASCULAR DISEASE

An emphasis on addressing sleep disturbances is critical for people with or at-risk for CVD, given its high prevalence and apparent negative consequences. Sleep screening and assessment focused on the most common contributors to sleep difficulties are critically important but currently not consistently provided in primary care settings, cardiology practices, hospitals, and cardiac rehabilitation settings where people with CVD receive health care. Protocols to guide assessment, treatment, and referral, as well as clinical management of patients who return to the primary care or cardiology practice after receiving specialized sleep assessment and treatment, may improve adherence to CVD treatment and successful outcomes.

The increased availability of electronic medical record (EMR) systems may facilitate screening with a focus on sleep duration, insomnia symptoms, and risk factors for sleep-disordered breathing—sleep abnormalities that are most common among patients with CVD. The use of such systems may facilitate the ability to identify participants who have signs and symptoms that indicate need for follow-up (e.g., snoring, excessive daytime sleepiness) with testing in a sleep disorders laboratory with ambulatory- or laboratory-based measurements, especially for those who evidence signs of sleep disordered breathing (see Chapter 3: Treatment of Breathing-Related Disorders, for more information). Although a full description of screening instruments is beyond the scope of this chapter, incorporating easy to use tools, such as the STOP-Bang Questionnaire[119] to screen for OSA and the Insomnia Severity Index[120] among others, into the EMR may be useful, especially for screening for patients who may already have risk factors for CVD, such as diabetes, obesity, or hypertension. Providing these structured and standard sleep assessment tools may facilitate uptake, given barriers including competing priorities, nonsleep specialist clinicians' lack of knowledge and awareness of sleep, and lack of time in busy clinic settings.

Although CVD-specific sleep screening instruments and methods are not needed for adults with this medical condition compared to others, alterations in sleep may cooccur with important cardiovascular indicators. For example, sleep disordered breathing may be associated with failure of blood pressure to drop at night (as is normal) or be associated with cardiac dysrhythmias during sleep. These observations suggest the importance of obtaining blood pressure and electrocardiographic measures during sleep.

There are numerous examples of opportunities to initiate sleep disorders screening, diagnosis, and treatment referral in cardiovascular populations in office,[121] hospital,[46,110,122–125] and cardiac rehabilitation settings.[126–128] Yet, assessment of sleep has still not been widely adopted in these settings, and the primary focus has been on sleep disordered breathing, despite evidence that other sleep disorders and sources of sleep deficiency may also be important to CVD and related outcomes (e.g., daytime function, symptoms, health-care resource utilization). Given evidence that sleep disorders may be more severe during exacerbation of CVD and differences in sensitivity and specificity of sleep questionnaires in different CVD populations,[121] the specific type of cardiovascular disorder and context for the screening efforts should be considered when evaluating symptoms and choosing measurement methods. For example, sleep disordered breathing or insomnia symptoms may be worse during acute exacerbation of HF with fluid overload. Sleep disturbance may also be poorer during acute care hospitalization due to environmental influences or changes in treatment compared to sleep in the home environment when health status is more stable. Nevertheless, signs of poor sleep during acute exacerbations may present an opportunity for future follow-up and evaluation. For example, previously undetected sleep disordered breathing may be observed in hospitalized patients and warrant follow-up evaluation.

INTERVENTIONS TO IMPROVE SLEEP IMPAIRMENTS IN PEOPLE WITH OR AT-RISK FOR CARDIOVASCULAR DISEASE

The associations of sleep impairments with various manifestations of CVD suggest the critical need for population-focused strategies to promote sleep health and prevent sleep deficiency across the lifespan. Focused clinical approaches may better identify, diagnose, and treat sleep deficiency, especially in people with or at-risk for CVD.

Population-Based Approaches

Strategies to raise awareness about the importance of sleep at the societal and community levels may contribute to increasing the adequacy of sleep and identifying and correcting sleep disorders, such as sleep-disordered breathing and insomnia. Recently accrued scientific evidence and sleep goals set forth for US Healthy People 2020 that sets forth specific health promotion and prevention goals for the population[129] focus on increasing sleep duration for youth and adults and identification and treatment of OSA. These efforts are focused on increasing public awareness about the negative consequences of sleep deficiency, such as the National Healthy Sleep Awareness Program. The National Sleep Foundation and the American Heart Association have begun to frame these messages regarding the specific effects on cardiovascular health. For example, the American Heart Association recently published a scientific statement recommending a public health campaign to incorporate sleep behavior as a means to promote cardiovascular health, including specific guidelines for sleep duration and integration of sleep assessment tools into public health and clinical settings.[23] Although the US Healthy People 2020 sleep goals and related activities will facilitate monitoring of improvements in sleep, further research will be needed to evaluate the effects of improvements in sleep on CVD health in the population.

Given that healthy sleep habits begin early in life, there is a pressing need for greater public awareness of the importance of obtaining sufficient sleep among infants and children that may lead to lifelong cardiovascular health. Improving sleep duration in children

110 — 5. CARDIOVASCULAR DISEASE

and adolescents may contribute not only to sleep extension but also to improved lifelong cardiovascular health, although cardiovascular health has not always been included in the rationale for these efforts. There is a need for research focused on the short-term (e.g., biomarkers) and longer term metabolic and cardiovascular outcomes of changes in sleep beginning early in life to focus on prevention of the negative cardiovascular consequences of sleep deficiency. Promotion of sleep health is likely to be a valuable addition to traditional efforts focused on cardiovascular risk factors.[23]

Clinical Approaches Targeting Sleep to Reduce Cardiovascular Disease Risk and Outcomes in People With Chronic Cardiovascular Disease

Despite increased understanding of the scope of sleep impairments and sleep disorders relative to CVD and available treatments, evidence of the efficacy of some potential sleep treatments is incomplete, and decision-making in these cases must be made based on clinical reasoning rather than comprehensive scientific evidence. For example, extending sleep duration seems to be a logical approach to addressing the cardiovascular abnormalities associated with short sleep duration. Although there have been several studies that considered the effects of sleep extension on cognitive function and other outcomes, to our knowledge, few studies have evaluated its cardiovascular effects. For example, in a study of a 6-week sleep extension intervention with 22 participants who had prehypertension or stage 1 hypertension, investigators randomized participants to sleep extension or sleep maintenance groups. Average 24-hour systolic and diastolic beat-to-beat blood pressures decreased significantly ($P < .05$) by an average of 14 and 8 mmHg, respectively, from baseline to after sleep extension. Although blood pressure reduction did not differ significantly between groups, this may be a result of low statistical power.[130] There is a need for further research on the effects of behavioral interventions to promote sleep extension on cardiovascular outcomes, especially blood pressure and biomarkers reflecting CVD risk.[7]

Insomnia treatment generally consists of hypnotic medication and/or behavioral treatment. Hypnotic medications do not present specific cardiovascular risks but may present risk for daytime cognitive effects, and possibly falls, especially in older adults who also have cardiovascular conditions and other comorbidities or functional limitations. Patients with insomnia and HF reported fear of addiction and negative daytime effects.[107] Cognitive-behavioral therapy for insomnia (CBT-I) is efficacious with similar short-term, but more durable, effects on insomnia than pharmacological interventions without the adverse consequences of hypnotic agents.[131] There is considerable evidence documenting the efficacy of CBT-I in insomnia and sleep characteristics in people with insomnia comorbid with medical or psychiatric conditions,[132] but few studies have examined its effects specifically in CVD patients. Several studies included mixed groups of older adults with CVD and other comorbidities and led to improvements in insomnia and sleep quality.[133–138]

However, in a recent feasibility study among patients with stable HF, the effects of CBT-I that includes sleep restriction, stimulus control, sleep hygiene, cognitive therapy, and progressive muscle relaxation based on standard CBT-I approaches were assessed. CBT-I led to large improvements in insomnia severity and fatigue at posttreatment and

II. SLEEP DISORDERS IN SPECIFIC MEDICAL CONDITIONS

sustained improvements in fatigue at 6 months after completing treatment, compared to an attention control condition consisting of standard HF self-management education that did not experience improvements.[139,140] Notably, the estimated incidence rates of rehospitalization at 12 months were 13.2% among the CBT-I group, compared with 21.1% in the control group, while controlling for symptoms at baseline and clinical and demographic characteristics (OR = 0.32, CI = 0.04–2.46). Nine percent of those with insomnia remission, compared with 27% of nonremitted patients, experienced hospital readmission. However, probably due to the small sample, the effects were not statistically significant.[140] Based on these promising effects, a fully powered randomized controlled trial (RCT) to examine the sustained effects of CBT-I on sleep, symptoms, daytime function, health resource utilization, and event-free survival is now underway.[141]

In addition to its effects on insomnia and sleep characteristics, CBT-I has promising effects on CVD biomarkers. A trial of CBT-I, compared with Tai Chi Chih (TCC) and a sleep seminar control among older adults (95% patients with CVD),[136,137,142] documented CBT-I-related improvements in inflammation, as measured by C-reactive protein (CRP), a measure associated with CVD risk, at 16 months compared to baseline.[138] CBT-I improved composite metabolic and cardiovascular risk factors (i.e., high-density lipoprotein, low-density lipoprotein, triglycerides, hemoglobin A1c, glucose, insulin, CRP, and fibrinogen).[137] Both CBT-I and TCC reduced the chance of being in a high-risk metabolic group at 16 months, but only CBT-I improved risk at 4 months. CBT-I also reduced monocyte production of inflammatory cytokines (all $Ps < .05$) and pro-inflammatory gene expression.[136] Among patients with stable HF and insomnia, improvements in sleep quality and fatigue from baseline to post-CBT-I correlated with increases in the ratio of diurnal to nocturnal urinary cortisol [($r = 0.29$, $P = .08$) and ($r = 0.54$, $P = .0007$), respectively]. The ratio of diurnal/nocturnal urinary epinephrine correlated with wake after sleep onset ($r = -0.52$, $P = .004$). However, there was no statistically significant effect of CBT-I on these outcomes. While these findings suggest the importance of CBT effects on the HPA axis and other CVD risk factors, CBT-I did not improve heart rate variability, a marker of autonomic function.[143]

Together, these findings suggest the potential benefits of CBT-I on significant biomarkers of CVD risk. However, further research is needed to confirm these effects and evaluate the long-term impact on clinical outcomes of CVD, as well as function, quality of life, and morbidity and mortality.

Treatment of Sleep-Disordered Breathing

Continuous positive airway pressure (CPAP) treatment is the mainstay intervention for OSA. CPAP acutely improves blood pressure and cardiac variability.[144] Results of a recent systematic review revealed that CPAP reduced 24-hour ambulatory systolic (2.32 mmHg; 95% CI, 3.65–1.00) and diastolic blood pressure (1.98 mmHg; 95% CI, 2.82–1.14). There were larger improvements in nocturnal than daytime blood pressure, and improvements in blood pressure were greater in people with resistant hypertension and those who were already taking antihypertensive medications.[145,146] Although some investigators and clinicians describe these effects as modest, this may be due to failure to detect the effect of CPAP on brief surges in blood pressure that may have deleterious effects.[147] Despite improvements in some intermediary CVD risk factors in past studies, a recent metaanalysis of moderate-to-large-sized CPAP trials revealed no effect on adverse cardiovascular events, hospitalization, or all-cause or CVD-related death.[148] While concerns about the extent of adherence to CPAP therapy have plagued interpretation about negative trials, the authors concluded that poor adherence did not explain the outcomes. Although both central and OSA are common and have negative consequences among patients with HF and the effects of CPAP have been extensively studied in these patients, there is little evidence of its long-term effects on cardiac outcomes or mortality.[149] Despite this, many people suffer from daytime sleepiness, cognitive deficits, and problems with safety and functional performance that warrant treatment of OSA, and CPAP therapy has well documented benefits for this purpose. A critical concern for people using CPAP therapy is adherence to treatment.

Investigators found that weight loss improved respiratory events associated with OSA[150] but are insufficient to normalize them.[151] However, given that obesity is a shared risk factor for both OSA and CVD, weight loss may contribute to improvement in both outcomes. Unknown is the extent to which improvements in OSA through weight loss might lead to reductions in death or adverse cardiovascular events.

AREAS FOR FUTURE RESEARCH

As reviewed in this chapter, there is a fairly large body of epidemiological evidence that suggests associations between various attributes of sleep disturbance (e.g., short sleep duration, insomnia symptoms) and sleep-disordered breathing and cardiovascular risk. Many of these studies focused on individual sleep disorders (e.g., sleep disordered breathing) and did not consider multiple attributes of sleep or comorbid sleep disorders that may explain outcomes and studies have also not considered subpopulations that may be at particular risk. In addition, with the exception of studies of the effects of positive airway pressure treatment delivered via a number of approaches on CVD outcomes, few studies have evaluated the effects of treatment of sleep disturbance on CVD outcomes. Therefore, the following research directions are suggestions to advance this body of knowledge.

1. Examine the relative contributions of sleep disorders (e.g., insomnia, sleep disordered breathing, RLS), sleep duration, and sleep fragmentation and combined sleep characteristics to the development and exacerbation of CVD.
2. Examine the extent to which subpopulations defined by genomic factors, age, race, sex, and socioeconomic status vary in CVD risk based on characteristics of sleep disturbance.
3. Examine the effects of behavioral and pharmacological treatments that address insomnia and sleep duration on prevention of CVD and reduction in negative outcomes among people with chronic CVD and their biobehavioral mechanisms.
4. Examine the effectiveness of treatment for sleep disordered breathing including the role of adherence.

5. Identify and disseminate best practices for effective screening and referral for sleep disturbance in community and clinical settings that serve patients with or at-risk for CVD.
6. Examine the cost-effectiveness of treatment of sleep disturbance in relation to people with or at-risk for CVD.

CLINICAL VIGNETTE

A male patient has had New York Heart Association Functional Classification III HF for 3 years. His left ventricular ejection fraction is 35%, and he has a pacemaker and an implanted cardioverter defibrillator for preventive reasons. He has type II diabetes, hypertension and experienced a myocardial infarction 4 years ago. He is adherent to evidence-based pharmacological HF treatment. Upon a routine primary care visit, health history revealed that he snored loudly and had pauses in breathing every night, especially while lying in the supine position. He also reported difficulty falling asleep and frequent awakening during the night. He reported falling asleep while sitting and reading in the afternoon but has not fallen asleep while driving.

The clinician referred the patient for evaluation of sleep disordered breathing. In-lab polysomnography revealed a respiratory disturbance index (RDI) of 35/hour in the supine position and 30/hour while side-lying. All of the respiratory events are obstructive in nature and were not associated with electrocardiographic abnormalities. The patient underwent a titration with CPAP with reduction of the RDI to 4 events/hour. He reported improvement in sleepiness after using CPAP for 4 weeks but continued to report difficulty falling asleep and staying asleep at night.

The patient returned to the sleep clinic where his Insomnia Severity Index score was 15, indicating moderate insomnia. He was referred to and completed four individual sessions of CBT-I with a behavioral health provider. After completion of CBT-I, his Insomnia Severity Index score decreased to 4, which falls in the normal range, and he reported a large improvement in fatigue during everyday activities. This effect was sustained over 6 months of follow-up.

CONCLUSIONS

There is significant but sometimes conflicting evidence of the importance of sleep disturbances, including short sleep duration, insomnia, sleep disordered breathing, and RLS to CVD and growing evidence of the effects of treatment of sleep disturbance on cardiovascular outcomes. Although more research is needed, this evidence suggests the need to screen people with CVD for sleep disturbance and to promote healthy sleep in the general population to reduce cardiovascular risk. Improved access to treatment of sleep disturbance may improve population outcomes.

114   5. CARDIOVASCULAR DISEASE

KEY PRACTICE POINTS

- Sleep disturbance is common among people with cardiovascular conditions and an important risk factor for these conditions.
- Sleep disturbance may contribute to exacerbation of CVD, morbidity, mortality, poor functional performance, and decrements in quality of life.
- Patients with CVD should be routinely assessed for signs and symptoms of sleep-disordered breathing (snoring, pauses in breathing during sleep, and excessive daytime sleepiness); insomnia symptoms (difficulty initiating sleep, maintaining sleep, or waking too early in the morning accompanied by daytime dysfunction); and sleep duration longer or shorter than 7–8 hours.
- Sleep disorders warrant referral to specialized sleep services and treatment.
- Treatment of sleep disturbance may reduce the onset or progression of CVD, improve daytime symptoms (e.g., fatigue, sleepiness), functional performance, and quality of life. It may also improve morbidity and health-care resource utilization.

References

1. Roth GA, Huffman MD, Moran AE, et al. Global and regional patterns in cardiovascular mortality from 1990 to 2013. *Circulation.* 2015;132(17):1667–1678.
2. Association AH. *Heart Disease and Stroke Statistics – 2017 At Glance.* 2017.
3. Colten HR, Altevogt BM, eds. *Sleep Disorders and Sleep Deprivation: An Unmet Public Health Problem.* Washington, DC: National Academies Press; 2006.
4. Watson NF, Badr MS, Belenky G, et al. Recommended amount of sleep for a healthy adult: a joint consensus statement of the American Academy of Sleep Medicine and Sleep Research Society. *Sleep.* 2015;38(6):843–844.
5. Liu Y, Wheaton AG, Chapman DP, Cunningham TJ, Lu H, Croft JB. Prevalence of healthy sleep duration among adults—United States, 2014. *Morb Mortal Wkly Rep.* 2016;65(6):137–141.
6. Altman NG, Izci-Balserak B, Schopfer E, et al. Sleep duration versus sleep insufficiency as predictors of cardiometabolic health outcomes. *Sleep Med.* 2012;13(10):1261–1270.
7. Itani O, Jike M, Watanabe N, Kaneita Y. Short sleep duration and health outcomes: a systematic review, meta-analysis, and meta-regression. *Sleep Med.* 2017;32:246–256.
8. Cappuccio FP, Cooper D, D'Elia L, Strazzullo P, Miller MA. Sleep duration predicts cardiovascular outcomes: a systematic review and meta-analysis of prospective studies. *Eur Heart J.* 2011;32(12):1484–1492.
9. Jike M, Itani O, Watanabe N, Buysse DJ, Kaneita Y. Long sleep duration and health outcomes: A systematic review, meta-analysis and meta-regression. *Sleep Med. Rev.* 2018;39:25–36.
10. Meng L, Zheng Y, Hui R. The relationship of sleep duration and insomnia to risk of hypertension incidence: a meta-analysis of prospective cohort studies. *Hypertens Res.* 2013;36(11):985–995.
11. Fernandez-Mendoza J, He F, Vgontzas AN, Liao D, Bixler EO. Objective short sleep duration modifies the relationship between hypertension and all-cause mortality. *J Hypertens.* 2017;35(4):830–836.
12. Lusardi P, Mugellini A, Preti P, Zoppi A, Derosa G, Fogari R. Effects of a restricted sleep regimen on ambulatory blood pressure monitoring in normotensive subjects. *Am J. Hypertens.* 1996;9(5):503–505.
13. Tochikubo O, Ikeda A, Miyajima E, Ishii M. Effects of insufficient sleep on blood pressure monitored by a new multibiomedical recorder. *Hypertension.* 1996;27(6):1318–1324.
14. Gangwisch JE, Heymsfield SB, Boden-Albala B, et al. Sleep duration as a risk factor for diabetes incidence in a large U.S. sample. *Sleep.* 2007;30(12):1667–1673.
15. Liu Y, Wheaton AG, Chapman DP, Croft JB. Sleep duration and chronic diseases among U.S. adults age 45 years and older: evidence from the 2010 Behavioral Risk Factor Surveillance System. *Sleep.* 2013;36 (10):1421–1427.
16. Ayas NT, White DP, Manson JE, et al. A prospective study of sleep duration and coronary heart disease in women. *Arch Intern Med.* 2003;163(2):205–209.

REFERENCES

17. Wannamethee SG, Papacosta O, Lennon L, Whincup PH. Self-reported sleep duration, napping, and incident heart failure: prospective associations in the British Regional Heart Study. *J Am Geriatr Soc*. 2016;64 (9):1845−1850.
18. Kayrak M, Gul EE, Aribas A, et al. Self-reported sleep quality of patients with atrial fibrillation and the effects of cardioversion on sleep quality. *Pacing Clin Electrophysiol*. 2013;36(7):823−829.
19. van den Berg JF, Tulen JH, Neven AK, et al. Sleep duration and hypertension are not associated in the elderly. *Hypertension*. 2007;50(3):585−589.
20. Gangwisch JE. A review of evidence for the link between sleep duration and hypertension. *Am J Hypertens*. 2014;27(10):1235−1242.
21. Rogers A, Necola O, Sexias A, et al. Resistant hypertension and sleep duration among blacks with metabolic syndrome. *J Sleep Disord Treat Care*. 2016;5(4).
22. Tobaldini E, Costantino G, Solbiati M, et al. Sleep, sleep deprivation, autonomic nervous system and cardiovascular diseases. *Neurosci Biobehav Rev*. 2017;74(Pt B):321−329.
23. St-Onge MP, Grandner MA, Brown D, et al. Sleep duration and quality: impact on lifestyle behaviors and cardiometabolic health: a scientific statement from the American Heart Association. *Circulation*. 2016;134(18): e367−e386.
24. Grandner MA, Buxton OM, Jackson N, Sands-Lincoln M, Pandey A, Jean-Louis G. Extreme sleep durations and increased C-reactive protein: effects of sex and ethnoracial group. *Sleep*. 2013;36(5):769−779E.
25. Matthews KA, Strollo Jr PJ, Hall M, et al. Associations of Framingham risk score profile and coronary artery calcification with sleep characteristics in middle-aged men and women: Pittsburgh SleepSCORE study. *Sleep*. 2011;34(6):711−716.
26. Lee JH, Park SK, Ryoo JH, et al. Sleep duration and quality as related to left ventricular structure and function. *Psychosom Med*. 2018;80(1):78−86.
27. Yang H, Haack M, Gautam S, Meier-Ewert HK, Mullington JM. Repetitive exposure to shortened sleep leads to blunted sleep-associated blood pressure dipping. *J Hypertens*. 2017;35(6):1187−1194.
28. Krueger PM, Friedman EM. Sleep duration in the United States: a cross-sectional population-based study. *Am J Epidemiol*. 2009;169(9):1052−1063.
29. Grandner MA, Drummond SP. Who are the long sleepers? Towards an understanding of the mortality relationship. *Sleep Med Rev*. 2007;11(5):341−360.
30. Prather AA, Vogelzangs N, Penninx BW. Sleep duration, insomnia, and markers of systemic inflammation: results from the Netherlands Study of Depression and Anxiety (NESDA). *J Psychiatr Res*. 2015;60:95−102.
31. Lanfranchi PA, Pennestri MH, Fradette L, Dumont M, Morin CM, Montplaisir J. Nighttime blood pressure in normotensive subjects with chronic insomnia: implications for cardiovascular risk. *Sleep*. 2009;32(6):760−766.
32. Suka M, Yoshida K, Sugimori H. Persistent insomnia is a predictor of hypertension in Japanese male workers. *J Occup Health*. 2003;45(6):344−350.
33. Haaramo P, Rahkonen O, Hublin C, Laatikainen T, Lahelma E, Lallukka T. Insomnia symptoms and subsequent cardiovascular medication: a register-linked follow-up study among middle-aged employees. *J Sleep Res*. 2014;23(3):281−289.
34. Javaheri S, Redline S. Insomnia and risk of cardiovascular disease. *Chest*. 2017;152(2):435−444.
35. Vgontzas AN, Fernandez-Mendoza J, Liao D, Bixler EO. Insomnia with objective short sleep duration: the most biologically severe phenotype of the disorder. *Sleep Med Rev*. 2013;17(4):241−254.
36. Montag SE, Knutson KL, Zee PC, et al. Association of sleep characteristics with cardiovascular and metabolic risk factors in a population sample: the Chicago Area Sleep Study. *Sleep Health*. 2017;3(2):107−112.
37. Johansson JK, Kronholm E, Jula AM. Variability in home-measured blood pressure and heart rate: associations with self-reported insomnia and sleep duration. *J Hypertens*. 2011;29(10):1897−1905.
38. Zhuang J, Zhan Y, Zhang F, et al. Self-reported insomnia and coronary heart disease in the elderly. *Clin Exp Hypertens*. 2016;38(1):51−55.
39. Sivertsen B, Lallukka T, Salo P, et al. Insomnia as a risk factor for ill health: results from the large population-based prospective HUNT Study in Norway. *J Sleep Res*. 2014;23(2):124−132.
40. Canivet C, Nilsson PM, Lindeberg SI, Karasek R, Ostergren PO. Insomnia increases risk for cardiovascular events in women and in men with low socioeconomic status: a longitudinal, register-based study. *J Psychosom Res*. 2014;76(4):292−299.
41. Westerlund A, Bellocco R, Sundstrom J, Adami HO, Akerstedt T, Trolle Lagerros Y. Sleep characteristics and cardiovascular events in a large Swedish cohort. *Eur J Epidemiol*. 2013;28(6):463−473.

II. SLEEP DISORDERS IN SPECIFIC MEDICAL CONDITIONS

42. Clark AJ, Salo P, Lange T, et al. Onset of impaired sleep and cardiovascular disease risk factors: a longitudinal study. *Sleep*. 2016;39(9):1709–1718.
43. Sands-Lincoln M, Loucks EB, Lu B, et al. Sleep duration, insomnia, and coronary heart disease among postmenopausal women in the Women's Health Initiative. *J Womens Health*. 2013;22(6):477–486.
44. Laugsand LE, Strand LB, Platou C, Vatten LJ, Janszky I. Insomnia and the risk of incident heart failure: a population study. *Eur Heart J*. 2014;35(21):1382–1393.
45. Ingelsson E, Lind L, Arnlov J, Sundstrom J. Sleep disturbances independently predict heart failure in overweight middle-aged men. *Eur. J. Heart Fail*. 2007;9(2):184–190.
46. Kanno Y, Yoshihisa A, Watanabe S, et al. Prognostic significance of insomnia in heart failure. *Circ J*. 2016;80 (7):1571–1577.
47. Strand LB, Laugsand LE, Dalen H, Vatten L, Janszky I. Insomnia and left ventricular function—an echocardiography study. *Scand Cardiovasc J*. 2016;50(3):187–192.
48. Basta M, Chrousos GP, Vela-Bueno A, Vgontzas AN. Chronic insomnia and stress system. *Sleep Med Clin*. 2007;2(2):279–291.
49. Kapsimalis F, Basta M, Varouchakis G, Gourgoulianis K, Vgontzas A, Kryger M. Cytokines and pathological sleep. *Sleep Med*. 2008;9(6):603–614.
50. Vgontzas AN, Bixler EO, Lin H-M, et al. Chronic insomnia is associated with nyctohemeral activation of the hypothalamic–pituitary–adrenal axis: clinical implications. *J Clin Endocrinol Metab*. 2001;86(8):3787–3794.
51. Floam S, Simpson N, Nemeth E, Scott-Sutherland J, Gautam S, Haack M. Sleep characteristics as predictor variables of stress systems markers in insomnia disorder. *J Sleep Res*. 2015;24(3):296–304.
52. Motivala SJ. Sleep and inflammation: psychoneuroimmunology in the context of cardiovascular disease. *Ann Behav Med*. 2011;42(2):141–152.
53. Gay CL, Zak RS, Lerdal A, Pullinger CR, Aouizerat BE, Lee KA. Cytokine polymorphisms and plasma levels are associated with sleep onset insomnia in adults living with HIV/AIDS. *Brain Behav Immun*. 2015;47:58–65.
54. Vgontzas AN, Chrousos GP. Sleep, the hypothalamic–pituitary–adrenal axis, and cytokines: multiple interactions and disturbances in sleep disorders. *Endocrinol Metab Clin North Am*. 2002;31(1):15–36.
55. Laugsand LE, Vatten LJ, Bjorngaard JH, Hveem K, Janszky I. Insomnia and high-sensitivity C-reactive protein: the HUNT study, Norway. *Psychosom Med*. 2012;74(5):543–553.
56. Backhaus J, Junghanns K, Hohagen F. Sleep disturbances are correlated with decreased morning awakening salivary cortisol. *Psychoneuroendocrinology*. 2004;29(9):1184–1191.
57. Abell JG, Shipley MJ, Ferrie JE, Kivimaki M, Kumari M. Recurrent short sleep, chronic insomnia symptoms and salivary cortisol: a 10-year follow-up in the Whitehall II study. *Psychoneuroendocrinology*. 2016;68:91–99.
58. Fernandez-Mendoza J, Vgontzas AN, Calhoun SL, et al. Insomnia symptoms, objective sleep duration and hypothalamic–pituitary–adrenal activity in children. *Eur J Clin Invest*. 2014;44(5):493–500.
59. Castro-Diehl C, Diez Roux AV, Redline S, Seeman T, Shrager SE, Shea S. Association of sleep duration and quality with alterations in the hypothalamic–pituitary–adrenocortical axis: the multi-ethnic study of atherosclerosis (MESA). *J Clin Endocrinol Metab*. 2015;100(8):3149–3158.
60. Redeker NS, Jeon S, Pacelli J, Anderson G. Sleep disturbance, sleep-related symptoms, and biological rhythms in heart failure patients who have insomnia. *Sleep*. 2014;37:A248–A249.
61. Irwin M, Clark C, Kennedy B, Christian Gillin J, Ziegler M. Nocturnal catecholamines and immune function in insomniacs, depressed patients, and control subjects. *Brain Behav Immun*. 2003;17(5):365–372.
62. Vgontzas AN, Tsigos C, Bixler EO, et al. Chronic insomnia and activity of the stress system: a preliminary study. *J Psychosom Res*. 1998;45(1):21–31.
63. Mitchell HA, Weinshenker D. Good night and good luck: norepinephrine in sleep pharmacology. *Biochem Pharmacol*. 2010;79(6):801–809.
64. Dodds KL, Miller CB, Kyle SD, Marshall NS, Gordon CJ. Heart rate variability in insomnia patients: a critical review of the literature. *Sleep Med Rev*. 2017;33:88–100.
65. Lin CL, Tsai YH, Yeh MC. The relationship between insomnia with short sleep duration is associated with hypercholesterolemia: a cross-sectional study. *J Adv Nurs*. 2016;72(2):339–347.
66. Routledge FS, Dunbar SB, Higgins M, et al. Insomnia symptoms are associated with abnormal endothelial function. *J Cardiovasc Nurs*. 2017;32(1):78–85.
67. Spielman AJ, Caruso LS, Glovinsky PB. A behavioral perspective on insomnia treatment. *Psychiatr Clin North Am*. 1987;10(4):541–553.
68. Gangwisch JE, Malaspina D, Posner K, et al. Insomnia and sleep duration as mediators of the relationship between depression and hypertension incidence. *Am J Hypertens*. 2010;23(1):62–69.

REFERENCES

69. Punjabi NM. The epidemiology of adult obstructive sleep apnea. *Proc Am Thorac Soc.* 2008;5(2):136–143.
70. Peppard PE, Young T, Barnet JH, Palta M, Hagen EW, Hla KM. Increased prevalence of sleep-disordered breathing in adults. *Am J Epidemiol.* 2013;177(9):1006–1014.
71. Pepin JL, Borel AL, Tamisier R, Baguet JP, Levy P, Dauvilliers Y. Hypertension and sleep: overview of a tight relationship. *Sleep Med Rev.* 2014;18(6):509–519.
72. Selim BJ, Koo BB, Qin L, et al. The association between nocturnal cardiac arrhythmias and sleep-disordered breathing: the DREAM study. *J Clin Sleep Med.* 2016;12(6):829–837.
73. Kwon Y, Misialek JR, Duprez D, et al. Association between sleep disordered breathing and electrocardiographic markers of atrial abnormalities: the MESA study. *Europace.* 2016;.
74. Gami AS, Pressman G, Caples SM, et al. Association of atrial fibrillation and obstructive sleep apnea. *Circulation.* 2004;110(4):364–367.
75. Shah NA, Yaggi HK, Concato J, Mohsenin V. Obstructive sleep apnea as a risk factor for coronary events or cardiovascular death. *Sleep Breath.* 2010;14(2):131–136.
76. Shahar E, Whitney CW, Redline S, et al. Sleep-disordered breathing and cardiovascular disease: cross-sectional results of the Sleep Heart Health Study. *Am J Respir Crit Care Med.* 2001;163(1):19–25.
77. Shamsuzzaman ASM, Gersh BJ, Somers VK. Obstructive sleep apnea: implications for cardiac and vascular disease. *JAMA.* 2003;290(14):1906–1914.
78. Somers VK, White DP, Amin R, et al. Sleep apnea and cardiovascular disease: an American Heart Association/American College of Cardiology Foundation Scientific Statement from the American Heart Association Council for High Blood Pressure Research Professional Education Committee, Council on Clinical Cardiology, Stroke Council, and Council on Cardiovascular Nursing. *J Am Coll Cardiol.* 2008;52(8):686–717.
79. Selim B, Won C, Yaggi HK. Cardiovascular consequences of sleep apnea. *Clin Chest Med.* 2010;31(2):203–220.
80. Gottlieb DJ, Somers VK, Punjabi NM, Winkelman JW. Restless legs syndrome and cardiovascular disease: a research roadmap. *Sleep Med.* 2017;31:10–17.
81. Olaithe M, Bucks RS. Executive dysfunction in OSA before and after treatment: a meta-analysis. *Sleep.* 2013;36(9):1297–1305.
82. Quan SF, Wright R, Baldwin CM, et al. Obstructive sleep apnea-hypopnea and neurocognitive functioning in the Sleep Heart Health Study. *Sleep Med.* 2006;7(6):498–507.
83. Righi CG, Martinez D, Goncalves SC, et al. Influence of high risk of obstructive sleep apnea on adherence to antihypertensive treatment in outpatients. *J Clin Hypertens.* 2017;19(5):534–539.
84. Riegel B, Lee CS, Ratcliffe SJ, et al. Predictors of objectively measured medication nonadherence in adults with heart failure. *Circ Heart Fail.* 2012;5(4):430–436.
85. Riegel B, Moelter ST, Ratcliffe SJ, et al. Excessive daytime sleepiness is associated with poor medication adherence in adults with heart failure. *J Card Fail.* 2011;17(4):340–348.
86. Phillips B, Hening W, Britz P, Mannino D. Prevalence and correlates of restless legs syndrome: results from the 2005 National Sleep Foundation Poll. *Chest.* 2006;129(1):76–80.
87. Winkelman JW, Shahar E, Sharief I, Gottlieb DJ. Association of restless legs syndrome and cardiovascular disease in the Sleep Heart Health Study. *Neurology.* 2008;70(1):35–42.
88. Cholley-Roulleau M, Chenini S, Beziat S, Guiraud L, Jaussent I, Dauvilliers Y. Restless legs syndrome and cardiovascular diseases: a case-control study. *PLoS One.* 2017;12(4):e0176552.
89. Sasai T, Matsuura M, Inoue Y. Change in heart rate variability precedes the occurrence of periodic leg movements during sleep: an observational study. *BMC Neurol.* 2013;13:139.
90. Ferrillo F, Beelke M, Canovaro P, et al. Changes in cerebral and autonomic activity heralding periodic limb movements in sleep. *Sleep Med.* 2004;5(4):407–412.
91. Bansil P, Kuklina EV, Merritt RK, Yoon PW. Associations between sleep disorders, sleep duration, quality of sleep, and hypertension: results from the National Health and Nutrition Examination Survey, 2005 to 2008. *J Clin Hypertens.* 2011;13(10):739–743.
92. Shafer H, Koehler U, Ploch T, Peter JH. Sleep-related myocardial ischemia and sleep structure in patients with obstructive sleep apnea and coronary heart disease. *Chest.* 1997;111:387–393.
93. Taylor DJ, Mallory LJ, Lichstein KL, Durrence HH, Riedel BW, Bush AJ. Comorbidity of chronic insomnia with medical problems. *Sleep.* 2007;30(2):213–218.
94. Redeker NS, Jeon S, Muench U, Campbell D, Walsleben J, Rapoport DM. Insomnia symptoms and daytime function in stable heart failure. *Sleep.* 2010;33(9):1210–1216.
95. Redeker NS, Stein S. Characteristics of sleep in patients with stable heart failure versus a comparison group. *Heart Lung.* 2006;35(4):252–261.

II. SLEEP DISORDERS IN SPECIFIC MEDICAL CONDITIONS

96. Redeker NS, Muench U, Zucker MJ, et al. Sleep disordered breathing, daytime symptoms, and functional performance in stable heart failure. *Sleep*. 2010;33(4):551–560.
97. Ludka O, Stepanova R, Vyskocilova M, et al. Sleep apnea prevalence in acute myocardial infarction—the Sleep Apnea in Post-acute Myocardial Infarction Patients (SAPAMI) Study. *Int J Cardiol*. 2014;176(1):13–19.
98. Ludka O, Stepanova R, Sert-Kuniyoshi F, Spinar J, Somers VK, Kara T. Differential likelihood of NSTEMI vs STEMI in patients with sleep apnea. *Int J Cardiol*. 2017;248:64–68.
99. Javaheri S. Sleep disorders in systolic heart failure: a prospective study of 100 male patients. The final report. *Int J Cardiol*. 2006;106(1):21–28.
100. Redeker NS, Mason DJ, Wykpisz E, Glica B. Women's patterns of activity over 6 months after coronary artery bypass surgery. *Heart Lung*. 1995;24(6):502–511.
101. Cole CS, McSweeney JC, Cleves MA, Armbya N, Bliwise DL, Pettey CM. Sleep disturbance in women before myocardial infarction. *Heart Lung*. 2012;41(5):438–445.
102. Redeker NS, Ruggiero J, Hedges C. Patterns and predictors of sleep pattern disturbance after cardiac surgery. *Res Nurs Health*. 2004;27(4):217–224.
103. Randerath W, Verbraecken J, Andreas S, et al. Definition, discrimination, diagnosis and treatment of central breathing disturbances during sleep. *Eur Respir J*. 2017;49(1).
104. Redeker NS, Mason DJ, Wykpisz E, Glica B. Sleep patterns in women after coronary artery bypass surgery. *Appl Nurs Res*. 1996;9(3):115–122.
105. Edell-Gustafsson UM, Hetta JE. Fragmented sleep and tiredness in males and females one year after percutaneous transluminal coronary angioplasty (PTCA). *J Adv Nurs*. 2001;34(2):203–211.
106. Redeker NS, Tamburri L, Howland CL. Prehospital correlates of sleep in patients hospitalized with cardiac disease. *Res Nurs Health*. 1998;21(1):27–37.
107. Andrews LK, Coviello J, Hurley E, Rose L, Redeker NS. "I'd eat a bucket of nails if you told me it would help me sleep:" perceptions of insomnia and its treatment in patients with stable heart failure. *Heart Lung*. 2013;42(5):339–345.
108. Redeker NS, Ruggiero JS, Hedges C. Sleep is related to physical function and emotional well-being after cardiac surgery. *Nurs Res*. 2004;53(3):154–162.
109. Mansukhani MP, Bellolio MF, Kolla BP, Enduri S, Somers VK, Stead LG. Worse outcome after stroke in patients with obstructive sleep apnea: an observational cohort study. *J Stroke Cerebrovasc Dis*. 2011;20(5):401–405.
110. Won CH, Chun HJ, Chandra SM, Sarinas PS, Chitkara RK, Heidenreich PA. Severe obstructive sleep apnea increases mortality in patients with ischemic heart disease and myocardial injury. *Sleep Breath*. 2013;17(1):85–91.
111. Araujo CM, Solimene MC, Grupi CJ, Genta PR, Lorenzi-Filho G, Da Luz PL. Evidence that the degree of obstructive sleep apnea may not increase myocardial ischemia and arrhythmias in patients with stable coronary artery disease. *Clinics*. 2009;64(3):223–230.
112. Wong JK, Maxwell BG, Kushida CA, et al. Obstructive sleep apnea is an independent predictor of postoperative atrial fibrillation in cardiac surgery. *J Cardiothorac Vasc Anesth*. 2015;29(5):1140–1147.
113. van Oosten EM, Hamilton A, Petsikas D, et al. Effect of preoperative obstructive sleep apnea on the frequency of atrial fibrillation after coronary artery bypass grafting. *Am J Cardiol*. 2014;113(6):919–923.
114. Wang H, Parker JD, Newton GE, et al. Influence of obstructive sleep apnea on mortality in patients with heart failure. *J Am Coll Cardiol*. 2007;49(15):1625–1631.
115. Calvin AD, Somers VK, van der Walt C, Scott CG, Olson LJ. Relation of natriuretic peptide concentrations to central sleep apnea in patients with heart failure. *Chest*. 2011;140(6):1517–1523.
116. Khayat R, Abraham W, Patt B, et al. Central sleep apnea is a predictor of cardiac readmission in hospitalized patients with systolic heart failure. *J Card Fail*. 2012;18(7):534–540.
117. Skomro R, Silva R, Alves R, Figueiredo A, Lorenzi-Filho G. The prevalence and significance of periodic leg movements during sleep in patients with congestive heart failure. *Sleep Breath*. 2009;13(1):43–47.
118. Yatsu S, Kasai T, Suda S, et al. Impact on clinical outcomes of periodic leg movements during sleep in hospitalized patients following acute decompensated heart failure. *Circ J*. 2017;81(4):495–500.
119. Prasad KT, Sehgal IS, Agarwal R, Nath Aggarwal A, Behera D, Dhooria S. Assessing the likelihood of obstructive sleep apnea: a comparison of nine screening questionnaires. *Sleep Breath*. 2017;21(4):909–917.

120. Bastien CH, Vallieres A, Morin CM. Validation of the Insomnia Severity Index as an outcome measure for insomnia research. *Sleep Med*. 2001;2(4):297−307.
121. Nunes FS, Danzi-Soares NJ, Genta PR, Drager LF, Cesar LA, Lorenzi-Filho G. Critical evaluation of screening questionnaires for obstructive sleep apnea in patients undergoing coronary artery bypass grafting and abdominal surgery. *Sleep Breath*. 2015;19(1):115−122.
122. Shear TC, Balachandran JS, Mokhlesi B, et al. Risk of sleep apnea in hospitalized older patients. *J Clin Sleep Med*. 2014;10(10):1061−1066.
123. Sharma S, Mather PJ, Chowdhury A, et al. Sleep Overnight monitoring for apnea in patients hospitalized with heart failure (SOMA-HF Study). *J Clin Sleep Med*. 2017;13(10):1185−1190.
124. Szymanski FM, Filipiak KJ, Hrynkiewicz-Szymanska A, Karpinski G, Opolski G. Clinical characteristics of patients with acute coronary syndrome at high clinical suspicion for obstructive sleep apnea syndrome. *Hellenic J Cardiol*. 2013;54(5):348−354.
125. Grigg-Damberger M. Why a polysomnogram should become part of the diagnostic evaluation of stroke and transient ischemic attack. *J Clin Neurophysiol*. 2006;23(1):21−38.
126. Loo G, Chua AP, Tay HY, Poh R, Tai BC, Lee CH. Sleep-disordered breathing in cardiac rehabilitation: prevalence, predictors, and influence on the six-minute walk test. *Heart Lung Circ*. 2016;25(6):584−591.
127. Marzolini S, Sarin M, Reitav J, Mendelson M, Oh P. Utility of screening for obstructive sleep apnea in cardiac rehabilitation. *J Cardiopulm Rehabil Prev*. 2016;36(6):413−420.
128. Jafari B. Rehabilitation of cardiovascular disorders and sleep apnea. *Sleep Med Clin*. 2017;12(2):193−203.
129. U.S. Office of Disease Prevention and Health Promotion. *Sleep Health Objectives*. <https://www.healthypeople.gov/2020/topics-objectives/topic/sleep-health/objectives>;2017 Accessed 17.06.09.
130. Haack M, Serrador J, Cohen D, Simpson N, Meier-Ewert H, Mullington JM. Increasing sleep duration to lower beat-to-beat blood pressure: a pilot study. *J Sleep Res*. 2013;22(3):295−304.
131. Morin CM, Bootzin RR, Buysse DJ, Edinger JD, Espie CA, Lichstein KL. Psychological and behavioral treatment of insomnia: update of the recent evidence (1998−2004). *Sleep*. 2006;29(11):1398−1414.
132. Smith MT, Huang MI, Manber R. Cognitive behavior therapy for chronic insomnia occurring within the context of medical and psychiatric disorders. *Clin Psychol Rev*. 2005;25(5):559−592.
133. Rybarczyk B, Stepanski E, Fogg L, Lopez M, Barry P, Davis A. A placebo-controlled test of cognitive-behavioral therapy for comorbid insomnia in older adults. *J Consult Clin Psychol*. 2005;73(6):1164−1174.
134. Rybarczyk B, Lopez M, Benson R, Alsten C, Stepanski E. Efficacy of two behavioral treatment programs for comorbid geriatric insomnia. *Psychol Aging*. 2002;17(2):288−298.
135. Conley S, Redeker NS. Cognitive behavioral therapy for insomnia in the context of cardiovascular conditions. *Curr Sleep Med Rep*. 2015;1(3):157−165.
136. Irwin MR, Olmstead R, Breen EC, et al. Cognitive behavioral therapy and Tai Chi Reverse cellular and genomic markers of inflammation in late-life insomnia: a randomized controlled trial. *Biol Psychiatry*. 2015;78 (10):721−729.
137. Carroll JE, Seeman TE, Olmstead R, et al. Improved sleep quality in older adults with insomnia reduces biomarkers of disease risk: pilot results from a randomized controlled comparative efficacy trial. *Psychoneuroendocrinology*. 2015;55:184−192.
138. Irwin MR, Olmstead R, Carrillo C, et al. Cognitive behavioral therapy vs. Tai Chi for late life insomnia and inflammatory risk: a randomized controlled comparative efficacy trial. *Sleep*. 2014;37(9):1543−1552.
139. Redeker NS, Jeon S, Andrews L, Cline J, Jacoby D, Mohsenin V. Feasibility and efficacy of a self-management intervention for insomnia in stable heart failure. *J Clin Sleep Med*. 2015;11(10):1109−1119.
140. Redeker NS, Jeon S, Andrews L, Cline J, Pacelli J, Jacoby D. Cognitive behavioral therapy for insomnia has sustained effects on daytime symptoms and hospitalization in patients with stable heart failure. *Sleep*. 2013;36:A297.
141. Redeker NS, Knies AK, Hollenbeak C, et al. Cognitive behavioral therapy for insomnia in stable heart failure: protocol for a randomized controlled trial. *Contemp Clin Trials*. 2017;55:16−23.
142. Cho HJ, Seeman TE, Kiefe CI, Lauderdale DS, Irwin MR. Sleep disturbance and longitudinal risk of inflammation: moderating influences of social integration and social isolation in the Coronary Artery Risk Development in Young Adults (CARDIA) study. *Brain Behav Immun*. 2015;46:319−326.
143. Jarrin DC, Chen IY, Ivers H, Lamy M, Vallieres A, Morin CM. Nocturnal heart rate variability in patients treated with cognitive-behavioral therapy for insomnia. *Health Psychol*. 2016;35(6):638−641.

144. Kufoy E, Palma JA, Lopez J, et al. Changes in the heart rate variability in patients with obstructive sleep apnea and its response to acute CPAP treatment. *PLoS One*. 2012;7(3):e33769.
145. Hu X, Fan J, Chen S, Yin Y, Zrenner B. The role of continuous positive airway pressure in blood pressure control for patients with obstructive sleep apnea and hypertension: a meta-analysis of randomized controlled trials. *J Clin Hypertens*. 2015;17(3):215–222.
146. Feldstein CA. Blood pressure effects of CPAP in nonresistant and resistant hypertension associated with OSA: a systematic review of randomized clinical trials. *Clin Exp Hypertens*. 2016;38(4):337–346.
147. Floras JS. Obstructive sleep apnea syndrome, continuous positive airway pressure and treatment of hypertension. *Eur J Pharmacol*. 2015;763(Pt A):28–37.
148. Yu J, Zhou Z, McEvoy RD, et al. Association of positive airway pressure with cardiovascular events and death in adults with sleep apnea: a systematic review and meta-analysis. *JAMA*. 2017;318(2):156–166.
149. Naughton MT, Kee K. Sleep apnoea in heart failure: to treat or not to treat? *Respirology*. 2017;22(2):217–229.
150. Anandam A, Akinnusi M, Kufel T, Porhomayon J, El-Solh AA. Effects of dietary weight loss on obstructive sleep apnea: a meta-analysis. *Sleep Breath*. 2013;17(1):227–234.
151. Araghi MH, Chen YF, Jagielski A, et al. Effectiveness of lifestyle interventions on obstructive sleep apnea (OSA): systematic review and meta-analysis. *Sleep*. 2013;36(10):1553–1562. 1562A-1562E.

CHAPTER

6

Lung Diseases

Lauren Tobias[1] and Christine Won[2]

[1]Department of Internal Medicine, Yale University School of Medicine, New Haven, CT,
United States [2]Yale University School of Medicine, Yale Centers for Sleep Medicine,
New Haven, CT, United States

OUTLINE

Introduction	122	Prevalence, Predictors, and Consequences	132	
Respiratory Physiology During Normal Sleep	122	Treatment of Sleep-Related Hypoxemia and Hypoventilation in Cystic Fibrosis	134	
Sleep Disorders in Chronic Obstructive Pulmonary Disease	123	Sleep in Interstitial Lung Disease	134	
Sleep-Related Hypoxemia and Hypoventilation in Chronic Obstructive Pulmonary Disease	124	Sleep-Related Respiratory Changes in Interstitial Lung Disease	134	
Chronic Obstructive Pulmonary Disease—Obstructive Sleep Apnea Overlap Syndrome	125	Sleep Architectural Disturbances in Interstitial Lung Disease	135	
Chronic Obstructive Pulmonary Disease and Insomnia	128	Comorbid Sleep Disorders in Interstitial Lung Disease	136	
Restless Legs Syndrome in Chronic Obstructive Pulmonary Disease	129	Sleep in Musculoskeletal Disorders	136	
		Description, Predictors, and Consequences	136	
Asthma and Sleep Disorders	130	Treatment of Sleep Disorders in Musculoskeletal Disorders	137	
Nocturnal Asthma	130			
Asthma and Sleep Quality	132	Sleep in Chronic Respiratory Failure Due to Neuromuscular Disease	137	
Asthma—Obstructive Sleep Apnea Overlap Syndrome	132	Myotonic Dystrophy	137	
Sleep in Cystic Fibrosis	132	Neurodegenerative Diseases	138	

Handbook of Sleep Disorders in Medical Conditions
DOI: https://doi.org/10.1016/B978-0-12-813014-8.00006-8

© 2019 Elsevier Inc. All rights reserved.

Obesity Hypoventilation Syndrome	139	Sleep in Pulmonary Hypertension	141
Definition, Prevalence, Predictors, and Consequences	*139*	Sleep After Lung Transplantation	143
		Conclusion	143
Treatment of Obesity Hypoventilation Syndrome	*141*	References	144

INTRODUCTION

Sleep disorders commonly coexist with respiratory disorders and may contribute to worse symptom control and outcomes in these patients. Sleep is associated with decreased ventilator responses to hypoxemia and hypercapnia, decreased upper airway and intercostal muscle tone, and reductions in minute ventilation. These alterations may be aggravated in patients with underlying lung disease, who are more susceptible to small fluctuations in gas exchange. In addition to the more common disorders such as asthma and chronic obstructive pulmonary disease (COPD), it is important to consider the role of sleep disorders in less commonly encountered pulmonary disorders such as pulmonary hypertension (PH), restrictive lung disease, obesity hypoventilation syndrome (OHS), and cystic fibrosis (CF). In general, the evidence for each of the respiratory disorders for which there is sufficient data available suggests that the presence of a comorbid sleep disorder carries a worse prognosis than the underlying respiratory disorder alone.

Therefore, there should be a low threshold for sleep evaluation and testing in patients with pulmonary disorders, particularly if awake hypoxemia or symptoms of sleep-disordered breathing (SDB) are present.

RESPIRATORY PHYSIOLOGY DURING NORMAL SLEEP

Sleep alters the process of breathing in fundamental ways. In healthy adults, sleep is associated with a depression in both the sensitivity and responsiveness to both hypoxemia and hypercapnia.[1] In normal individuals, this sleep-related ventilatory depression may cause mild elevations in the partial pressure of carbon dioxide ($PaCO_2$) on the order of 3–8 mm Hg, as well as reductions in partial pressure of oxygen (PaO_2) by 3–10 mm Hg. Elevations in $PaCO_2$ arise from reduction in minute ventilation, which occur primarily as a result of reductions in tidal volume with only a modest corresponding increase in respiratory rate.[2] Together, this results in a reduction in minute ventilation by approximately 6% during nonrapid eye movement (NREM) and 16% during rapid eye movement (REM) sleep.[3] During wakefulness, increases in $PaCO_2$ serve as a potent stimulus to brain stem chemoreceptors to increase respiratory drive, but this ventilatory response is diminished during sleep.[4,5] The response to hypoxemia, on the other hand, is controlled by

SLEEP DISORDERS IN CHRONIC OBSTRUCTIVE PULMONARY DISEASE

TABLE 6.1 Effects of Sleep on Breathing

- Blunted ventilatory responses to hypoxemia and hypercapnia (in rapid eye movement > nonrapid eye movement)
- Decreased tonic drive to pharyngeal muscles → reduction in upper airway muscle tone → increased airway resistance → increased mechanical load and potential to cause arousals
- Positional changes in respiratory mechanics: Supine positioning may reduce functional residual capacity by 20%−50% (exacerbated during rapid eye movement-associated atonia of intercostal muscles) and predispose to upper airway collapse
- Fall in minute ventilation in response to decreased metabolism and decreased chemosensitivity to O_2 and CO_2 (lower tidal volumes without commensurate increase in respiratory rate)
- Reduced responses to increased airways resistance compared with wakefulness
- More regular respiratory pattern in nonrapid eye movement sleep than wakefulness; variable respiratory patterns during rapid eye movement

peripheral chemoreceptors in the aortic and carotid bodies. Normal patients will experience only a slight, if any, decrease in oxygen saturation (SpO_2) as a result of these changes. Patients with pulmonary disease however may experience exaggerations of the normal gas exchange abnormalities observed during sleep. Abnormalities such as hypoxemia and hypercapnia are often most prominent during REM sleep, when ventilatory responses from the respiratory control center are blunted. Sleep is also associated with a reduction in drive to the upper airway dilator muscles, which may cause inspiratory airflow limitation in those with a predisposition.[4] Effects of sleep on control of breathing are summarized in Table 6.1 and Fig. 6.1.

SLEEP DISORDERS IN CHRONIC OBSTRUCTIVE PULMONARY DISEASE

COPD is a respiratory disorder characterized by chronic airflow obstruction caused by small airways disease and parenchymal lung destruction. Estimated to affect 10% of the general population, COPD is a leading cause of morbidity, health care utilization, and currently the third-ranked cause of death in the United States.[6] The prevalence of COPD is expected to increase over time given continued exposure to risk factors including aging of the world's population and smoking and air pollution. The cardinal symptoms of COPD include dyspnea, chronic cough, and sputum production. While early definitions of COPD differentiated between different subtypes (i.e., emphysema vs chronic bronchitis), the current definition no longer makes this distinction, recognizing that there is significant overlap in presentation between individual patients.[7] Pulmonary function tests (PFTs) are used to determine the severity of airflow limitation as measured by the forced expiratory volume in 1 second (FEV_1) and forced vital capacity. COPD management consists of inhaled pharmacologic therapy (e.g., long-acting β-agonists and muscarinic agents, inhaled glucocorticoids), smoking cessation, and supplemental oxygen in some patients. A key goal in managing COPD is attention to the multiple comorbid conditions that can contribute to worsened symptoms and mortality, including coronary artery disease, heart failure, metabolic syndrome, skeletal muscle dysfunction, anxiety and depression, and gastroesophageal reflux (GERD).

II. SLEEP DISORDERS IN SPECIFIC MEDICAL CONDITIONS

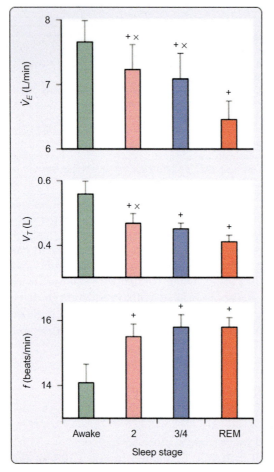

FIGURE 6.1 Changes in ventilation and lung volumes during sleep. Minute ventilation (V_E), tidal volume (V_T), and breathing frequency (f) during wakefulness and different sleep stages are illustrated. V_E is reduced during NREM sleep, with a further reduction in REM sleep. *NREM*, Nonrapid eye movement; *REM*, rapid eye movement. *Source: From Kryger MH, Roth T, Dement WC. Principles and Practice of Sleep Medicine, 6th ed. Philadelphia: Elsevier; 2017. Reproduced with permission from the ©ERS 2013. European Respiratory Review September 2013;22(129):365–375; http://dx.doi.org/10.1183/09059180.00003213.*

Sleep disturbance is also a common comorbidity in COPD, with patients reporting a range of symptoms including exacerbation of dyspnea, cough, wheezing, and insomnia.[8,9] Patients with COPD are at increased risk of several types of SDB, including sleep-related hypoxemia, alveolar hypoventilation, and coexisting sleep apnea (termed "overlap syndrome"). Clinicians likely underestimate the prevalence of sleep disruption in their patients with COPD and may not routinely assess patients for sleep quality, despite its relationship to quality of life.[10]

Sleep-Related Hypoxemia and Hypoventilation in Chronic Obstructive Pulmonary Disease

Nocturnal hypoxemia is common even among those COPD patients whose daytime arterial SpO_2 is normal; isolated nighttime desaturation is estimated to affect over 50% of the patients overall.[11] The presence of daytime gas exchange abnormalities are somewhat

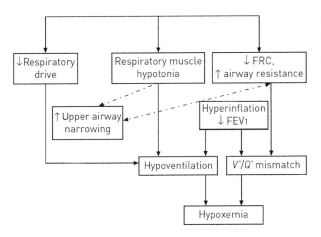

FIGURE 6.2 Mechanisms contributing to sleep-related hypoxemia in patients with COPD. Sleep affects multiple facets of respiration that impact hypoxemia. *COPD*, Chronic obstructive pulmonary disease. *Source: Reproduced with permission from the © ERS 2013. European Respiratory Review Sep 2013, 22 (129) 365–375; http://dx.doi.org/10.1183/09059180.00003213.*

predictive of the tendency to desaturate during sleep in patients with COPD[12] and the risk of sleep-related hypoxemia is related to underlying COPD severity.[13] However, it has been reported that up to 70% of patients with daytime saturations between 90% and 95% experience significant nocturnal desaturation.[11] Nocturnal hypoxemia may be associated with arousals and sleep fragmentation.[14]

There are several proposed mechanisms of hypoxemia in patients with COPD (Fig. 6.2). First, patients on the steeper portion of the oxyhemoglobin dissociation curve would be expected to experience greater reductions in SpO_2 during sleep, especially during REM sleep. Furthermore, the physiologic hypoventilation that occurs with sleep may be accentuated in patients with COPD as a result of lung hyperinflation leading to reductions in diaphragmatic efficiency, increased upper airway resistance due to loss of upper pharyngeal muscle tone, and a reduction in responsiveness to hypercapnia. One study demonstrated that minute ventilation may decrease by as much as 32% during REM sleep and 16% during NREM sleep in patients with COPD.[15]

Nearly half of all patients with COPD and daytime hypercapnia will exhibit alveolar hypoventilation during sleep.[16,17] In patients whose COPD is milder or who have no evidence of awake hypercapnia, hypoventilation may be most apparent during REM sleep. Hypoventilation during sleep is defined as an increase in $PaCO_2$ above what would be expected in a normal patient. The ICSD-3 defines sleep-related hypoventilation as a $PaCO_2$ of >50 mm Hg, or a rise in $PaCO_2$ of >10 mm Hg from baseline, lasting for at least 10 minutes. The presence of hypoventilation may result in an increased frequency of hypercapnic arousals, which may occur when $PaCO_2$ levels exceed 15 mm Hg above baseline values.[1]

Chronic Obstructive Pulmonary Disease–Obstructive Sleep Apnea Overlap Syndrome

Description, Predictors, and Consequences

The impetus for characterizing patients as having COPD–OSA (obstructive sleep apnea) "overlap syndrome" originated from a belief that their prognosis was worse than for patients

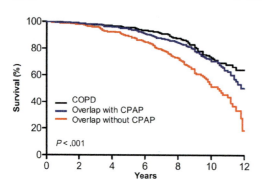

FIGURE 6.3 Patients with COPD/OSA overlap syndrome whose OSA is untreated have a higher mortality than those with either COPD alone or overlap syndrome on CPAP. *COPD*, Chronic obstructive pulmonary disease; *CPAP*, continuous positive airway pressure; *OSA*, obstructive sleep apnea. *Source: Reprinted with permission of the American Thoracic Society. Copyright © 2019 American Thoracic Society. Marin JM, Soriano JB, Carrizo SJ, Boldova A, Celli BR. Outcomes in patients with chronic obstructive pulmonary disease and obstructive sleep apnea. Am J Respir Crit Care Med. 2010;182(3):325–331. The American Journal of Respiratory and Critical Care Medicine is an official journal of the American Thoracic Society.*

with either disorder alone. Initially described over 30 years ago,[18] subsequent studies have supported this supposition. Patients with overlap syndrome have more pronounced sleep-related hypoxemia and hypercapnia than patients with either OSA or COPD alone.[19] Patients with the overlap syndrome have a 1.7 times greater relative risk of suffering severe COPD exacerbations, COPD-related hospitalizations, and death (Fig. 6.3[20]) compared to those with COPD alone. As a population, this group tends to be more obese and have more comorbidities than their counterparts with either disorder alone.

The description of a patient as having "overlap syndrome" rests on the presence of criteria for both conditions independently. Although it is not completely clear that these two disorders coexist more often than would be predicted by chance, several factors could explain why they might be plausibly associated. First, there are features of COPD that might increase the prevalence of OSA in this population, including frequent steroid use causing an increase in neck circumference, weight gain, and edema of both the pharynx and lower extremities; upper airway edema resulting from cor pulmonale; decreased exercise capacity contributing to obesity; and generalized muscle weakness predisposing to collapsibility of the upper airway.[16] Similarly, there are factors seen in OSA that could explain an increased prevalence of COPD in this population, including the presence of systemic inflammation resulting in inflammation of the airways themselves, oxidative stress leading to parenchymal lung destruction, worse GERD from intrathoracic pressure swings present in untreated OSA, and poor sleep quality contributing to daytime sleepiness and mood disturbances which could contribute to smoking or difficulty quitting. Finally, several conditions could reasonably contribute to the presence of both disorders, including smoking, obesity, allergic rhinitis, and GERD.

Overweight and obese patients with COPD/OSA overlap syndrome may be particularly susceptible to abnormalities in gas exchange during sleep as a result of reductions in ventilatory responses. The normal response to airway obstruction during apneas and hypopneas in patients with OSA is to increase diaphragmatic muscle effort in order to overcome increased resistance in the upper airway; patients with COPD may be unable to mount such effort, however, as a result of lung hyperinflation and resulting mechanical disadvantage to the diaphragm. Furthermore, patients with COPD may be slower to respond to obstructive airway events, with slower and longer apneas, greater alterations of PaO_2 and $PaCO_2$ during events, and failure to return to baseline PaO_2 and $PaCO_2$ levels between respiratory events (Fig. 6.4).

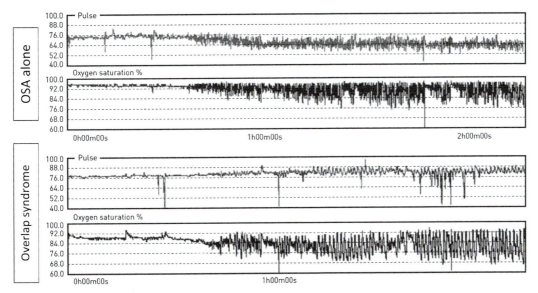

FIGURE 6.4 Oxygenation in OSA versus COPD/OSA overlap syndrome. In the patient with overlap syndrome, there is persistent desaturation without return to baseline between apneic episodes. *COPD*, Chronic obstructive pulmonary disease; *OSA*, obstructive sleep apnea. *Source: Reproduced with permission from the © ERS 2013. European Respiratory Review Sep 2013, 22 (129) 365–375; http://dx.doi.org/10.1183/09059180.00003213.*

The overlap syndrome results from complex pathophysiology of the upper and lower airways and lung parenchyma, and therefore, home sleep testing is neither validated, nor recommended by the American Academy of Sleep Medicine for diagnosis of sleep apnea in COPD patients.[21] Attended polysomnography (PSG) involves continuous monitoring of SpO_2 and CO_2 levels either by end-tidal CO_2 or transcutaneous ($tcPCO_2$) monitoring, which allows for more accurate identification and categorization of sleep-related breathing disorders in COPD as well as other lung diseases. For similar reasons, in-laboratory PSG for titration of continuous positive airway pressure (CPAP) therapy is preferred over automated machines. It is worth noting that, although widely available, nocturnal oximetry is probably an insufficiently sensitive screen for the presence of SDB in most patients with lung disease, particularly in patients with milder sleep apnea severity or where there is concern for hypoventilation.[22,23]

Treatment of Patients With COPD and COPD–OSA Overlap Syndrome

Therapeutic management of COPD/OSA overlap syndrome consists of optimizing management of the individual conditions following clinical guidelines. CPAP is considered first-line treatment for patients with OSA. CPAP has shown to improve COPD-related outcomes such as COPD exacerbations, hospitalizations, and mortality in patients with overlap syndrome.[20,24] The benefit of CPAP therapy in overlap syndrome may be greatest for hypercapnic patients since CPAP therapy improves mortality in these patients but not in patients with normocapnia.[25] However, recent studies suggest wake time hypercapnia and hypoxemia predict early CPAP failure [defined as persistent hypoxemia, intolerance to pressures, or persistent obstructive events with apnea/hypopnea index (AHI) remaining >15 events per hour], and that some of these patients may do better on bilevel positive airway pressure (PAP) therapy. Other predictors of early CPAP failure in this group are obesity and comorbid disease.

Currently, the decision about whether CPAP versus bilevel PAP is more appropriate remains a case-by-case decision. Patients in whom hypercapnia is present or suspected are generally recommended to undergo laboratory titration, ideally with carbon dioxide monitoring. Supplemental oxygen may be considered if hypoxemia persists despite achieving target minute ventilation. Supplemental oxygen is generally prescribed under the following conditions: (1) an arterial PaO_2 ≤ 55 mm Hg or an arterial SpO_2 $\leq 88\%$ for a total of at least 5 minutes during sleep or (2) a decrease in arterial SpO_2 by >5% for a total of at least 5 minutes during sleep associated with signs or symptoms attributed to hypoxemia (cor pulmonale, "P" pulmonale on electrocardiogram, PH, or erythrocytosis).

For patients with isolated COPD, noninvasive positive-pressure ventilation (NIPPV) has clear benefit in the management of acute-on-chronic hypercapnic respiratory failure, but its role as a long-term therapy for chronic, stable COPD remains controversial. While several early trials failed to show a benefit, insufficient sample size and follow-up duration may have contributed to negative results.[26] A more recent trial using higher airway pressures showed a survival benefit for NIPPV users in this population.[27] From the standpoint of medical insurers in the United States, initiation of NIPPV in patients with severe COPD typically requires the following criteria: (1) OSA has been ruled out; (2) awake $PaCO_2$ ≥ 52 mm Hg; and (3) sleep oximetry shows SpO_2 of 88% or less for 5 minutes or more while breathing supplemental O_2 at 2 L/min or at the patient's prescribed fraction of inspired oxygen.

Chronic Obstructive Pulmonary Disease and Insomnia

Prevalence, Predictors, and Assessment

Chronic insomnia is a significant public health problem associated with decreased quality of life that is estimated to affect 10% of the adult population.[28,29] COPD is associated with an increased prevalence of insomnia, which is present in up to one-third of patients.[30] Insomnia is associated with adverse health outcomes, daytime sleepiness, and poor quality of life.[31] Current tobacco use and sadness/anxiety are associated with increased prevalence of insomnia, while oxygen use is associated with a lower prevalence of insomnia.[31]

Several mechanisms have been postulated to contribute to this relationship, including (1) the presence of respiratory symptoms such as cough and sputum production; (2) nocturnal dyspnea that worsens in a supine position; (3) medications used to treat COPD including β-agonists; (4) smoking and either the sympathetic activation from nicotine or its withdrawal; (5) the presence of restless legs syndrome (RLS), which is seen with increased frequency in patients with COPD; (6) comorbid psychiatric disorders such as depression and anxiety, both of which occur in high rates among patients with COPD; and (7) increased sympathetic activity in COPD patients as is suggested by the elevated markers of systemic inflammation and oxidative stress in this population.[16] While some data suggest that the presence of nocturnal hypoxemia may be correlated with likelihood of insomnia, it is not clear that supplemental oxygen improves sleep quality.[31–33]

In COPD patients who complain of nighttime symptoms, practitioners may consider administering the recently developed nighttime symptoms of COPD instrument. This validated self-completed electronic daily diary measures nighttime symptom occurrence and severity, nocturnal awakening due to COPD symptoms, and nighttime rescue medication use in patients with COPD and may illuminate the degree to which COPD itself is contributing to a patient's insomnia.[34]

Treatment of Insomnia in Chronic Obstructive Pulmonary Disease

Standard treatment for insomnia in the general population includes cognitive-behavioral therapy for insomnia (CBT-I) and pharmacologic agents. A small study of patients with COPD and comorbid insomnia suggested a similarly favorable impact of CBT-I as in patients with other chronic diseases.[35] Studies have suggested benzodiazepines effectively improve objective sleep quality; however, caution is recommended in prescribing certain benzodiazepines to patients with underlying COPD as they may decrease alveolar ventilation, blunt the arousal response, and worsen SDB.[36] A safer option may be the class of nonbenzodiazepine sedative-hypnotics, which appear to have no adverse effects on gas exchange in patients with COPD nor carry a risk of worsening SDB.[37] A new orexin receptor antagonist, suvorexant, has also been tested in COPD patients and found not to cause nocturnal hypoxemia nor worsen AHI.[38] The subset of patients with COPD and comorbid insomnia may also benefit from a focus on symptoms that are exacerbated at night such as cough, sputum production, and nocturnal dyspnea. Few studies of pharmacologic agents for COPD have used sleep disturbance as a primary outcome. However, two recent trials showed a favorable impact of dual bronchodilator therapy on nighttime symptom severity.[39,40]

Given the efficacy of low-dose opioids to manage refractory breathlessness in patients with severe pulmonary disease including COPD, it is worth mentioning the impact of this treatment on SDB and sleep quality. Clinicians are often hesitant to prescribe medications that could suppress respiration, reduce arousals, or precipitate central apneas, to patients with underlying lung disease, and at high doses, these effects have been demonstrated.[41] However, low-dose, sustained-release opioids appear to have a limited impact on breathing and may in fact lead to improvements in sleep quality in COPD patients.[42,43]

Restless Legs Syndrome in Chronic Obstructive Pulmonary Disease

Description, Prevalence, Risk Factors, and Consequences

RLS (also called Willis–Ekbom disease) is a clinical diagnosis made based on the presence of four cardinal symptoms, including (1) an irresistible urge to move the limbs, often associated with paresthesias or dysesthesias; (2) worsening of symptoms at rest; (3) partial or temporary relief of symptoms with activity; and (4) worsening of symptoms in

130 6. LUNG DISEASES

the evening and at night.[44] Up to one-third of the patients with stable COPD report symptoms consistent with RLS, with most reporting moderate/severe symptoms[45–47] and the prevalence increases in those experiencing acute exacerbations of COPD.[48] Predictors of RLS include low creatinine levels (a marker of reduced muscle mass) and low ferritin (a marker of iron stores known to correlate with RLS severity in the general population as well).[45] Patients with RLS report worse sleep quality, fatigue, and more symptoms of depression (for which patients with COPD are already at increased risk). Although controversial, it has been suggested that RLS may lead to adverse cardiovascular consequences. Given that cardiovascular disease is a well-established cause of morbidity and mortality in patients with COPD,[49] reduction in cardiovascular risk may be another compelling reason to address RLS symptoms in this population.

Treatment of Restless Legs Syndrome in Chronic Obstructive Pulmonary Disease

Management of RLS in patients with COPD is similar to that in the general population and includes avoidance of triggers and pharmacologic therapy. Medications (including antiemetics, antipsychotics, antidepressants, and diphenhydramine), caffeine, alcohol, and sleep deprivation may worsen RLS symptoms. Pharmacologic agents to treat RLS include iron supplementation in those with low ferritin levels, antiseizure medications (particularly gabapentin, pregabalin), and dopamine agonists; opioids and benzodiazepines may be used in particularly severe cases refractory to other medical therapy.

ASTHMA AND SLEEP DISORDERS

Asthma is among the most common respiratory diseases and affects patients across the lifespan, with an estimated prevalence of 8% in the United States across all ages.[50] Asthma is characterized by symptoms of intermittent dyspnea, cough, and wheezing brought on by characteristic triggers, relieved with bronchodilator medications, and with demonstration of variable expiratory airflow obstruction on PFTs. Pathologically, asthma is a disorder of chronic airways inflammation and bronchial hyperresponsiveness. Nearly half of all asthma-related deaths occur during nighttime hours; therefore, the factors that contribute to nocturnal asthma are of major public health importance.[51]

Nocturnal Asthma

Many patients with asthma report worsening asthma symptoms at night.[52,53] Nearly half of all after-hours calls from asthmatic patients occur between the hours of 11 p.m. and 7 a.m.[54] and half of deaths due to asthma exacerbations occur between 6:00 p.m. and 3:00 a.m.[51] Asthma symptoms may be amplified at night because of normal circadian changes in ventilation, airway responsiveness and inflammation, mucociliary clearance, ventilator responses to hypercapnia and hypoxia, and hormone levels. For example, plasma cortisol and histamine levels exhibit circadian variations with a nadir for cortisol at midnight and peak level for histamine at 4:00 a.m. These changes in hormone and mediator levels have temporal relationships with peak expiratory flow rate (PEFR) in

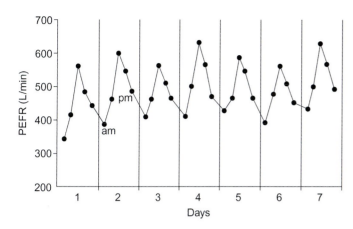

FIGURE 6.5 Diurnal variation in peak expiratory flow rate in an asthmatic patient over time. Peak expiratory flow rate typically nadirs in the early-morning hours. Several studies have demonstrated circadian changes in flow rates and airway resistance. PEFR, Peak expiratory flow rate. *Source: From Khan WH, Mohsenin V, D'Ambrosio CM. Sleep in asthma. Clin Chest Med. 2014;35 (3):483–493.*

TABLE 6.2   Factors Contributing to Pathophysiology of Nocturnal Asthma

- Increased airway resistance and diminished flow rates
- Decreased inspiratory muscle activity
- Decreased functional residual capacity
- Airway cooling
- Enhanced bronchial hyperreactivity
- Increased parasympathetic tone
- Bronchial inflammation
- Increased circulating histamine

Adapted from Ref. [52].

asthmatic subjects. Fig. 6.5 demonstrates the diurnal variation in PEFR as an example of one predisposing factor to circadian exacerbations of asthma. Other physiologic changes that exhibit circadian physiology and tend to worsen asthma at night are summarized in Table 6.2.[52]

Sleep-related, as opposed to circadian-related, changes in respiration may also contribute to nocturnal asthma. The sleep state is characterized by a reduction in inspiratory muscle tonic activity leading to physiologic reductions in functional residual capacity, PEFR, minute ventilation, and tidal volume, which is more pronounced in REM than NREM sleep. These sleep-related changes in lung volumes are more exaggerated in asthmatics and may predispose to nocturnal hypoxemia and hypercapnia. In a study of shift workers, sleep, and not the time of day, was responsible for the increased airway resistance seen in asthmatics.[55] Sleep may worsen bronchoreactivity as a result of increased parasympathetic tone and decreased nonadrenergic and noncholinergic discharge.[56] Sleep also leads to increase in airway secretions as a result of cough suppression. Comorbid GERD may worsen during sleep. Acid irritation of the lower esophagus may cause vagally mediated reflex bronchoconstriction in addition to direct airway inflammation by microaspiration. Finally, supine positioning itself reduces lung volumes and may contribute to nocturnal asthma symptoms.

Asthma and Sleep Quality

Patients with nocturnal asthma may report frequent nocturnal awakenings and arousals and worse sleep quality as a result of exacerbation of symptoms of bronchoconstriction, including breathlessness, cough, and wheezing. As many as 75% of the asthma patients may experience asthma symptoms during sleep at least once a week.[57] Perhaps as a result of these symptoms, the prevalence of insomnia in patients with asthma is significantly higher than among nonasthmatics (47% vs 37% in one study[58]), with increased asthma severity associated with higher likelihood of insomnia. Poor sleep quality and sleep fragmentation appear to be associated with impaired daytime cognitive function and quality of life in asthmatics.[59] Furthermore, improvements in sleepiness and sleep quality are associated with improvements in quality of life as well as improvement in asthma control. A recent study suggested that the relationship between asthma and insomnia may be bidirectional, showing that the presence of chronic insomnia was associated with a threefold risk for the development of incident asthma.[60] In general, nocturnal asthma symptoms should be approached similarly to the disease generally. This multipronged approach entails (1) avoidance of triggers (e.g., nocturnal allergen exposure); (2) optimizing pharmacotherapy (inhaled glucocorticoids, β-agonists, and leukotriene receptor antagonists) and ensuring appropriate inhaler technique and adherence; and (3) management of comorbidities (GERD, sleep apnea, and rhinitis).

Asthma—Obstructive Sleep Apnea Overlap Syndrome

Epidemiologic studies have suggested that the prevalence of OSA is increased in patients with asthma, which has led the Global Initiative for Asthma to recommend consideration of OSA when evaluating patients with asthma.[61] Data from the Wisconsin cohort, a prospective longitudinal cohort study, have demonstrated that asthma is associated with increased risk of new-onset OSA. Patients with asthma had a 1.39 relative risk of incident OSA compared to those without asthma,[62] and patients with longer standing asthma were at greater risk. Data also suggest a negative impact of OSA on asthma control.[63] In children with asthma, SDB has been linked to a 3.6-fold increase in odds for progression to severe asthma at 1 year. Treatment of OSA with CPAP may improve nocturnal and daytime symptoms of comorbid asthma.[64] The proposed mechanism by which CPAP improves asthma symptoms includes elimination of chronic upper airway irritation from snoring and repeated apneas which may lead to neutrally mediated reflex bronchoconstriction. CPAP also eliminates intermittent hypoxemia due to OSA, which may also potentiate bronchoconstriction.

SLEEP IN CYSTIC FIBROSIS

Prevalence, Predictors, and Consequences

Sleep quality is reported to be poor in as many as 50% of patients with CF. Worse sleep is predicted by severity of underlying lung disease[65] and presence of nighttime symptoms.[66] Common sleep complaints include sleep-onset insomnia, frequent awakenings, nighttime cough, snoring, and excessive daytime sleepiness (EDS). Despite commonly experiencing sleep symptoms, CF patients are less likely to seek medical attention for their sleep problems compared to other medical conditions, and sleep complaints are most often addressed by their pulmonologists.[67] CF patients often report significantly improved sleep quality after a hospitalization, or after therapy for CF exacerbation with chest physiotherapy, antibiotics, and nutritional support.[68] Other factors hypothesized to contribute to poor sleep quality in this population include fear and anxiety of death as well as the need for nocturnal therapies.[69]

Objective sleep measurements in CF patients confirm poor sleep quality on PSG with reduced sleep efficiency, reduced REM sleep, increased arousals, and increased wake time after sleep onset. These PSG findings in CF adults are associated with more severe pulmonary disease, lower FEV_1, and the presence of nocturnal hypoxemia.[70,71] In children with CF, however, objective sleep disruption is observed even in mild-to-moderate lung disease and in the absence of significant hypoxemia or hypercapnia. Multiple sleep latency testing (MLST) of adult CF patients confirm complaints of EDS with short mean sleep latency times.[72]

Nocturnal hypoxemia often precedes daytime hypoxemia in most lung diseases. Interestingly, the degree of hypoxemia during sleep in CF patients is often more exaggerated than during exercise. As with other lung diseases, the sleep state lends itself to hypoxemia and hypoventilation due to reduced tidal volumes, worsening ventilation and perfusion mismatch, increased airway resistance, and altered ventilatory chemoreflexes. There is no single clinical, radiographic, or spirometric parameter that reliably predicts nocturnal hypoxemia for CF patients. However, nocturnal hypoxemia is generally suspected when FEV_1 is $<64\%$, or when daytime resting SpO_2 is $<93\%$.[73]

Sleep is normally characterized by hypoventilation related to decreased respiratory drive and respiratory muscle activation. These physiologic processes are exaggerated in CF due to increased airway resistance, secretions, and hyperinflation. In addition, children with CF have been shown to exhibit abnormalities in gas exchange and increased respiratory load during sleep compared with normal controls, even after adjustment for age and body mass index (BMI).[70] Nocturnal hypercapnia in CF may be present without daytime hypoxemia or hypercapnia, or even without nocturnal hypoxemia. Moreover, initiation of supplemental oxygen in hypoxemic CF patients may lead to hypercapnia in some instances due to suppression of hypoxemic drive.

CF is frequently associated with nasal and sinus disease, including nasal obstruction due to polyps, chronic discharge, or mucosal edema, often resulting in chronic mouth breathing. OSA has been found to be prevalent in as many as 56% of children with CF.[74] Factors associated with OSA in CF children are similar to the general pediatric population including tonsillar hypertrophy and jaw malocclusion. In addition, chronic rhinosinusitis places CF children at increased risk for OSA. In adolescent and adult patients with CF, the occurrence of OSA has consistently shown to be rare, possibly due to the generally malnourished condition of this population.[75]

Treatment of Sleep-Related Hypoxemia and Hypoventilation in Cystic Fibrosis

Although sleep disturbances are associated with hypoxemia in CF patients, the effect on sleep architecture of supplemental oxygen, CPAP, or NIPPV has proven to be inconsistent. Supplemental oxygen is recommended by consensus guidelines for hypoxemic CF patients, though studies have yet to show a clear mortality benefit. Supplemental oxygen may decrease PH, but most studies show no effect on cognition, exacerbations, disease progression, quality of life, or sleep quality.[76] Supplemental oxygen is not without harm, as it may pose psychologic impact, logistical issues, and restrict mobility, to a generally young patient population. Moreover, some studies have shown supplemental oxygen may suppress hypoxemic respiratory drive and result in hypercapnia.[77]

A recent Cochrane review examining the role of noninvasive ventilation for patients with CF found that NIPPV may improve gas exchange during sleep to an extent greater than oxygen therapy alone in patients with moderate-to-severe obstructive disease.[78] In addition to improving nocturnal hypercapnia and hypoxemia, NIPPV may improve work of breathing, symptoms of dyspnea and headaches, as well as quality of life and exercise performance.[79] NIPPV however has no measurable effect on sleep quality or sleep architecture, and its impact on exacerbations and disease progression remains unclear. There is currently no consensus about when to start NIPPV in CF, but decisions may be guided by symptoms of dyspnea or headaches in the setting of nocturnal hypercapnia or hypoxemia. CF patients with end-stage lung disease have shown to generally tolerate NIPPV; however, CF patients with milder disease may complain of worsening sleep fragmentation. Adherence rates to NIPPV in this population have been described to be approximately 60% for greater than 4-hour nightly use.[79,80]

SLEEP IN INTERSTITIAL LUNG DISEASE

Interstitial lung disease (ILD) results in a restrictive ventilatory defect due to inflammation and fibrosis of lung parenchyma. Restrictive ventilator defects that arise outside of the lung parenchyma include chest wall deformity (e.g., kyphoscoliosis) and neuromuscular disease. Fatigue is a common complaint among patients with restrictive lung disease, part of which may be sleep disturbance due to nocturnal hypoxemia, increased work of breathing, and comorbid sleep disorders.

Sleep-Related Respiratory Changes in Interstitial Lung Disease

Patients with ILD often have rapid, shallow breathing during wakefulness as a result of decreased lung compliance, and a sensation of dyspnea due to afferent stimulation from vagal receptors.[81] Some studies suggest individuals with ILD may experience persistently elevated respiratory rates with shorter inspiratory and expiratory phases during sleep. This persistent tachypnea into sleep state may be attributed to the persistence of vagal-mediated reflexes that lead to hyperventilation during wakefulness.[82,83] The pattern of breathing may also relate to greater inefficiency and greater work of breathing during sleep. Patients with ILD require more excursion and effort to produce a given tidal volume during sleep.[84]

In contrast, others have demonstrated a reduction in hyperventilatory responses during sleep in patients with ILD.[85] Studies show that, even when hypoxemia is resolved with supplemental oxygen, ILD patients continue to hyperventilate compared to healthy individuals, suggesting a respiratory stimulatory effect that is independent of hypoxemia during wakefulness.[86] However, during sleep, and most notably during slow wave sleep, ILD patients reduce their respiratory rate to a comparable degree as healthy individuals. Similarly, the transcutaneous $PaCO_2$ (partial pressure of transcutaneous carbon dioxide) is reduced in ILD subjects during wakefulness, but there are no differences during sleep, suggesting an amelioration of hyperventilatory responses during sleep.[87]

Sleep-related hypoxemia is common in ILD and is associated with worse clinical outcomes. Respiratory muscles with the exception of the diaphragm become atonic during REM sleep. While these normal physiologic changes are associated with only minor gas exchange abnormalities in healthy individuals, they may lead to profound hypoventilation and hypoxemia in patients with restrictive lung disease. REM-related hypoxemia is often more severe than resting and exertional daytime hypoxemia.[82,83] Therefore, if an ILD patient is hypoxemic during exercise, nocturnal hypoxemia should be suspected. Average SpO_2 desaturations have been described on the order of approximately 9% during sleep compared to approximately 4% in controls.[84] ILD patients may experience SpO_2 nadir of 10%−16% compared to 3%−6% in healthy controls. ILD patients have been described to spend approximately 17% of the total sleep time with $SpO_2 < 85\%$, and more than a third of ILD patients may spend $>10\%$ of the sleep with $SpO_2 < 90\%$. In one study, the desaturation index, defined as the number of desaturation events $>4\%$, was shown to be independently associated with mortality.[88] Similarly, in a study of female patients with lymphangioleiomyomatosis, nocturnal hypoxemia (defined as 10% total sleep time with $SpO_2 < 90\%$) occurred in 56% of the patients, and sleep time spent with $SpO_2 < 90\%$ could be predicted by low diffusion capacity and FEV_1, and increased residual volume to total lung capacity ratio.[89] Noninvasive ventilation has been shown to improve SpO_2 as well as sleep quality in patients with chronic respiratory failure and restrictive lung disease.

Sleep Architectural Disturbances in Interstitial Lung Disease

Disrupted sleep and frequent arousals are common in patients with ILD. Their sleep is often characterized by increased arousals, increased stage N1 and stage N2 sleep, frequent stage shifts, and deficient stage N3 and REM sleep.[83] REM sleep onset is also often delayed, and 65% of the ILD patients may not achieve any slow wave sleep. These architectural disruptions are more pronounced in ILD patients with daytime hypoxemia. Similar patterns of sleep disruption have been demonstrated in patients with lung fibrosis awaiting lung transplantation.[90] There are a number of potential mechanisms contributing to sleep fragmentation in ILD patients, such as nocturnal hypoxemia, hypercapnia, cough, chest pain related to esophageal dysmotility, nocturnal or supine reflux, and other respiratory reflexes.[91] Finally, comorbid sleep disorders have been described with increasing frequency in patients with ILD.[92]

Comorbid Sleep Disorders in Interstitial Lung Disease

OSA has been shown to be common in ILD. As many as 68% of the nonobese individuals with ILD related to idiopathic pulmonary fibrosis (IPF), sarcoidosis, or scleroderma may have OSA.[93] OSA in this group is generally mild, with more than half of the patients having predominantly REM-related sleep apnea. Those with radiographically more severe ILD had worse AHI and oxygen desaturations. Unlike in the general population, OSA severity does not correlate with BMI in those with comorbid IPF.[94] However, the degree of restriction as measured by total lung capacity and impairment in diffusion capacity on PFTs do correlate with OSA severity.[95]

Comorbid sleep apnea in this patient population may worsen sleep quality as well as impact disease progression. Sleep efficiency, slow wave, and REM sleep are reduced, while arousal index is increased.[93] GERD is common in ILD due to the effects of lung fibrosis on intrathoracic pressure, diaphragm dysfunction, and lower esophageal sphincter tone.[96] GERD may exacerbate respiratory symptoms of dyspnea and cough, and may promote inflammation and progression of fibrotic lung disease.[97] OSA has also been implicated in GERD among healthy individuals due to large intrathoracic pressure swings generated by obstructive apneic episodes, though OSA has yet to be implicated in promoting GERD in ILD patients.[98–100] Finally, OSA confers an increased risk of PH when comorbid with ILD, much like OSA does for COPD, and treatment of OSA with CPAP therapy improves pulmonary artery pressures in ILD patients.

Other sleep disorders observed with increased frequency in ILD are periodic limb movements in sleep (PLMS) and RLS. As expected, these conditions worsen subjective sleep quality, worsen sleep efficiency, and increase arousal index in a dose-dependent manner with the number of limb movements per hour during sleep.[91]

SLEEP IN MUSCULOSKELETAL DISORDERS

Description, Predictors, and Consequences

Severe chest wall deformity may lead to ineffective respiratory muscle mechanics leading to increased work of breathing and easy fatigability. In contrast to patients with restrictive parenchymal lung disease, patients with extrapulmonary restriction often develop reduced chemo-responses to breathe and may hypoventilate during sleep and wakefulness. Hypoventilation with associated atelectasis and ventilation/perfusion mismatches become particularly problematic during supine positioning and REM sleep, when accessory respiratory muscles are atonic.[101] Persistent hypoventilation during sleep may lead to chronic hypercapnia and hypoxemia, pulmonary arterial vasoconstriction and remodeling, and eventual cor pulmonale.[102] Hypoxemia and hypercapnia themselves impair respiratory muscle function, increasing fatigability, and accelerating the development of respiratory failure. A wide variety of abnormal breathing patterns has been reported during sleep, including periodic breathing, central and obstructive apneas.[103,104] The cause of apneas is unclear but may be due to sleep-related reduction in muscle tone in accessory muscles especially during REM, and alterations in control of breathing.

Patients with musculoskeletal deformities complain of disrupted nocturnal sleep, severe nocturnal and/or morning headaches, and symptoms of EDS. The headaches may be due to the rise in $PaCO_2$ during sleep. Sleep in these patients has been characterized by frequent arousals associated with hypoxemia and hypercapnia. Stage N3 sleep and REM sleep as a proportion of total sleep time are reduced, while stage N1 and N2 are increased.[101] These sleep disturbances may contribute to fatigue in these patients and worsen respiratory efforts. In fact, decreased sleep duration has been proposed as a risk factor for the development of musculoskeletal thoracic cage abnormality, specifically degenerative lumbar scoliosis.[105] Sleep deprivation has been implicated in lowered bone mineral density and increased interleukin-1, which in turn has been implicated in intervertebral disk degeneration and osteoporosis.

Treatment of Sleep Disorders in Musculoskeletal Disorders

Nocturnal hypoxemia and hypercapnia may be treated with NIPPV.[106] NIPPV may reduce work of breathing and allow respiratory muscles to rest during sleep with theoretical improvement in daytime function. In addition, NIPPV provides positive end-expiratory pressure which improves alveolar recruitment and oxygenation, as well as eliminates obstructive apneas and hypopneas. Supplemental oxygen for those with sustained desaturations may prevent pulmonary vascular hypoxic vasoconstriction and PH. Tracheostomy with positive-pressure ventilation is reserved for severe restrictive lung physiology and chronic respiratory failure. To date, there are little data regarding the effect on sleep these treatments may have in musculoskeletal restrictive lung disease.

SLEEP IN CHRONIC RESPIRATORY FAILURE DUE TO NEUROMUSCULAR DISEASE

Neuromuscular disease may lead to significant restrictive ventilatory defect, hypoventilation, and respiratory failure. Many patients with neuromuscular disease die from respiratory failure due to severe diaphragmatic weakness or abnormalities in respiratory control. In individuals with respiratory muscle weakness, sleep is often disrupted with shorter total sleep time, frequent arousals, increased N1 sleep, and a reduction in REM sleep. SDB and nocturnal desaturations are common and most severe during REM sleep since the diaphragm is the only working respiratory muscle during REM sleep. Daytime respiratory function is weakly associated with nocturnal desaturation. Noninvasive ventilation improves sleep quality and breathing in patients with respiratory muscle weakness. However, the optimal timing of initiation and its role in progressive neuromuscular diseases is unclear.

Myotonic Dystrophy

Myotonic dystrophy type 1 (DM1) is the most common adult-onset form of muscular dystrophy, although its prevalence varies markedly across different geographic regions. DM1 is the neuromuscular condition with the most identified sleep disorders including hypersomnia, central and OSAs, RLS, PLMS, and REM sleep dysregulation.[107] EDS is reported in 70%–80% of the patients and is the most frequent nonmuscular complaint in DM1. Different sleep-related findings may mimic several sleep disorders, including idiopathic hypersomnia and narcolepsy without cataplexy. For example, patients with DM1 may have short sleep latency and sleep-onset REM periods on MLST, similar to that observed in narcolepsy.[108] However, the pathophysiologic mechanism for EDS and REM intrusion is likely distinct from that of narcolepsy since hypocretin levels in cerebrospinal fluid are not found to be abnormally low nor do these patients have defects in hypocretin receptors. While to some degree subjective and objective daytime sleepiness may be associated with muscular impairment and sleep disturbance, it appears that EDS in DM1 patients is caused by a primarily central process rather than by sleep fragmentation or comorbid sleep disorders. EDS, for example, tends to persist despite successful treatment of SDB in DM1 patients.[109]

Myotonic dystrophy type 2 (DM2), also called proximal myotonic myopathy, is rarer than DM1 and generally manifests with milder signs and symptoms. Descriptions of sleep in DM2 are less common than those reported for DM1. PSG data in DM2 patients show increased arousals, decreased sleep efficiency, and increased alpha–delta sleep. Many patients also have OSA and paradoxical breathing during REM sleep.[110] Patients are also noted to have dream enactment without atonia during REM consistent with REM behavioral disorder (RBD).[107] MLST shows reduced mean sleep latencies without sleep-onset REM periods. Patients commonly complain of EDS, poor sleep quality, fatigue, and RLS complaints.[111] The pathogenesis of sleep disturbance in DM2 patients remains unknown. OSA may be related to upper airway muscle weakness and myotonia. In addition, abnormal central control of breathing and sleep–wake state due to brainstem abnormality may contribute to SDB, insomnia, and RBD. Whether sleep dysfunction in DM2 is of the same characteristic and severity as in DM1 is unknown.

Neurodegenerative Diseases

Sleep disorders are common in neurodegenerative diseases such as Parkinson's disease, multiple system atrophy (MSA), and amyotrophic lateral sclerosis (ALS). Cell loss in the respiratory center of the brain stem, and dysfunction of bulbar and diaphragmatic muscles may increase the risk for SDB in MSA and ALS. The most common SDB in MSA is stridor, while in ALS, diaphragmatic weakness and nocturnal hypoventilation is more commonly experienced. As with other restrictive lung diseases, these patients are particularly vulnerable to having apneas and nocturnal hypoxemia during REM sleep when the diaphragm remains the only active respiratory muscle. Moreover, ALS patients may experience progressive bulbar palsy with motor neuron degeneration that specifically impairs the glossopharyngeal nerve, vagus nerve, and hypoglossal nerve. In addition to difficulty with chewing, talking, and swallowing, patients may experience weak palatal control and tongue protrusion and as a result are at increased risk for OSA. Nocturnal hypoxemia in this group may be associated with cognitive dysfunction, including poor memory and recall. Moreover, stridor and nocturnal hypoventilation in MSA and ALS are associated with increased mortality.

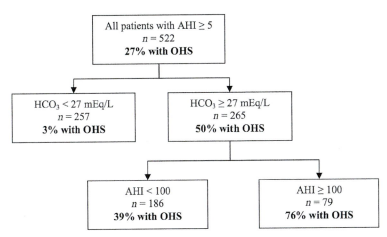

FIGURE 6.6 Decision tree to screen for OHS in OSA. The presence of an elevated serum bicarbonate level (≥27 mEq/L), reflecting metabolic compensation for a chronic respiratory acidosis, is a sensitive but not specific marker for OHS in the obese patient with OSA. *OHS*, Obesity hypoventilation syndrome; *OSA*, obstructive sleep apnea. *Source: From Babak Mokhlesi, Aiman Tulaimat, Ilja Faibussowitsch et al., Obesity hypoventilation syndrome: prevalence and predictors in patients with obstructive sleep apnea, Sleep and Breath, 2007 Jun;11(2):117−24.*

Fatigue is a common symptom reported in ALS. ALS patients have difficulties staying asleep, increased nocturia, and frequent nocturnal cramps. Other sleep disorders that have been described in ALS patients include RBD and RLS.[47,112] Sleep efficiency is poor, and ALS patients with sleep disturbances have worse functional disability as measured by the ALS Functional Rating Scale-Revised and Pittsburgh Sleep Quality Index, worse EDS as measured by Epworth Sleepiness Scale, and greater depression on Beck Depression Inventory.[113,114] When disease progresses and a patient becomes "locked in," they may lose their circadian rhythm for heart rate and body temperature, and have greater fragmentation of stage N3 sleep without change in total stage N3 time.[115]

NIPPV has shown to prolong survival in ALS of the bulbar dysfunction phenotype and to help ALS patients improve oxygenation during sleep (Fig. 6.6).[116] However, whether NIPPV improves sleep quality, sleep efficiency, arousal index, and sleep architecture in ALS patients is unclear.[117] Meanwhile, diaphragmatic pacing in ALS patients have been shown to improve sleep efficiency and decrease arousal index, as well as treat REM-related apneas and hypopneas. More research is necessary to investigate the impact of respiratory support on sleep.[118]

OBESITY HYPOVENTILATION SYNDROME

Definition, Prevalence, Predictors, and Consequences

OHS is defined as the triad of obesity (BMI > 30 kg/m^2, or >28 kg/m^2 in Asian population), daytime alveolar hypoventilation (PaCO$_2$ > 45 mm Hg), and SDB[119] (Table 6.3). The prevalence of OHS increases linearly with the degree of obesity[120] with a

6. LUNG DISEASES

TABLE 6.3   Definition of Obesity Hypoventilation Syndrome

Obesity
- Body mass index $\geq 30 \, kg/m^2$

Chronic hypoventilation
- Awake daytime hypercapnia ($PCO_2 \geq 45$ mm Hg, $PO_2 < 70$ mm Hg)
- Possible role of serum venous bicarbonate or calculated capillary blood gas bicarbonate >27 mEq/L

Sleep-disordered breathing
- Obstructive sleep apnea (apnea/hypopnea index ≥ 5 events/hour)
- Nonobstructive sleep hypoventilation (Apnea/hypopnea index < 5 events/hour, PCO_2 increases by ≥ 7 mm Hg during sleep or oxygen saturation $\leq 88\%$ for at least 5 min without obstructive respiratory events)

Exclusion of alternative causes of hypoventilation
- Severe obstructive pulmonary disease
- Severe interstitial lung disease
- Severe chest wall disorders (e.g., kyphoscoliosis)
- Severe hypothyroidism
- Neuromuscular disease
- Congenital hypoventilation syndromes

prevalence as high as 25% in those whose BMI exceeds $40 \, kg/m^2$.[120] Although often unrecognized, OHS is associated with excess mortality and multiple comorbidities, including risk for PH, right heart failure, atrial fibrillation, and hypercapnic respiratory failure.[121] While the vast majority (approximately 90%) of patients with OHS have concomitant OSA, a minority have SDB in the form of nocturnal hypoventilation. A decision tree to aid in screening OSA patients for comorbid OHS is shown in Fig. 6.7.[122]

Obesity itself increases the load on the respiratory system by several mechanisms. Excess adiposity causes a reduction in lung volumes (total lung capacity, vital capacity, residual volume) and the degree of restrictive lung physiology becomes more profound with increasingly severe obesity. Obese patients breathe on the less compliant portion of the thoracic volume pressure curve which means that expiratory flow limitation can occur at normal tidal volumes, predisposing to the development of dynamic hyperinflation and atelectasis.[123] Obese patients have an increased total work of breathing at rest; morbidly obese patients have been shown to have fivefold higher oxygen consumption than those with normal weight.[124,125]

Central respiratory depression may result from maladaptation to SDB. In some individuals, the postapnea hyperventilation period is insufficient to completely eliminate the $PaCO_2$ accumulated during an apneic event. As a result, $PaCO_2$ accumulates steadily during the night.[126,127] The kidneys may retain bicarbonate during sleep briefly to buffer the ongoing respiratory academia. If the kidney's rate of bicarbonate excretion is slow, metabolic alkalemia may result at the end of the sleep period, and the patient will tend to hypoventilate during wakefulness to compensate. Many factors are hypothesized to contribute to the insufficient interapnea hyperventilation periods. In patients with eucapnic SDB, acute hypoxemia is a known respiratory stimulant and may contribute to the increased ventilatory drive at the termination of a respiratory event. Abrupt awakenings from sleep are associated with increased sympathetic signaling, which also stimulates

II. SLEEP DISORDERS IN SPECIFIC MEDICAL CONDITIONS

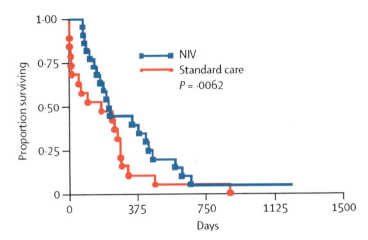

FIGURE 6.7 Noninvasive ventilation has been shown to prolong survival by a median of 205 days in patients with amyotophic lateral sclerosis. Subgroup analysis demonstrated survival benefit only in the subgroup of patients without severe bulbar dysfunction. *Source: Reprinted with permission from Elsevier Bourke SC, Tomlinson M, Williams TL, Bullock RE, Shaw PJ, Gibson GJ. Effects of noninvasive ventilation on survival and quality of life in patients with amyotrophic lateral sclerosis: a randomised controlled trial. Lancet Neurol. 2006;5 (2):140−7.*

postapneic hyperventilation. However, sustained hypoxemia is a neurocognitive depressant. In addition, chronic hypoxemia may impair respiratory load sensation and impair respiratory arousal from sleep, leading to exaggerated apneic episodes and inadequate postevent ventilatory responses. Additional factors that may impair postapnea ventilation include blunted sympathetic stimulation or response from arousals, and inadequate upper airway mechanoreceptor activation leading to persistently increased upper airway resistance upon awakening.[128]

Treatment of Obesity Hypoventilation Syndrome

Treatment of SDB with CPAP therapy or tracheostomy even without concomitant weight loss improves daytime hypoventilation in most OHS patients. The need for supplemental oxygen during wakefulness will decrease in those treated with CPAP. However, CPAP therapy does not reverse the underlying propensity for chronic hypoventilation, and hypercapnia will recur if OHS patients stop using it.

Several trials have sought to answer the question of whether CPAP versus NIPPV is a preferable mode in patients with OHS. Despite some theoretical appeal for NIPPV over CPAP, given its ability to specifically target ventilation, studies have shown no difference in the benefit of either modality with respect to several outcomes studied, including hospital admission, adherence, quality of life, or markers of gas exchange (improvements in $PaCO_2$, PaO_2, and serum bicarbonate).[129]

SLEEP IN PULMONARY HYPERTENSION

PH is defined as a mean pulmonary artery pressure exceeding 25 mm Hg and is categorized by the World Health Organization into five classes, including (1) primary pulmonary arterial hypertension; (2) PH due to left-sided heart disease; (3) PH due to chronic lung disease (including COPD and OSA); (4) chronic thromboembolic PH; and (5) PH due to miscellaneous etiologies. The gold standard for diagnosis of PH is right heart catheterization. Although uncommon even in severe OSA, PH is a potential consequence of longstanding SDB that appears related to the severity of sleep-related oxygen desaturation present.

OSA appears to be fairly common in patients with PH, with a prevalence higher than that of the general population, ranging from 27% to 71% across studies.[130,131] The presence of coexistent OSA may worsen underlying PH by several mechanisms: negative intrathoracic pressure changes causing right ventricular overload; intermittent hypoxemia causing constriction of pulmonary arterioles, endothelial dysfunction, and remodeling; and elevated sympathetic nervous system activity[132] (Fig. 6.8). Similarly, PH may contribute to the development of OSA by causing right heart failure with resulting fluid accumulation and rostral fluid shift from the lower extremities during the day to the pharynx at night.[133] Studies have shown that central sleep apnea is also seen with increased frequency in patients with PH.[134] When PH cooccurs with OSA, mean pulmonary artery pressure is generally only mildly elevated (20–52 mm Hg), unless underlying pulmonary or cardiac disease, or chronic daytime hypoxemia or hypercapnia, is present, in which case PH

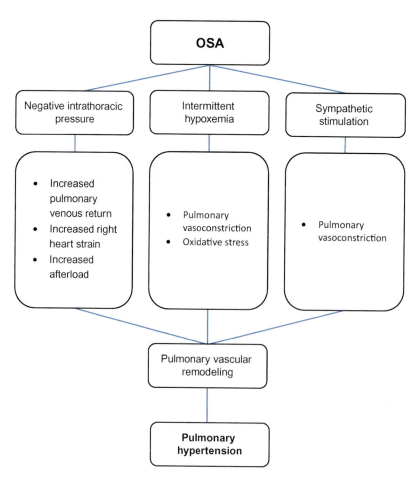

FIGURE 6.8 Pathways by which OSA may contribute to pulmonary hypertension. *OSA*, Obstructive sleep apnea.

may be severe. Predictors of PH in OSA are obesity, poor lung function, and degree and duration of hypoxemia and hypercapnia. PH does not appear to be associated with age, sex, or AHI.[135,136] Patients with both sleep apnea and PH have lower quality of life and higher mortality than those with OSA alone.[135,137] It is worth noting that many PH patients with OSA may have normal exercise oximetry and that even nocturnal oximetry is insensitive as a screen for OSA in this population, so full overnight PSG or home sleep testing is preferable.[134,138,139]

Studies examining the effect of CPAP on reducing pulmonary artery pressures have been limited by small sample sizes, but generally support a modest benefit.[140,141] Moreover, CPAP has been shown to improve serum markers of cardiac and vascular injury. Because the effect appears to be greater in more severe disease, the true benefit of identifying and treating OSA may apply to those with significant PH. Most studies use mean pulmonary artery pressure as their primary outcome; however, mean pulmonary artery pressure is not a survival correlate in PH. As PH progresses, the right ventricle fails and may cause mean pulmonary artery pressures to actually decrease over time. Further studies are required to evaluate the impact of CPAP therapy on cardiovascular morbidity and mortality as well as functional outcomes in patients with OSA and PH.

SLEEP AFTER LUNG TRANSPLANTATION

SDB is common in patients undergoing lung transplantation, estimated to affect nearly two-thirds of patients in some studies.[142,143] Although the improvements in oxygenation seen following lung transplantation may result in improvements in OSA in some patients, others may develop OSA newly, possibly as a result of the weight gain and obesity associated with immunosuppressive medication regimens. Metabolic syndrome after lung transplant is common, affecting 25% of recipients.[143] The limited data available suggests no association between the presence of SDB and posttransplant survival.[144]

CONCLUSION

Sleep disorders in the form of sleep disruption or fragmentation, insomnia, and SDB are common in patients with underlying lung disease. Often times, the physiologic changes in respiration become pathologic in those with compromised gas exchange capabilities due to airways, lung parenchymal, thoracic cage, or respiratory muscle deficiencies. Disrupted sleep or comorbid sleep disorders may impact cognitive function, quality of life, and morbidity or mortality in some lung diseases. Some comorbid sleep disorders may even affect progression of disease and disease-related symptoms. The threshold to evaluate and test for sleep disorders in patients with pulmonary disorders should be low, especially in the presence of awake hypoxemia or symptoms consistent with SDB. These patients often require in-laboratory sleep testing and titration for CPAP or nocturnal noninvasive ventilation, with supplemental oxygen where indicated, since these can be effective treatments for improving gas exchange, symptoms, and quality of life in certain conditions.

KEY PRACTICE POINTS

- Sleep disorders are common in patients with underlying lung disease and their recognition may lead to improvements in outcomes and quality of life. Severity of underlying lung disease (as assessed by pulmonary function testing, clinically, and radiographically) and presence of comorbid disorders must be considered.
- Overnight laboratory polysomnography (as opposed to unattended sleep testing) is recommended for patients with significant lung disease.
- The absence of daytime gas exchange abnormalities (hypoxemia and hypercapnia) does not exclude the possibility of sleep-related derangements in gas exchange.
- Sleep-related hypoxemia in patients with pulmonary disease has various etiologies depending on the underlying disease pathophysiology, but may include decrements in minute ventilation resulting from alveolar hypoventilation (especially during REM), reduced sensitivity to hypoxemia and hypercapnia, decreased tone and activity of accessory muscles of respiration, hyperinflation and diaphragmatic flattening, and upper airway resistance leading to flow limitation
- The relationship between COPD and OSA is likely bidirectional, with one condition aggravating or predisposing to the other, and CPAP is first-line therapy for patients with overlap syndrome.
- Nocturnal oximetry, while commonly ordered as a screening tool to evaluate for the presence of SDB in patients with underlying lung disease, is insensitive and risks underdiagnosis of concomitant OSA

References

1. Douglas NJ, White DP, Weil JV, Pickett CK, Zwillich CW. Hypercapnic ventilatory response in sleeping adults. *Am Rev Respir Dis.* 1982;126(5):758−762. Available from: https://doi.org/10.1164/arrd.1982.126.5.758.
2. Douglas NJ, White DP, Pickett CK, Weil JV, Zwillich CW. Respiration during sleep in normal man. *Thorax.* 1982;37(11):840−844.
3. Gould GA, Gugger M, Molloy J, Tsara V, Shapiro CM, Douglas NJ. Breathing pattern and eye movement density during REM sleep in humans. *Am Rev Respir Dis.* 1988;138(4):874−877. Available from: https://doi.org/10.1164/ajrccm/138.4.874.
4. Newton K, Malik V, Lee-Chiong T. Sleep and breathing. *Clin Chest Med.* 2014;35(3):451−456. Available from: https://doi.org/10.1016/j.ccm.2014.06.001.
5. Sowho M, Amatoury J, Kirkness JP, Patil SP. Sleep and respiratory physiology in adults. *Clin Chest Med.* 2014;35(3):469−481. Available from: https://doi.org/10.1016/j.ccm.2014.06.002.
6. Ford ES. Trends in mortality from COPD among adults in the United States. *Chest.* 2015;148(4):962−970. Available from: https://doi.org/10.1378/chest.14-2311.
7. GOLD. *Global Strategy for the Diagnosis, Management and Prevention of COPD.* Glob Initiat Chronic Obstr Lung Dis—GOLD; 2017. <http://goldcopd.org/gold-2017-global-strategy-diagnosis-management-prevention-copd/> Accessed 22.10.17.
8. Rennard S, Decramer M, Calverley PMA, et al. Impact of COPD in North America and Europe in 2000: subjects' perspective of confronting COPD international survey. *Eur Respir J.* 2002;20(4):799−805.
9. Agusti A, Hedner J, Marin JM, Barbé F, Cazzola M, Rennard S. Night-time symptoms: a forgotten dimension of COPD. *Eur Respir Rev.* 2011;20(121):183−194. Available from: https://doi.org/10.1183/09059180.00004311.

REFERENCES

10. Scharf S, Maimon N, Simon-Tuval T, Bernhard-Scharf B, Reuveni H, Tarasiuk A. Sleep quality predicts quality of life in chronic obstructive pulmonary disease. *Int J Chron Obstruct Pulmon Dis.* 2010;1. Available from: https://doi.org/10.2147/COPD.S15666.
11. Lewis CA, Fergusson W, Eaton T, Zeng I, Kolbe J. Isolated nocturnal desaturation in COPD: prevalence and impact on quality of life and sleep. *Thorax.* 2009;64(2):133–138. Available from: https://doi.org/10.1136/thx.2007.088930.
12. Mulloy E, McNicholas WT. Ventilation and gas exchange during sleep and exercise in severe COPD. *Chest.* 1996;109(2):387–394.
13. Sanders MH, Newman AB, Haggerty CL, et al. Sleep and sleep-disordered breathing in adults with predominantly mild obstructive airway disease. *Am J Respir Crit Care Med.* 2003;167(1):7–14. Available from: https://doi.org/10.1164/rccm.2203046.
14. Cormick W, Olson LG, Hensley MJ, Saunders NA. Nocturnal hypoxaemia and quality of sleep in patients with chronic obstructive lung disease. *Thorax.* 1986;41(11):846–854.
15. Becker HF, Piper AJ, Flynn WE, et al. Breathing during sleep in patients with nocturnal desaturation. *Am J Respir Crit Care Med.* 1999;159(1):112–118. Available from: https://doi.org/10.1164/ajrccm.159.1.9803037.
16. Budhiraja R, Siddiqi TA, Quan SF. Sleep disorders in chronic obstructive pulmonary disease: etiology, impact, and management. *J Clin Sleep Med: JCSM.* 2015;11(3):259–270. Available from: https://doi.org/10.5664/jcsm.4540.
17. O'Donoghue FJ, Catcheside PG, Ellis EE, et al. Sleep hypoventilation in hypercapnic chronic obstructive pulmonary disease: prevalence and associated factors. *Eur Respir J.* 2003;21(6):977–984.
18. Flenley DC. Sleep in chronic obstructive lung disease. *Clin Chest Med.* 1985;6(4):651–661.
19. McNicholas WT, Verbraecken J, Marin JM. Sleep disorders in COPD: the forgotten dimension. *Eur Respir Rev.* 2013;22(129):365–375. Available from: https://doi.org/10.1183/09059180.00003213.
20. Marin JM, Soriano JB, Carrizo SJ, Boldova A, Celli BR. Outcomes in patients with chronic obstructive pulmonary disease and obstructive sleep apnea. *Am J Respir Crit Care Med.* 2010;182(3):325–331. Available from: https://doi.org/10.1164/rccm.200912-1869OC.
21. Collop NA, Anderson WM, Boehlecke B, et al. Clinical guidelines for the use of unattended portable monitors in the diagnosis of obstructive sleep apnea in adult patients. Portable Monitoring Task Force of the American Academy of Sleep Medicine. *J Clin Sleep Med: JCSM.* 2007;3(7):737–747.
22. Cooper BG, Veale D, Griffiths CJ, Gibson GJ. Value of nocturnal oxygen saturation as a screening test for sleep apnoea. *Thorax.* 1991;46(8):586–588.
23. Chung F, Liao P, Elsaid H, Islam S, Shapiro CM, Sun Y. Oxygen desaturation index from nocturnal oximetry: a sensitive and specific tool to detect sleep-disordered breathing in surgical patients. *Anesth Analg.* 2012;114(5):993–1000. Available from: https://doi.org/10.1213/ANE.0b013e318248f4f5.
24. Stanchina ML, Welicky LM, Donat W, Lee D, Corrao W, Malhotra A. Impact of CPAP use and age on mortality in patients with combined COPD and obstructive sleep apnea: the overlap syndrome. *J Clin Sleep Med: JCSM.* 2013;9(8):767–772. Available from: https://doi.org/10.5664/jcsm.2916.
25. Jaoude P, El-Solh AA. Survival benefit of CPAP favors hypercapnic patients with the overlap syndrome. *Lung.* 2014;192(5):633–634. Available from: https://doi.org/10.1007/s00408-014-9614-5.
26. Struik FM, Lacasse Y, Goldstein R, Kerstjens HM, Wijkstra PJ. Nocturnal non-invasive positive pressure ventilation for stable chronic obstructive pulmonary disease. *Cochrane Database Syst Rev.* 2013;(6)CD002878. Available from: https://doi.org/10.1002/14651858.CD002878.pub2.
27. Köhnlein T, Windisch W, Köhler D, et al. Non-invasive positive pressure ventilation for the treatment of severe stable chronic obstructive pulmonary disease: a prospective, multicentre, randomised, controlled clinical trial. *Lancet Respir Med.* 2014;2(9):698–705. Available from: https://doi.org/10.1016/S2213-2600(14)70153-5.
28. Zammit GK, Weiner J, Damato N, Sillup GP, McMillan CA. Quality of life in people with insomnia. *Sleep.* 1999;22(suppl 2):S379–S385.
29. Ohayon M. Epidemiological study on insomnia in the general population. *Sleep.* 1996;19(3 suppl):S7–S15.
30. Valipour A, Lavie P, Lothaller H, Mikulic I, Burghuber OC. Sleep profile and symptoms of sleep disorders in patients with stable mild to moderate chronic obstructive pulmonary disease. *Sleep Med.* 2011;12(4):367–372. Available from: https://doi.org/10.1016/j.sleep.2010.08.017.

31. Budhiraja R, Parthasarathy S, Budhiraja P, Habib MP, Wendel C, Quan SF. Insomnia in patients with COPD. *Sleep*. 2012;35(3):369−375. Available from: https://doi.org/10.5665/sleep.1698.
32. Goldstein RS, Ramcharan V, Bowes G, McNicholas WT, Bradley D, Phillipson EA. Effect of supplemental nocturnal oxygen on gas exchange in patients with severe obstructive lung disease. *N Engl J Med*. 1984;310 (7):425−429. Available from: https://doi.org/10.1056/NEJM198402163100704.
33. McKeon JL, Murree-Allen K, Saunders NA. Supplemental oxygen and quality of sleep in patients with chronic obstructive lung disease. *Thorax*. 1989;44(3):184−188.
34. Hareendran A, Palsgrove AC, Mocarski M, et al. The development of a patient-reported outcome measure for assessing nighttime symptoms of chronic obstructive pulmonary disease. *Health Qual Life Outcomes*. 2013;11 (1):104. Available from: https://doi.org/10.1186/1477-7525-11-104.
35. Kapella M, Herdegen, Perlis, et al. Cognitive behavioral therapy for insomnia comorbid with COPD is feasible with preliminary evidence of positive sleep and fatigue effects. *Int J Chron Obstruct Pulmon Dis*. 2011;625. Available from: https://doi.org/10.2147/COPD.S24858.
36. Lu X-M, Zhu J-P, Zhou X-M. The effect of benzodiazepines on insomnia in patients with chronic obstructive pulmonary disease: a meta-analysis of treatment efficacy and safety. *Int J Chron Obstruct Pulmon Dis*. 2016;11:675−685. Available from: https://doi.org/10.2147/COPD.S98082.
37. Steens RD, Pouliot Z, Millar TW, Kryger MH, George CF. Effects of zolpidem and triazolam on sleep and respiration in mild to moderate chronic obstructive pulmonary disease. *Sleep*. 1993;16(4):318−326.
38. Sun H, Palcza J, Rosenberg R, et al. Effects of suvorexant, an orexin receptor antagonist, on breathing during sleep in patients with chronic obstructive pulmonary disease. *Respir Med*. 2015;109(3):416−426. Available from: https://doi.org/10.1016/j.rmed.2014.12.010.
39. D'Urzo AD, Rennard SI, Kerwin EM, et al. Efficacy and safety of fixed-dose combinations of aclidinium bromide/formoterol fumarate: the 24-week, randomized, placebo-controlled AUGMENT COPD study. *Respir Res*. 2014;15:123. Available from: https://doi.org/10.1186/s12931-014-0123-0.
40. Mahler DA, Decramer M, D'Urzo A, et al. Dual bronchodilation with QVA149 reduces patient-reported dyspnoea in COPD: the BLAZE study. *Eur Respir J*. 2014;43(6):1599−1609. Available from: https://doi.org/10.1183/09031936.00124013.
41. Rose AR, Catcheside PG, McEvoy RD, et al. Sleep disordered breathing and chronic respiratory failure in patients with chronic pain on long term opioid therapy. *J Clin Sleep Med: JCSM*. 2014;10(8):847−852. Available from: https://doi.org/10.5664/jcsm.3950.
42. Wang D, Somogyi AA, Yee BJ, et al. The effects of a single mild dose of morphine on chemoreflexes and breathing in obstructive sleep apnea. *Respir Physiol Neurobiol*. 2013;185(3):526−532. Available from: https://doi.org/10.1016/j.resp.2012.11.014.
43. Barnes H, McDonald J, Smallwood N, Manser R. Opioids for the palliation of refractory breathlessness in adults with advanced disease and terminal illness. *Cochrane Database Syst Rev*. 2016;3:CD011008. Available from: https://doi.org/10.1002/14651858.CD011008.pub2.
44. Allen RP, Picchietti DL, Garcia-Borreguero D, et al. Restless legs syndrome/Willis−Ekbom disease diagnostic criteria: updated International Restless Legs Syndrome Study Group (IRLSSG) consensus criteria—history, rationale, description, and significance. *Sleep Med*. 2014;15(8):860−873. Available from: https://doi.org/10.1016/j.sleep.2014.03.025.
45. Cavalcante AGM, de Bruin PFC, de Bruin VMS, et al. Restless legs syndrome, sleep impairment, and fatigue in chronic obstructive pulmonary disease. *Sleep Med*. 2012;13(7):842−847. Available from: https://doi.org/10.1016/j.sleep.2012.03.017.
46. Kaplan Y, Inonu H, Yilmaz A, Ocal S. Restless legs syndrome in patients with chronic obstructive pulmonary disease. *Can J Neurol Sci*. 2008;35(3):352−357.
47. Lo Coco D, Mattaliano A, Lo Coco A, Randisi B. Increased frequency of restless legs syndrome in chronic obstructive pulmonary disease patients. *Sleep Med*. 2009;10(5):572−576. Available from: https://doi.org/10.1016/j.sleep.2008.04.014.
48. Aras G, Kadakal F, Purisa S, Kanmaz D, Aynaci A, Isik E. Are we aware of restless legs syndrome in COPD patients who are in an exacerbation period? Frequency and probable factors related to underlying mechanism. *COPD*. 2011;8(6):437−443. Available from: https://doi.org/10.3109/15412555.2011.623737.
49. Maclay JD, MacNee W. Cardiovascular disease in COPD: mechanisms. *Chest*. 2013;143(3):798−807. Available from: https://doi.org/10.1378/chest.12-0938.

REFERENCES

50. Akinbami LJ, Moorman JE, Liu X. Asthma prevalence, health care use, and mortality: United States, 2005–2009. *Natl Health Stat Rep*. 2011;32:1–14.
51. Robertson CF, Rubinfeld AR, Bowes G. Deaths from asthma in Victoria: a 12-month survey. *Med J Aust*. 1990;152(10):511–517.
52. Khan WH, Mohsenin V, D'Ambrosio CM. Sleep in asthma. *Clin Chest Med*. 2014;35(3):483–493. Available from: https://doi.org/10.1016/j.ccm.2014.06.004.
53. Douglas NJ, Flenley DC. Breathing during sleep in patients with obstructive lung disease. *Am Rev Respir Dis*. 1990;141(4 Pt 1):1055–1070. Available from: https://doi.org/10.1164/ajrccm/141.4_Pt_1.1055.
54. Horn CR, Clark TJ, Cochrane GM. Is there a circadian variation in respiratory morbidity? *Br J Dis Chest*. 1987;81(3):248–251.
55. Clark TJ, Hetzel MR. Diurnal variation of asthma. *Br J Dis Chest*. 1977;71(2):87–92.
56. Smith DR, Lee-Chiong T. Respiratory physiology during sleep. *Sleep Med Clin*. 2008;3(4):497–503. Available from: https://doi.org/10.1016/j.jsmc.2008.07.002.
57. PubMed—NCBI. Epidemiology of nocturnal asthma. <https://www.ncbi.nlm.nih.gov/pubmed/?term = Turner-Warwick%2C + Am + J + Med%2C + 1988> Accessed 30.10.17.
58. Sundbom F, Lindberg E, Bjerg A, et al. Asthma symptoms and nasal congestion as independent risk factors for insomnia in a general population: results from the GA(2)LEN survey. *Allergy*. 2013;68(2):213–219. Available from: https://doi.org/10.1111/all.12079.
59. Mastronarde JG, Wise RA, Shade DM, Olopade CO, Scharf SM. American Lung Association Asthma Clinical Research Centers. Sleep quality in asthma: results of a large prospective clinical trial. *J Asthma*. 2008;45 (3):183–189. Available from: https://doi.org/10.1080/02770900801890224.
60. Brumpton B, Mai X-M, Langhammer A, Laugsand LE, Janszky I, Strand LB. Prospective study of insomnia and incident asthma in adults: the HUNT study. *Eur Respir J*. 2017;49(2):1601327. Available from: https://doi.org/10.1183/13993003.01327-2016.
61. GINA-2016-main-report_tracked.pdf. <http://ginasthma.org/wp-content/uploads/2016/04/GINA-2016-main-report_tracked.pdf>Accessed 20.10.17.
62. Teodorescu M, Barnet JH, Hagen EW, Palta M, Young TB, Peppard PE. Association between asthma and risk of developing obstructive sleep apnea. *JAMA*. 2015;313(2):156–164. Available from: https://doi.org/10.1001/jama.2014.17822.
63. Teodorescu M, Broytman O, Curran-Everett D, et al. Obstructive sleep apnea risk, asthma burden, and lower airway inflammation in adults in the severe asthma research program (SARP) II. *J Allergy Clin Immunol Pract*. 2015;3(4). 566.e1-575.e1. doi:10.1016/j.jaip.2015.04.002.
64. Chan CS, Woolcock AJ, Sullivan CE. Nocturnal asthma: role of snoring and obstructive sleep apnea. *Am Rev Respir Dis*. 1988;137(6):1502–1504. Available from: https://doi.org/10.1164/ajrccm/137.6.1502.
65. Milross MA, Piper AJ, Norman M, et al. Subjective sleep quality in cystic fibrosis. *Sleep Med*. 2002;3 (3):205–212.
66. Erdem E, Ersu R, Karadag B, et al. Effect of night symptoms and disease severity on subjective sleep quality in children with non-cystic-fibrosis bronchiectasis. *Pediatr Pulmonol*. 2011;46(9):919–926. Available from: https://doi.org/10.1002/ppul.21454.
67. Sheehan J, Massie J, Hay M, et al. The natural history and predictors of persistent problem behaviours in cystic fibrosis: a multicentre, prospective study. *Arch Dis Child*. 2012;97(7):625–631. Available from: https://doi.org/10.1136/archdischild-2011-301527.
68. Dobbin CJ, Bartlett D, Melehan K, Grunstein RR, Bye PTP. The effect of infective exacerbations on sleep and neurobehavioral function in cystic fibrosis. *Am J Respir Crit Care Med*. 2005;172(1):99–104. Available from: https://doi.org/10.1164/rccm.200409-1244OC.
69. Silva AM, Descalço A, Salgueiro M, et al. Respiratory sleep disturbance in children and adolescents with cystic fibrosis. *Rev Port Pneumol*. 2016;22(4):202–208. Available from: https://doi.org/10.1016/j.rppnen.2016.02.007.
70. Paranjape SM, McGinley BM, Braun AT, Schneider H. Polysomnographic markers in children with cystic fibrosis lung disease. *Pediatrics*. 2015;136(5):920–926. Available from: https://doi.org/10.1542/peds.2015-1747.
71. Jankelowitz L, Reid KJ, Wolfe L, Cullina J, Zee PC, Jain M. Cystic fibrosis patients have poor sleep quality despite normal sleep latency and efficiency. *Chest*. 2005;127(5):1593–1599. Available from: https://doi.org/10.1378/chest.127.5.1593.

72. Dancey DR, Tullis ED, Heslegrave R, Thornley K, Hanly PJ. Sleep quality and daytime function in adults with cystic fibrosis and severe lung disease. *Eur Respir J*. 2002;19(3):504–510.
73. de Castro-Silva C, de Bruin VMS, Cavalcante AGM, Bittencourt LRA, de Bruin PFC. Nocturnal hypoxia and sleep disturbances in cystic fibrosis. *Pediatr Pulmonol*. 2009;44(11):1143–1150. Available from: https://doi.org/10.1002/ppul.21122.
74. Ramos RTT, Salles C, Gregório PB, et al. Evaluation of the upper airway in children and adolescents with cystic fibrosis and obstructive sleep apnea syndrome. *Int J Pediatr Otorhinolaryngol*. 2009;73(12):1780–1785. Available from: https://doi.org/10.1016/j.ijporl.2009.09.037.
75. Perin C, Fagondes SC, Casarotto FC, Pinotti AFF, Menna Barreto SS, Dalcin Pde TR. Sleep findings and predictors of sleep desaturation in adult cystic fibrosis patients. *Sleep Breath Schlaf Atm*. 2012;16(4):1041–1048. Available from: https://doi.org/10.1007/s11325-011-0599-5.
76. Elphick HE, Mallory G. Oxygen therapy for cystic fibrosis. *Cochrane Database Syst Rev*. 2013;(7)CD003884. Available from: https://doi.org/10.1002/14651858.CD003884.pub4.
77. Gozal D. Nocturnal ventilatory support in patients with cystic fibrosis: comparison with supplemental oxygen. *Eur Respir J*. 1997;10(9):1999–2003.
78. Moran F, Bradley JM, Piper AJ. Non-invasive ventilation for cystic fibrosis. *Cochrane Database Syst Rev*. 2017;2:CD002769. Available from: https://doi.org/10.1002/14651858.CD002769.pub5.
79. Young AC, Wilson JW, Kotsimbos TC, Naughton MT. Randomised placebo controlled trial of non-invasive ventilation for hypercapnia in cystic fibrosis. *Thorax*. 2008;63(1):72–77. Available from: https://doi.org/10.1136/thx.2007.082602.
80. Fauroux B, Burgel P-R, Boelle P-Y, et al. Practice of noninvasive ventilation for cystic fibrosis: a nationwide survey in France. *Respir Care*. 2008;53(11):1482–1489.
81. American Thoracic Society. Dyspnea. Mechanisms, assessment, and management: a consensus statement. *Am J Respir Crit Care Med*. 1999;159(1):321–340. Available from: https://doi.org/10.1164/ajrccm.159.1.ats898.
82. Bye PT, Issa F, Berthon-Jones M, Sullivan CE. Studies of oxygenation during sleep in patients with interstitial lung disease. *Am Rev Respir Dis*. 1984;129(1):27–32. Available from: https://doi.org/10.1164/arrd.1984.129.1.27.
83. Perez-Padilla R, West P, Lertzman M, Kryger MH. Breathing during sleep in patients with interstitial lung disease. *Am Rev Respir Dis*. 1985;132(2):224–229. Available from: https://doi.org/10.1164/arrd.1985.132.2.224.
84. Hira HS, Sharma RK. Study of oxygen saturation, breathing pattern and arrhythmias in patients of interstitial lung disease during sleep. *Indian J Chest Dis Allied Sci*. 1997;39(3):157–162.
85. McNicholas WT, Coffey M, Fitzgerald MX. Ventilation and gas exchange during sleep in patients with interstitial lung disease. *Thorax*. 1986;41(10):777–782.
86. Shea SA, Winning AJ, McKenzie E, Guz A. Does the abnormal pattern of breathing in patients with interstitial lung disease persist in deep, non-rapid eye movement sleep? *Am Rev Respir Dis*. 1989;139(3):653–658. Available from: https://doi.org/10.1164/ajrccm/139.3.653.
87. PubMed—NCBI. Oxygen desaturation during sleep and exercise in patients with interstitial lung disease. <https://www.ncbi.nlm.nih.gov/pubmed/?term = Midgren + B%2C + Hansson + L%2C + Eriksson + L%2C + Airikkala + P%2C + Elmqvist + D. + Oxygen + desaturation + during + sleep + and + exercise + in + patients + with + interstitial + lung + disease. + Thorax + 1987> Accessed 25.10.17.
88. Corte TJ, Wort SJ, Talbot S, et al. Elevated nocturnal desaturation index predicts mortality in interstitial lung disease. *Sarcoidosis Vasc Diffuse Lung Dis*. 2012;29(1):41–50.
89. Medeiros P, Lorenzi-Filho G, Pimenta SP, Kairalla RA, Carvalho CRR. Sleep desaturation and its relationship to lung function, exercise and quality of life in LAM. *Respir Med*. 2012;106(3):420–428. Available from: https://doi.org/10.1016/j.rmed.2011.12.008.
90. Pascual N, Jurado B, Rubio JM, Santos F, Lama R, Cosano A. Respiratory disorders and quality of sleep in patients on the waiting list for lung transplantation. *Transplant Proc*. 2005;37(3):1537–1539. Available from: https://doi.org/10.1016/j.transproceed.2005.02.043.
91. Prado GF, Allen RP, Trevisani VMF, Toscano VG, Earley CJ. Sleep disruption in systemic sclerosis (scleroderma) patients: clinical and polysomnographic findings. *Sleep Med*. 2002;3(4):341–345.
92. Schiza S, Mermigkis C, Margaritopoulos GA, et al. Idiopathic pulmonary fibrosis and sleep disorders: no longer strangers in the night. *Eur Respir Rev*. 2015;24(136):327–339. Available from: https://doi.org/10.1183/16000617.00009114.

REFERENCES

149

93. Pihtili A, Bingol Z, Kiyan E, Cuhadaroglu C, Issever H, Gulbaran Z. Obstructive sleep apnea is common in patients with interstitial lung disease. *Sleep Breath Schlaf Atm*. 2013;17(4):1281−1288. Available from: https://doi.org/10.1007/s11325-013-0834-3.

94. Lancaster LH, Mason WR, Parnell JA, et al. Obstructive sleep apnea is common in idiopathic pulmonary fibrosis. *Chest*. 2009;136(3):772−778. Available from: https://doi.org/10.1378/chest.08-2776.

95. Mermigkis C, Stagaki E, Tryfon S, et al. How common is sleep-disordered breathing in patients with idiopathic pulmonary fibrosis? *Sleep Breath Schlaf Atm*. 2010;14(4):387−390. Available from: https://doi.org/10.1007/s11325-010-0336-5.

96. Raghu G, Freudenberger TD, Yang S, et al. High prevalence of abnormal acid gastro-oesophageal reflux in idiopathic pulmonary fibrosis. *Eur Respir J*. 2006;27(1):136−142. Available from: https://doi.org/10.1183/09031936.06.00037005.

97. Zhang XJ, Bonner A, Hudson M, Canadian Scleroderma Research Group, Baron M, Pope J. Association of gastroesophageal factors and worsening of forced vital capacity in systemic sclerosis. *J Rheumatol*. 2013;40(6):850−858. Available from: https://doi.org/10.3899/jrheum.120705.

98. Shepherd KL, James AL, Musk AW, Hunter ML, Hillman DR, Eastwood PR. Gastro-oesophageal reflux symptoms are related to the presence and severity of obstructive sleep apnoea. *J Sleep Res*. 2011;20(1 Pt 2):241‑ 249. Available from: https://doi.org/10.1111/j.1365-2869.2010.00843.x.

99. Shepherd K, Hillman D, Holloway R, Eastwood P. Mechanisms of nocturnal gastroesophageal reflux events in obstructive sleep apnea. *Sleep Breath Schlaf Atm*. 2011;15(3):561−570. Available from: https://doi.org/10.1007/s11325-010-0404-x.

100. Pillai M, Olson AL, Huie TJ, et al. Obstructive sleep apnea does not promote esophageal reflux in fibrosing interstitial lung disease. *Respir Med*. 2012;106(7):1033−1039. Available from: https://doi.org/10.1016/j.rmed.2012.03.014.

101. Guilleminault C, Kurland G, Winkle R, Miles LE. Severe kyphoscoliosis, breathing, and sleep: the "Quasimodo" syndrome during sleep. *Chest*. 1981;79(6):626−630.

102. Mezon BL, West P, Israels J, Kryger M. Sleep breathing abnormalities in kyphoscoliosis. *Am Rev Respir Dis*. 1980;122(4):617−621. Available from: https://doi.org/10.1164/arrd.1980.122.4.617.

103. Esau SA. Hypoxic, hypercapnic acidosis decreases tension and increases fatigue in hamster diaphragm muscle in vitro. *Am Rev Respir Dis*. 1989;139(6):1410−1417. Available from: https://doi.org/10.1164/ajrccm/139.6.1410.

104. Kiyan E, Okumus G, Cuhadaroglu C, Deymeer F. Sleep apnea in adult myotonic dystrophy patients who have no excessive daytime sleepiness. *Sleep Breath Schlaf Atm*. 2010;14(1):19−24. Available from: https://doi.org/10.1007/s11325-009-0270-6.

105. Li H, Liang C, Shen C, Li Y, Chen Q. Decreased sleep duration: a risk of progression of degenerative lumbar scoliosis. *Med Hypotheses*. 2012;78(2):244−246. Available from: https://doi.org/10.1016/j.mehy.2011.10.036.

106. Gonzalez C, Ferris G, Diaz J, Fontana I, Nuñez J, Marín J. Kyphoscoliotic ventilatory insufficiency: effects of long-term intermittent positive-pressure ventilation. *Chest*. 2003;124(3):857−862.

107. Romigi A, Albanese M, Placidi F, et al. Sleep disorders in myotonic dystrophy type 2: a controlled polysomnographic study and self-reported questionnaires. *Eur J Neurol*. 2014;21(6):929−934. Available from: https://doi.org/10.1111/ene.12226.

108. Ciafaloni E, Mignot E, Sansone V, et al. The hypocretin neurotransmission system in myotonic dystrophy type 1. *Neurology*. 2008;70(3):226−230. Available from: https://doi.org/10.1212/01.wnl.0000296827.20167.98.

109. Dauvilliers YA, Laberge L. Myotonic dystrophy type 1, daytime sleepiness and REM sleep dysregulation. *Sleep Med Rev*. 2012;16(6):539−545. Available from: https://doi.org/10.1016/j.smrv.2012.01.001.

110. Chokroverty S, Bhat S, Rosen D, Farheen A. REM behavior disorder in myotonic dystrophy type 2. *Neurology*. 2012;78(24):2004. Available from: https://doi.org/10.1212/WNL.0b013e318259e28c.

111. Lam EM, Shepard PW, St Louis EK, et al. Restless legs syndrome and daytime sleepiness are prominent in myotonic dystrophy type 2. *Neurology*. 2013;81(2):157−164. Available from: https://doi.org/10.1212/WNL.0b013e31829a340f.

112. Ebben MR, Shahbazi M, Lange DJ, Krieger AC. REM behavior disorder associated with familial amyotrophic lateral sclerosis. *Amyotroph Lateral Scler*. 2012;13(5):473−474. Available from: https://doi.org/10.3109/17482968.2012.673172.

113. Lo Coco D, La Bella V. Fatigue, sleep, and nocturnal complaints in patients with amyotrophic lateral sclerosis. *Eur J Neurol*. 2012;19(5):760−763. Available from: https://doi.org/10.1111/j.1468-1331.2011.03637.x.

II. SLEEP DISORDERS IN SPECIFIC MEDICAL CONDITIONS

114. Lo Coco D, Mattaliano P, Spataro R, Mattaliano A, La Bella V. Sleep—wake disturbances in patients with amyotrophic lateral sclerosis. *J Neurol Neurosurg Psychiatry*. 2011;82(8):839–842. Available from: https://doi.org/10.1136/jnnp.2010.228007.

115. Soekadar SR, Born J, Birbaumer N, et al. Fragmentation of slow wave sleep after onset of complete locked-in state. *J Clin Sleep Med: JCSM*. 2013;9(9):951–953. Available from: https://doi.org/10.5664/jcsm.3002.

116. Bourke SC, Tomlinson M, Williams TL, Bullock RE, Shaw PJ, Gibson GJ. Effects of non-invasive ventilation on survival and quality of life in patients with amyotrophic lateral sclerosis: a randomised controlled trial. *Lancet Neurol*. 2006;5(2):140–147. Available from: https://doi.org/10.1016/S1474-4422(05)70326-4.

117. Katzberg HD, Selegiman A, Guion L, et al. Effects of noninvasive ventilation on sleep outcomes in amyotrophic lateral sclerosis. *J Clin Sleep Med: JCSM*. 2013;9(4):345–351. Available from: https://doi.org/10.5664/jcsm.2586.

118. Gonzalez-Bermejo J, Morélot-Panzini C, Salachas F, et al. Diaphragm pacing improves sleep in patients with amyotrophic lateral sclerosis. *Amyotroph Lateral Scler*. 2012;13(1):44–54. Available from: https://doi.org/10.3109/17482968.2011.597862.

119. Mokhlesi B. Obesity hypoventilation syndrome: a state-of-the-art review. *Respir Care*. 2010;55(10):1347–1362 [discussion 1363–1365].

120. Balachandran JS, Masa JF, Mokhlesi B. Obesity hypoventilation syndrome epidemiology and diagnosis. *Sleep Med Clin*. 2014;9(3):341–347. Available from: https://doi.org/10.1016/j.jsmc.2014.05.007.

121. Pierce AM, Brown LK. Obesity hypoventilation syndrome: current theories of pathogenesis. *Curr Opin Pulm Med*. 2015;21(6):557–562. Available from: https://doi.org/10.1097/MCP.0000000000000210.

122. Mokhlesi B, Tulaimat A, Faibussowitsch I, Wang Y, Evans AT. Obesity hypoventilation syndrome: prevalence and predictors in patients with obstructive sleep apnea. *Sleep Breath Schlaf Atm*. 2007;11(2):117–124. Available from: https://doi.org/10.1007/s11325-006-0092-8.

123. Hillman D, Singh B, McArdle N, Eastwood P. Relationships between ventilatory impairment, sleep hypoventilation and type 2 respiratory failure. *Respirol Carlton Vic*. 2014;19(8):1106–1116. Available from: https://doi.org/10.1111/resp.12376.

124. Sharp JT, Henry JP, Sweany SK, Meadows WR, Pietras RJ. The total work of breathing in normal and obese men. *J Clin Invest*. 1964;43:728–739. Available from: https://doi.org/10.1172/JCI104957.

125. Kress JP, Pohlman AS, Alverdy J, Hall JB. The impact of morbid obesity on oxygen cost of breathing (VO (2RESP)) at rest. *Am J Respir Crit Care Med*. 1999;160(3):883–886. Available from: https://doi.org/10.1164/ajrccm.160.3.9902058.

126. Berger KI, Ayappa I, Sorkin IB, Norman RG, Rapoport DM, Goldring RM. Postevent ventilation as a function of CO_2 load during respiratory events in obstructive sleep apnea. *J Appl Physiol*. 2002;93(3):917–924. Available from: https://doi.org/10.1152/japplphysiol.01082.2001.

127. Ayappa I, Berger KI, Norman RG, Oppenheimer BW, Rapoport DM, Goldring RM. Hypercapnia and ventilatory periodicity in obstructive sleep apnea syndrome. *Am J Respir Crit Care Med*. 2002;166(8):1112–1115. Available from: https://doi.org/10.1164/rccm.200203-212OC.

128. Berger K, Goldring R, Rapoport D. Obesity hypoventilation syndrome. *Semin Respir Crit Care Med*. 2009;30 (03):253–261. Available from: https://doi.org/10.1055/s-0029-1222439.

129. Howard ME, Piper AJ, Stevens B, et al. A randomised controlled trial of CPAP versus non-invasive ventilation for initial treatment of obesity hypoventilation syndrome. *Thorax*. 2017;72(5):437–444. Available from: https://doi.org/10.1136/thoraxjnl-2016-208559.

130. Dumitrascu R, Tiede H, Eckermann J, et al. Sleep apnea in precapillary pulmonary hypertension. *Sleep Med*. 2013;14(3):247–251. Available from: https://doi.org/10.1016/j.sleep.2012.11.013.

131. Minic M, Granton JT, Ryan CM. Sleep disordered breathing in group 1 pulmonary arterial hypertension. *J Clin Sleep Med: JCSM*. 2014;10(3):277–283. Available from: https://doi.org/10.5664/jcsm.3528.

132. Kholdani C, Fares WH, Mohsenin V. Pulmonary hypertension in obstructive sleep apnea: is it clinically significant? A critical analysis of the association and pathophysiology. *Pulm Circ*. 2015;5(2):220–227. Available from: https://doi.org/10.1086/679995.

133. Yumino D, Redolfi S, Ruttanaumpawan P, et al. Nocturnal rostral fluid shift: a unifying concept for the pathogenesis of obstructive and central sleep apnea in men with heart failure. *Circulation*. 2010;121 (14):1598–1605. Available from: https://doi.org/10.1161/CIRCULATIONAHA.109.902452.

REFERENCES

151

134. Schulz R, Baseler G, Ghofrani HA, Grimminger F, Olschewski H, Seeger W. Nocturnal periodic breathing in primary pulmonary hypertension. *Eur Respir J.* 2002;19(4):658–663.
135. Chaouat A, Weitzenblum E, Krieger J, Ifoundza T, Oswald M, Kessler R. Association of chronic obstructive pulmonary disease and sleep apnea syndrome. *Am J Respir Crit Care Med.* 1995;151(1):82–86. Available from: https://doi.org/10.1164/ajrccm.151.1.7812577.
136. PubMed—NCBI. Daytime pulmonary hypertension in patients with obstructive sleep apnea syndrome. <https://www.ncbi.nlm.nih.gov/pubmed/?term = Weitzenblum + Daytime + pulmonary + hypertension + in + patients + with + obstructive + sleep + apnea + syndrome. + Am + Rev + Respir + Dis + 1988%3B> Accessed 25.10.17.
137. Minai OA, Ricaurte B, Kaw R, et al. Frequency and impact of pulmonary hypertension in patients with obstructive sleep apnea syndrome. *Am J Cardiol.* 2009;104(9):1300–1306. Available from: https://doi.org/10.1016/j.amjcard.2009.06.048.
138. Minai OA, Pandya CM, Golish JA, et al. Predictors of nocturnal oxygen desaturation in pulmonary arterial hypertension. *Chest.* 2007;131(1):109–117. Available from: https://doi.org/10.1378/chest.06-1378.
139. Ulrich S, Fischler M, Speich R, Bloch KE. Sleep-related breathing disorders in patients with pulmonary hypertension. *Chest.* 2008;133(6):1375–1380. Available from: https://doi.org/10.1378/chest.07-3035.
140. Nahmias J, Lao R, Karetzky M. Right ventricular dysfunction in obstructive sleep apnoea: reversal with nasal continuous positive airway pressure. *Eur Respir J.* 1996;9(5):945–951.
141. Arias MA, García-Río F, Alonso-Fernández A, Martínez I, Villamor J. Pulmonary hypertension in obstructive sleep apnoea: effects of continuous positive airway pressure: a randomized, controlled cross-over study. *Eur Heart J.* 2006;27(9):1106–1113. Available from: https://doi.org/10.1093/eurheartj/ehi807.
142. Naraine VS, Bradley TD, Singer LG. Prevalence of sleep disordered breathing in lung transplant recipients. *J Clin Sleep Med: JCSM.* 2009;5(5):441–447.
143. Sommerwerck U, Kleibrink BE, Kruse F, et al. Predictors of obstructive sleep apnea in lung transplant recipients. *Sleep Med.* 2016;21:121–125. Available from: https://doi.org/10.1016/j.sleep.2016.01.005.
144. Malouf MA, Milrose MA, Milross MA, et al. Sleep-disordered breathing before and after lung transplantation. *J Heart Lung Transplant.* 2008;27(5):540–546. Available from: https://doi.org/10.1016/j.healun.2008.01.021.

II. SLEEP DISORDERS IN SPECIFIC MEDICAL CONDITIONS

CHAPTER

7

Obesity, Diabetes, and Metabolic Syndrome

Sundeep Shenoy[1], Azizi Seixas[2] and Michael A. Grandner[3]

[1]Division of Cardiovascular Medicine, University of Virginia, Charlottesville, VA, United States [2]Department of Population Health and Psychiatry, New York University Langone Health, New York, NY, United States [3]Sleep and Health Research Program, Department of Psychiatry, University of Arizona, Tucson, AZ, United States

OUTLINE

Introduction	154	Definition and Components of Metabolic Syndrome	161
Sleep and Overweight–Obesity	154	Sleep Duration and Components of Metabolic Syndrome	162
Definition and Public Health Relevance	154	Insomnia and Components of Metabolic Syndrome	163
Insufficient Sleep and Obesity	155		
Insomnia and Obesity	155	Obstructive Sleep Apnea and Components of Metabolic Syndrome	164
Obstructive Sleep Apnea and Obesity	156		
Sleep, Appetite, and Food Intake	157	Assessment of Sleep Disturbances	165
Sleep and Diabetes	158	Treatment of Sleep Disturbances	166
Definition and Public Health Relevance	158	Areas for Future Research	166
Insufficient Sleep and Diabetes	160	Conclusion	167
Insomnia and Diabetes	160	References	168
Obstructive Sleep Apnea and Diabetes	160		
Sleep and Metabolic Syndrome	161		

Handbook of Sleep Disorders in Medical Conditions
DOI: https://doi.org/10.1016/B978-0-12-813014-8.00007-X

153

© 2019 Elsevier Inc. All rights reserved.

INTRODUCTION

Sleep has emerged as an important factor for cardiometabolic health [obesity, diabetes, and metabolic syndrome (MetS)]. In particular, sleep duration, insomnia, and obstructive sleep apnea (OSA) are all implicated in risk for obesity, diabetes, and MetS. A recent position statement identified sleep disruption as a key metabolic risk factor.[1] This review will address how each of these three sleep conditions is related to each of these metabolic outcomes.

SLEEP AND OVERWEIGHT–OBESITY

Definition and Public Health Relevance

Definition and Diagnosis

Obesity is defined in terms of body mass and represents excessive or abnormal weight relative to a person's height. The World Health Organization defines overweight as a body mass index (BMI) greater than 25 kg/m^2 but less than 30 kg/m^2, while obesity is classified as a BMI greater than 30 kg/m^2. See Table 7.1 for a description of BMI categories.

Etiology

Generally, obesity results when there is a prolonged state of energy imbalance—an imbalance between the consumption and expenditure of calories. This energy balance over time results in excess accumulation of body weight. However, obesity is a complex interplay of genetic and environmental factors. About 60 genetic markers have been implicated in obesity of which 32 are considered to be highly associated with obesity. Individuals with a high-obesity risk genetic profile may weigh 7–9 kg more than their non-high-risk counterparts, independent of obesity status.[2] An area of interest in the causation of obesity is the role of diet and lifestyle during the perinatal period. Parental obesity, maternal smoking, gestational diabetes are some of the factors that have been associated with long-term effects on obesity.[3]

TABLE 7.1   Classification of Obesity Among Adults

Underweight	<18.5
Normal	18.5–24.9
Overweight	25–29.9
Obese	>30
Obese class I	30–34.9
Obese class II	34.9–39.9
Obese class III	>40

Adapted from the World Health Organization fact sheets and definitions on 'Overweight' and 'Obesity—http://www.who.int/mediacentre/factsheets/fs311/en/'.

Public Health Relevance

Obesity is associated with numerous chronic medical conditions such as diabetes mellitus (DM) type 2 and MetS and has a significant psychosocial and economic burden to individuals. The estimated annual cost in the United States (US) attributed to obesity approximately 10 years ago was about $147 billion, with an average obese individual spending $1429 (42%) higher on healthcare than a person with normal weight,[4] and these numbers are likely higher today. These may also be underestimates, since they do not account for costs of medical and surgical treatment of obesity.

Insufficient Sleep and Obesity

Short sleep, insufficient sleep, and sleep loss are terms that are often used interchangeably in the medical literature, yet they may reflect different phenomena. Short sleep denotes a habitual sleep of 6 hours or less, while sleep insufficiency is a reduction of sleep time to the degree that results in an adverse outcome.[5] Sleep deficiency represents a combination of insufficient sleep duration, poor quality, and/or circadian misalignment, all resulting in deleterious health consequences. There is a growing concern about the high rates of insufficient sleep duration throughout the world.[6]

Several studies have demonstrated a relationship between insufficient sleep and obesity (for reviews, see [6–16]). In a recent metaanalysis of studies linking short sleep duration and incident obesity, Wu et al.[17] found that habitual short sleepers were 45% more likely to be obese compared to average sleepers [odds ratio (OR) = 1.45, 95% confidence interval (CI) (1.25–1.67)].

Insufficient sleep duration in both men and women has been associated with obesity. In a self-reported study of 1.1 million individuals, sleep duration of more than 8 hours or less than 6 hours was associated with increased mortality and BMI.[18] Similarly, a study from Québec, Canada showed that, in comparison with 7–8 hours of sleep, both 5–6 hours and greater than 9 hours were associated with obesity. The ORs were 1.69 for 5–6 hours of sleep and 1.38 for those who slept greater than 9 hours.[19]

Sleep insufficiency is associated with excessive daytime fatigue, changes in the neurohormonal secretions, which in turn can trigger excess caloric intake resulting in obesity.

Some factors may dictate aspects of this relationship, though. For example, studies using nationally representative data from the National Health and Nutrition Examination Survey (NHANES) found that the relationship between sleep duration and BMI depends on race/ethnicity[20] as well as age.[21] For example, younger adults who need more sleep demonstrate a relationship where more sleep is associated with lower BMIs, but in middle age, a "U"-shaped relationship emerges and both short and long sleeps are associated with higher BMIs.

Insomnia and Obesity

Insomnia is not the same as short sleep. Insomnia is a clinical condition represented by (1) difficulty falling asleep or resuming sleep after awakening (e.g., difficulty initiating sleep, difficulty maintaining sleep, and early morning awakenings) and (2) resultant daytime impairment. This can be expressed as insomnia symptoms, or in the context of insomnia disorder, a clinical condition where persistent difficulty exists at least 3 nights per week and has persisted for at least 3 months.[22] Insomnia is characterized by a long sleep latency (>30 minutes), long wake time after sleep onset (>30 minutes), and/or early morning awakening, all resulting in impairment in day-to-day functioning of an individual or significant psychological distress.[5] It may or may not exist in the context of short sleep. The association between obesity and insomnia is complex, and further research is needed to determine the role of insomnia in obesity. It is possible that insomnia may be a risk factor for obesity, but that obesity may also represent a risk factor for insomnia via sympathetic activation, inflammation, or other pathways. This complexity is likely due to the involvement of variables such as lack of unanimity on the definitions of "short sleep" and "insomnia" in the studies, the interplay of other comorbid conditions such as OSA, restless leg syndrome, and psychosocial conditions. Also, it is possible that insomnia results in hyperarousal, which in turn can alter the physiology of weight gain.

Several studies have documented an association between insomnia (variably defined) and obesity. For example, data from the 2006 Behavioral Risk Factor Surveillance System (BRFSS) showed that US adults who reported difficulty sleeping at least 3 nights per week were 35% more likely to be obese.[23] Although these effects were attenuated after adjustment for age, sex, race/ethnicity, education, income, marital status, employment, census region, mental health, health insurance, access to healthcare, smoking, and alcohol use, individuals with insomnia were still 14% at greater odds of obesity, compared to individuals with no insomnia. The relationship between insomnia and obesity in these studies is equivocated by the lack of accounting for the confounding effects of short sleep duration.[24]

Obstructive Sleep Apnea and Obesity

OSA and obesity commonly coexist, and obesity is the strongest risk factor for the development of OSA. The prevalence of OSA has augmented with the increasing prevalence of obesity. In a large prospective study of 690 participants, a 10% weight gain resulted in 32% increase in apnea–hypopnea index (AHI), and a 10% weight loss led to a 26% improvement of the AHI.[25] Similarly, a prospective study from Sao Paulo, Brazil showed that, amongst the 1042 individuals who underwent a polysomnography, obese individuals were 10 times more likely to have OSA.[26] Peppard et al.[27] looked at the Wisconsin sleep cohort study data and showed an increasing trend in the prevalence of OSA between 1988–94 and 2007–10. See Fig. 7.1 for a depiction of sleep apnea prevalence by BMI categories.

Obesity results in anatomical changes in the neck and chest which may contribute to the development of OSA. Increased fat deposition in the neck and tongue base results in airway narrowing, especially when the individual is asleep. Obesity also results in mechanical compression of the chest and reduction in lung volumes and respiratory system compliance. OSA is a condition that is due to periodic closure of the airways resulting in intermittent episodes of hypoxia. These recurrent episodes of hypoxia induce arousal, which in a single night can be as high as 100 times. The consequences of these repeated

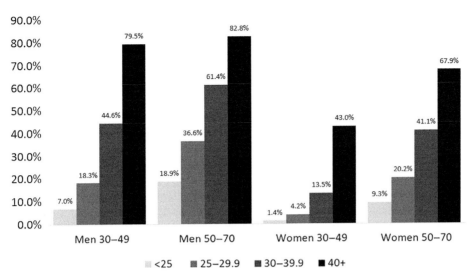

FIGURE 7.1 Prevalence of OSA by age group and BMI category. *BMI*, Body mass index; *OSA*, obstructive sleep apnea. *Adapted from Peppard, Paul E., et al. "Increased prevalence of sleep-disordered breathing in adults." American Journal of Epidemiology 177.9 (2013): 1006–1014.*

arousals are not only excessive daytime somnolence, decreased neurocognitive performance but also long-term consequences such as increased cardiovascular complications, hypertension, and MetS. There is considerable evidence to suggest that OSA has a causal relationship with these chronic medical conditions, and the presence of obesity magnifies the severity of these conditions.

Obesity hypoventilation syndrome (OHS) is another condition that is frequently overlooked in the obese population and is essentially hypoventilation at the alveolar level during awake and sleep states. This condition is often seen in individuals with BMI > 50 and sometimes in people with BMI > 30. OHS usually coexists with OSA in these individuals.

Sleep, Appetite, and Food Intake

Short sleep and insufficient sleep are associated with poor dietary habits. Matricciani et al.[28], in their systematic review of data, highlight that over the last 100 years, there has been a decline in the sleep duration of at least 1 hour amongst children and adolescents. A recent metaanalysis of all randomized controlled trials revealed that partial sleep deprivation was associated with weight gain.[29] Several epidemiological studies revealed similar findings implicating a role of sleep hygiene in food intake, appetite, and overall weight regulation.[30] The data is robust in the pediatric population in whom a linear relationship is noted between weight gain and sleep deprivation. Sleep deprivation in children can alter hypothalamic mechanisms that regulate appetite. Partial sleep deprivation for even a single night can result in increased food intake on the subsequent day.[31] Long-term effects of sleep deprivation were investigated by Patel et al.[32] who found that habitual sleep duration less than 5 hours was associated with a BMI $2.5 \, \text{kg/m}^2$ greater in men and $1.8 \, \text{kg/m}^2$ in women compared to those who slept an average of 7–8 hours at night.

Subcutaneous fat releases a hormone called leptin, which to a large part regulates satiety relative to food intake and has a role in obesity. Complementary to leptin, ghrelin is an amino acid peptide that stimulates hunger and is produced by the stomach. Both leptin and ghrelin interact with the orexin system and work in the central nervous system and influence appetite regulation in the hypothalamus. Taheri et al.,[33] in their study of 1024 participants, noticed that sleep duration of less than 5 hours per night was associated with at least a 15.5% reduction of leptin and 14.9% increase of ghrelin levels independent of the BMI, when compared to individuals who slept for at least 8 hours. This patterns of hormonal changes are likely to induce overeating and over time a positive energy balance resulting in obesity. Sleep deprivation is also associated with the release of stress hormones and cortisol which can further change the caloric intake. Data from neuroimagining studies also suggests that sleep deprivation is associated with a hedonic drive to increase caloric intake.[34]

Several laboratory studies have demonstrated that acute sleep deprivation can result in increased caloric intake. St-Onge et al.[35] showed that sleep restriction to 5 hours was associated with consumption of approximately 500 more kcal/day compared to a well-rested condition. Other studies have found similar results. In a study by Markwald et al.,[36] this increase in caloric intake was shown to be only during the evening/night, especially during the period when the individual would normally be asleep. This finding was replicated in a more diverse sample in a large study by Spaeth et al.[37,38] In addition, it showed that this increased caloric intake was associated with about 1 kg of weight gain across 5 days. In another laboratory study, when individuals were sleep restricted, they consumed more calorie-dense snacks.[39] Thus, sleep loss may lead to increased caloric consumption in the evening or at night, especially from energy-dense sources, resulting in weight gain.[40]

SLEEP AND DIABETES

Definition and Public Health Relevance

Definition and Diagnosis

DM affects 12%–14% of the US population.[41] Type 2 DM (T2DM) type comprises 90%–95% of this cohort.[42] Most discussion around the relationship of DM and sleep refers to T2DM. T2DM is diagnosed based on fasting plasma glucose, a 2-hour plasma glucose level post a 75 g glucose challenge, or the hemoglobin A1c% (HbA1C). See Table 7.2 for criteria for diabetes. Polyuria, polydipsia, polyphagia are all symptoms of hyperglycemia associated with T2DM. The persistent hyperglycemia may result in complications such as retinopathy, vision loss, neuropathy, impaired healing of tissue, and increased susceptibility to infections. Patients with T2DM often have comorbid conditions such as hypertension, dyslipidemia and together significantly increases the risk of the main cardiovascular events such as myocardial infarction, stroke, and sudden cardiac death.

TABLE 7.2 Diagnosis of Diabetes

Method 1	Fasting plasma glucose of 126 mg/dL (7.0 mmol/L)
Method 2	Plasma glucose of 200 mg/dL (11.1 mmol/L) following a 2-h oral glucose tolerance test performed using 75 g anhydrous glucose dissolved in water
Method 3	Hemoglobin A1c of 6.5%
Method 4	A random plasma glucose of 200 mg/dL in a patient with classic symptoms of hyperglycemia or a hyperglycemic crisis

Adapted from: American Diabetes Association. "2. Classification and Diagnosis of Diabetes." Diabetes Care 40. Supplement 1 (2017): S11–S24.

Etiology

T2DM is characterized by hyperglycemia and a combination of impaired insulin secretion and insulin resistance. A combination of genetic and environmental factors contributes to the development of T2DM. The overall result is a deficiency of insulin at the end organ resulting in impaired protein, carbohydrate, fat metabolism, and hyperglycemia. Over time, the hyperglycemia-induced cell damage results in end organ failure; this includes damage to the pancreatic beta cells which results in reduced insulin secretion further exacerbating the hyperglycemia. In the recent years, the increase in the incidence of T2DM has paralleled the incidence of obesity, partially because adipose tissue secretes leptin, adiponectin, and tumor necrosis factor-alpha, which all contribute to insulin resistance. Also, individuals with a strong family history of diabetes have a twofold higher chance of developing diabetes in their lifetime.[43]

Public Health Relevance

In the US, the Centers for Disease Control and Prevention estimates that total medical costs and lost work attributable to diabetes are as high as $245 billion with an average diabetic spending twice the amount on healthcare per year compared to those without diabetes. It is estimated that a person diagnosed with diabetes at age 40 spends an excess of $211,400 over his/her lifetime compared to his/her nondiabetic counterpart. This excess lifetime cost lowers with an increasing age of diagnosis.[44] What adds complexity to this model is that there are at least 79 million people with prediabetes in the US, and the overall burden of this disease is likely going to be higher in the future.

Globally, diabetes is a leading cause of renal failure, lower limb amputations, and preventable blindness. Diabetes is more prevalent among ethnic minorities such as the Native Americans, African-Americans, and the Hispanics/Latinos compared to non-Hispanic Whites. Data from the National Health Interview Survey (NHIS) shows that the incidence of diabetes had doubled between 1980 and 2008 with a plateauing between 2008 and 2012.[42] However, the increasing incidence of obesity and diabetes in the adolescent population is very concerning. It is estimated that about one-third of adolescents in the US are either overweight or obese. With obesity being one of the strongest risk factors for the development of diabetes, there has been a 30% increase in the prevalence of T2DM in the adolescent population.[45]

Insufficient Sleep and Diabetes

Recent metaanalyses indicate that insufficient sleep duration is an important risk factor for the development of T2DM, with relative risks of up to 1.48 (95% CI: 1.25−1.76) associated with sleep duration of 5 hours or less per night.[46] Also, a recent metaanalysis found that short sleep duration was associated with HbA1C, with a weighted mean difference of 0.23% (95% CI: 0.10−0.36)[47] higher. Cardiometabolic consequences of short sleep duration may be relevant for adults with prediabetes as well. For example, short sleep duration (≤5 hours) was associated with a doubling of the odds of prediabetes (95% CI: 1.00−4.22) compared to individuals who reported 7 hours of sleep.[48]

The relationship between sleep duration and T2DM has been documented to be stronger among men than women.[49] The influence of race is less clear. Both Black/African-American and non-Hispanic White short sleepers (≤5 hours) were found to be 91% more likely to have T2DM versus those getting 6−8 hours.[50] In contrast, both Jackson et al.[51] and Grandner et al.[20] found that the relationship between short sleep duration and diabetes was stronger among non-Hispanic Whites than Blacks/African-Americans. In sum, the aforementioned evidence indicates that the relationship between short sleep duration and diabetes and diabetes outcomes may be mediated by unhealthy glucose levels and insulin resistance.

Insomnia and Diabetes

Few studies have examined insomnia as a risk factor for T2DM (or, likewise, T2DM as a risk factor for insomnia). Cross-sectional data from the 2006 BRFSS shows that difficulty sleeping at least 3 nights per week was associated with a 54% increased likelihood of having diabetes.[23] This was attenuated to a 18% increased risk after adjusting for age, sex, race/ethnicity, education, income, marital status, census region, mental health, health insurance, healthcare access, smoking, and alcohol use. In another cross-sectional study, the odds of T2DM were two−threefold higher among those with both insomnia and short sleep (<6 hours in the laboratory), compared to those with neither.[71] Those with insomnia who slept more than 6 hours in the laboratory did not show elevated risk for T2DM.

Obstructive Sleep Apnea and Diabetes

The prevalence of OSA among T2DM patients is exceedingly high. Foster et al.[52] documented a prevalence of 86% in obese diabetic patients, and Aronsohn et al.[53] reported a 77% prevalence among diabetic patients. Brooks et al.[54] found a slightly lower prevalence of 70%. The overlapping prevalence between these two conditions is notable and alarming. One previous study found that among T2DM patients, only about 18% had received a diagnosis of OSA, suggesting high rates of underdiagnosis.[55] At the population level, epidemiologic studies have also shown that T2DM is prevalent among patients with OSA.[56−58]

SLEEP AND METABOLIC SYNDROME

Definition and Components of Metabolic Syndrome

Definition and Diagnosis

MetS is a condition that represents the combination of central (abdominal) obesity, impaired glucose tolerance, dyslipidemia, low-density lipoprotein (LDL) and high-density lipoprotein (HDL) or elevated triglycerides, and high blood pressure. "Syndrome X" and "insulin resistance syndrome" are other names given to this condition.[59] There are several definitions of MetS, which have led to complexity in data collection and comparison. The main difference among these definitions is the indicator of central obesity. The most commonly accepted set of guidelines was published in a joint consensus statement from the International Diabetes Federation Task Force on Epidemiology and Prevention, National Heart, Lung, and Blood Institute, American Heart Association, World Heart Federation, International Atherosclerosis Society, and International Association for the Study of Obesity in 2009.[60] See Table 7.3 for MetS criteria.

Etiology

MetS is a cluster of conditions that together carry an increased risk for T2DM and cardiovascular disease. Several studies including two metaanalyses of these investigations have conferred a greater risk of cardiovascular and all-cause mortality from MetS.[61,62] Obesity and insulin resistance are likely the preceding events to the development of the MetS. Although a strong genetic component behind MetS is suspected, the lack of a unified definition, differences in lifestyle factors, as well as lack of genetic studies that have looked at MetS as a whole over individual components make it difficult to draw any clear conclusions.

MetS is an inflammatory, prothrombotic state, and elevated levels of inflammatory markers such as interleukin-6 and C-related peptides have been noted. The role of these elevated inflammatory markers in the pathogenesis of MetS and the clinical significance of the measurement of these markers is unclear. However, it is postulated that this chronic

TABLE 7.3   Diagnosis of Metabolic Syndrome

Measure	Criteria/Cut Point
Elevated waist circumference	At least 35 in. (women) or 40 in. (men)
High triglycerides	150 mg/dL or higher or on medication to lower triglycerides
Low high-density lipoproteins cholesterol	Less than 50 mg/dL for women or 40 mg/dL for men or on medication to raise High-density lipoproteins cholesterol
High blood pressure	130/85 mmHg or higher
High fasting glucose	Fasting plasma glucose of 100 mg/dL or higher

Adapted from https://www.nhlbi.nih.gov/health/health-topics/topics/ms/diagnosis.

7. OBESITY, DIABETES, AND METABOLIC SYNDROME

low-grade inflammation is the underlying pathology of MetS that drives the individual components of this complex syndrome.[63] In addition, individuals with MetS have an increased risk for other condition such as fatty liver, asthma, sleep disturbances, and polycystic ovarian disease.[59]

Public Health Relevance

MetS is a global problem, with prevalence ranging from 10% to 84%.[63] This wide variability is due to the differences in the definitions of MetS, region studied, population characteristics, and environment (urban vs rural, clinic vs community, etc.).[63] In the US, the prevalence has increased over time. Data from the NHANES 2003–06 revealed that about 34.0% of the adults met the criteria for MetS, as defined using the National Cholesterol Education Program/Adult Treatment Panel III guidelines, compared to 25.3% in a period of 1988–94.[64] The prevalence increased with age, male sex, and BMI, with the obese male being 32 times more likely to have MetS. Wilson et al.[65] found that even a 2.25 kg (5 pound) weight gain over 16 years was associated with a considerable increase in the risk for MetS and atherosclerosis. It is estimated that for every 11 cm increase in waist circumference, the odds of developing MetS are 1.7 times higher.[66] Early recognition and prompt focus on risk factor control that includes aggressive lifestyle changes, management of diet, and weight reduction is required for the management of MetS.

Sleep Duration and Components of Metabolic Syndrome

Hypertension

Altman et al.[67] found that habitual short sleep (6 hours or less per night) was associated with an increased risk of hypertension. Similarly, in an analysis of the 2004–05 NHIS data, Buxton and Marcelli[68] found that short sleep duration was associated with increased hypertension prevalence. Using the 2007–08 NHANES data, Grandner et al.[20] showed an association between short sleep (6 hours or less) and elevated risk for self-reported history of hypertension but not recorded blood pressure. A recent metaanalysis by Meng et al.[69] combined the results of several prospective studies and found that habitual short sleep duration (variously defined) was associated with an approximately 20% increased likelihood of incident hypertension.

Insulin Resistance

Insulin resistance has also been documented in individuals with short sleep duration. Chaput et al.[70] noted that sleeping less than 6 hours per night was associated with an elevated risk for MetS as defined by the American Heart Association/National Heart, Lung, and Blood Institute criteria with an OR of 1.76 compared to individuals sleeping for 7–8 hours. Data from the NHANES noticed a "U"-shaped relationship between sleep duration and risk for development of insulin resistance and T2DM with risk worse in the non-Hispanic White group compared to Blacks/African-Americans.[51] Hall et al.[71] found similar results where short and long duration of sleep increased the risk of MetS by 45% compared to those who slept for 7–8 hours. This "U"-shaped relationship was noted not only in the development of insulin resistance and diabetes but also general cardiovascular disease.[61, 72]

Grandner et al.,[20] using 2007−08 NHANES data, found that both habitual short and long sleep were associated with impaired fasting glucose, but these relationships were no longer significant after adjustment for demographics, socioeconomics, smoking, and caffeine use.

Dyslipidemia

Several cross-sectional studies from around the world as well as longitudinal studies have shown the deleterious effects of short sleep on the lipid profile in terms of reduced HDL and elevated total cholesterol levels.[73−75] Data from the ARIRANG study revealed that, in 2579 individuals followed for 2.6 years, short sleepers (less than 6 hours) had higher odds of having obesity, low HDL, and elevated triglycerides.[76] History of diagnosis of hypercholesterolemia was associated with habitual short sleep duration in studies by Altman et al.[67] and Grandner et al.[20] Further, the study by Grandner et al. found that habitual short sleep duration of less than 5 hours per night was associated with elevated plasma cholesterol. This population also carried an increased risk of developing hypertension and impaired fasting glucose. Studies on sleep disturbances and dyslipidemia have obtained conflicting results in the pediatric population.[75,77] It is not entirely clear if abnormalities in the lipid panel are due to obesity or sleep insufficiency.[78]

Sleep Duration and Other Unhealthy Behaviors

Habitual short sleep duration is associated with other unhealthy behaviors that may increase risk for MetS. For example, habitual short sleep duration is associated with a sedentary lifestyle.[79,80] Sedentary behavior is also associated with perceived insufficient sleep,[81] general sleep disturbances,[82] and evening chronotype.[80,83] Smoking is related to both short sleep duration[80] and poor sleep quality.[84] Further, this relationship may interact with chronotype, such that short sleepers may be more likely to smoke, but this is especially the case among evening types.[80,83] Individuals with sleep difficulties are also more likely to consume excessive alcohol.[85,86]

Insomnia and Components of Metabolic Syndrome

Evidence linking insomnia to MetS is limited and inconsistent.[87−89] One potential way that insomnia could represent a MetS risk factor is depression. Insomnia is a well-recognized risk factor for the development of depression.[90−93] Individuals with insomnia often have associated dysphoric mood, decreased functioning during wakefulness and are at high risk for having reduced quality of life. Depression, likewise, is a well-characterized risk factor for MetS.[94,95] Depression may serve a causal role (decreased self-care, decreased motivation, decreased emotion regulation leading to unhealthy eating, etc.), may be a consequence (poor health, lack of energy, cognitive complications, and negative view of self and the world), or a correlate of MetS (both potentially partially caused by inflammation, lack of healthy sleep, poor diet, etc.).

Hypertension

A longitudinal study of more than 8000 participants reported that insomnia, as defined by difficulty in the initiation and maintenance of sleep, was associated with an increased risk of developing hypertension.[96] Similar results were reported in cross-sectional studies.[97] A 10-year follow-up study in patients with insomnia reported a higher prevalence of hypertension in patients with insomnia, although no statistical association was noted in the multivariate analysis.[98] It is postulated that hypertension in insomniacs is likely due to increased activation of the hypothalamic pituitary adrenal axis.[99]

Insulin Resistance

Even a single night's reduction of sleep from 12 to 4 hours in an experimental setting has been shown to increase insulin resistance.[100] Vgontzas et al.[101] in their study of 1742 individuals noted that the odds of developing diabetes were approximately 200% higher in people with insomnia with less than <5 hours of sleep per night compared to people without insomnia and sleeping more than 6 hours. This increased risk was independent of other comorbidities frequently associated with insomnia or diabetes. However, Gottlieb et al.[102], in a cross-sectional study of 1486 individuals, found no increased risk of impaired glucose tolerance or diabetes in patients with insomnia. Also, an effective pharmacologic treatment of insomnia has been shown to improve glycemic controls in T2DM as evidenced by the improvement in their HbA1c.[103]

Insomnia and Lipid Disorders

There is mixed data on the association of insomnia and dyslipidemia. Data from the NHANES study found no association of insomnia with dyslipidemia (low HDL or elevated LDL). However, individuals using sleeping pills were noted to have a 118% risk of having an elevated LDL.[104] Similar findings were observed in a prospective cohort study of 3430 adults from China.[105] A large cross-sectional study of 10,054 participants from China showed that frequent insomnia was associated with a slightly increased risk of dyslipidemia; however, the findings were limited to women with elevated total cholesterol. No significant associations were obtained between insomnia symptoms and HDL, LDL, or triglycerides amongst men or women. An important confound, though, was that about 40% of the sample had OSA comorbid with insomnia.[106]

Obstructive Sleep Apnea and Components of Metabolic Syndrome

OSA increases the risk of developing MetS. The risk of developing each feature of Mets increases proportionally to the OSA severity, independent of the BMI.[107] OSA independent of the obesity has shown to worsen insulin resistance and increase the risk for T2DM with the converse relation holding true as well.[52] The International Diabetes Federation recommends screening for OSA in all patients with T2DM, while the American Diabetes Association recommends active treatment of OSA when the conditions coexist.[108] This relationship has been confirmed by several population-based, longitudinal, and cross-sectional studies.[58,97,109,110] Also, glycemic control is worse in individuals with OSA than their counterparts without OSA. A similar trend has also been noted amongst pediatric and adolescent population[111]; however, the relationship is not as conclusive as in the adults with several studies showing no risk of insulin resistance in children with OSA.[112,113]

Hypertension

OSA is the most common cause of drug-resistant hypertension, and the prevalence in this population is as high as 83%. OSA is linked to hypertension independent of the BMI. Several cross-sectional and longitudinal studies have noted this association.[25,114,115] The use of a continuous positive airway pressure (CPAP) for at least 12 weeks was associated with improvement in the control of blood pressure in individuals with drug-resistant hypertension. A recent metaanalysis showed that this improvement was more pronounced in individuals with severe OSA.[116] Also, better compliance with the CPAP resulted in better blood pressure control.[117]

Dyslipidemia

The proposed mechanism linking OSA and dyslipidemia is that hypoxia activates the sterol regulatory element binding protein-1 and stearoyl-coenzyme A desaturase-1, which are involved in the triglycerides and phospholipids biosynthesis. Animal models have shown an association between intermittent hypoxia and dyslipidemia leading to atherosclerosis.[118, 119] In humans, although a similar effect is possible, data on the association of OSA and dyslipidemia is limited. Most of the studies are cross-sectional and nonrandomized. Among these studies, OSA is often associated with lower HDL and higher triglycerides in OSA, with smaller relationships to LDL and total cholesterol.[120−125] The use of CPAP has shown to improve the lipid profile. However, this evidence stems from nonrandomized trials and small randomized trials of short duration.[126,127] Again, data in children is mixed with some studies showing a direct relationship between OSA and abnormal lipids, while others demonstrate obesity as the predominant culprit for dyslipidemia.[113,128]

ASSESSMENT OF SLEEP DISTURBANCES

It is estimated that approximately 50% of the population has some form of sleep disruption (e.g., sleep disorder, poor sleep hygiene, or circadian disorders). However, a large portion of these individuals are unaware (1) if their sleep disruption is due to a primary or secondary cause, such as a sleep disorder, stress/lifestyle, chronic health condition, or medication and/or (2) of the adverse health consequences of habitual sleep disturbances.

To increase awareness and knowledge about sleep disturbances and corollary health risks, several assessment tools have been developed. These assessment tools can be classified into two categories, subjective/self-report/observational and objective/device-based assessments. Subjective assessment tools are often generally brief self-report scales; examples include the insomnia severity index[129] for insomnia or the STOP-BANG questionnaire[130] for sleep apnea. These questionnaires can be used for screening purposes. Other questionnaires may also capture sleep behaviors and parameters such as the sleep timing questionnaire,[131] the Epworth Sleepiness Scale,[132] or the Pittsburgh sleep quality index.[133] Prospective self-report, such as sleep diaries,[134] can provide useful information about sleep continuity. Objective assessment tools (e.g., actigraphy[135] and polysomnography) can often reliably estimate sleep parameters, such as total sleep time, sleep latency, sleep efficiency, and wake after sleep onset, etc. Measures that use electroencephalography may also assess sleep architecture (stages). Subjective assessment tools are considered easier to administer and capture insomnia well but are limited by recall bias, while objective assessment tools, especially polysomnography, are preferred for assessing physiologic sleep. Regarding assessment of sleep in patients with MetS, it is important to assess for common sleep disorders, as well as risk for sleep disorders such as OSA and insomnia.

TREATMENT OF SLEEP DISTURBANCES

Targeting sleep may be a promising opportunity to improve treatment of obesity and other cardiometabolic disease risk factors such as diabetes and hypertension, given the above literature showing relationships to sleep. Interventions include treatments aimed at treating sleep disorders and those aimed at improving subclinical sleep problems. Regarding treatments for sleep disorders, empirically supported treatments exist for both insomnia and OSA (as well as other disorders outside the scope of this chapter). For insomnia, the recommended first-line treatment is cognitive behavioral therapy for insomnia,[136] and short-term hypnotic medications are also recommended as a potential second-line treatment.[137] Regarding OSA, continuous positive airway pressure is the recommended first-line therapy, particularly in MetS,[138] though other treatment options exist, including positional therapy,[139] oral appliances,[140] surgery,[141] and other approaches. Interventions for general sleep habits usually include sleep hygiene[85] or recommendations of at least 7 hours of sleep.[142-144] Lifestyle-based sleep interventions provide valuable recommendations to improve and optimize sleep and include consistent wake—sleep schedules, limiting stimulants before bed, avoid exercise and eating immediately before bedtime, establish sleep-conducive environment, and avoid naps. Studies indicate that if sleep disturbances are caught and treated early, consistently, and efficiently, it can either stave off or better manage adverse health consequences. In addition, existing interventions for promoting weight loss and/or cardiometabolic health (e.g., weight loss, exercise, smoking cessation, medical treatment) should consider sleep as a component of these interventions, since treating sleep problems may facilitate treatment and ignoring sleep problems may attenuate the effectiveness of treatments.

AREAS FOR FUTURE RESEARCH

Sleep has emerged as an important metabolic risk factor. Specifically, short sleep duration, insomnia, and sleep apnea are all associated with obesity, diabetes, and MetS. It is possible that the pathways implicated in these relationships overlap across conditions, but it is also likely that there are some unique pathways specific to each of these sleep disturbances. Further complicating matters is that these conditions are often comorbid. Future research is needed to better understand the mechanisms linking sleep disturbances to obesity and diabetes. Although several pathways have been implicated, it is not clear which is most relevant and which exerts causal control. Research is needed to clarify the mechanisms and causal chains. Additional work is also warranted to determine how to incorporate sleep health into obesity/diabetes prevention efforts and what the added benefits are of adding sleep components to obesity and diabetes prevention programs. In the future, it will also be important to identify and refine sleep-based interventions for the purpose of reducing obesity and metabolic risk. Although guidelines for healthy sleep and treatment of sleep disorders exist, there are still gaps that need to be addressed: (1) which interventions work to reduce metabolic risk; (2) how can these interventions be optimized to maximize metabolic benefits; (3) how can these interventions be successfully disseminated; and (4) how can these interventions be adequately tailored to meet the needs of at-risk groups, including racial/ethnic minorities and the socioeconomically disadvantaged. It is possible that sleep represents an important modifiable risk factor for metabolic disease.

CONCLUSION

MetS, and the related problems of obesity, cardiovascular disease, and T2DM, are major public health problems. Habitual short sleep duration, insomnia, and OSA are all prevalent in the population and represent important considerations in understanding MetS and related conditions. A growing literature is demonstrating that these sleep disturbances may represent important risk factors for the presence of and, in some cases, the development of cardiometabolic disease. Further, they may also represent a consequence of these problems, making the relationship bidirectional. It is still unknown whether sleep interventions may represent an opportunity to prevent or reduce the impact of MetS, obesity, and other conditions, but preliminary work suggests that not only can sleep interventions improve outcomes, but also they are likely more feasible than other interventions such as diet and exercise. Future research needs to more fully elucidate the causal mechanisms linking sleep to metabolic risk, which will aid in the development of new interventions.

KEY PRACTICE POINTS

- Short sleep duration is associated with weight gain and incident and prevalent obesity, as well as hypertension, diabetes, and other components of the MetS.
- Insomnia is also a risk factor for cardiometabolic disease risk, though mechanisms are less clear and the relationship may overlap with that of short sleep; individuals with insomnia combined with short sleep may be at highest risk.
- OSA is a well-characterized risk factor for weight gain, obesity, diabetes, high blood pressure, and other aspects of cardiac and cardiometabolic risk.
- Treatments for sleep disorders are promising approaches to improving treatments for cardiometabolic disease risk.
- Improving sleep duration in those achieving insufficient sleep may reduce cardiometabolic disease risk and may assist in the treatment of obesity and other disorders.

References

1. Arble DM, Bass J, Behn CD, et al. Impact of sleep and circadian disruption on energy balance and diabetes: a summary of workshop discussions. *Sleep*. 2015;38:1849−1860.
2. Speliotes EK, Willer CJ, Berndt SI, et al. Association analyses of 249,796 individuals reveal 18 new loci associated with body mass index. *Nat Genet*. 2010;42:937−948.
3. Dabelea D, Harrod CS. Role of developmental overnutrition in pediatric obesity and type 2 diabetes. *Nutr Rev*. 2013;71(Suppl 1):S62−S67.
4. Finkelstein EA, Trogdon JG, Cohen JW, Dietz W. Annual medical spending attributable to obesity: payer-and service-specific estimates. *Health Aff (Millwood)*. 2009;28:w822−w831.
5. Grandner MA, Patel NP, Gehrman PR, Perlis ML, Pack AI. Problems associated with short sleep: bridging the gap between laboratory and epidemiological studies. *Sleep Med Rev*. 2010;14:239−247.
6. Grandner MA, Seixas A, Shetty S, Shenoy S. Sleep duration and diabetes risk: population trends and potential mechanisms. *Curr Diab Rep*. 2016;16:106.
7. Itani O, Jike M, Watanabe N, Kaneita Y. Short sleep duration and health outcomes: a systematic review, meta-analysis, and meta-regression. *Sleep Med*. 2017;32:246−256.
8. Grandner MA, Alfonso-Miller P, Fernandez-Mendoza J, Shetty S, Shenoy S, Combs D. Sleep: important considerations for the prevention of cardiovascular disease. *Curr Opin Cardiol*. 2016;31:551−565.
9. Adenekan B, Pandey A, McKenzie S, Zizi F, Casimir GJ, Jean-Louis G. Sleep in America: role of racial/ethnic differences. *Sleep Med Rev*. 2013;17:255−262.
10. St-Onge MP. The role of sleep duration in the regulation of energy balance: effects on energy intakes and expenditure. *J Clin Sleep Med*. 2013;9:73−80.
11. Reiter RJ, Tan DX, Korkmaz A, Ma S. Obesity and metabolic syndrome: association with chronodisruption, sleep deprivation, and melatonin suppression. *Ann Med*. 2012;44:564−577.
12. Morselli LL, Guyon A, Spiegel K. Sleep and metabolic function. *Pflugers Arch*. 2012;463:139−160.
13. Knutson KL. Does inadequate sleep play a role in vulnerability to obesity? *Am J Hum Biol*. 2012;24:361−371.
14. Akinnusi ME, Saliba R, Porhomayon J, El-Solh AA. Sleep disorders in morbid obesity. *Eur J Intern Med*. 2012;23:219−226.
15. Penev PD. Update on energy homeostasis and insufficient sleep. *J Clin Endocrinol Metab*. 2012;97:1792−1801.
16. Guidolin M, Gradisar M. Is shortened sleep duration a risk factor for overweight and obesity during adolescence? A review of the empirical literature. *Sleep Med*. 2012;13:779−786.
17. Wu Y, Zhai L, Zhang D. Sleep duration and obesity among adults: a meta-analysis of prospective studies. *Sleep Med*. 2014;15:1456−1462.
18. Kripke DF, Garfinkel L, Wingard DL, Klauber MR, Marler MR. Mortality associated with sleep duration and insomnia. *Arch Gen Psychiatry*. 2002;59:131−136.
19. Chaput JP, Despres JP, Bouchard C, Tremblay A. Short sleep duration is associated with reduced leptin levels and increased adiposity: results from the Quebec family study. *Obesity (Silver Spring)*. 2007;15:253−261.
20. Grandner MA, Chakravorty S, Perlis ML, Oliver L, Gurubhagavatula I. Habitual sleep duration associated with self-reported and objectively determined cardiometabolic risk factors. *Sleep Med*. 2014;15:42−50.
21. Grandner MA, Schopfer EA, Sands-Lincoln M, Jackson N, Malhotra A. Relationship between sleep duration and body mass index depends on age. *Obesity (Silver Spring)*. 2015;23:2491−2498.
22. American Academy of Sleep Medicine. *International Classification of Sleep Disorders*. 3rd ed Darien, IL: American Academy of Sleep Medicine; 2014.
23. Grandner MA, Jackson NJ, Pak VM, Gehrman PR. Sleep disturbance is associated with cardiovascular and metabolic disorders. *J Sleep Res*. 2012;21:427−433.
24. Cronlein T. Insomnia and obesity. *Curr Opin Psychiatry*. 2016;29:409−412.
25. Peppard PE, Young T, Palta M, Skatrud J. Prospective study of the association between sleep-disordered breathing and hypertension. *N Engl J Med*. 2000;342:1378−1384.
26. Tufik S, Santos-Silva R, Taddei JA, Bittencourt LR. Obstructive sleep apnea syndrome in the Sao Paulo Epidemiologic Sleep Study. *Sleep Med*. 2010;11:441−446.
27. Peppard PE, Young T, Barnet JH, Palta M, Hagen EW, Hla KM. Increased prevalence of sleep-disordered breathing in adults. *Am J Epidemiol*. 2013;177:1006−1014.
28. Matricciani L, Olds T, Petkov J. In search of lost sleep: secular trends in the sleep time of school-aged children and adolescents. *Sleep Med Rev*. 2012;16:203−211.

REFERENCES

29. Capers PL, Fobian AD, Kaiser KA, Borah R, Allison DB. A systematic review and meta-analysis of randomized controlled trials of the impact of sleep duration on adiposity and components of energy balance. *Obes Rev.* 2015;16:771–782.
30. Taheri S. The link between short sleep duration and obesity: we should recommend more sleep to prevent obesity. *Arch Dis Child.* 2006;91:881–884.
31. Brondel L, Romer MA, Nougues PM, Touyarou P, Davenne D. Acute partial sleep deprivation increases food intake in healthy men. *Am J Clin Nutr.* 2010;91:1550–1559.
32. Patel SR, Blackwell T, Redline S, et al. The association between sleep duration and obesity in older adults. *Int J Obes (Lond).* 2008;32:1825–1834.
33. Taheri S, Lin L, Austin D, Young T, Mignot E. Short sleep duration is associated with reduced leptin, elevated ghrelin, and increased body mass index. *PLoS Med.* 2004;1:e62.
34. St-Onge MP, Wolfe S, Sy M, Shechter A, Hirsch J. Sleep restriction increases the neuronal response to unhealthy food in normal-weight individuals. *Int J Obes (Lond).* 2014;38:411–416.
35. St-Onge MP, Roberts AL, Chen J, et al. Short sleep duration increases energy intakes but does not change energy expenditure in normal-weight individuals. *Am J Clin Nutr.* 2011;94:410–416.
36. Markwald RR, Melanson EL, Smith MR, et al. Impact of insufficient sleep on total daily energy expenditure, food intake, and weight gain. *Proc Natl Acad Sci USA.* 2013;110:5695–5700.
37. Spaeth AM, Dinges DF, Goel N. Sex and race differences in caloric intake during sleep restriction in healthy adults. *Am J Clin Nutr.* 2014;100:559–566.
38. Spaeth AM, Dinges DF, Goel N. Effects of experimental sleep restriction on weight gain, caloric intake, and meal timing in healthy adults. *Sleep.* 2013;36:981–990.
39. Nedeltcheva AV, Kilkus JM, Imperial J, Kasza K, Schoeller DA, Penev PD. Sleep curtailment is accompanied by increased intake of calories from snacks. *Am J Clin Nutr.* 2009;89:126–133.
40. Shechter A, Grandner MA, St-Onge MP. The role of sleep in the control of food intake. *Am J Lifestyle Med.* 2014;8:371–374.
41. Menke A, Casagrande S, Geiss L, Cowie CC. Prevalence of and trends in diabetes among adults in the United States, 1988–2012. *JAMA.* 2015;314:1021–1029.
42. Geiss LS, Wang J, Cheng YJ, et al. Prevalence and incidence trends for diagnosed diabetes among adults aged 20 to 79 years, United States, 1980–2012. *JAMA.* 2014;312:1218–1226.
43. Scott RA, Langenberg C, Sharp SJ, et al. The link between family history and risk of type 2 diabetes is not explained by anthropometric, lifestyle or genetic risk factors: the EPIC-InterAct study. *Diabetologia.* 2013;56:60–69.
44. Zhuo X, Zhang P, Barker L, Albright A, Thompson TJ, Gregg E. The lifetime cost of diabetes and its implications for diabetes prevention. *Diabetes Care.* 2014;37:2557–2564.
45. Dabelea D, Mayer-Davis EJ, Saydah S, et al. Prevalence of type 1 and type 2 diabetes among children and adolescents from 2001 to 2009. *JAMA.* 2014;311:1778–1786.
46. Anothaisintawee T, Reutrakul S, Van Cauter E, Thakkinstian A. Sleep disturbances compared to traditional risk factors for diabetes development: systematic review and meta-analysis. *Sleep Med Rev.* 2015;30:11–24.
47. Lee SW, Ng KY, Chin WK. The impact of sleep amount and sleep quality on glycemic control in type 2 diabetes: a systematic review and meta-analysis. *Sleep Med Rev.* 2016;.
48. Engeda J, Mezuk B, Ratliff S, Ning Y. Association between duration and quality of sleep and the risk of prediabetes: evidence from NHANES. *Diabet Med.* 2013;30:676–680.
49. Mallon L, Broman JE, Hetta J. High incidence of diabetes in men with sleep complaints or short sleep duration: a 12-year follow-up study of a middle-aged population. *Diabetes Care.* 2005;28:2762–2767.
50. Zizi F, Pandey A, Murrray-Bachmann R, et al. Race/ethnicity, sleep duration, and diabetes mellitus: analysis of the National Health Interview Survey. *Am J Med.* 2012;125:162–167.
51. Jackson CL, Redline S, Kawachi I, Hu FB. Association between sleep duration and diabetes in black and white adults. *Diabetes Care.* 2013;36:3557–3565.
52. Foster GD, Sanders MH, Millman R, et al. Obstructive sleep apnea among obese patients with type 2 diabetes. *Diabetes Care.* 2009;32:1017–1019.
53. Aronsohn RS, Whitmore H, Van Cauter E, Tasali E. Impact of untreated obstructive sleep apnea on glucose control in type 2 diabetes. *Am J Respir Crit Care Med.* 2010;181:507–513.
54. Brooks B, Cistulli PA, Borkman M, et al. Obstructive sleep apnea in obese noninsulin-dependent diabetic patients: effect of continuous positive airway pressure treatment on insulin responsiveness. *J Clin Endocrinol Metab.* 1994;79:1681–1685.

II. SLEEP DISORDERS IN SPECIFIC MEDICAL CONDITIONS

55. Heffner JE, Rozenfeld Y, Kai M, Stephens EA, Brown LK. Prevalence of diagnosed sleep apnea among patients with type 2 diabetes in primary care. *CHEST J*. 2012;141:1414−1421.
56. Reichmuth KJ, Austin D, Skatrud JB, Young T. Association of sleep apnea and type II diabetes: a population-based study. *Am J Respir Crit Care Med*. 2005;172:1590−1595.
57. Marshall NS, Wong KK, Phillips CL, Liu PY, Knuiman MW, Grunstein RR. Is sleep apnea an independent risk factor for prevalent and incident diabetes in the Busselton Health Study? *J Clin Sleep Med: JCSM*. 2009;5:15−20.
58. Lindberg E, Theorell-Haglöw J, Svensson M, Gislason T, Berne C, Janson C. Sleep apnea and glucose metabolism: a long-term follow-up in a community-based sample. *CHEST J*. 2012;142:935−942.
59. Grundy SM, Brewer Jr. HB, Cleeman JI, Smith Jr. SC, Lenfant C. Definition of metabolic syndrome: report of the National Heart, Lung, and Blood Institute/American Heart Association conference on scientific issues related to definition. *Circulation*. 2004;109:433−438.
60. Alberti KG, Eckel RH, Grundy SM, et al. Harmonizing the metabolic syndrome: a joint interim statement of the International Diabetes Federation Task Force on Epidemiology and Prevention; National Heart, Lung, and Blood Institute; American Heart Association; World Heart Federation; International Atherosclerosis Society; and International Association for the Study of Obesity. *Circulation*. 2009;120:1640−1645.
61. Ford ES. Habitual sleep duration and predicted 10-year cardiovascular risk using the pooled cohort risk equations among US adults. *J Am Heart Assoc*. 2014;3:e001454.
62. Gami AS, Somers VK. Obstructive sleep apnoea, metabolic syndrome, and cardiovascular outcomes. *Eur Heart J*. 2004;25:709−711.
63. Kaur J. A comprehensive review on metabolic syndrome. *Cardiol Res Pract*. 2014;2014:943162.
64. Moore JX, Chaudhary N, Akinyemiju T. Metabolic syndrome prevalence by race/ethnicity and sex in the United States, National Health and Nutrition Examination Survey, 1988−2012. *Prev Chronic Dis*. 2017;14:E24.
65. Wilson PW, Kannel WB, Silbershatz H, D'Agostino RB. Clustering of metabolic factors and coronary heart disease. *Arch Intern Med*. 1999;159:1104−1109.
66. Palaniappan L, Carnethon MR, Wang Y, et al. Predictors of the incident metabolic syndrome in adults: the Insulin Resistance Atherosclerosis Study. *Diabetes Care*. 2004;27:788−793.
67. Altman NG, Izci-Balserak B, Schopfer E, et al. Sleep duration versus sleep insufficiency as predictors of cardiometabolic health outcomes. *Sleep Med*. 2012;13:1261−1270.
68. Buxton OM, Marcelli E. Short and long sleep are positively associated with obesity, diabetes, hypertension, and cardiovascular disease among adults in the United States. *Soc Sci Med*. 2010;71:1027−1036.
69. Meng L, Zheng Y, Hui R. The relationship of sleep duration and insomnia to risk of hypertension incidence: a meta-analysis of prospective cohort studies. *Hypertens Res*. 2013;36:985−995.
70. Chaput JP, McNeil J, Despres JP, Bouchard C, Tremblay A. Short sleep duration as a risk factor for the development of the metabolic syndrome in adults. *Prev Med*. 2013;57:872−877.
71. Hall MH, Muldoon MF, Jennings JR, Buysse DJ, Flory JD, Manuck SB. Self-reported sleep duration is associated with the metabolic syndrome in midlife adults. *Sleep*. 2008;31:635−643.
72. Sabanayagam C, Zhang R, Shankar A. Markers of sleep-disordered breathing and metabolic syndrome in a multiethnic sample of US adults: results from the National Health and Nutrition Examination Survey 2005−2008. *Cardiol Res Pract*. 2012;2012:630802.
73. Choi KM, Lee JS, Park HS, Baik SH, Choi DS, Kim SM. Relationship between sleep duration and the metabolic syndrome: Korean National Health and Nutrition Survey 2001. *Int J Obes (Lond)*. 2008;32:1091−1097.
74. Bjorvatn B, Sagen IM, Oyane N, et al. The association between sleep duration, body mass index and metabolic measures in the Hordaland Health Study. *J Sleep Res*. 2007;16:66−76.
75. Gangwisch JE, Malaspina D, Babiss LA, et al. Short sleep duration as a risk factor for hypercholesterolemia: analyses of the National Longitudinal Study of Adolescent Health. *Sleep*. 2010;33:956−961.
76. Kim JY, Yadav D, Ahn SV, et al. A prospective study of total sleep duration and incident metabolic syndrome: the ARIRANG study. *Sleep Med*. 2015;16:1511−1515.
77. Berentzen NE, Smit HA, Bekkers MB, et al. Time in bed, sleep quality and associations with cardiometabolic markers in children: the Prevention and Incidence of Asthma and Mite Allergy birth cohort study. *J Sleep Res*. 2014;23:3−12.
78. Rey-Lopez JP, de Carvalho HB, de Moraes AC, et al. Sleep time and cardiovascular risk factors in adolescents: the HELENA (Healthy Lifestyle in Europe by Nutrition in Adolescence) study. *Sleep Med*. 2014;15:104−110.

REFERENCES

79. Xiao Q, Keadle SK, Hollenbeck AR, Matthews CE. Sleep duration and total and cause-specific mortality in a large US cohort: interrelationships with physical activity, sedentary behavior, and body mass index. *Am J Epidemiol.* 2014;180:997−1006.

80. Patterson F, Malone SK, Lozano A, Grandner MA, Hanlon AL. Smoking, screen-based sedentary behavior, and diet associated with habitual sleep duration and chronotype: data from the UK Biobank. *Ann Behav Med.* 2016;.

81. Grandner MA, Jackson NJ, Izci-Balserak B, et al. Social and behavioral determinants of perceived insufficient sleep. *Front Neurol.* 2015;6:112.

82. Grandner MA, Patel NP, Perlis ML, et al. Obesity, diabetes, and exercise associated with sleep-related complaints in the American population. *J Public Health.* 2011;19:463−474.

83. Patterson F, Malone SK, Grandner MA, Lozano A, Perkett M, Hanlon A. Interactive effects of sleep duration and morning/evening preference on cardiovascular risk factors. *Eur J Public Health.* 2017;.

84. Jaehne A, Unbehaun T, Feige B, Lutz UC, Batra A, Riemann D. How smoking affects sleep: a polysomnographical analysis. *Sleep Med.* 2012;13:1286−1292.

85. Irish LA, Kline CE, Gunn HE, Buysse DJ, Hall MH. The role of sleep hygiene in promoting public health: a review of empirical evidence. *Sleep Med Rev.* 2015;22:23−36.

86. Hoevenaar-Blom MP, Spijkerman AM, Kromhout D, Verschuren WM. Sufficient sleep duration contributes to lower cardiovascular disease risk in addition to four traditional lifestyle factors: the MORGEN study. *Eur J Prev Cardiol.* 2014;21:1367−1375.

87. Jennings JR, Muldoon MF, Hall M, Buysse DJ, Manuck SB. Self-reported sleep quality is associated with the metabolic syndrome. *Sleep.* 2007;30:219−223.

88. Hall MH, Okun ML, Sowers M, et al. Sleep is associated with the metabolic syndrome in a multi-ethnic cohort of midlife women: the SWAN Sleep Study. *Sleep.* 2012;35:783−790.

89. Kazman JB, Abraham PA, Zeno SA, Poth M, Deuster PA. Self-reported sleep impairment and the metabolic syndrome among African Americans. *Ethn Dis.* 2012;22:410−415.

90. Spiegelhalder K, Regen W, Nanovska S, Baglioni C, Riemann D. Comorbid sleep disorders in neuropsychiatric disorders across the life cycle. *Curr Psychiatry Rep.* 2013;15:364.

91. Baglioni C, Riemann D. Is chronic insomnia a precursor to major depression? Epidemiological and biological findings. *Curr Psychiatry Rep.* 2012;14:511−518.

92. Baglioni C, Battagliese G, Feige B, et al. Insomnia as a predictor of depression: a meta-analytic evaluation of longitudinal epidemiological studies. *J Affect Disord.* 2011;135:10−19.

93. Baglioni C, Spiegelhalder K, Lombardo C, Riemann D. Sleep and emotions: a focus on insomnia. *Sleep Med Rev.* 2010;14:227−238.

94. Mezick EJ, Hall M, Matthews KA. Are sleep and depression independent or overlapping risk factors for cardiometabolic disease? *Sleep Med Rev.* 2011;15:51−63.

95. de Melo LGP, Nunes SOV, Anderson G, et al. Shared metabolic and immune-inflammatory, oxidative and nitrosative stress pathways in the metabolic syndrome and mood disorders. *Prog Neuropsychopharmacol Biol Psychiatry.* 2017;78:34−50.

96. Suka M, Yoshida K, Sugimori H. Persistent insomnia is a predictor of hypertension in Japanese male workers. *J Occup Health.* 2003;45:344−350.

97. Elmasry A, Lindberg E, Berne C, et al. Sleep-disordered breathing and glucose metabolism in hypertensive men: a population-based study. *J Intern Med.* 2001;249:153−161.

98. Janson C, Lindberg E, Gislason T, Elmasry A, Boman G. Insomnia in men-a 10-year prospective population based study. *Sleep.* 2001;24:425−430.

99. Vgontzas AN, Tsigos C, Bixler EO, et al. Chronic insomnia and activity of the stress system: a preliminary study. *J Psychosom Res.* 1998;45:21−31.

100. Donga E, van Dijk M, van Dijk JG, et al. Partial sleep restriction decreases insulin sensitivity in type 1 diabetes. *Diabetes Care.* 2010;33:1573−1577.

101. Vgontzas AN, Liao D, Pejovic S, Calhoun S, Karataraki M, Bixler EO. Insomnia with objective short sleep duration is associated with type 2 diabetes: a population-based study. *Diabetes Care.* 2009;32:1980−1985.

102. Gottlieb DJ, Punjabi NM, Newman AB, et al. Association of sleep time with diabetes mellitus and impaired glucose tolerance. *Arch Intern Med.* 2005;165:863−867.

II. SLEEP DISORDERS IN SPECIFIC MEDICAL CONDITIONS

103. Garfinkel D, Zorin M, Wainstein J, Matas Z, Laudon M, Zisapel N. Efficacy and safety of prolonged-release melatonin in insomnia patients with diabetes: a randomized, double-blind, crossover study. *Diabetes Metab Syndr Obes*. 2011;4:307–313.
104. Vozoris NT. Insomnia symptoms are not associated with dyslipidemia: a population-based study. *Sleep*. 2016;39:551–558.
105. Chien KL, Chen PC, Hsu HC, et al. Habitual sleep duration and insomnia and the risk of cardiovascular events and all-cause death: report from a community-based cohort. *SLEEP*. 2010;33:177–184.
106. Vozoris NT. Sleep apnea-plus: prevalence, risk factors, and association with cardiovascular diseases using United States population-level data. *Sleep Med*. 2012;13:637–644.
107. Peled N, Kassirer M, Shitrit D, et al. The association of OSA with insulin resistance, inflammation and metabolic syndrome. *Respir Med*. 2007;101:1696–1701.
108. Westlake K, Polak J. Screening for obstructive sleep apnea in type 2 diabetes patients—questionnaires are not good enough. *Front Endocrinol (Lausanne)*. 2016;7:124.
109. Botros N, Concato J, Mohsenin V, Selim B, Doctor K, Yaggi HK. Obstructive sleep apnea as a risk factor for type 2 diabetes. *Am J Med*. 2009;122:1122–1127.
110. Fredheim JM, Rollheim J, Omland T, et al. Type 2 diabetes and pre-diabetes are associated with obstructive sleep apnea in extremely obese subjects: a cross-sectional study. *Cardiovasc Diabetol*. 2011;10:84.
111. Koren D, Gozal D, Philby MF, Bhattacharjee R, Kheirandish-Gozal L. Impact of obstructive sleep apnoea on insulin resistance in nonobese and obese children. *Eur Respir J*. 2016;47:1152–1161.
112. Kaditis AG, Alexopoulos EI, Damani E, et al. Obstructive sleep-disordered breathing and fasting insulin levels in nonobese children. *Pediatr Pulmonol*. 2005;40:515–523.
113. Tauman R, O'Brien LM, Ivanenko A, Gozal D. Obesity rather than severity of sleep-disordered breathing as the major determinant of insulin resistance and altered lipidemia in snoring children. *Pediatrics*. 2005;116: e66–e73.
114. Nieto FJ, Young TB, Lind BK, et al. Association of sleep-disordered breathing, sleep apnea, and hypertension in a large community-based study. Sleep Heart Health Study. *JAMA*. 2000;283:1829–1836.
115. Hla KM, Young TB, Bidwell T, Palta M, Skatrud JB, Dempsey J. Sleep apnea and hypertension. A population-based study. *Ann Intern Med*. 1994;120:382–388.
116. Alajmi M, Mulgrew AT, Fox J, et al. Impact of continuous positive airway pressure therapy on blood pressure in patients with obstructive sleep apnea hypopnea: a meta-analysis of randomized controlled trials. *Lung*. 2007;185:67–72.
117. Haentjens P, Van Meerhaeghe A, Moscariello A, et al. The impact of continuous positive airway pressure on blood pressure in patients with obstructive sleep apnea syndrome: evidence from a meta-analysis of placebo-controlled randomized trials. *Arch Intern Med*. 2007;167:757–764.
118. Li J, Thorne LN, Punjabi NM, et al. Intermittent hypoxia induces hyperlipidemia in lean mice. *Circ Res*. 2005;97:698–706.
119. Li J, Savransky V, Nanayakkara A, Smith PL, O'Donnell CP, Polotsky VY. Hyperlipidemia and lipid peroxidation are dependent on the severity of chronic intermittent hypoxia. *J Appl Physiol*. 1985;2007(102):557–563.
120. Coughlin SR, Mawdsley L, Mugarza JA, Calverley PM, Wilding JP. Obstructive sleep apnoea is independently associated with an increased prevalence of metabolic syndrome. *Eur Heart J*. 2004;25:735–741.
121. Czerniawska J, Bielen P, Plywaczewski R, et al. [Metabolic abnormalities in obstructive sleep apnea patients]. *Pneumonol Alergol Pol*. 2008;76:340–347.
122. Kono M, Tatsumi K, Saibara T, et al. Obstructive sleep apnea syndrome is associated with some components of metabolic syndrome. *Chest*. 2007;131:1387–1392.
123. McArdle N, Hillman D, Beilin L, Watts G. Metabolic risk factors for vascular disease in obstructive sleep apnea: a matched controlled study. *Am J Respir Crit Care Med*. 2007;175:190–195.
124. Tan KC, Chow WS, Lam JC, et al. HDL dysfunction in obstructive sleep apnea. *Atherosclerosis*. 2006;184:377–382.
125. Drager LF, Jun J, Polotsky VY. Obstructive sleep apnea and dyslipidemia: implications for atherosclerosis. *Curr Opin Endocrinol Diabetes Obes*. 2010;17:161–165.
126. Robinson GV, Pepperell JC, Segal HC, Davies RJ, Stradling JR. Circulating cardiovascular risk factors in obstructive sleep apnoea: data from randomised controlled trials. *Thorax*. 2004;59:777–782.

REFERENCES

127. Drager LF, Bortolotto LA, Figueiredo AC, Krieger EM, Lorenzi GF. Effects of continuous positive airway pressure on early signs of atherosclerosis in obstructive sleep apnea. *Am J Respir Crit Care Med.* 2007;176:706–712.

128. Alexopoulos EI, Gletsou E, Kostadima E, et al. Effects of obstructive sleep apnea severity on serum lipid levels in Greek children with snoring. *Sleep Breath.* 2011;15:625–631.

129. Bastien CH, Vallieres A, Morin CM. Validation of the Insomnia Severity Index as an outcome measure for insomnia research. *Sleep Med.* 2001;2:297–307.

130. Farney RJ, Walker BS, Farney RM, Snow GL, Walker JM. The STOP-Bang equivalent model and prediction of severity of obstructive sleep apnea: relation to polysomnographic measurements of the apnea/hypopnea index. *J Clin Sleep Med.* 2011;7:459–465.

131. Monk TH, Buysse DJ, Kennedy KS, Pods JM, DeGrazia JM, Miewald JM. Measuring sleep habits without using a diary: the sleep timing questionnaire. *Sleep.* 2003;26:208–212.

132. Johns MW. A new method for measuring daytime sleepiness: the Epworth sleepiness scale. *Sleep.* 1991;14:540–545.

133. Buysse DJ, Reynolds 3rd CF, Monk TH, Berman SR, Kupfer DJ. The Pittsburgh Sleep Quality Index: a new instrument for psychiatric practice and research. *Psychiatry Res.* 1989;28:193–213.

134. Carney CE, Buysse DJ, Ancoli-Israel S, et al. The consensus sleep diary: standardizing prospective sleep self-monitoring. *Sleep.* 2012;35:287–302.

135. Ancoli-Israel S, Martin JL, Blackwell T, et al. The SBSM guide to actigraphy monitoring: clinical and research applications. *Behav Sleep Med.* 2015;13(Suppl 1):S4–S38.

136. Qaseem A, Kansagara D, Forciea MA, Cooke M, Denberg TD. Clinical Guidelines Committee of the American College of P. Management of Chronic Insomnia Disorder in Adults: A Clinical Practice Guideline From the American College of Physicians. *Ann Intern Med.* 2016;.

137. Sateia MJ, Buysse DJ, Krystal AD, Neubauer DN, Heald JL. Clinical practice guideline for the pharmacologic treatment of chronic insomnia in adults: an American Academy of Sleep Medicine Clinical Practice Guideline. *J Clin Sleep Med.* 2017;13:307–349.

138. Bertisch S, Patel SR. CPAP for obstructive sleep apnea and the metabolic syndrome. *N Engl J Med.* 2012;366:963–964. author reply 5-6.

139. de Vries N, Ravesloot M, van Maanen JP. *Positional Therapy in Obstructive Sleep Apnea.* Cham: Springer; 2015.

140. Lam B, Sam K, Lam JC, Lai AY, Lam CL, Ip MS. The efficacy of oral appliances in the treatment of severe obstructive sleep apnea. *Sleep Breath.* 2011;15:195–201.

141. Freedman N. Treatment of obstructive sleep apnea syndrome. *Clin Chest Med.* 2010;31:187–201.

142. Watson NF, Badr MS, Belenky G, et al. Recommended amount of sleep for a healthy adult: a joint consensus statement of the American Academy of Sleep Medicine and Sleep Research Society. *Sleep.* 2015;38:843–844.

143. St-Onge MP, Grandner MA, Brown D, et al. Sleep duration and quality: impact on lifestyle behaviors and cardiometabolic health: a scientific statement from the American Heart Association. *Circulation.* 2016;.

144. Hirshkowitz M, Whiton K, Alpert SM, et al. National sleep foundation's updated sleep duration recommendations: final report. *Sleep Health.* 2015;1:233–243.

CHAPTER

8

Cancer

Josée Savard[1,2,3]

[1]School of Psychology, Université Laval, Québec, QC, Canada [2]CHU de Québec-Université Laval Research Center, Québec, QC, Canada [3]Université Laval Cancer Research Center, Québec, QC, Canada

OUTLINE

Introduction	176	Obstructive Sleep Apnea	185
Brief Overview of the Medical Condition	176	Periodic Limb Movement in Sleep/Restless Legs Syndrome	186
Epidemiology of Sleep Disorders in Cancer	177	Assessment of Sleep Difficulties in Cancer	186
Insomnia	178	Screening	186
Obstructive Sleep Apnea	179	Clinical Interview	187
Periodic Limb Movement in Sleep/Restless Legs Syndrome	180	Sleep Diary	187
		Questionnaires	187
		Polysomnography	187
Sleep Difficulties in Special Populations	181	Actigraphy	187
Advanced Cancer and Palliative Care	181	Treatment of Sleep Difficulties in Cancer	188
Survivors of Childhood Cancer	181	Treatment of Insomnia	188
Etiology of Sleep Disorders in Cancer	182	Treatment of Other Sleep Disorders	191
Insomnia	182	Clinical Vignette	192
Obstructive Sleep Apnea	183		
Periodic Limb Movement in Sleep/Restless Legs Syndrome	184	Areas for Future Research	193
Consequences of Sleep Disorders in Cancer	184	Conclusion	193
Insomnia	184	References	194

Handbook of Sleep Disorders in Medical Conditions
DOI: https://doi.org/10.1016/B978-0-12-813014-8.00008-1

© 2019 Elsevier Inc. All rights reserved.

INTRODUCTION

Once a neglected area of research, cancer-related sleep disturbances have received increased attention in the past 10–15 years. Over the years, it has been demonstrated that sleep difficulties are extremely common in patients with cancer, before, during, and after their treatment. Although many more clinical trials have been published on the treatment of insomnia in general, cancer-related sleep disturbances in general appear to be highly treatable. This chapter will provide information that clinicians need to know in order to routinely screen and assess sleep disorders in their clinical practice and to treat them efficaciously.

BRIEF OVERVIEW OF THE MEDICAL CONDITION

Cancer is now the first cause of death in many Western countries (e.g., Canada, Western Europe)[1,2] and is second after cardiovascular disease in the United States and worldwide.[2] In 2012 there were 14 million new cases of cancer and 8.2 million cancer-related deaths worldwide.[3] The global incidence of cancer is expected to rise to 21 million within the next 20 years, while the cancer deaths are expected to reach 13 million.[4] Overall, one in six deaths is due to cancer.

Breast cancer is by far the most common malignancy in women (accounting for 25% of all cancers in 2012) in both more and less developed countries. In men, lung cancer is the most common type of cancer worldwide (representing 16.7% of all cancers in 2012), but prostate cancer is the most common in developed countries such as Canada, the United States, and Western Europe. The most common causes of cancer deaths are, in order, lung, liver, colorectal, stomach, and breast cancer.[4]

There are more than 200 cancer types, each with its own risk factors, treatment protocols, and prognosis. Cancers are multifactorial diseases and thus originate from a combination of factors including a genetic vulnerability and three types of external factors: physical (e.g., ultraviolet), chemical (e.g., cigarette smoking, asbestos), and biological carcinogens (e.g., infections such as the human papilloma virus). Alcohol use, unhealthy diet, physical inactivity/sedentary lifestyle, and hormonal factors are other significant risk factors for cancer.[4] The incidence of cancer increases significantly with aging, which explains in large part its augmentation worldwide.

The selection of cancer treatments depends heavily on cancer type and stage, as well as on other biological indicators (e.g., Oncotype DX test for breast cancer). It also varies as a function of patients' age and general health and functional status. A combination of treatments is often used when the cancer is in its early stage, with a curative intent. A treatment with a palliative intent is administered when the cancer is advanced (noncurable). Removal of the tumor by surgery is the primary treatment for several solid cancers (e.g., breast, gastrointestinal, head, and neck) whereas chemotherapy is the primary treatment for hematological cancers (e.g., leukemia, lymphoma). Chemotherapy is a systemic treatment that aims to kill cancer cells throughout the body. It is used before (i.e., neoadjuvant treatment) or after (i.e., adjuvant treatment) cancer surgery. Regimens are composed of a single or a combination of drugs, whose administration is repeated over several cycles (i.e., treatment followed by a period of rest), usually of 2–3 weeks, over a period of about 3–6 months. Radiation therapy is another treatment that can be used as a neoadjuvant or an adjuvant treatment. This localized treatment can also be offered alone (e.g., localized prostate cancer). It can be administered externally via linear accelerators or internally, for example, through the placement of radioactive seeds that are left in place permanently (i.e., brachytherapy). It is a local treatment, administered at the site of the tumor, to destroy the tumor and possibly remaining cancer cells surrounding the tumor after it has been surgically excised. Hormone therapy is another treatment that is used for hormone-dependent cancers such as breast and prostate cancer. It can consist of surgically removing hormone-producing glands or organs, of using radiation therapy to destroy or damage hormone-producing tissue and of using medications that block hormone secretion or their binding to hormone receptors (e.g., antiestrogens for breast cancer, antiandrogens for prostate cancer). Immunotherapy and targeted therapies are promising newer treatments that are increasingly investigated and used.

Oncological treatments have numerous side effects and these are likely to be multiplied or accentuated when a combination of treatments is used. Side effects of surgery depend on the tumor site and its size but almost always involve some levels of pain. Nausea, vomiting, alopecia, low white blood count, and fatigue are common side effects of chemotherapy. Some drugs (e.g., taxanes) are also likely to induce peripheral neuropathic pain. Fatigue is one of the most common side effects of radiation therapy. Skin problems are also frequent during and after this treatment. Both chemotherapy and hormone therapy administered in women are likely to induce premature menopause (and infertility) or aggravate preexisting menopausal symptoms such as hot flashes. Pain is another undesirable effect of hormone therapy. In men, androgen-deprivation therapy is likely to cause sexual dysfunction (e.g., loss of libido, erectile difficulties), hot flashes, osteoporosis, bone fractures, fatigue, and gynecomastia. Cognitive impairments are other possible side effects not only of chemotherapy (the so-called chemo-brain) but also of hormone therapy.

EPIDEMIOLOGY OF SLEEP DISORDERS IN CANCER

Sleep difficulties have been shown to be highly prevalent in cancer patients. When several symptoms are assessed, sleep disturbance is the first or at least among the first three most burdensome symptoms reported by patients.[5–8] Studies assessing the prevalence of specific sleep disorders [e.g., insomnia, obstructive sleep apnea (OSA)] in the context of cancer are lacking.[9] Indeed, previous studies have mainly investigated the presence of unspecific sleep difficulties, with assessment methods that do not distinguish across specific sleep disorders [e.g., self-report scales such as the *Pittsburgh Sleep Quality Index* (PSQI)]. Insomnia and OSA are the two specific sleep disorders that have been most frequently documented. The following will describe the available epidemiological evidence by specific sleep disorder, although it should be recognized that a differential diagnosis using a structured clinical interview or polysomnography (PSG) has generally not been done.

Insomnia

To the best of our knowledge, our study is the only one that has evaluated the prevalence of insomnia using a structured clinical interview and specific diagnostic criteria.[10,11] A total of 962 patients with mixed cancer sites participated in this longitudinal study and were recruited at their preoperative visit. Patients were excluded if they self-reported having received a diagnosis of or were currently being treated for a sleep disorder other than insomnia. Hence, no PSG was used to detect other sleep disorders. Participants were administered the *Insomnia Interview Schedule*[12] at baseline (before or just after their surgery with a curative intent), and 2, 6, 10, 16, 20, and 24 months later. An insomnia syndrome was diagnosed when patients met the following criteria: (1) subjective complaint of sleep difficulties; (2) sleep onset latency or wake after sleep onset greater than 30 minutes; (3) occurring at least three nights per week; (4) for a duration of 1 month or greater; and when they were associated with impaired daytime functioning or marked distress. Participants using a hypnotic medication three nights or more per week were also categorized in the insomnia syndrome group. Patients were categorized in the insomnia symptoms group when they had a complaint of sleep difficulties but did not meet the criteria for an insomnia syndrome or when they reported using a hypnotic medication one or two nights per week. The remaining patients were considered good sleepers.

Results showed that, overall, 59% of the patients had insomnia at baseline, including the 28% with an insomnia syndrome. This prevalence rate of insomnia syndrome is at least two to three times higher than in the general population.[13–15] Insomnia rates steadily declined over the course of the study but there were still 36% of the patients who had insomnia (21% with a syndrome; 15% with symptoms) at the 18-month evaluation. Women with breast (42%–69%) and gynecological cancer (33%–68%) showed higher prevalence rates of insomnia as compared to men with prostate cancer (25%–39%). During the course of the study, 14.4% of the participants had a first incidence of insomnia and 19.5% experienced a relapse of insomnia, for a total incidence rate of 31.8%. The general persistence rate was 50.7%, which was defined as having insomnia symptoms or an insomnia syndrome on two consecutive study time points, while the general remission rate was 45.8%. Among the patients who had an insomnia syndrome at baseline, 37.6% continued to meet the diagnostic criteria on each and every subsequent time point of the study, thus indicating that insomnia syndrome is a particularly enduring condition in this population.

Another large-scale longitudinal study ($N = 823$) assessed the prevalence of insomnia symptoms in patients with mixed cancer sites receiving chemotherapy.[16] Insomnia was assessed using the six insomnia items of the *Hamilton Depression Inventory*, a self-report scale. The authors were nonetheless able to categorize patients into the same three categories: insomnia syndrome, insomnia symptoms, and good sleepers. In this case, the insomnia syndrome was defined as follows: (1) difficulty falling asleep, difficulty staying asleep, and/or early morning awakenings of at least 30 minutes and for at least three nights a week, for 2 weeks. The impact of sleep difficulties on patients' functioning and distress could not be assessed, hence giving a more inclusive definition of insomnia, and the presence of other sleep disorders was not used as an exclusion criterion. Patients with some sleep difficulties but not meeting the frequency or duration criteria were categorized in the insomnia symptoms group, while patients with a score of 0 on all items were considered good sleepers. At day 7 of cycle 1 of chemotherapy, 36.6% of the patients had insomnia symptoms and 43.0% met the criteria for an insomnia syndrome, for a total prevalence of 79.6%. Women with breast cancer and patients treated for lung cancer showed the highest prevalence rates, while those with alimentary tract cancers had the lowest ones. Insomnia symptoms persisted at cycle 2 in 60% of the patients. A total of 34.6% of the patients developed insomnia symptoms or an insomnia syndrome at cycle 2 (incidence).

Only a few small-scale studies have reported information on PSG-assessed sleep architecture of cancer patients. In a preliminary report published in 1985 of 14 patients receiving radiation therapy for lung cancer, no difference was found between self-declared good ($n = 9$) versus poor sleepers ($n = 5$) on PSG indices of sleep latency, rapid eye movement (REM) latency, and percentage of REM time spent in Stage I versus Stage II sleep.[17] The only significant between-groups difference was on delta sleep with good sleepers having an average of 8.9 minutes compared to 5.5 minutes in poor sleepers. Later, a study of 3282 men and women with a variety of medical conditions revealed no difference on PSG sleep variables between patients with and without cancer.[18] A more recent prospective study ($N = 26$; 77% with breast cancer) found no significant effect of chemotherapy on sleep architecture and sleep continuity as assessed with PSG, with the exception of an increase of about 20 minutes in the total time spent asleep between the pre- and the postchemotherapy evaluations.[19] Overall, investigations have failed in consistently identifying particularities in sleep architecture of cancer patients.

Obstructive Sleep Apnea

The association between OSA and cancer has received an increased attention in recent years. Early cross-sectional studies conducted in small convenience samples found that OSA was more prevalent in cancer patients than what could be expected in the general population. Head and neck cancer is the most likely to lead to OSA because of the narrowing of the upper-airway or neural alterations produced by the tumor itself or its localized treatment. This partial obstruction may be functional during the day but collapses during sleep. Obesity would be less influential in this population.[20]

The first study by Friedman et al.[21] conducted in 24 patients successfully treated for a head and neck cancer revealed that 91.7% of them had OSA, of which 72.7% had clinically defined symptoms of OSA. This is much greater than the prevalence rate of OSA in the general population (approximately 9.0% of male and 4.0% of female middle-aged individuals). These results are consistent with those of another study of 17 patients with head and neck cancer which showed that 76.0% of them had OSA.[22] Interestingly, the mean size of the tumor was not related with the apnea-hypopnea index (AHI). There is some evidence suggesting that the type of tumor can influence the risk of developing OSA. In a study of 31 patients, those with a hypopharynx or larynx tumor and those with a transient tracheostomy most often had a pathological $AHI \geq 20$.[23] Cancer treatments such as surgery and radiation therapy also appear to play a significant role in increasing risks of OSA. The retrospective investigation by Qian et al.[24] found that patients who received surgery for a head and neck cancer had a significantly higher prevalence of moderate-to-severe OSA

(AHI $\geq$ 15) in the postoperative period (11 out of 15 patients; 73.3%), compared to a non-surgical group who received chemoradiation only (3 out of 9 patients; 33.3%). Surgery can increase OSA risk by altering pharyngeal musculature and restricting upper-airway compliance with bulky flap reconstruction.[20] Another study found that, while the severity of apnea and hypopnea events and snoring decreased in most patients treated with radiotherapy for a head and neck cancer, the AHI increased in 8 of the 18 patients.[25] The more recent prospective case series of 16 patients who underwent radiation therapy for head and neck cancer showed that 50% had OSA assessed objectively and marginally significant results suggested that the presence of OSA was associated with a shorter time interval since the completion of this treatment and a gastrostomy tube placement during radiation.[26] The effect of radiation therapy on OSA risk is believed to be multifactorial including chronic mucosal edema, decreased elasticity, and increased fibrosis, thus causing poor pharyngeal constriction and dysphagia.[20]

Rates of OSA have also been assessed in patients with other types of cancer, although to a much lesser extent. For instance, a recent study conducted in 160 women undergoing abdominal surgery for gynecological cancer revealed that 45.0% were obese and 50.0% had OSA assessed with portable sleep oximetry.[27] A small-scale study of 11 patients operated for an intracranial tumor found that 6 patients demonstrated signs of OSA and 1 had mixed obstructive and central sleep apnea.[28] After the surgery, the mean AHI significantly decreased from 23.3 to 8.1.

In sum, the prevalence of OSA appears to be higher in cancer patients than in the general population, especially in patients with head and neck cancer because of the direct effect of the tumor and its treatment on the upper-airway tract. However, large epidemiological studies with representative samples are needed to confirm these results. Moreover, the design of previous studies (e.g., cross-sectional, retrospective) precludes from concluding on the extent to which OSA is a consequence of cancer or a preexisting condition. Recent research efforts have even investigated whether OSA is a risk factor for cancer, a literature that will be discussed later in this chapter ("Consequences of Sleep Disorders in Cancer" section).

Periodic Limb Movement in Sleep/Restless Legs Syndrome

A few studies have evaluated rates of periodic limb movement in sleep (PLMS) and restless legs syndrome (RLS) in cancer patients. An early study found higher rates of PLMS in breast cancer patients (60%) than in patients with insomnia (25%) and healthy controls (22%).[29] More recently, Reinsel et al.[30] found that 12 of 26 breast cancer survivors (on average 54.0 months posttreatment) who underwent a sleep laboratory assessment had a PLMS index $\geq$ 15. Interestingly, women who had a PSQI score $\geq$ 10, which was used to indicate the presence of clinical insomnia, had a higher rate of PLMS symptoms compared to good sleepers. Although PSQI is not a specific measure of insomnia, this could indicate a high comorbidity between insomnia and PLMS in breast cancer survivors.

Scarce data also suggest that RLS is common in cancer patients. Of the 173 patients undergoing chemotherapy who were assessed in one study, 58.8% had sleep disturbances on the PSQI (score $\geq$ 5) and 20% screened positive for RLS, which is higher than what is expected in the general population.[31] A significant correlation was found between the two. Clearly, further research is needed to delineate the role of PLMS and RLS on the occurrence of sleep complaints and their evolution during the cancer care trajectory.

SLEEP DIFFICULTIES IN SPECIAL POPULATIONS

Advanced Cancer and Palliative Care

Available data suggest that sleep impairments are even more prevalent in the context of advanced cancer and palliative care with prevalence rates between 50% and 75%.[32–34] Again, authors have rarely assessed the prevalence of specific sleep disorders. An exception is the recent study by our group which was conducted among 55 community-dwelling cancer patients who had significant functioning limitations and were receiving palliative care.[35] Sections of the *Duke Structured Interview for Sleep Disorders*[36] pertaining to diagnoses of insomnia, hypersomnolence, OSA, RLS, PLMS, environmental sleep disorder, and circadian rhythm sleep–wake disorders were administered. Results indicated that 68.6% of the sample had at least one type of sleep–wake difficulty at a syndromal or a subsyndromal level. Fourteen (27.5%) patients met the diagnostic criteria for an insomnia disorder and 13 (25.5%) had a hypersomnolence disorder, the two most commonly encountered disorders. These findings suggest that many palliative care patients have sleep–wake difficulties occurring around the clock. The high prevalence of hypersomnolence in this population appears to be due in large part to the usage of pain medications. Among the patients having a hypersomnolence disorder, 10 (71.4%) reported that their excessive daytime sleepiness occurred or aggravated after initiating or changing dosage of their pain medication. Indeed, daytime sleepiness is a common side effect of opioids such as morphine. Recent data using PSG in 28 patients admitted to a palliative care service also suggest that opioids lead to respiratory disturbance, predominantly of an obstructive type, which may also contribute to the excessive daytime sleepiness.[37]

Survivors of Childhood Cancer

Although no large-scale epidemiological findings exist on the prevalence of specific sleep disorders, sleep–wake disturbances are also highly frequent in adult survivors of childhood cancer. This issue has particularly been investigated in survivors of brain cancer. In that population, sleep difficulties could be the direct consequence of brain injury from the tumor, hydrocephalus, brain surgery, and cranial radiation therapy, and through the indirect role of chemotherapy or complications of cancer (e.g., seizures, obesity).[38] High rates of OSA and excessive daytime sleepiness have been found. For instance, a survey of 153 children and adolescents at least 5 years' postdiagnosis and 2 years' posttreatment showed that one-third of adolescents and one-fifth of children reported excessive daytime sleepiness. This problem was associated with obesity. Another study of 31 patients referred for a PSG assessment found that the most common diagnosis was OSA (14 patients).[39] Again, the presence of sleep disorders was associated with obesity. Insomnia also appears to be common in this population. The study by Zhou et al.[40]

revealed that 26% of the sample had a sleep efficiency lower than 85% on the PSQI, which was used as an indicator of insomnia symptoms. Insomnia symptoms were more frequent in women although the difference with men was not significant. In another investigation, 78 adult survivors of childhood brain tumors were found to be 2.7 times more likely to take more than 30 minutes to fall asleep than a population-based comparison group.[41] However, no significant difference was found between these two groups on the mean PSQI total score. Factors associated with poor sleep quality were the female sex, radiation to the hypothalamic–pituitary axis, and obesity.

In summary, large-scale prospective studies have shown that insomnia is a highly prevalent and enduring condition in cancer patients. Although the high prevalence of insomnia is well established, future studies, using PSG or at least validated subjective instruments (e.g., *Duke Structured Interview for Sleep Disorders*), should better distinguish insomnia from other sleep disorders that may lead to an insomnia complaint. Findings on other sleep disorders are scarcer but available data suggest that cancer patients, at least with certain types of malignancy and when at an advanced stage, also display high rates of OSA and hypersomnolence. Although larger epidemiological studies are needed in order to assess the presence of these disorders in more representative study samples, clinicians working in oncology should remain alert to the possible presence of these disorders in their patients.

ETIOLOGY OF SLEEP DISORDERS IN CANCER

As in the general population, the etiology of sleep disorders in cancer is most commonly multifactorial.

Insomnia

Spielman's model, also known as the 3P model,[42] is extremely helpful to explain the development and persistence of cancer-related insomnia. This model delineates three types of risk factors for insomnia: predisposing, precipitating, and perpetuating factors. *Predisposing factors* are those that increase one's vulnerability to experience insomnia in his/her lifetime. They include demographic and psychological variables. As in the general population, women are at a higher risk to experience insomnia during their cancer care trajectory.[10] However, while older age is associated with increased risk for insomnia in the general population, it is younger age that is related to a greater risk in the cancer context.[16,43–45] This could be due to higher levels of psychological distress typically experienced in younger cancer patients and to the presence of more severe sleep-interfering somatic symptoms such as nocturnal hot flashes. There is also evidence suggesting that a hyperarousability trait, that is the tendency to be easily physiologically and cognitively aroused, is a predisposing factor of cancer-related insomnia.[10] Other possible predisposing factors include a personal and family history of insomnia and of psychological disorders (e.g., anxiety and depression disorders).

Precipitating factors are variables that trigger the onset of insomnia, which includes not only the successive stressful events that characterize the cancer experience (e.g., cancer diagnosis, initiation of each cancer treatment, hospitalization, medical follow-ups, cancer recurrence, terminal disease, and end of life) but also many somatic symptoms induced by the cancer itself or its treatment. For instance, as already mentioned, chemotherapy and hormone therapy (e.g., tamoxifen, arimidex) induce a premature menopause in younger women or aggravation of preexisting menopausal symptoms through deficiency in estrogen levels. There is now strong evidence showing a relationship between nocturnal hot flashes and sleep disturbances in breast cancer.[46–48] Similarly, hormone therapy (e.g., goserelin) or orchiectomy in men with prostate cancer engenders nocturnal hot flashes that may impair sleep.[49] Radiation therapy is also associated with increased sleep difficulties, but to a lesser extent than hormone therapy.[50] Other somatic symptoms that are likely to interfere with sleep are pain, nocturia (for instance, following surgical or radiation treatment for prostate cancer), dyspnea, nausea, and vomiting.[51–53] Interestingly, although fatigue is most commonly perceived as a consequence of sleep disturbances, there is some evidence showing that it can also be a risk factor for insomnia.[54] This could be attributable to changes in sleep routines (e.g., more daytime napping) that fatigued patients tend to introduce, which desynchronize their circadian rhythms in the long term.

Finally, the *perpetuating factors* are those that contribute to maintaining insomnia over time. While the persistence of somatic symptoms that played a role in precipitating insomnia initially (e.g., chronic pain) may also contribute in the sustainment of insomnia over time, behavioral and cognitive factors are believed to be the most influential perpetuating factors. These include maladaptive sleep behaviors that the person developed in order to get some rest, alleviate fatigue, and catch up on sleep loss such as spending too much time in bed and napping. While these behaviors appear to be a logical response to fatigue and sleep disturbances and are often reinforced by health-care providers and caregivers, they eventually desynchronize the sleep–wake cycle and alter the association (conditioning) that should exist between the bedroom environment and sleep. Cognitive factors such as dysfunctional beliefs about sleep and monitoring of sleep-monitoring threats (i.e., the tendency to selectively attend to salient internal or external threat cues related to sleep) also play a role in maintaining the problem over time.[10] Cancer patients tend to entertain the same dysfunctional beliefs about sleep as healthy individuals (e.g., "I need 8 hours of sleep to function well during the day") but they also have some specific ones especially with regard to the possible impact of their sleep difficulties on cancer progression (e.g., "If I don't sleep well, my cancer will come back"). Together, such beliefs significantly increase performance anxiety and preoccupations related to sleep during daytime and as bedtime approaches ("I really need to sleep tonight"), thus making the problem even worse.

Obstructive Sleep Apnea

Except for patients with head and neck cancer in whom specific risk factors for OSA have been identified (see earlier), risk factors for cancer-related OSA appear to be the same as in the general population (e.g., male sex, aging, obesity, craniofacial and upper-airway structure, and medical comorbidity). Weight excess appears to be particularly influential. Several patients are overweight or obese at cancer diagnosis. Obesity is even increasingly recognized as a significant risk factor for several types of cancer (e.g., breast, ovary, colorectal, and gastric).[55] Moreover, some cancer treatments may induce weight gain. For instance, a recent meta-analysis of 25 studies indicated that women receiving chemotherapy for breast cancer gain 2.7 kg on average.[56] Hormone therapy does not appear to lead to significant weight gain.[57] On the other hand, the prevalence of OSA increases during menopause in healthy women. Again, weight gain as well as changes in the distribution of adipose tissue (accumulating in the upper part of the body and visceral fat) seem to explain this association but hormonal mechanisms are also possible.[58,59]

Periodic Limb Movement in Sleep/Restless Legs Syndrome

To our knowledge, risk factors of PLMS and RLS have not been studied specifically in cancer patients. A few studies have investigated the relationships between the menopausal transition and PLMS incidence with inconclusive results. It has recently been concluded that the increased prevalence of PLMS (and RLS) after menopause is more related to aging.[58]

In summary, the etiology of sleep disorders in cancer patients is complex. While the etiology of cancer-related insomnia is getting clearer, more research is needed to identify risk factors of OSA and PLMS that are specific to cancer.

CONSEQUENCES OF SLEEP DISORDERS IN CANCER

Insomnia

Although often overlooked in clinical practice, insomnia is not a trivial problem. Untreated insomnia frequently becomes chronic[60] and, as such, it is very likely to be associated with several negative consequences on patients' daily functioning and quality of life.[61–63] Specific consequences of insomnia in the cancer context have received little attention, but they are likely to be the same as, if not worse than, those documented in primary insomnia. Fatigue is the number one complaint of patients with chronic insomnia,[64] a symptom that is commonly associated with cancer treatment, even in the absence of sleep difficulties. Other possible consequences include a higher risk of subsequently developing depressive and anxiety disorders,[65,66] and a greater incidence of several physical conditions such as hypertension, pain, and permanent work disability.[67–70]

In the specific context of cancer, there is some evidence linking insomnia to negative medical outcomes. We recently found that patients with an insomnia syndrome, diagnosed with a structured interview, were at a significantly higher risk of developing infectious episodes during the cancer care trajectory in general[71] and after chemotherapy for breast or gynecological cancer more specifically.[72] Poor sleep has also been hypothesized to be associated with a worse prognosis. A large prospective epidemiological study revealed that, among women who developed breast cancer, those who had regular sleep difficulties (as assessed with a single question) had a 49% increased risk of all-cause mortality during the study's follow-up as compared to those who reported little to none. Also, women with a sleep duration ≥9 hours postdiagnosis were more at risk of breast cancer and all-cause mortality compared to those who slept 8 hours.[73] Results of another recent epidemiological study suggested that short sleep duration (6 hours) and worse sleep quality were associated with the development of more aggressive breast tumors.[74] There is also evidence showing that short or long sleep duration and poor sleep assessed subjectively or with actigraphy are associated with higher mortality and progression and poor treatment response in patients with advanced cancer.[75–77] These links between poor sleep and cancer could be due to impaired immune functioning, decreased melatonin levels, which may have an antiproliferative role, and impaired circadian rhythms, which may be involved in DNA repair.[76,78] However, these findings need to be interpreted with great caution as sleep is typically assessed at only one point in time in these studies and, therefore, such measures do not take into account changes that occur over time on sleep duration and quality, as well as the persistence of sleep difficulties. In addition, crude measures of sleep quality (single items) have often been used. It is also important to keep in mind that short sleep duration is not a typical feature of insomnia which is more characterized by low sleep efficiency. Hence, the degree to which these findings apply to (chronic) insomnia or other sleep disorders is unknown.

Obstructive Sleep Apnea

Negative consequences of OSA in the general population are well known and include a higher risk of all-cause mortality, especially of cardiovascular disease.[79] There is mounting literature suggesting that OSA may also be associated with increased cancer incidence and mortality as well. However, early studies are characterized by many weaknesses including the fact that they were not originally designed to examine this question and had a retrospective design, the inclusion of highly selected patients (referred for sleep studies), the lack of objective assessment of OSA in control patients, a small number of cancer cases, and the failure to control for other cancer risk factors.[80,81] Two meta-analyses of this literature yielded contradictory results: one indicated that sleep disordered breathing/OSA was related to a significantly increased risk for cancer, even after adjusting for some cancer risk factors,[82] while the other one found no such effect.[83] The recent epidemiological study by Gozal et al.[84] using a large nationally representative database (approximately 5.6 million individuals) is worth describing. It revealed that OSA was not a significant risk factor of cancer when all cancer diagnoses were combined; it was associated with an increased risk for pancreatic and kidney cancer but with a lower risk for colorectal, breast, and prostate cancer. Moreover, among patients with cancer, the presence of OSA was not associated with an increased risk of metastatic disease or mortality. This is consistent with results of another nationwide registry study conducted in Sweden which revealed no association of OSA with cancer mortality.[85] Further research is needed to confirm the OSA–cancer relationships and to better understand the mechanisms through which OSA could augment cancer risk. These include inflammation and sympathetic nervous system activity.[81]

Periodic Limb Movement in Sleep/Restless Legs Syndrome

To our knowledge, no study has specifically assessed the impact of PLMS or RLS specifically in cancer.

ASSESSMENT OF SLEEP DIFFICULTIES IN CANCER

Screening

Sleep disturbances are typically underdiagnosed and undertreated in cancer care; hence, actively screening for these problems on a routine basis is an essential first step to ensure that they are appropriately dealt with. Cancer centers around the world have increasingly implemented a routine screening procedure of psychological distress in their clinics. In Canada, which has taken a leading role in implementing such screening, the *Edmonton Symptom Assessment System* (ESAS)[86] and the *Canadian Problem Checklist* (CPC)[87,88] are used along with the *Distress Thermometer*[89,90] to screen for psychological distress and other psychological, physical, practical, information, and spiritual issues that may affect cancer patients. These tools are administered at several critical times during patients' care trajectory.

The original version of the ESAS included no specific insomnia item but patients could use the "Other Problem" item to report their sleep difficulties. The ESAS also contains an item assessing drowsiness rated on a "0" to "10" scale; however, although this symptom can sometimes be a consequence of insomnia, it is more specific to other sleep disorders (e.g., OSA) and could also be due to some medications (e.g., opioids). With regard to the CPC, it contains a sleep item, which is a box that can be checked if the problem is present. Unfortunately, a study conducted by our team showed that the ESAS-other problem item was ineffective to detect clinical levels of insomnia when using a score of eight or greater on the *Insomnia Severity Index* (ISI) as the standard criterion, since none of the patients used this item to report their sleep difficulties. Moreover, neither the CPC-sleep nor the ESAS-drowsiness item provided an optimal screen for clinical insomnia when used alone. An alternative that was found to be valuable was to use a positive answer on the CPC-sleep item OR a score of 2 or greater on the ESAS-drowsiness item, a combination of criteria that yielded a sensitivity of 84.2% and a specificity of 69.7% to detect clinical insomnia. Another option is to add a specific sleep item to the ESAS, rated from 0 ("no trouble sleeping") to 10 ("worst sleep imaginable"), as was previously done in the context of advanced cancer.[91] We recently showed that, again using a score of 8 or greater on the ISI as the standard criterion for clinical insomnia, a score of 2 or higher on the ESAS-sleep item was the one that showed the best screening indices (sensitivity: 87.1%; specificity: 76.1%; positive predictive value: 72.7%; negative predictive value: 89.0%) (Savard et al., in preparation).

Clinical Interview

A positive screening of sleep disturbance should ideally be followed by a more in-depth evaluation which will often include a clinical interview. This will allow clinicians to assess aspects such as the exact nature of sleep disturbance, its duration, its trigger factors, the person's sleep history, as well as the impact cancer had on sleep (e.g., preexisting insomnia that aggravated with cancer). The *Insomnia Interview Schedule*[12] could be very useful for that purpose. A semistructured interview such as the *Duke Structured Interview for Sleep Disorders*[36] could also be used to assess the diagnostic criteria of the main sleep disorders for differential diagnosis.

Sleep Diary

A daily sleep diary such as the *Consensus Sleep Diary*[92] is an extremely useful tool to assess the daily variations in various sleep parameters assessed subjectively (e.g., sleep onset latency, wake after sleep onset, and sleep efficiency). It is usually filled out every day for 1 or 2 weeks to establish a baseline before the introduction of an intervention and at posttreatment to assess its effects. It is also helpful—if not necessary—to have it completed by the patient during cognitive–behavioral therapy to monitor progress and to guide the application of sleep restriction (see description later).

Questionnaires

No questionnaire has been developed to assess sleep disturbances specifically in cancer patients. The ISI,[12,93] a measure of insomnia severity, and the PSQI,[94] a more general measure of sleep problems, are the most commonly used self-report scales in the cancer literature. In cancer, a score of 8 or greater on the ISI has been found to indicate the presence of clinical levels of insomnia, while of score of 15 or greater is indicative of an insomnia disorder.[95] For the PSQI, a psychometric study suggested that a score of 8 or greater, rather than the usual cutoff score of 5, was the optimal cutoff score for detecting sleep problems in cancer patients,[96] but this recommendation has not been followed consistently across studies.

Polysomnography

An all-night PSG assessment is required when the clinician suspects the presence of a sleep disorder other than insomnia such as OSA and PLMS.

Actigraphy

Actigraphy has increasingly been used in research over the past years to objectively quantify sleep parameters. It is especially useful to characterize the 24-hour rest–activity patterns.

TREATMENT OF SLEEP DIFFICULTIES IN CANCER

Treatment of Insomnia

Pharmacotherapy is, by far, the most commonly used treatment by cancer patients to deal with insomnia. It is also often the first treatment initiated, which is not consistent with current clinical guidelines recommending cognitive–behavioral therapy for insomnia (CBT-I) as the first-line intervention for chronic insomnia in cancer.[97,98] Yet, while the efficacy and safety of hypnotics have not been investigated specifically in cancer patients, the efficacy of CBT-I is now well established in this population.

Pharmacotherapy

A recent epidemiological study conducted in France revealed that 52% of patients received a psychotropic drug prescription within the first year following their cancer diagnosis, with anxiolytics and hypnotics being the most frequently prescribed classes of medications.[99] Hypnotics were prescribed to 25.1% of cancer patients during the same timeframe, compared to only 9.9% in matched individuals with no cancer, a rate that decreased to only 17.4% during the second year postdiagnosis. Another study conducted in 1984 among Canadian cancer survivors found that 41% of them had received a prescription for hypnotic medication since their cancer diagnosis (within the past 10 years), with 23% who were still using one.[100] The mean duration of use in current users was 58.1 months, which considerably exceeds the limit duration of 4–6 weeks recommended by the National Institutes of Health.[101] A prolonged nightly usage of hypnotic medications should be avoided as much as possible given the risks for the development of tolerance and of physical and psychological dependence. However, a short-term or occasional usage of hypnotics may well be appropriate when patients have to cope with acute stress (e.g., upcoming surgery and follow-up appointment with the oncologist, corticosteroid-induced insomnia during chemotherapy).

There is a large variety of over-the-counter products that may be tried by cancer patients looking for alternatives to prescribed medications. Among them, melatonin is receiving a growing interest. A recent 2-week double-blind clinical trial of 50 patients with various types of cancer (stages I–IV) and sleep complaints for more than a month showed significantly larger improvements of insomnia symptoms in patients receiving melatonin (3 mg) compared to the placebo group.[102] No adverse effect was reported. However, these promising preliminary results need to be replicated in larger studies with longer follow-ups. More generally, the efficacy and safety of natural products need to be ascertained.

Cognitive–Behavioral Therapy

A fairly large body of evidence now supports the efficacy of CBT-I in cancer patients. A meta-analysis of clinical trials conducted up to 2014 and including 8 studies (752 participants) revealed significantly greater improvements of various sleep parameters in CBT-I patients as compared to controls.[103] Relative to control patients, posttreatment effects of CBT-I patients were of a medium-to-large magnitude on sleep onset latency, wake after sleep onset, and self-reported insomnia severity. These effects were found to be well maintained at 6-month follow-up. More commonly, the standard face-to-face format was used but there are also recent findings supporting the utility of self-administered formats (e.g., video and web based).[11,104–107] This is of great importance given the extremely limited access patients have to CBT-I in most cancer settings.

Despite the unique features of cancer-related insomnia (e.g., strong association with somatic symptoms), more commonly, it is the standard CBT-I protocol that has been tested in previous clinical trials. Given that treatment effects appear to be of a similar magnitude as in healthy individuals, it suggests that this protocol does not need much adaptation to be efficacious in this population. This is coherent with CBT-I's rationale to target the behavioral and cognitive factors that are believed to be the most contributive to the maintenance of insomnia over time, rather than the factors that triggered it initially (e.g., hot flashes, pain). It would nonetheless be interesting to investigate whether treatment effects can be enhanced when targeting cancer-related factors in the intervention.

There are still some clinical particularities that the clinicians may want to take into account. First, it may be difficult for cancer patients, especially when in the midst of their treatment, to totally avoid napping given the fatigue they experience. In our work, we recommend patients who need to nap to sleep no longer than 60 minutes and no later than 3:00 p.m. The goal is to limit the negative impact of napping on the quality and continuity of the subsequent night. We also provide psychoeducation on the fact that poor sleep is not the only contributor to cancer-related fatigue and, as such, sleeping is not the sole strategy that can be used to alleviate it. In brief, it is possible to get some rest without sleeping (e.g., by doing a relaxing activity).

Sleep restriction may also be challenging for many patients, again given their cancer-related fatigue. The first step of sleep restriction is to encourage patients to reduce their time spent in bed to their actual sleep duration for a week (ideally based on information from a daily sleep diary; e.g., not more than 6 hours spent in bed if sleep duration is 6 hours/night on average). As their sleep efficiency improves over the subsequent weeks, their time in bed is increased until the optimal duration is found. As would be done in resistant patients with no medical comorbidity, the weekly sleep window may be relaxed and negotiated with cancer patients to a point where they feel confident to adhere to it, as a greater adherence will eventually result in increased beneficial effects.[108]

As already stated, there is one specific erroneous belief that cancer patients commonly entertain about their insomnia that warrants further discussion, that is the belief that insomnia will increase their risk of cancer recurrence. It is frequent for patients with insomnia in general to fear the impact their insomnia may have on their health but, in cancer patients, this anxiety is greatly enhanced. Entertaining this kind of cognition throughout the day and as bedtime approaches will inevitably accentuate performance anxiety to sleep, which will make the sleep problem even worse. In our work, we challenge the validity of this belief using Socratic questioning (e.g., "Is there any evidence that this belief is true/untrue?"; "What are the most realistic consequences of insomnia?"). This helps patients understand that cancer is a multifactorial disease, that insomnia alone is very unlikely to lead to a cancer recurrence, and that worrying about the possible consequences of insomnia and catastrophizing it only aggravate sleep difficulties.

Treating patients who are at an advanced stage of the disease, particularly when bedridden, poses another challenge for clinicians using CBT-I.[109] For these patients, some adaptations to the stimulus control instructions will need to be done. For instance, patients with mobility issues will be instructed, when awake at night, to sit in a chair near the bed if possible or to sit in their bed rather than getting out of the bedroom. Recently, our team developed a cognitive–behavioral and environmental intervention for sleep–wake difficulties in palliative care patients which also emphasizes the importance for patients to remain as much exposed as possible to natural daylight (spend time outside if possible, remain in the brightest room of their home, open the curtains), while limiting their exposure to light and noise during the night.[109] As mentioned earlier, patients with advanced cancer receiving palliative care frequently experience excessive daytime sleepiness,[35] a problem that may be counteracted to some extent by increasing exposure to daylight. A feasibility study of this adapted protocol revealed promising findings for patients with insomnia symptoms but these need to be replicated in a larger and controlled study.[109]

Mindfulness-Based Stress Reduction

Mindfulness-based stress reduction (MBSR) promotes the progressive acquisition of mindfulness that is the ability to be present in the moment, with acceptance and nonjudgmental awareness. Over recent years, MBSR approaches have significantly gained in popularity and have been promoted as an effective alternative to treat various psychological conditions associated with cancer. However, the available evidence for treating cancer-related insomnia specifically is mixed. While early uncontrolled studies found encouraging results,[110,111] more recent randomized controlled trials (RCTs) revealed either no effect of MBSR on subjective sleep estimates (PSQI; sleep diary) at posttreatment[112] or no sustainment of sleep quality improvements at a 12-month follow-up.[113] A critical point when assessing the efficacy of alternative interventions is to demonstrate that they are as efficacious as more established treatments such as CBT-I. In a noninferiority RCT conducted in patients with mixed cancer sites, Garland, Carlson, Stephens, Antle, Samuels, and Campbell[114] observed that a mindfulness-based cancer recovery (MBCR) intervention (an adaptation of MBSR) was significantly inferior to CBT-I in reducing insomnia severity at posttreatment but was not significantly inferior at the 3-month follow-up. Interestingly, the attrition rate was much larger in the MBCR group. The authors rightly concluded that CBT-I remains the treatment of choice for cancer-related insomnia because of its more rapid and durable effects.

Exercise

The numerous benefits of regular exercise in oncology are well recognized (e.g., improved functioning and fitness, reduction of cancer treatment side effects, improved quality of life), but the question as to whether exercise interventions can treat efficaciously cancer-related insomnia remains unanswered at this time. A number of RCTs have investigated the effects of exercise programs on sleep, but sleep was rarely the main outcome measure and, most importantly, none was conducted in patients with clinical levels of insomnia. Another challenge when interpreting the findings is the lack of consistency across studies in terms of type of exercise, number of sessions/weeks, and the type of supervision provided. This body of literature (21 studies, including 17 RCTs) was examined in a recent systematic review and meta-analysis.[115] Although the qualitative review revealed a positive effect of exercise interventions on sleep in nearly half of the studies (48%), the meta-analysis of RCTs showed no significant superiority of these interventions relative to controls on subjective (PSQI and *General Sleep Disturbance Scale*) or objective (actigraphy-recorded sleep onset latency and sleep efficiency) sleep variables. Using a non-inferiority research design, our research team recently compared the efficacy of a 6-week home-based exercise program to a 6-week video-based CBT-I in 41 patients with clinical levels of insomnia.[116] Results indicated that both interventions produced significant sleep improvements at posttreatment. The exercise intervention was statistically inferior to CBT-I in diminishing ISI scores at posttreatment but not inferior at 3- and 6-month follow-ups. Again, it was concluded that CBT-I remains the treatment of choice for cancer-related insomnia. However, more comparative research is needed given some methodological particularities of this study (e.g., small sample, usage of two home-based interventions) and challenges that were encountered (e.g., minimal difference between exercise and CBT-I groups on activity level during treatment, thus suggesting a contamination effect).

Bright Light Therapy

There is some evidence, mostly from small-scale RCTs, supporting the usefulness of bright light therapy to improve cancer symptoms that are commonly related to sleep disturbances such as fatigue[117–119] and depression.[120] For sleep disturbances per se, the available evidence is sparse. To our knowledge, only one recent preliminary RCT conducted in 44 patients with clinical levels of fatigue has examined this question.[121] Compared to a dim-red light condition, a 30-minute morning exposure to bright white light for 4 weeks was associated with a greater increase in the proportion of patients with a sleep efficiency of 85 or greater as assessed with actigraphy. However, none of the subjective or other objective sleep parameters differed across groups. Another study of breast cancer patients ($N = 39$) found that a morning administration of bright light therapy protected from sleep−wake desynchronization during chemotherapy.[122] Although this assumption warrants investigation, such effects are of a particular interest for patients with significant circadian disruptions such as patients receiving palliative care.[123] More research is needed to assess the efficacy of this type of intervention on sleep impairments in general in oncology, as well as in patients with advanced malignancies specifically.

Treatment of Other Sleep Disorders

There is a paucity of empirical data supporting the efficacy of treatments for sleep disorders other than insomnia in the context of cancer. Until it is proven otherwise, empirically based treatments for sleep disorders in general are assumed to be appropriate courses of action in cancer patients [e.g., continuous positive airway pressure (CPAP) for OSA]. Some particularities are nonetheless possible. In patients with head and neck cancer, it has been shown that adherence to CPAP treatment was a significant issue, which could of course translate into decreased efficacy.[20] This could be due to aggravation of postradiation mouth dryness, a symptom that can be alleviated with humidification and fluid intake.[20] With regard to efficacy, a case series of 14 patients with OSA suggested beneficial effects associated with CPAP therapy and/or modafinil.[124] Clearly, these preliminary results need replication.

CLINICAL VIGNETTE

The patient is a 45-year-old woman who has been suffering from occasional insomnia for many years. After she was diagnosed with breast cancer, 1 year ago, her insomnia became more frequent and interfered more significantly with her daily functioning. At the interview, she describes that her sleep first deteriorated around the time of the diagnosis because of the anxiety she experienced about cancer in general and the upcoming surgery. At that time, she had a lot of difficulties falling asleep because of all of her worries ("What will happen to me?"; "Am I going to die?"). Later, after beginning her chemotherapy, she started having frequent and intense hot flashes throughout the 24-hour period and which awakened her in the middle of the night. She did not consult at that time because she thought that her insomnia would go away by itself after the completion of her cancer treatment, but this had not been the case. In the sleep diary, it is observed that the patient goes to bed early (she says that she does so to make sure to rest and get enough sleep), takes on average 60 minutes to fall asleep, and is awakened two times/night for a total duration of around 60 minutes. On average, out of 480 minutes (8 hours) she spends in bed, she sleeps 360 minutes (6 hours) for a sleep efficiency of only 75%. She also naps approximately 45 minutes in midafternoon. She complains of severe fatigue, especially since her chemotherapy, but has no excessive daytime sleepiness and denies heavy snoring and having abnormal sensations and movements in her legs. A diagnosis of chronic insomnia disorder is established.

A standard CBT-I is offered to the patient, combining sleep restriction, stimulus control, cognitive restructuring, and sleep hygiene. Because the patient still experiences a great deal of fatigue, she is extremely reluctant to restrict her time in bed to her actual total sleep time of 6 hours and to stop napping. After a discussion with the patient about aspects that would make sleep restriction easier for her (e.g., plan a social activity in the morning), the therapist and the patient collaboratively establish the sleep window at 6.5 hours for the first week. The patient decides to delay her bedtime to 12:30 a.m. and to wake up at 7:00 a.m. For naps, the therapist recommends limiting them as much as possible but, when the patient feels that a nap is necessary, it should be taken before 3:00 p.m. and for a duration shorter than 60 minutes. At the next session, the patient's sleep diary reveals no clear improvement. The patient even has the impression that her sleep has deteriorated given the narrowing of the time she could spend in bed. During the discussion, it becomes clear that the patient has inconsistently adhered to some behavioral strategies including the instruction to get out of bed when awake and to get up at the same time every morning even on weekends and regardless of how much sleep she was able to get. The rationale of these strategies is further emphasized (i.e., to reinforce the association between the bedroom and sleep, to resynchronize the biological clock). To do so, the therapist empathically agrees with the patient that these strategies are difficult to apply but stresses their importance. The therapist explains that like in a recipe, all ingredients of CBT-I are needed to be successful and that selecting only the ones that seem easiest (e.g., sleep hygiene) will not make much of a change. Practical tips are also discussed to help the patient get out of bed when awake (e.g., leave a blanket on the couch) and to get up at the same time during weekends (e.g., use an alarm, plan an activity with a friend).

At the next session, the patients' sleep diary reveals an important improvement of sleep efficiency. Given that it now exceeds 85%, the sleep window is extended to 7 hours for the upcoming week. The patient decides to go to bed 30 minutes earlier given the severe sleepiness she has experienced, particularly in the hour preceding bedtime. During the course of the program, the sleep window is further extended until it is proven that extending it further leads to a sleep deterioration.

Another critical aspect of the program is to help the patient challenge the validity of her belief that not sleeping well would increase her risk of having a cancer recurrence. Socratic questioning is used as described above. One particularity here is that the patient had read on the Internet that it was proven that short sleep and insomnia are associated with immunosuppression; insomnia could therefore increase the risk of cancer recurrence. To address that, the therapist informs the patient that it is true that insomnia and poor sleep have been found to be associated with some immune alterations and an increased risk for infections (e.g., common cold). The therapist also explains that, however, it is still unknown whether immune changes that are associated with poor sleep are of a nature, duration, and intensity to really have an impact on a complex, multifactorial, disease such as cancer. Conversely, the negative impact that worrying about the possible consequences of insomnia has on her sleep is clear. The idea here is not to deny that insomnia has some negative impact but to help the patient focus on what helps her to sleep better.

AREAS FOR FUTURE RESEARCH

Several areas for future research were identified throughout this chapter. Overall, the epidemiology, etiology, consequences, and treatment of sleep disorders other than insomnia need to be further investigated in cancer. For insomnia, the gold standard is CBT-I and it is now a priority to make it more accessible to patients through the utilization, for instance, of self-administered interventions and stepped care models.[125] Implementation studies are therefore needed to investigate whether it is feasible to integrate this type of intervention in the real world. Other nonpharmacological interventions are available to reduce sleep disturbances (e.g., MBSR, exercise, bright light therapy). Although preliminary data are available to support their usefulness, larger and more rigorous studies are needed to conclude about their efficacy, especially comparative trials with CBT-I.

CONCLUSION

Sleep disturbances are highly prevalent in cancer patients. When left untreated, which is the rule rather than the exception, sleep difficulties tend to become chronic and to bring with them many negative consequences. Clinicians need to better screen and assess these difficulties in their patients, in order to offer them an appropriate intervention. For sleep disorders other than insomnia, the same treatments as in the general population should be used (e.g., CPAP for OSA), although their efficacy has very rarely been tested specifically in that context. For insomnia, the efficacy of CBT-I is now well established in this specific population. Despite this progress, insomnia remains largely overlooked in clinical practice and, when a treatment is initiated, it is often a pharmacological

194   8. CANCER

intervention that is offered with all the limitations that characterize this approach. The accessibility to CBT-I is an important issue but self-administered formats have been developed and found to be efficacious in clinical trials. Such approaches now need to be implemented in clinical care. Other nonpharmacological alternatives can be used (e.g., MBSR, exercise, bright light therapy) although their efficacy and applications need to be confirmed.

KEY PRACTICE POINTS

Assessment of Sleep Disturbances in Patients With Cancer

- Cancer patients underreport their sleep difficulties.
- Screening is a critical first step to better manage sleep difficulties in this population.
- Simple one-item tools can be used for screening, followed by a short clinical interview and other self-report scales to determine the nature, duration, severity, and contributing factors of patients' sleep impairments, as well as their antecedents.
- Referral to a specialized sleep disorder clinic is needed for PSG assessment when there is suspicion of a sleep disorder other than insomnia.

Treatment of Sleep Disturbances in Patients With Cancer

- Although efficacy data are lacking and unless otherwise proven, standard treatments should be used for sleep disorders other than insomnia (e.g., CPAP for OSA).
- A prolonged nightly usage of hypnotic medications should be avoided (no longer than 4–6 weeks).
- Occasional or short-term usage of hypnotics may be appropriate to better cope with acute stressors (e.g., upcoming surgery).
- CBT-I should be the first-line treatment for cancer-related insomnia, especially when chronic.
- Self-administered forms of CBT-I are effective alternatives to standard face-to-face CBT-I.
- The standard CBT-I protocol is efficacious in cancer although there are some slight adaptations that may need to be made to enhance patients' acceptability and adherence, as well as treatment effects.
- Some other types of intervention (e.g., exercise, MBSR, bright light therapy) may be attempted although additional research is warranted to establish their efficacy for clinical insomnia and their applications.

References

1. Canadian Cancer Society. Canadian cancer statistics. <http://www.cancer.ca>; 2018.
2. Townsend N, Wilson L, Bhatnagar P, et al. Cardiovascular disease in Europe: epidemiological update 2016. *Eur Heart J.* 2016;37(42):3232–3245.
3. Ferlay J, Soerjomataram I, Dikshit R, et al. Cancer incidence and mortality worldwide: sources, methods and major patterns in GLOBOCAN 2012. *Int J Cancer.* 2015;136(5):E359–E386.

REFERENCES

4. World Health Organization. Cancer. <http://www.who.int/en/news-room/fact-sheets/detail/cancer>; 2018.

5. Mao JJ, Armstrong K, Bowman MA, et al. Symptom burden among cancer survivors: impact of age and comorbidity. *J. Am Board Fam Med*. 2007;20(5):434−443.

6. Hong F, Blonquist TM, Halpenny B, et al. Patient-reported symptom distress, and most bothersome issues, before and during cancer treatment. *Patient Relat Outcome Meas*. 2016;7:127−135.

7. Cleeland CS, Zhao F, Chang VT, et al. The symptom burden of cancer: evidence for a core set of cancer-related and treatment-related symptoms from the Eastern Cooperative Oncology Group Symptom Outcomes and Practice Patterns study. *Cancer*. 2013;119(24):4333−4340.

8. Ataseven B, Findte J, Harter P, et al. Change of patient perceptions of chemotherapy side effects in breast and ovarian cancer patients. *Annu Conf Eur Soc Med Oncol*. 2017;.

9. ESMO. *Annals of Oncology*. 2017;28(Suppl. 5):v605−v649.

10. Savard J, Villa J, Ivers H, et al. Prevalence, natural course, and risk factors of insomnia comorbid with cancer over a 2-month period. *J Clin Oncol*. 2009;27(31):5233−5239.

11. Savard J, Villa J, Simard S, et al. Feasibility of a self-help treatment for insomnia comorbid with cancer. *Psychooncology*. 2011;20:1013−1019.

12. Morin CM. *Insomnia: Psychological Assessment and Management*. New York: The Guilford Press; 1993.

13. Mai E, Buysse DJ. Insomnia: prevalence, impact, pathogenesis, differential diagnosis, and evaluation. *Sleep Med Clin*. 2008;3(2):167−174.

14. Ohayon MM. Prevalence of DSM-IV diagnostic criteria of insomnia: distinguishing insomnia related to mental disorders from sleep disorders. *J Psychol Res*. 1997;31(3):333−346.

15. Ohayon MM. Prevalence and comorbidity of sleep disorders in general population. *Rev Prat*. 2007;57 (14):1521−1528.

16. Palesh OG, Roscoe JA, Mustian KM, et al. Prevalence, demographics, and psychological associations of sleep disruption in patients with cancer: University of Rochester Cancer Center−Community Clinical Oncology Program. *J Clin Oncol*. 2010;28(2):292−298.

17. Silberfarb PM, Hauri PJ, Oxman TE, et al. Insomnia in cancer patients. *Soc Sci Med*. 1985;20(8):849−850.

18. Budhiraja R, Roth T, Hudgel DW, et al. Prevalence and polysomnographic correlates of insomnia comorbid with medical disorders. *Sleep*. 2011;34(7):859−867.

19. Roscoe JA, Perlis ML, Pigeon WR, et al. Few changes observed in polysomnographic-assessed sleep before and after completion of chemotherapy. *J Psychosom Res*. 2011;71(6):423−428.

20. Zhou J, Jolly S. Obstructive sleep apnea and fatigue in head and neck cancer patients. *Am J Clin Oncol*. 2015;411−414.

21. Friedman M, Landsberg R, Pryor S, et al. The occurrence of sleep-disordered breathing among patients with head and neck cancer. *Laryngoscope*. 2001;111(11 Pt 1):1917−1919.

22. Payne RJ, Hier MP, Kost KM, et al. High prevalence of obstructive sleep apnea among patients with head and neck cancer. *J Otolaryngol*. 2005;34(5):304−311.

23. Steffen A, Graefe H, Gehrking E, et al. Sleep apnoea in patients after treatment of head neck cancer. *Acta Otolaryngol*. 2009;129(11):1300−1305.

24. Qian W, Haight J, Poon I, et al. Sleep apnea in patients with oral cavity and oropharyngeal cancer after surgery and chemoradiation therapy. *Otolaryngol Head Neck Surg*. 2010;143(2):248−252.

25. Lin HC, Friedman M, Chang HW, et al. Impact of head and neck radiotherapy for patients with nasopharyngeal carcinoma on sleep-related breathing disorders. *JAMA Otolaryngol Head Neck Surg*. 2014;140 (12):1166−1172.

26. Huyett P, Kim S, Johnson JT, et al. Obstructive sleep apnea in the irradiated head and neck cancer patient. *Laryngoscope*. 2017;127(11):2673−2677.

27. Bamgbade OA, Khaw RR, Sawati RS, et al. Obstructive sleep apnea and postoperative complications among patients undergoing gynecologic oncology surgery. *Int J Gynaecol Obstet*. 2017;138(1):69−73.

28. Pollak L, Shpirer I, Rabey JM, et al. Polysomnography in patients with intracranial tumors before and after operation. *Acta Neurol Scand*. 2004;109(1):56−60.

29. Silberfarb PM, Hauri PJ, Oxman TE, et al. Assessment of sleep in patients with lung cancer and breast cancer. *J Clin Oncol*. 1993;11(5):997−1004.

30. Reinsel RA, Starr TD, O'Sullivan B, et al. Polysomnographic study of sleep in survivors of breast cancer. *J Clin Sleep Med*. 2015;11(12):1361−1370.

31. Saini A, Berruti A, Ferini-Strambi L, et al. Restless legs syndrome as a cause of sleep disturbances in cancer patients receiving chemotherapy. *J Pain Symptom Manage*. 2013;46(1):56–64.
32. Yennurajalingam S, Chisholm G, Palla SL, et al. Self-reported sleep disturbance in patients with advanced cancer: frequency, intensity, and factors associated with response to outpatient supportive care consultation—a preliminary report. *Palliat Support Care*. 2015;13(2):135–143.
33. Sela RA, Watanabe S, Nekolaichuk CL. Sleep disturbances in palliative cancer patients attending a pain and symptom control clinic. *Palliat Support Care*. 2005;3(1):23–31.
34. Gibbins J, McCoubrie R, Kendrick AH, et al. Sleep–wake disturbances in patients with advanced cancer and their family carers. *J Pain Symptom Manage*. 2009;38(6):860–870.
35. Bernatchez MS, Savard J, Savard MH, et al. Sleep–wake difficulties in community-dwelling cancer patients receiving palliative care: subjective and objective assessment. *Palliat Support Care*. 1–11 (in press).
36. Edinger JD, Kirby A, Lineberger M, et al. *Duke Structured Interview for Sleep Disorders*. Durham, NC: University Medical Center; 2004.
37. Good P, Pinkerton R, Bowler S, et al. Impact of opioid therapy on sleep and respiratory patterns in adults with advanced cancer receiving palliative care. *J Pain Symptom Manage*. 2018;55(3):962–967.
38. Kaleyias J, Manley P, Kothare SV. Sleep disorders in children with cancer. *Semin Pediatr Neurol*. 2012;19(1):25–34.
39. Mandrell BN, Wise M, Schoumacher RA, et al. Excessive daytime sleepiness and sleep-disordered breathing disturbances in survivors of childhood central nervous system tumors. *Pediatr Blood Cancer*. 2012;58(5):746–751.
40. Zhou ES, Manley PE, Marcus KJ, et al. Medical and psychosocial correlates of insomnia symptoms in adult survivors of pediatric brain tumors. *J Pediatr Psychol*. 2016;41(6):623–630.
41. Nolan VG, Gapstur R, Gross CR, et al. Sleep disturbances in adult survivors of childhood brain tumors. *Qual Life Res*. 2013;22(4):781–789.
42. Spielman AJ, Glovinsky P. Case studies in insomnia. In: Hauri PJ, ed. *The Varied Nature of Insomnia*. New York: Plenum Press; 1991:1–15.
43. Davidson JR, MacLean AW, Brundage MD, et al. Sleep disturbance in cancer patients. *Soc Sci Med*. 2002;54:1309–1321.
44. Desai K, Mao JJ, Su I, et al. Prevalence and risk factors for insomnia among breast cancer patients on aromatase inhibitors. *Support Care Cancer*. 2013;21(1):43–51.
45. Mercadante S, Aielli F, Adile C, et al. Sleep disturbances in patients with advanced cancer in different palliative care settings. *J Pain Symptom Manage*. 2015;50(6):786–792.
46. Savard MH, Savard J, Trudel-Fitzgerald C, et al. Changes in hot flashes are associated with concurrent changes in insomnia symptoms among breast cancer patients. *Menopause*. 2011;18:985–993.
47. Savard MH, Savard J, Caplette-Gingras A, et al. Relationship between objectively recorded hot flashes and sleep disturbances among breast cancer patients: Investigating hot flash characteristics other than frequency. *Menopause*. 2013;20(10):997–1005.
48. Hanisch LJ, Gooneratne NS, Soin K, et al. Sleep and daily functioning during androgen deprivation therapy for prostate cancer. *Eur J Cancer Care (Engl)*. 2011;20(4):549–554.
49. Savard J, Ivers H, Savard MH, et al. Cancer treatments and their side effects are associated with aggravation of insomnia: results of a longitudinal study. *Cancer*. 2015;121(10):1703–1711.
50. Costa AR, Fontes F, Pereira S, et al. Impact of breast cancer treatments on sleep disturbances—a systematic review. *Breast*. 2014;23(6):697–709.
51. Sharma N, Hansen CH, O'Connor M, et al. Sleep problems in cancer patients: prevalence and association with distress and pain. *Psychooncology*. 2013;21(9):1003–1009.
52. Palesh OG, Collie K, Batiuchok D, et al. A longitudinal study of depression, pain, and stress as predictors of sleep disturbance among women with metastatic breast cancer. *Biol Psychol*. 2007;75(1):37–44.
53. Shi Q, Giordano SH, Lu H, et al. Anastrozole-associated joint pain and other symptoms in patients with breast cancer. *J Pain*. 2013;14(3):290–296.
54. Trudel-Fitzgerald C, Savard J, Ivers H. Which symptoms come first? Exploration of temporal relationships between cancer-related symptoms over an 18-month period. *Ann Behav Med*. 2013;45(3):329–337.
55. Colditz GA, Peterson LL. Obesity and cancer: evidence, impact, and future directions. *Clin Chem*. 2018;64(1):154–162.

REFERENCES

56. van den Berg MM, Winkels RM, de Kruif JT, et al. Weight change during chemotherapy in breast cancer patients: a meta-analysis. *BMC Cancer*. 2017;17(1):259.
57. Nyrop KA, Deal AM, Lee JT, et al. Weight changes in postmenopausal breast cancer survivors over 2 years of endocrine therapy: a retrospective chart review. *Breast Cancer Res Treat*. 2017;162(2):375–388.
58. Baker FC, Joffe H, Lee KA. Sleep and menopause. In: Kryger MH, Roth T, eds. *Principles and Practice of Sleep Medicine*. Philadelphia, PA: Elsevier; 2017:1553–1563.
59. Eichling PS, Sahni J. Menopause related sleep disorders. *J Clin Sleep Med*. 2005;1(3):291–300.
60. Savard J, Ivers H, Villa J, et al. Natural course of insomnia comorbid with cancer: an 18-month longitudinal study. *J Clin Oncol*. 2011;29(26):3580–3586.
61. Daley M, Morin CM, LeBlanc M, et al. Insomnia and its relationship to health-care utilization, work absenteeism, productivity and accidents. *Sleep Med*. 2009;10:427–438.
62. Chevalier H, Los F, Boichut D, et al. Evaluation of severe insomnia in the general population: results of a European multinational survey. *J Psychopharmacol*. 1999;13(4, suppl 1):S21–S24.
63. Zammit G, Weiner J, Damato N, et al. Quality of life in people with insomnia. *Sleep*. 1999;22(suppl 2):S379–385.
64. Lichstein KL, Means MK, Noe SL, et al. Fatigue and sleep disorders. *Behav Res Ther*. 1997;35(8):733–740.
65. Breslau N, Roth T, Rosenthal L, et al. Sleep disturbance and psychiatric disorders: a longitudinal epidemiological study of young adults. *Biol Psychiatry*. 1996;39:411–418.
66. Ford DE, Kamerow DB. Epidemiologic study of sleep disturbances and psychiatric disorders: an opportunity for prevention? *JAMA*. 1989;262(11):1479–1484.
67. Morphy H, Dunn KM, Lewis M, et al. Epidemiology of insomnia: a longitudinal study in a UK population. *Sleep*. 2007;30(3):274–280.
68. Suka M, Yoshida K, Sugimori H. Persistent insomnia is a predictor of hypertension in Japanese male workers. *J Occup Health*. 2003;45(6):344–350.
69. Sivertsen B, Overland S, Neckelmann D, et al. The long-term effect of insomnia on work disability: the HUNT-2 historical cohort study. *Am J Epidemiol*. 2006;163(11):1018–1024.
70. Sivertsen B, Lallukka T, Salo P, et al. Insomnia as a risk factor for ill health: results from the large population-based prospective HUNT Study in Norway. *J Sleep Res*. 2014;23(2):124–132.
71. Ruel S, Savard J, Ivers H. Insomnia and self-reported infections in cancer patients: an 18-month longitudinal study. *Health Psychol*. 2015;34(10):983–991.
72. Ruel S, Ivers H, Savard MH, et al. Insomnia, immunity and self-reported infections in patients with chemotherapy for breast or gynaecological cancer: results from a longitudinal study [submitted].
73. Trudel-Fitzgerald C, Zhou ES, Poole EM, et al. Sleep and survival among women with breast cancer: 30 years of follow-up within the Nurses' Health Study. *Br J Cancer*. 2017;116(9):1239–1246.
74. Soucise A, Vaughn C, Thompson CL, et al. Sleep quality, duration, and breast cancer aggressiveness. *Breast Cancer Res Treat*. 2017;164(1):169–178.
75. Innominato PF, Spiegel D, Ulusakarya A, et al. Subjective sleep and overall survival in chemotherapy-naive patients with metastatic colorectal cancer. *Sleep Med*. 2015;16(3):391–398.
76. Collins KP, Geller DA, Antoni M, et al. Sleep duration is associated with survival in advanced cancer patients. *Sleep Med*. 2017;32:208–212.
77. Palesh O, Aldridge-Gerry A, Zeitzer JM, et al. Actigraphy-measured sleep disruption as a predictor of survival among women with advanced breast cancer. *Sleep*. 2014;37(5):837–842.
78. Vaughn CB, Freudenheim JL, Nie J, et al. Sleep and breast cancer in the Western New York Exposures and Breast Cancer (WEB) Study. *J Clin Sleep Med*. 2018;14(1):81–86.
79. Young T, Finn L, Peppard PE, et al. Sleep disordered breathing and mortality: eighteen-year follow-up of the Wisconsin sleep cohort. *Sleep*. 2008;31(8):1071–1078.
80. Martinez-Garcia MA, Campos-Rodriguez F, Barbe F. Cancer and OSA: current evidence from human studies. *Chest*. 2016;150(2):451–463.
81. Owens RL, Gold KA, Gozal D, et al. Sleep and breathing … and cancer? *Cancer Prev Res (Phila)*. 2016;9 (11):821–827.
82. Palamaner GSS, Kumar AA, Cheskin LJ, et al. Association between sleep-disordered breathing, obstructive sleep apnea, and cancer incidence: a systematic review and meta-analysis. *Sleep Med*. 2015;16(10):1289–1294.
83. Zhang XB, Peng LH, Lyu Z, et al. Obstructive sleep apnoea and the incidence and mortality of cancer: a meta-analysis. *Eur J Cancer Care (Engl)*. 2017;26(2):1–8.

84. Gozal D, Ham SA, Mokhlesi B. Sleep apnea and cancer: analysis of a nationwide population sample. *Sleep.* 2016;39(8):1493–1500.
85. Rod NH, Kjeldgard L, Akerstedt T, et al. Sleep apnea, disability pensions, and cause-specific mortality: a Swedish nationwide register linkage study. *Am J Epidemiol.* 2017;186(6):709–718.
86. Bruera E, Kuehn N, Miller MJ, et al. The Edmonton Symptom Assessment System (ESAS): a simple method for the assessment of palliative care patients. *J Palliat Care.* 1991;7(2):6–9.
87. Ashbury FD, Findlay H, Reynolds B, et al. A Canadian survey of cancer patients' experiences: are their needs being met? *J Pain Symptom Manage.* 1998;16(5):298–306.
88. Fitch MI, Porter HB, Page BD. *Supportive Care Framework: A Foundation for Person-Centered Care.* Pembroke, ON: Pappin Communications; 2008.
89. Jacobsen PB, Donovan KA, Trask PC, et al. Screening for psychologic distress in ambulatory cancer patients. *Cancer.* 2005;103(7):1494–1502.
90. Roth AJ, Kornblith AB, Batel-Copel L, et al. Rapid screening for psychologic distress in men with prostate carcinoma: a pilot study. *Cancer.* 1998;82:1904–1908.
91. Delgado-Guay M, Yennurajalingam S, Parsons H, et al. Association between self-reported sleep disturbance and other symptoms in patients with advanced cancer. *J Pain Symptom Manage.* 2011;41(5):819–827.
92. Carney CE, Buysse DJ, Ancoli-Israel S, et al. The consensus sleep diary: standardizing prospective sleep self-monitoring. *Sleep.* 2012;35(2):287–302.
93. Bastien CH, Vallières A, Morin CM. Validation of the Insomnia Severity Index as an outcome measure for insomnia research. *Sleep Med.* 2001;2(4):297–307.
94. Buysse DJ, Reynolds III CF, Monk TH, et al. The Pittsburgh Sleep Quality Index: a new instrument for psychiatric practice and research. *Psychiatry Res.* 1989;28:193–213.
95. Savard MH, Savard J, Simard S, et al. Empirical validation of the Insomnia Severity Index in cancer patients. *Psychooncology.* 2005;14(6):429–441.
96. Carpenter JS, Andrykowski MA. Psychometric evaluation of the Pittsburgh Sleep Quality Index. *J Psychosom Res.* 1998;45(1):5–13.
97. Howell D, Oliver TK, Keller-Olaman S, et al. A pan-Canadian practice guideline: prevention, screening, assessment, and treatment of sleep disturbances in adults with cancer. *Support Care Cancer.* 2012;.
98. Howell D, Oliver TK, Keller-Olaman S, et al. Sleep disturbance in adults with cancer: a systematic review of evidence for best practices in assessment and management for clinical practice. *Ann Oncol.* 2014;25(4):791–800.
99. Verger P, Cortaredona S, Tournier M, et al. Psychotropic drug dispensing in people with and without cancer in France. *J Cancer Surviv.* 2017;11(1):92–101.
100. Casault L, Savard J, Ivers H, et al. Utilization of hypnotic medication in the context of cancer: predictors and frequency of use. *Support Care Cancer.* 2012;20(6):1203–1210.
101. National Institutes of Health. NIH State-of-the-Science Conference Statement on manifestations and management of chronic insomnia in adults. *Natl Inst Health State Sci Conf.* 2005;1–22.
102. Kurdi MS, Muthukalai SP. The efficacy of oral melatonin in improving sleep in cancer patients with insomnia: a randomized double-blind placebo-controlled study. *Indian J Palliat Care.* 2016;22(3):295–300.
103. Johnson JA, Rash JA, Campbell TS, et al. A systematic review and meta-analysis of randomized controlled trials of cognitive behavior therapy for insomnia (CBT-I) in cancer survivors. *Sleep Med Rev.* 2016;27:20–28.
104. Ritterband LM, Bailey ET, Thorndike FP, et al. Initial evaluation of an Internet intervention to improve the sleep of cancer survivors with insomnia. *Psychooncology.* 2012;21:695–705.
105. Savard J, Ivers H, Savard MH, et al. Is a video-based cognitive–behavioral therapy for insomnia as efficacious as a professionally-administered treatment in breast cancer? Results of a randomized controlled trial. *Sleep.* 2014;37(8):1305–1314.
106. Savard J, Ivers H, Savard MH, et al. Long-term effects of two formats of cognitive–behavioral therapy for insomnia comorbid with breast cancer. *Sleep.* 2016;39(4):813–823.
107. Zachariae R, Amidi A, Damholdt MF, et al. Internet-delivered cognitive–behavioral therapy for insomnia in breast cancer survivors: a randomized controlled trial. *J Natl Cancer Inst.* 2018;110(8):880–887.
108. Zhou ES, Suh S, Youn S, et al. Adapting cognitive–behavior therapy for insomnia in cancer patients. *Sleep Med Res.* 2017;8(2):51–61.

REFERENCES

109. Bernatchez MS, Savard J, Savard MH, et al. Feasibility of a cognitive–behavioral and environmental intervention for sleep–wake difficulties in community-dwelling cancer patients receiving palliative care. *Cancer Nurs.* (in press).

110. Carlson LE, Garland SN. Impact of mindfulness-based stress reduction (MBSR) on sleep, mood, stress and fatigue symptoms in cancer outpatients. *Int J Behav Med.* 2005;12(4):278–285.

111. Shapiro SL, Bootzin RR, Figueredo AJ, et al. The efficacy of mindfulness-based stress reduction in the treatment of sleep disturbance in women with breast cancer: an exploratory study. *J Psychosom Res.* 2003;54(1):85–91.

112. Lengacher CA, Reich RR, Paterson CL, et al. The effects of mindfulness-based stress reduction on objective and subjective sleep parameters in women with breast cancer: a randomized controlled trial. *Psychooncology.* 2015;24(4):424–432.

113. Andersen SR, Wurtzen H, Steding-Jessen M, et al. Effect of mindfulness-based stress reduction on sleep quality: results of a randomized trial among Danish breast cancer patients. *Acta Oncol.* 2013;52(2):336–344.

114. Garland SN, Carlson LE, Stephens AJ, et al. Mindfulness-based stress reduction compared with cognitive behavioral therapy for the treatment of insomnia comorbid with cancer: a randomized, partially blinded, noninferiority trial. *J Clin Oncol.* 2014;32(5):449–457.

115. Mercier J, Savard J, Bernard P. Exercise interventions to improve sleep in cancer patients: a systematic review and meta-analysis. *Sleep Med Rev.* 2017;36:43–56.

116. Mercier J, Ivers H, Savard J. A non-inferiority randomized controlled trial comparing a home-based aerobic exercise program to a self-administered cognitive–behavioral therapy for insomnia in cancer patients. *Sleep.* in press;41(10):1–15.

117. Redd WH, Valdimarsdottir H, Wu LM, et al. Systematic light exposure in the treatment of cancer-related fatigue: a preliminary study. *Psychooncology.* 2014;23(12):1431–1434.

118. Ancoli-Israel S, Rissling M, Neikrug A, et al. Light treatment prevents fatigue in women undergoing chemotherapy for breast cancer. *Support Care Cancer.* 2012;20(6):1211–1219.

119. Johnson JA, Garland SN, Carlson LE, et al. Bright light therapy improves cancer-related fatigue in cancer survivors: a randomized controlled trial. *J Cancer Surviv.* 2018;12(2):206–215.

120. Desautels C, Savard J, Ivers H, et al. Treatment of depressive symptoms in patients with breast cancer: a randomized controlled trial comparing cognitive therapy and bright light therapy. *Health Psychol.* 2018;37(1):1–13.

121. Wu LM, Amidi A, Valdimarsdottir H, et al. The effect of systematic light exposure on sleep in a mixed group of fatigued cancer survivors. *J Clin Sleep Med.* 2018;14(1):31–39.

122. Neikrug AB, Rissling M, Trofimenko V, et al. Bright light therapy protects women from circadian rhythm desynchronization during chemotherapy for breast cancer. *Behav Sleep Med.* 2012;10(3):202–216.

123. Bernatchez MS, Savard J, Ivers H. Disruptions in sleep–wake cycles in community-dwelling cancer patients receiving palliative care and their correlates. *Chronobiol Int.* 2018;35(1):49–62.

124. Crowley RK, Woods C, Fleming M, et al. Somnolence in adult craniopharyngioma patients is a common, heterogeneous condition that is potentially treatable. *Clin Endocrinol (Oxf).* 2011;74(6):750–755.

125. Espie CA. Stepped care: a health technology solution for delivering cognitive behavioral therapy as a first line insomnia treatment. *Sleep.* 2009;32(12):1549–1558.

CHAPTER

9

Chronic Pain

Caitlin B. Murray[1] and Tonya M. Palermo[1,2]

[1]Center for Child Health, Behavior, and Development, Seattle Children's Research Institute, Seattle, WA, United States [2]Department of Anesthesiology & Pain Medicine, University of Washington School of Medicine, Seattle, WA, United States

OUTLINE

Introduction	201	Sleep Assessment	207
		Clinical Interview	*207*
Sleep Deficiency in Chronic Pain		*Questionnaires*	*208*
Conditions	202	*Sleep Diaries*	*209*
Osteoarthritis (OA)	*203*	*Actigraphy*	*209*
Juvenile Idiopathic Arthritis (JIA)	*203*	*Polysomnography*	*210*
Cancer	*203*		
Sickle Cell Disease	*204*	**Sleep Treatment**	210
Chronic Headache	*204*		
Fibromyalgia	*204*	**Conclusions and Areas for Future**	
		Research	212
Sleep and Pain: Interconnection and			
Shared Mechanisms	205	**References**	213
Interconnection Between Sleep and Pain	*205*		
Shared Mechanisms	*205*		

INTRODUCTION

Chronic pain refers to pain that recurs or persists beyond the expected time of healing from illness, injury, or surgical procedures (International Association for the Study of Pain[1]). It is a complex, multidimensional experience that varies considerably in intensity,

Handbook of Sleep Disorders in Medical Conditions
DOI: https://doi.org/10.1016/B978-0-12-813014-8.00009-3

© 2019 Elsevier Inc. All rights reserved.

location, frequency, and functional impact for each person. Individuals can experience chronic pain in association with a medical illness such as cancer or sickle cell disease (SCD), or in the absence of an identifiable physical illness or injury, as with fibromyalgia and migraine headaches.

Chronic pain is a common, debilitating, and costly health issue that may develop at any age. Although prevalence rates vary due to differences in definitions of chronic pain, epidemiological research indicates that between 11% and 38% of children and adolescents report chronic pain, most commonly musculoskeletal, abdominal, and headache pain.[2] Further, there are emerging data to support the persistence of childhood pain into adulthood,[2,3] with adult chronic pain prevalence rates similar to those of childhood (19%−30%[4,5]). The most common causes of chronic pain in adulthood are musculoskeletal conditions—predominantly back pain and joint pain.[6]

Chronic pain is associated with substantial burden and impact including physical limitations, social impairment and psychological distress,[7,8] lost work and school productivity,[9,10] and high health-care costs.[9,11] Chronic pain also occurs in association with a number of physical and psychiatric morbidities (e.g., depression[12]). Sleep deficiency—particularly insomnia—is one of the most common and debilitating comorbidities presenting in the vast majority of children and adults with chronic pain conditions (see reviews by Tang[8] and Allen et al.[13]). Sleep deficiency is independently linked to pervasive impairments in physical health, cognitive function, and mood disturbance.[13–16] The negative additive impact of sleep deficiency therefore represents a dramatic increase in suffering for individuals with chronic pain.

Addressing sleep deficiency in individuals with chronic pain is a clinical and research priority. Yet sleep issues are often underidentified and undertreated, perhaps due to the competing demands associated with chronic pain assessment and management. The purpose of this chapter is to review common presentations of sleep deficiency in chronic pain conditions, highlight shared pathophysiological and cognitive−affective factors underlying the sleep−pain association, and provide an overview of current sleep assessment and treatment strategies. We conclude by highlighting several clinical priorities and avenues for future research to address sleep deficiency in individuals with chronic pain.

SLEEP DEFICIENCY IN CHRONIC PAIN CONDITIONS

Both subjective report and objective sleep measures [e.g., polysomnography (PSG)] have been used to document the high occurrence of sleep deficiency in children and adults with pain conditions. Insomnia symptoms (difficulties initiating and maintaining sleep) are the most common cause of sleep deficiency in children and adults with chronic pain. Sleep disorders such as sleep disordered breathing (SDB) and periodic limb movement may also present more frequently in the context of certain pain conditions. We organize this section according to major categories of pain conditions to describe common presentations of sleep deficiencies and sleep disorders in children and adults with chronic pain.

Osteoarthritis (OA)

OA is the most common form of arthritis among adults, with an increasing prevalence associated with older age and obesity. The pathophysiology of this condition involves the breakdown of joint cartilage and changes in the underlying bone and soft tissue. Research indicates that 25%–50% of individuals with OA experience nighttime joint pain[17,18] that is highly comorbid with sleep deficiency—particularly insomnia.[19–21] For example, in a study of 429 individuals with OA, sleep deficiencies were common and included a high prevalence of difficulties with sleep maintenance (81%), followed by frequent early morning awakenings (51%), and prolonged sleep onset (31%[21]). Studies using PSG have further revealed that individuals with OA spend more time in light sleep compared to healthy controls [i.e., nonrapid eye movement (NREM) Stage 1[22]].

Juvenile Idiopathic Arthritis (JIA)

Pain is a common experience for children and adolescents with JIA due to recurring, unpredictable episodes of acute inflammation that characterize this rheumatic condition. In fact, an estimated 20% of youth with JIA may experience lifelong pain and disability.[23] Youth with JIA also experience significant sleep deficiency. A recent systematic review found that children with JIA experience greater difficulties with sleep maintenance and excessive daytime sleepiness compared to otherwise healthy children.[24] Studies using objective measures of sleep including PSG and actigraphy provide further evidence of sleep deficiencies in this condition. Specifically, on PSG, these youth demonstrate sleep fragmentation, less time in slow wave sleep, and more alpha/delta activity during NREM.[25,26] Results of a recent study utilizing actigraphy also found that young children aged 2–3 years newly diagnosed with JIA experience significantly less total sleep time and lower sleep efficiency compared to their otherwise healthy peers.[27]

Cancer

Pain is one of the most common and distressing symptoms reported by individuals with cancer. Cancer pain may be secondary to treatment side effects (e.g., mucositis) or recurring medical procedures.[28–30] Tumor-related pain may also occur with delayed diagnosis and with advanced disease. Sleep deficiency and fatigue are also reported to be significant problems across cancer diagnoses[31] and may persist into survivorship. The most common self-reported sleep deficiencies in adult and pediatric oncology patients include difficulties falling and staying asleep, reduced sleep duration, and irregular sleep schedules.[32–34] While cancer-related pain may disrupt sleep, thereby exacerbating fatigue, other salient contributors of sleep and fatigue include central nervous system involvement, specific treatments (e.g., corticosteroids, radiation), environmental or schedule changes, frequent hospitalizations, and stress and anxiety related to having a life- threatening illness.[35,36] Fatigue is an important consideration across pain conditions, yet cancer-related fatigue has been the focus of research attention due to its profound effects on patients'

Sickle Cell Disease

SCD is a common genetic blood disorder associated with recurring vaso-occlusive pain episodes. Individuals with SCD may also experience chronic pain with or without contributory disease factors.[39] SCD is associated with significant sleep deficiency including an increased rate of SDB and insomnia symptoms. SDB affects 40% of children and young adults with SCD[40,41] and may be particularly important to assess due to the potential for SDB-related oxygen desaturation to trigger or exacerbate sickle cell crises. Insomnia related to difficulties with sleep onset and maintenance is also highly prevalent in individuals with this condition.[42] In one study of 8–12 year olds with SCD utilizing diary report, children showcased difficulties falling asleep and maintaining sleep more than 30% of days.[43] While the majority of sleep research has focused on children and adolescents with SCD, recent studies focused on adults with this condition have revealed similarly high rates of SDB and insomnia.[44–46]

Chronic Headache

Chronic headache pain is frequent among children and adults and refers to migraine or tension headaches occurring more than 15 days per month. Migraine headache is characterized by throbbing or pulsating pain on one or both sides of the head and is typically associated with visual disturbances and nausea. Tension-type headache pain is usually mild to moderate in intensity and accompanied by bilateral pressure or tightening quality around the head. Insomnia symptoms are frequent in children and adults with chronic headache pain, occurring in one-half to two-thirds of patients (see Almoznino, Benoliel, Sharav, and Haviv[47] for a review). Sleep-related bruxism (e.g., repetitive jaw activity related to chewing/grinding of teeth) and restless leg syndrome have also been identified in individuals with tension-type and migraine headache, respectively.[48,49] Morning headache has also been associated with SDB.[50]

Fibromyalgia

Fibromyalgia is characterized by chronic widespread musculoskeletal pain, stiffness, and fatigue. Although typically considered to be an adult condition, fibromyalgia also occurs in children and adolescents. Sleep deficiency is highly prevalent in this pain condition: an estimated 70%–80% of patients with fibromyalgia report insomnia symptoms (sleep onset and maintenance difficulties) and nonrestorative sleep.[51–53] See Spaeth, Rizzi, and Sarzi-Puttini[54] for a review. PSG data have further revealed that children and adults with fibromyalgia show decreased total sleep time, poor sleep efficiency, increased arousals, periodic limb movements, and disturbed electroencephalogram (EEG) frequency during slow-wave sleep.[51,55,56]

SLEEP AND PAIN: INTERCONNECTION AND SHARED MECHANISMS

Existing research establishing the high cooccurrence of sleep deficiency in individuals with chronic pain has led to great clinical and scientific interest in unraveling the relationship between sleep and pain. In particular, advancements in research over the past decade have clarified our understanding of temporal pathways and shared mechanisms of sleep and pain.

Interconnection Between Sleep and Pain

The interconnection between sleep and pain is often cited as a bidirectional relationship, such that pain interferes with sleep onset and maintenance and sleep deficiency amplifies pain sensation. Indeed, the majority (60%) of individuals with chronic pain have identified painful symptoms as the sole cause of their sleep disruption.[57] Well-designed laboratory research using healthy volunteers has further revealed that the induction of nociceptive stimuli during sleep leads to transient arousals and brief awakenings.[58,59] Several prospective studies have also found that pain intensity is a predictor of sleep deficiency in chronic pain populations, yet this finding has not been consistent,[60,61] and the impact may be small or nonsignificant when controlling for psychological and cognitive factors (e.g., attention to pain[62]).

Comparatively, there are more robust data to support for the impact of sleep on pain. Micro-longitudinal studies investigating daily temporal relationships have revealed that poor nighttime sleep is a strong predictor of next-day pain intensity in several chronic pain populations (see Finan, Goodin, and Smith[63] for review). Experimental research has also highlighted the significant impact of sleep deficiency on impairments in pain-inhibitory function in individuals with chronic pain.[64-66] Moreover, prospective studies conducted over longer time intervals (several months to years) have revealed that sleep deficiency is associated with the development of new-onset chronic pain as well as worsening prognosis in individuals with existing pain conditions.[63]

Overall, the accumulation of clinical literature over the past decade suggests that the effect of sleep on pain appears stronger than that of pain on sleep. Little is known about the mechanisms that underlie the nature, etiology, or clinical significance of this bidirectional interaction. There are likely both nonspecific and specific effects of sleep on pain, and mechanisms may vary by type of pain condition. Active research areas include shared neural/glial mechanisms of sleep and pain, the effect of poor sleep on nociceptive signals, inflammatory modulators, central arousal, and cognitive/affective modulators which may reveal new insights into this relationship.

Shared Mechanisms

To deepen understanding of the cooccurrence of chronic pain and sleep deficiency and to guide assessment and treatment strategies, we review current evidence of pathophysiological and cognitive—affective mechanisms that may enhance risk for symptom amplification and maintenance of these highly comorbid conditions.

Dopamine. Altered dopamine signaling has been identified as one potential mechanism that may explain the overlap between chronic pain and insomnia (see Finan and Smith[67] for a review). Dopamine regulation is integral to the maintenance of arousal states, and abnormalities in this system may promote sleep-interfering arousal states. In support of this theory, individuals with widespread chronic pain including fibromyalgia have altered dopamine metabolite concentrations in the cerebrospinal fluid.[68] However, further research is needed to better understand the effect of dopaminergic dysregulation on pain modulation. While research in this area is still in its infancy, dopamine abnormalities may perpetuate insomnia and pain symptoms, and persistent symptoms may serve to further impair dopaminergic function in negative cycle.[67]

Inflammation. Experimental research has also identified the effects of sleep deficiency on inflammatory processes in animal models and healthy adult populations.[69] Inflammatory markers including prostaglandins and proinflammatory cytokines have been shown to sensitize nociceptors, likely leading to pain amplification.[70–72] In one of the first studies to link sleep deficiency-induced inflammatory markers to pain, Haack et al.[70] found that prolonged sleep deprivation (10 days, 4 hours/night) elevated proinflammatory cytokine (Interleukin 6), which was strongly associated with increased self-reported pain ratings in healthy adults.

Cognitive factors. Key cognitive constructs highly relevant to chronic pain may fuel central arousal and disrupt sleep. Presleep cognitive arousal, characterized by excessive mental activity or rumination at nighttime, has been consistently linked to insomnia in individuals with chronic pain.[73–75] Individuals with comorbid chronic pain and insomnia commonly experience highly distressing sleep-interfering cognitions about their health, pain, and related functional impairments. For example, a study by Smith et al. [57] found that individuals with chronic pain who experienced a higher proportion of negative pain-related thoughts prior to sleep (vs sleep-related cognitions) had the greatest difficulties with sleep onset.

Pain catastrophizing, characterized by exaggerated negative thoughts about pain, has also been implicated in the development and severity of both insomnia and chronic pain through heightened somatic hypervigilance and the intrusive and distressing nature of catastrophizing thoughts.[75–77] More generally, dysfunctional beliefs could amplify cognitive arousal and promote poor sleep practices,[78] including unhelpful beliefs about pain (*This pain will never get better*), sleep (*I will never get a good night of sleep*), and their relationship (*I will only be able to sleep again when my pain goes away*).

Mood. A number of studies have implicated mood disturbance—particularly depression—as a prominent comorbidity and potential mechanism linking chronic pain and sleep deficiency. Sleep disruptions may result in depressed mood, which can impact pain perception through increased physiologic or cognitive arousal. Alternatively, pain may mediate the sleep—mood relationship such that disrupted sleep increases pain severity and consequently exacerbates negative cognitions and affectivity.[79] There is further evidence that depressed mood may moderate or strengthen the relationship between sleep and pain. Research using daily diaries in pediatric and adult chronic pain populations indicate that the association between sleep quality and next-day pain severity is stronger in patients with higher (vs lower) levels of depressive and negative affectivity.[80,81] Together, insomnia, chronic pain, and depression form an "unhappy triad" of symptoms with deleterious economic and functional consequences for the sufferer.[79] While prospective and micro-longitudinal research may help to clarify the temporal complexities of sleep, pain, and mood, it is likely that each symptom has the potential to mediate or moderate the others in a vicious cycle. Research identifying common pathophysiological substrates of insomnia, chronic pain, and depression—such as dopamine[67]—may elucidate vulnerability factors for the development and maintenance of this complex symptom triad.

SLEEP ASSESSMENT

Evidence of the high prevalence and deleterious impact of sleep deficiency on pain sensitivity and functional disability highlights the importance of routine assessment of sleep in individuals with chronic pain. The assessment of sleep is the first step necessary in identifying sleep deficiency and incorporating sleep interventions into a comprehensive pain management program. Consider the following case example, which will be revisited throughout this section:

Chloe is a 19-year-old Caucasian female with chronic daily low back pain that began after a motor vehicle accident. Since Chloe transitioned out of her parents' home to begin college, she has experienced increased depressive symptoms including irritability, feelings of isolation, and hopelessness. She constantly worries about the impact of pain on her ability to maintain new friendships and keep up with a highly challenging college workload. Chloe complains of persistent difficulties initiating and maintaining sleep. She also endorses poor sleep habits reflective of college lifestyle factors, such as going to bed late to hang out with friends in her dorm or attend social events.

Despite its importance, sleep assessment can be challenging. Sleep is a broad and multifaceted domain reflecting a variety of behaviors, patterns, disorders, and symptoms. A multi-method approach using a combination of measurement systems is often necessary to capture the different domains and components of sleep. Ultimately, the chosen method or combination of assessment tools depends on the structure of the treatment setting, unique characteristics of the patient and his or her chronic pain condition, and the intended function of assessment (e.g., screening vs treatment monitoring).

For children and adults with chronic pain, an evaluation of sleep may begin with simple screening questions to identify disturbances in sleep patterns, nighttime awakenings, and sleep onset and maintenance. A more detailed clinical interview can follow to obtain a clearer picture of sleep. When feasible and available, subjective measures (questionnaires, sleep diaries) and actigraphy can provide useful information to supplement clinical interview. Sleep assessment may also involve a referral for PSG if an underlying physiological sleep disorder is suspected. Below, we review sleep assessment methods commonly used in individuals with chronic pain.

Clinical Interview

A detailed clinical interview includes evaluation of the following areas: (1) sleep schedule and bedtime routine; (2) sleep habits and physical environment (e.g., presence of media in the bedroom); (3) the nature, frequency, and duration of sleep complaints including nocturnal behaviors (e.g., sleep latency, nighttime awakenings, discomfort or pain during the night, symptoms of SDB including snoring and morning headaches); (4) maladaptive daytime consequences and behaviors (e.g., fatigue, napping as a means to cope or escape from pain); (5) mood, affect, and cognitions about sleep; (6) family history of sleep problems; and (7) current prescriptions and nonprescription medications (to consider their effects on sleep; e.g., opioids). As an example, the following was obtained during clinical interview with the patient, Chloe:

Chloe reports that it typically takes 2 hours to fall asleep due to nighttime worry and difficulty finding a comfortable position. During the week, she attempts to go to sleep around 12:00 a.m. and usually falls asleep well after 2:00 a.m. She typically wakes up 2—3 times during the night (between 3:00 a.m. and 4:00 a.m.) and is unable to fall back asleep for 20—30 minutes. She sleeps until 10:00 a.m. during the week unless she has a morning class. On the weekend, she has a later bedtime and sleeps until the early afternoon. Chloe has a very difficult time waking up in the morning to attend classes. Poor concentration and motivation have impacted her ability to complete her course work. She is currently failing several classes, which has recently resulted in academic probation. Chloe has not received any health-care services for over a year because she is not sure which clinics will allow her to use her parents' insurance.

To further assess her symptoms of insomnia, additional interview prompts may focus on pain- and sleep-related maladaptive cognitive—affective and behavioral responses. Questions about thoughts at bedtime can be used to identify dysfunctional beliefs about sleep (e.g., "I must get 8 hours of sleep to feel refreshed and function well the next day"), high levels of fear related to pain or disease symptoms, and cognitive rumination or catastrophizing about pain prior to sleep onset (e.g., "My pain is never going to get better"). All patients complaining of insomnia symptoms should be screened for depression and anxiety, given their frequent cooccurrence. The interview can also be used to assess maladaptive behavioral responses and consequences of poor sleep and painful symptoms such as work or school absenteeism, social isolation, and use of excessive rest to alleviate symptoms. Consider the following:

On interview, Chloe describes several maladaptive thoughts and cognitions at nighttime. She frequently ruminates about her health and whether academic probation will require her to leave college and return home. Chloe expressed the belief that her physician may be missing an underlying medical diagnosis due to her continued pain and fatigue. She believes that her unremitting pain and fatigue will never improve. She feels hopeless and isolated about her situation. Chloe stopped exercising and has had difficulty reinitiating moderate-to-vigorous physical activity. She recently attempted to begin light jogging but stopped due to experiencing increased pain the following day and fear of aggravating her pain condition. She often lies in bed to watch TV or nap and will turn down invitations to engage in social events when she feels fatigued or pain exacerbations.

Questionnaires

Use of standardized questionnaires may be useful in complementing the clinical interview and for tracking treatment progress. A number of sleep questionnaires have been developed to assess specific sleep disturbances (e.g., insomnia) as well as sleep quality more broadly via multiple domains (e.g., onset latency, duration). While subjective sleep measures have not yet been developed specifically for individuals with chronic pain conditions, there are several existing measures that have shown excellent validity and feasibility in chronic pain samples.

Well-established measures of sleep in pediatric chronic pain include the *Children's Sleep Habits Questionnaire* (CSHQ[82]), the *Adolescent Sleep–Wake Scale* (ASWS), and the *Adolescent Sleep Habits Scale* (ASHS[83]). The CSHQ is a parent-report measure used to assess multiple aspects of sleep in school-age children (ages 4–10), including bedtime behavioral issues and symptoms of SDB.[82] The ASWS and ASHS are complementary measures of sleep quality and sleep habits, respectively. While reliable and relevant to pediatric pain, all three questionnaires are lengthy and potentially burdensome to complete. A short form (10-item) of the ASWS was recently developed that may be useful to integrate into quick-paced tertiary care settings.[84] For a comprehensive review of available sleep measurement tools in pediatric pain, see de la Vega and Miro.[85]

For adults with chronic pain, commonly used questionnaires include the *Pittsburg Sleep Quality Index* (PSQI[86]), the *Patient-Reported Outcomes Measurement Information System (PROMIS)-Sleep Disturbance* item bank,[87] and the *Insomnia Severity Index* (ISI[88,89]). The PSQI and PROMIS-Sleep Disturbance are easy to administer, psychometrically robust measures of global sleep quality. The ISI is a brief, seven-item instrument measuring perceived insomnia severity and has been increasingly used for the purposes of screening and documentation of treatment response. The ISI is easy to score with interpretable cut points (e.g., mild insomnia, moderate insomnia).

Further, self-report of cognitive and behavioral factors that perpetuate sleep deficiency may be beneficial to incorporate into questionnaire assessment. The *Presleep Arousal Scale*[90,91] and *Dysfunctional Beliefs and Attitudes about Sleep Scale*[89,92–94] have been used in adolescent and adult chronic pain populations to identify cognitive/physiological arousal prior to sleep onset and maladaptive sleep cognitions, respectively.

Sleep Diaries

Sleep diaries are subjective daily reports providing detailed information about sleep scheduling and patterns over prospective time periods (1–2 weeks). Diaries are completed each morning and include one-item questions about the previous nights' sleep onset, nighttime awakenings, sleep times, and sleep quality. Items related to medication use and sleep-related daytime symptoms (pain, fatigue, and mood) may also be included to provide additional insight for case formulation and intervention.

Actigraphy

Actigraphy has been used in pediatric and adult pain populations to provide an estimate of sleep patterns across multiple days/weeks. Actigraphy assessment relies on an accelerometer (a small, wristwatch-like device) worn in the home environment to record the presence or absence of movement as an approximation of sleep and wake cycles (see Horne and Biggs[95] for a review). While not indicated for diagnosis of insomnia or physiological sleep disorders, actigraphy may supplement patient-reported measures (questionnaires, diaries) and the clinical interview to identify disturbed sleep patterns. Actigraphy has also been used as a measure of adherence for behavioral sleep treatment and to modify sleep restriction schedules.[96] However, because actiwatches are costly and require specialized training, they are only typically available for use in sleep clinics or as part of research programs.

Polysomnography

PSG is considered the gold standard for identifying and quantifying sleep-related breathing and periodic limb disturbances. As part of clinical intake in certain pain conditions, it is important for clinicians to screen for SDB (e.g., in SCD and chronic headache) as well as periodic limb disturbances (e.g., in chronic headache and fibromyalgia). For example, clinicians may conduct a targeted assessment of excessive daytime sleepiness, poor sleep quality, heavy snoring, and gasping in children and young adults with SCD to facilitate a referral for PSG evaluation of SDB.

SLEEP TREATMENT

Specific treatment targeting improved sleep is an important consideration in the care of patients with chronic pain. Treatments may include education and instruction in healthy sleep habits, cognitive–behavioral therapy (CBT), physical therapy, breathing devices (continuous positive airway pressure—CPAP, or oral appliance), and medications (sleep facilitators, e.g., zolpidem; or antidepressants, e.g., trazodone, duloxetine; or neuroleptics, e.g., pregabalin[97]). For a review, see Cheatle et al.[98]

Overall, the empirical literature on treatment of sleep disorders in patients with chronic pain is very limited. However, because certain sleep disorders may be more common in individuals with chronic pain or may have specific management concerns in this population, there has been a great deal of clinical interest. For example, the management of SDB has been particularly discussed in relationship to chronic opioid therapy because of the link established between long-term opioid use and risk for central sleep apnea.[99] Several studies have examined the efficacy of noninvasive ventilation, particularly adaptive servo-ventilation for the treatment of opioid-associated SDB. Given that morning headache is a common feature of obstructive sleep apnea (OSA), there has been some consideration of CPAP treatment to alleviate both headache and OSA. However, mixed results have been reported with regards to headache improvement with CPAP.[50] Further studies are needed to more fully understand risk for OSA in individuals with chronic pain conditions as well as effective management strategies.

The most well-studied sleep intervention for individuals with chronic pain is behavioral treatment of insomnia. A number of studies have evaluated the efficacy of CBT for insomnia (CBT-I) encompassing education about sleep and sleep hygiene, stimulus control, relaxation techniques, and cognitive strategies in diverse patient populations including individuals with chronic pain. Sleep interventions can be a stand-alone treatment or included as a part of CBT for chronic pain.[100,101] Sleep strategies that are commonly integrated in these hybrid CBT treatments include modifying sleep habits (e.g., no phones or screens in bed, regular and relaxing bedtime routine), keeping a regular bedtime and wake time which allows for adequate duration of sleep, limiting naps, reducing negative thoughts about sleep, as well as teaching specific strategies to decrease insomnia symptoms. For a review, see Wu, Appleman, Salazar, and Ong.[102] CBT-I is recommended by the American Academy of Sleep Medicine as first-line treatment for adult insomnia.

CBT-I has been delivered without any major adaptations to multiple populations of adults with chronic pain showing robust effects. Recent metaanalyses conclude that CBT-I produces reliable and durable improvement in sleep in adults with cooccurring pain conditions including arthritis and fibromyalgia.[99,103,104] There is emerging evidence suggesting that changes in sleep as a result of behavioral treatment lead to changes in pain symptoms. For example, in a recent trial of CBT-I in patients with knee OA and insomnia, patients who received CBT-I had greater improvements in sleep and the baseline-to-posttreatment change predicted subsequent decreases in pain.[105] Moreover, CBT-I is a flexible treatment with evidence emerging that it can be delivered in brief form (i.e., in four sessions[106]), in group settings, and remotely through the internet with similar positive benefits.[107] Such adaptations to methods of treatment delivery will be necessary for broader dissemination as access to psychologists specialized in treating insomnia is extremely limited in most communities.

Although the treatment literature for child and adolescent insomnia interventions is more limited, there has been a recent uncontrolled trial in an adolescent pain population.[106] In this study, adolescents with a range of physical and psychiatric comorbidities (e.g., depression, chronic pain, anxiety, gastrointestinal problems) received a brief four-session CBT-I intervention. CBT-I was associated with treatment improvements in insomnia symptoms, sleep quality, sleep hygiene, presleep arousal, and sleep patterns as well as improvements in psychological symptoms and health-related quality of life. Returning to the case example:

Chloe saw a primary care provider through her university health clinic for complaints of insomnia and back pain. Chloe was referred to work with a psychologist for cognitive—behavioral intervention to address difficulties falling and staying asleep. The psychologist first provided education concerning adequate sleep and healthy sleep habits and helped Chloe establish a regular bedtime routine and schedule. Chloe was encouraged to engage in nonstimulating, relaxing activities as part of her bedtime routine and to limit electronic use.

The psychologist used sleep restriction and stimulus control interventions to address the long sleep onset delay. Based on Chloe's reported bedtimes in her sleep diary, a starting, temporary bedtime of 2:00 a.m. was selected with a wake time of 9:00 a.m. to provide enough time to attend morning classes. A wake time of 10:00 a.m. was selected for weekends. In order to establish a stronger connection with her bed and sleep, Chloe was instructed to restrict activities to sleeping only and to get out of bed and engage in a nonstimulating activity when she could not fall asleep within 20 minutes. She continued to complete a sleep diary which was reviewed at each session to allow for titrating her sleep window. As Chloe demonstrated success in falling asleep quickly at the stated bedtime, her sleep window was adjusted by 20 minutes (e.g., her next sleep window was set at 1:40 a.m.).

Cognitive therapy was introduced through guided discussion of how Chloe's worrisome presleep thoughts heightened emotional and cognitive arousal and interfered with sleep onset. She was encouraged to restrict worrying to 20 minutes between her morning classes to reduce ruminating thoughts at nighttime. Cognitive restructuring exercises were integrated through in-session and homework exercises which encouraged Chloe to challenge and replace her maladaptive, catastrophic thoughts about pain and sleep (e.g., "My pain and fatigue are causing me to fail out of school") with less distressing versions of these cognitions (e.g., "This pain and fatigue have made college extremely difficult, but every day I am taking small steps to get better"). Chloe was instructed to take a brisk walk to campus when she experienced worsening fatigue symptoms to serve the combined purpose of combating fatigue and increasing her physical activity level.

Chloe continued to work with the sleep clinic psychologist on these interventions over the following 8 weeks.

CONCLUSIONS AND AREAS FOR FUTURE RESEARCH

Children and adults with chronic pain commonly report sleep deficiency that may amplify pain severity and disability. While progress has been made to identify sleep deficiency in individuals with chronic pain and to begin to develop assessment and treatment approaches, there remain significant unanswered questions and unmet treatment needs.

Understanding mechanistic factors that explain the sleep—pain association will be an active area of investigation across basic, translational, and clinical studies. Our review highlights a number of crosscutting factors that may contribute to symptom amplification and maintenance of both sleep deficiency and chronic pain. Several cognitive—affective (e.g., mood disturbance and presleep arousal) and neurobiological mechanisms have been identified as playing a possible mechanistic role, and further research is needed to apply understanding and inform effective treatments. In particular, future research should aim to understand whether mechanisms that account for the sleep—pain relationship vary by type of pain condition.

There is also an overarching need to develop an integrated framework to enhance understanding of the etiology and maintenance of sleep deficiency in chronic pain. Application of a developmental and biopsychosocial framework to existing cognitive—behavioral models may enable recognition of understudied condition-related (e.g., surgery) and social-contextual (e.g., family functioning, developmental stage) factors that perpetuate sleep deficiency and pain. An integrated model of sleep deficiency in chronic pain may help to identify gaps in knowledge and ultimately guide novel screening and treatment strategies to simultaneously address pain and sleep.

Very limited empirical work has been conducted to evaluate sleep treatments in pain populations. The most promising sleep treatment to date is CBT-I. Evidence for improving sleep has accumulated in diverse adult pain populations. There is more limited research evaluating the efficacy and acceptability of this treatment for children, adolescents, and young adults and this is an important research priority. There are also limited data to suggest that improvements in sleep resulting from CBT-I lead to reductions in pain. More research on mechanisms of treatment effects is needed to inform the best way to treat comorbid pain and insomnia. Hybrid treatment approaches that combine the most potent sleep and pain interventions and include flexible components tailored to the individual patient (e.g., optional modules to manage fatigue) hold promise for improving comorbid insomnia and chronic pain. Additional priorities include utilizing innovative methodological approaches including multicenter randomized controlled trials and eHealth delivery methods (e.g., mobile apps) to increase access to efficacious sleep treatments.

KEY PRACTICE POINTS

- The most common cause of sleep deficiency in children and adults with chronic pain is insomnia (e.g., difficulties falling and staying asleep). Sleep disorders may also present more frequently in the context of certain pain conditions (e.g., Sleep Disordered Breathing in Sickle Cell Disease).
- Initial evaluation of sleep deficiency for individuals with chronic pain may begin with gathering information on sleep patterns, sleep habits, nocturnal behaviors (e.g., sleep latency, nighttime awakenings, pain during the night), mood, maladaptive cognitions about sleep and pain at nighttime, and maladaptive behavioral responses to poor sleep and painful symptoms (e.g., school or work absenteeism; excessive napping).
- Opioid and nonopioid medications should be reviewed carefully to identify possible disruption on sleep. Opioid pain medications can disrupt sleep and prevent patients from entering deep sleep after as little as one dose. Opioid pain medications can also cause sleep-related breathing disturbances. If medications that are associated with sleep disruption cannot be discontinued or switched, consideration should be given to adjusting the timing of administration. On the other hand, certain pain medications can improve sleep and might be prescribed for patients who have both a sleep disorder and a specific pain disorder. For a primer, clinicians can consult Marshansky et al.[97]
- CBT-I delivered to patients with comorbid insomnia and chronic pain produces clinically meaningful improvements in sleep symptoms.
- Due to physical inactivity and functional impairments often experienced by patients with chronic pain, treatment approaches that build appropriate schedules, routines, and increase physical activity may be a useful aspect of both sleep and pain management.

References

1. Merskey H, Bogduk N. *Classification of Chronic Pain*. 2nd ed. Seattle: IASP Press; 1994.
2. King S, Chambers CT, Huguet A, et al. The epidemiology of chronic pain in children and adolescents revisited: a systematic review. *Pain*. 2011;152(12):2729−2738. Available from: https://doi.org/10.1016/j.pain.2011.07.016.
3. Campo JV, Di Lorenzo C, Chiappetta L, et al. Adult outcomes of pediatric recurrent abdominal pain: do they just grow out of it? *Pediatrics*. 2001;108(1):E1. Available from: https://doi.org/10.1542/peds.108.1.e1.
4. Breivik H, Collett B, Ventafridda V, Cohen R, Gallacher D. Survey of chronic pain in Europe: prevalence, impact on daily life, and treatment. *Eur J Pain*. 2006;10(4):287−333. Available from: https://doi.org/10.1016/j.ejpain.2005.06.009.

5. Johannes CB, Le TK, Zhou X, Johnston JA, Dworkin RH. The prevalence of chronic pain in United States adults: results of an internet-based survey. *J Pain*. 2010;11(11):1230–1239. Available from: https://doi.org/10.1016/j.jpain.2010.07.002.
6. Smith BH, Torrance N. Epidemiology of chronic pain. In: McQuay HJ, Kalso EMR, eds. *Systematic Reviews in Pain Research: Methodology Refined*. 2008:233–246. <http://media.axon.es/pdf/90663.pdf> Accessed 27.09.17.
7. Cohen LL, Vowles KE, Eccleston C. The impact of adolescent chronic pain on functioning: disentangling the complex role of anxiety. *J Pain*. 2010;11(11):1039–1046. Available from: https://doi.org/10.1016/j.jpain.2009.09.009.
8. Tang NKY, Crane C. Suicidality in chronic pain: a review of the prevalence, risk factors and psychological links. *Psychol Med*. 2006;36(5):575. Available from: https://doi.org/10.1017/S0033291705006859.
9. Andrew R, Derry S, Taylor RS, Straube S, Phillips CJ. The costs and consequences of adequately managed chronic non-cancer pain and chronic neuropathic pain. *Pain Pract*. 2014;14(1):79–94. Available from: https://doi.org/10.1111/papr.12050.
10. Palermo TM. Impact of recurrent and chronic pain on child and family daily functioning: a critical review of the literature. *J Dev Behav Pediatr*. 2000;21(1):58–69. <http://www.ncbi.nlm.nih.gov/pubmed/10706352> Accessed September 28, 2017.
11. Groenewald CB, Essner BS, Wright D, Fesinmeyer MD, Palermo TM. The economic costs of chronic pain among a cohort of treatment-seeking adolescents in the United States. *J Pain*. 2014;15(9):925–933. Available from: https://doi.org/10.1016/j.jpain.2014.06.002.
12. Ohayon MM, Schatzberg AF. Chronic pain and major depressive disorder in the general population. *J Psychiatr Res*. 2010;44(7):454–461. Available from: https://doi.org/10.1016/j.jpsychires.2009.10.013.
13. Baglioni C, Battagliese G, Feige B, et al. Insomnia as a predictor of depression: a meta-analytic evaluation of longitudinal epidemiological studies. *J Affect Disord*. 2011;135(1–3):10–19. Available from: https://doi.org/10.1016/j.jad.2011.01.011.
14. Léger D, Partinen M, Hirshkowitz M, et al. Daytime consequences of insomnia symptoms among outpatients in primary care practice: EQUINOX international survey. *Sleep Med*. 2010;11(10):999–1009. Available from: https://doi.org/10.1016/j.sleep.2010.04.018.
15. Lovato N, Gradisar M. A meta-analysis and model of the relationship between sleep and depression in adolescents: recommendations for future research and clinical practice. *Sleep Med Rev*. 2014;18(6):521–529. Available from: https://doi.org/10.1016/j.smrv.2014.03.006.
16. Shochat T, Cohen-Zion M, Tzischinsky O. Functional consequences of inadequate sleep in adolescents: a systematic review. *Sleep Med Rev*. 2014;18(1):75–87. Available from: https://doi.org/10.1016/j.smrv.2013.03.005.
17. Power JD, Perruccio AV, Badley EM. Pain as a mediator of sleep problems in arthritis and other chronic conditions. *Arthritis Rheum*. 2005;53(6):911–919. Available from: https://doi.org/10.1002/art.21584.
18. Sasaki E, Tsuda E, Yamamoto Y, et al. Nocturnal knee pain increases with the severity of knee osteoarthritis, disturbing patient sleep quality. *Arthritis Care Res (Hoboken)*. 2014;66(7):1027–1032. Available from: https://doi.org/10.1002/acr.22258.
19. Allen KD, Renner JB, Devellis B, Helmick CG, Jordan JM. Osteoarthritis and sleep: the Johnston County Osteoarthritis Project. *J Rheumatol*. 2008;35(6):1102–1107. Available from: http://www.ncbi.nlm.nih.gov/pubmed/18484690.
20. Petrov ME, Goodin BR, Cruz-almeida Y, et al. Disrupted sleep is associated with altered pain processing by sex and ethnicity in knee osteoarthritis. *J Pain*. 2015;16(5):478–490. Available from: https://doi.org/10.1016/j.jpain.2015.02.004.
21. Wilcox S, Brenes GA, Levine D, Sevick MA, Shumaker SA, Craven T. Factors related to sleep disturbance in older adults experiencing knee pain or knee pain with radiographic evidence of knee osteoarthritis. *J Am Geriatr Soc*. 2000;48(10):1241–1251. Available from: https://doi.org/10.1111/j.1532-5415.2000.tb02597.x.
22. Leigh TJ, Hindmarch I, Bird HA, Wright V. Comparison of sleep in osteoarthritic patients and age and sex matched healthy controls. *Ann Rheum Dis*. 1988;47(1):40–42. Available from: http://www.ncbi.nlm.nih.gov/entrez/query.fcgi?cmd=Retrieve&db=PubMed&dopt=Citation&list_uids=3345103.
23. Gowdie PJ, Tse SML. Juvenile idiopathic arthritis. *Pediatr Clin North Am*. 2012;59(2):301–327. Available from: https://doi.org/10.1016/j.pcl.2012.03.014.

REFERENCES

24. Stinson JN, Hayden JA, Ahola Kohut S, et al. Sleep problems and associated factors in children with juvenile idiopathic arthritis: a systematic review. *Pediatr Rheumatol Online J*. 2014;12(1):1–12. Available from: https://doi.org/10.1186/1546-0096-12-19.

25. Lopes MC, Guilleminault C, Rosa A, Passarelli C, Roizenblatt S, Tufik S. Delta sleep instability in children with chronic arthritis. *Braz J Med Biol Res*. 2008;41(10):938–943. Available from: https://doi.ord/S0100-879X2008001000018 [pii] ET - 2008/11/26.

26. Zamir G, Press J, Tal A, Tarasiuk A. Sleep fragmentation in children with juvenile rheumatoid arthritis. *J Rheumatol*. 1998;25(6):1191–1197. Available from: https://www.ncbi.nlm.nih.gov/pubmed/9632085.

27. Yuwen W, Chen ML, Cain KC, Ringold S, Wallace CA, Ward TM. Daily sleep patterns, sleep quality, and sleep hygiene among parent–child dyads of young children newly diagnosed with juvenile idiopathic arthritis and typically developing children. *J Pediatr Psychol*. 2016;41(March):1–10. Available from: https://doi.org/10.1093/jpepsy/jsw007.

28. Twycross A, Parker R, Williams A, Gibson F. Cancer-related pain and pain management. *J Pediatr Oncol Nurs*. 2015;32(6):369–384. Available from: https://doi.org/10.1177/1043454214563751.

29. van den Beuken-van Everdingen MHJ, de Rijke JM, Kessels AG, Schouten HC, van Kleef M, Patijn J. Prevalence of pain in patients with cancer: a systematic review of the past 40 years. *Ann Oncol*. 2007;18 (9):1437–1449. Available from: https://doi.org/10.1093/annonc/mdm056.

30. Mercadante S. Cancer pain management in children. *Palliat Med*. 2004;18(7):654–662. Available from: https://doi.org/10.1191/0269216304pm945rr.

31. Walter LM, Nixon GM, Davey MJ, Downie PA, Horne RSC. Sleep and fatigue in pediatric oncology: a review of the literature. *Sleep Med Rev*. 2015;24:71–82. Available from: https://doi.org/10.1016/j.smrv.2015.01.001.

32. Wright M. Children receiving treatment for cancer and their caregivers: a mixed methods study of their sleep characteristics. *Pediatr Blood Cancer*. 2011;56(4):638–645. Available from: https://doi.org/10.1002/pbc.22732.

33. Palesh OG, Roscoe JA, Mustian KM, et al. Prevalence, demographics, and psychological associations of sleep disruption in patients with cancer: University of Rochester Cancer Center-community clinical oncology program. *J Clin Oncol*. 2010;28(2):292–298. Available from: https://doi.org/10.1200/JCO.2009.22.5011.

34. Rosen G, Brand SR. Sleep in children with cancer: case review of 70 children evaluated in a comprehensive pediatric sleep center. *Support Care Cancer*. 2011;19(7):985–994. Available from: https://doi.org/10.1007/s00520-010-0921-y.

35. Daniel LC, Schwartz LA, Mindell JA, Tucker CA, Barakat LP. Initial validation of the sleep disturbances in pediatric cancer model. *J Pediatr Psychol*. 2016;41(6):588–599. Available from: https://doi.org/10.1093/jpepsy/jsw008.

36. Rosen GM, Shor AC, Geller TJ. Sleep in children with cancer. *Curr Opin Pediatr*. 2008;20(6):676–681. Available from: https://doi.org/10.1097/MOP.0b013e328312c7ad.

37. Campos MPO, Hassan BJ, Riechelmann R, Del Giglio A. Cancer-related fatigue: a practical review. *Ann Oncol*. 2011;22(6):1273–1279. Available from: https://doi.org/10.1093/annonc/mdq458.

38. Brown JC, Huedo-Medina TB, Pescatello LS, Pescatello SM, Ferrer RA, Johnson BT. Efficacy of exercise interventions in modulating cancer-related fatigue among adult cancer survivors: a meta-analysis. *Cancer Epidemiol Biomarkers Prev*. 2011;20(1):123–133. Available from: https://doi.org/10.1158/1055-9965.EPI-10-0988.

39. Dampier C, Palermo TM, Darbari DS, Hassell K, Smith W, Zempsky W. AAPT diagnostic criteria for chronic sickle cell disease pain perspective: an evidence-based classification system for chronic SCD pain was constructed for the. *J Pain*. 2017;18:490–498. Available from: https://doi.org/10.1016/j.jpain.2016.12.016.

40. Rosen CL, Debaun MR, Strunk RC, et al. Obstructive sleep apnea and sickle cell anemia. *Pediatrics*. 2014;134 (2):273–281. Available from: https://doi.org/10.1542/peds.2013-4223.

41. Samuels MP, Stebbens VA, Davies SC, Picton-Jones E, Southall DP. Sleep related upper airway obstruction and hypoxaemia in sickle cell disease. *Arch Dis Child*. 1992;67(7):925–929. Available from: https://doi.org/10.1136/ADC.67.7.925.

42. Daniel LC, Grant M, Kothare SV, Dampier C, Barakat LP. Sleep patterns in pediatric sickle cell disease. *Pediatr Blood Cancer*. 2010;55(3):501–507. Available from: https://doi.org/10.1002/pbc.22564.

43. Valrie CR, Gil KM, Redding-Lallinger R, Daeschner C. Brief report: sleep in children with sickle cell disease: an analysis of daily diaries utilizing multilevel models. *J Pediatr Psychol*. 2007;32(7):857–861. Available from: https://doi.org/10.1093/jpepsy/jsm016.

44. Whitesell PL, Owoyemi O, Oneal P, et al. Sleep-disordered breathing and nocturnal hypoxemia in young adults with sickle cell disease. *Sleep Med*. 2016;22:47−49. Available from: https://doi.org/10.1016/j.sleep.2016.05.006.
45. Sharma S, Efird JT, Knupp C, et al. Sleep disorders in adult sickle cell patients. *J Clin Sleep Med*. 2015;11 (3):219−223. Available from: https://doi.org/10.5664/jcsm.4530.
46. Wallen GR, Minniti CP, Krumlauf M, et al. Sleep disturbance, depression and pain in adults with sickle cell disease. *BMC Psychiatry*. 2014;14(1):207. Available from: https://doi.org/10.1186/1471-244X-14-207.
47. Almoznino G, Benoliel R, Sharav Y, Haviv Y. Sleep disorders and chronic craniofacial pain: characteristics and management possibilities. *Sleep Med Rev*. 2016. Available from: https://doi.org/10.1016/j.smrv.2016.04.005.
48. Vendrame M, Kaleyias J, Valencia I, Legido A, Kothare SV. Polysomnographic findings in children with headaches. *Pediatr Neurol*. 2008;39(1):6−11. Available from: https://doi.org/10.1016/j.pediatrneurol.2008.03.007.
49. Esposito M, Parisi P, Miano S, Carotenuto M. Migraine and periodic limb movement disorders in sleep in children: a preliminary case-control study. *J Headache Pain*. 2013;14(1):57. Available from: https://doi.org/10.1186/1129-2377-14-57.
50. Stark CD, Stark RJ. Sleep and chronic daily headache. *Curr Pain Headache Rep*. 2015;19(1):468. Available from: https://doi.org/10.1007/s11916-014-0468-6.
51. Roizenblatt S, Tufik S, Goldenberg J, Pinto LR, Hilario MO, Feldman D. Juvenile fibromyalgia: clinical and polysomnographic aspects. *J Rheumatol*. 1997;24(3):579−585. Available from: http://www.ncbi.nlm.nih.gov/pubmed/9058669.
52. Belt NK, Kronholm E, Kauppi MJ. Sleep problems in fibromyalgia and rheumatoid arthritis compared with the general population. *Clin Exp Rheumatol*. 2009;27(1):35−41. Available from: http://www.ncbi.nlm.nih.gov/pubmed/19327227.
53. Wu Y, Chang L, Lee H. Sleep disturbances in fibromyalgia: a meta-analysis of case-control studies. *J Psychosom Res*. 2017;96(March):89−97. Available from: https://doi.org/10.1016/j.jpsychores.2017.03.011.
54. Spaeth M, Rizzi M, Sarzi-Puttini P. Fibromyalgia and sleep. *Best Pract Res Clin Rheumatol*. 2011;25(2):227−239. Available from: https://doi.org/10.1016/j.berh.2011.03.004.
55. Tayag-Kier CE, Keenan GF, Scalzi LV, et al. Sleep and periodic limb movement in sleep in juvenile fibromyalgia. *Pediatrics*. 2000;106(5):e70. Available from: https://doi.org/10.1542/peds.106.5.e70.
56. Stuifbergen AK, Phillips L, Carter P, Morrison J, Todd A. Subjective and objective sleep difficulties in women with fibromyalgia syndrome. *J Am Acad Nurse Pract*. 2010;22(10):548−556. Available from: https://doi.org/10.1111/j.1745-7599.2010.00547.x.
57. Smith MT, Perlis ML, Smith MS, Giles DE, Carmody TP. Sleep quality and presleep arousal in chronic pain. *J Behav Med*. 2000;23(1). < https://link-springer-com.flagship.luc.edu/content/pdf/10.1023%2FA%3A1005444719169.pdf > Accessed July 18, 2017.
58. Lavigne G, Brousseau M, Kato T, et al. Experimental pain perception remains equally active over all sleep stages. *Pain*. 2004;110(3):646−655. Available from: https://doi.org/10.1016/j.pain.2004.05.003.
59. Lavigne G, Zucconi M, Castronovo C, Manzini C, Marchettini P, Smirne S. Sleep arousal response to experimental thermal stimulation during sleep in human subjects free of pain and sleep problems. *Pain*. 2000;84 (2):283−290. Available from: https://doi.org/10.1016/S0304-3959(99)00213-4.
60. Lewandowski AS, Palermo TM, De la Motte S, Fu R. Temporal daily associations between pain and sleep in adolescents with chronic pain versus healthy adolescents. *Pain*. 2010;151(1):220−225. Available from: https://doi.org/10.1016/j.pain.2010.07.016.
61. Tang NKY, Goodchild CE, Sanborn AN, Howard J, Salkovskis PM. Deciphering the temporal link between pain and sleep in a heterogeneous chronic pain patient sample: a multilevel daily process study. *Sleep*. 2012;35 (5):675A−687A. Available from: https://doi.org/10.5665/sleep.1830.
62. Affleck G, Urrows S, Tennen H, Higgins P, Abeles M. Sequential daily relations of sleep, pain intensity, and attention to pain among women with fibromyalgia. *Pain*. 1996;68(2):363−368. Available from: https://doi.org/10.1016/S0304-3959(96)03226-5.
63. Finan PH, Goodin BR, Smith MT. The association of sleep and pain: an update and a path forward. *J Pain*. 2013;14(12):1539−1552. Available from: https://doi.org/10.1016/j.jpain.2013.08.007.
64. Roehrs T, Hyde M, Blaisdell B, Greenwald M, Roth T. Sleep loss and REM sleep loss are hyperalgesic. *Sleep*. 2006;29(2):145−151. Available from: https://doi.org/10.1093/sleep/29.2.145.

REFERENCES

65. Smith MT, Edwards RR, McCann UD, Haythornthwaite JA. The effects of sleep deprivation on pain inhibition and spontaneous pain in women. *Sleep.* 2007;30(4):494–505. < https://watermark.silverchair.com/api/watermark?token = AQECAHi208BE49Ooan9kkhW_Ercy7Dm3ZL_9Cf3qfKAc485ysgAAAdswggHXBgkqhkiG9w0BBwagggHIMIIBxAIBADCCAb0GCSqGSIb3DQEHATAeBglghkgBZQMEAS4wEQQMs-4ruI2SU2JrFd_NAgEQgIIBjoDGtMjnhK3ut4RVb1_DnYRdu4ttC2mpOcHozYwEvFrxc > Accessed September 28, 2017.

66. Irwin MR, Olmstead R, Carrillo C, et al. Sleep loss exacerbates fatigue, depression, and pain in rheumatoid arthritis. *Sleep.* 2012;35(4):537–543. Available from: https://doi.org/10.5665/sleep.1742.

67. Finan PH, Smith MT. The comorbidity of insomnia, chronic pain, and depression: dopamine as a putative mechanism. *Sleep Med Rev.* 2013;17(3):173–183. Available from: https://doi.org/10.1016/j.smrv.2012.03.003.

68. Legangneux E, Mora JJ, Spreux-Varoquaux O, et al. Cerebrospinal fluid biogenic amine metabolites, plasma-rich platelet serotonin and [3H]imipramine reuptake in the primary fibromyalgia syndrome. *Rheumatology.* 2001;40(3):290–296. Available from: https://doi.org/10.1093/rheumatology/40.3.290.

69. Mullington JM, Simpson NS, Meier-Ewert HK, Haack M. Sleep loss and inflammation. *Best Pract Res Clin Endocrinol Metab.* 2010;24(5):775–784. Available from: https://doi.org/10.1016/j.beem.2010.08.014.

70. Haack M, Sanchez E, Mullington JM. Elevated inflammatory markers in response to prolonged sleep restriction are associated with increased pain experience in healthy volunteers. *Sleep.* 2007;30(9):1145–1152. Available from: https://doi.org/10.1093/sleep/30.9.1145.

71. Haack M, Lee E, Cohen DA, Mullington JM. Activation of the prostaglandin system in response to sleep loss in healthy humans: potential mediator of increased spontaneous pain. *Pain.* 2009;145(1):136–141. Available from: https://doi.org/10.1016/j.pain.2009.05.029.

72. Watkins LR, Maier SF. Immune regulation of central nervous system functions: from sickness responses to pathological pain. *J Intern Med.* 2005;257(2):139–155. Available from: https://doi.org/10.1111/j.1365-2796.2004.01443.x.

73. Palermo TM, Wilson AC, Lewandowski AS, Toliver-Sokol M, Murray CB. Behavioral and psychosocial factors associated with insomnia in adolescents with chronic pain. *Pain.* 2011;152(1):89–94. Available from: https://doi.org/10.1016/j.pain.2010.09.035.

74. Tang NKY, Goodchild CE, Hester J, Salkovskis PM. Pain-related insomnia versus primary insomnia. *Clin J Pain.* 2012;28(5):428–436. Available from: https://doi.org/10.1097/AJP.0b013e31823711bc.

75. Byers HD, Lichstein KL, Thorn BE. Cognitive processes in comorbid poor sleep and chronic pain. *J Behav Med.* 2016;39(2):233–240. Available from: https://doi.org/10.1007/s10865-015-9687-5.

76. Campbell CM, Buenaver LF, Finan P, et al. Sleep, pain catastrophizing, and central sensitization in knee osteoarthritis patients with and without insomnia. *Arthritis Care Res.* 2015;67(10). Available from: https://doi.org/10.1002/acr.22609.

77. Buenaver LF, Quartana PJ, Grace EG, et al. Evidence for indirect effects of pain catastrophizing on clinical pain among myofascial temporomandibular disorder participants: the mediating role of sleep disturbance. *Pain.* 2012;153(6):1159–1166. Available from: https://doi.org/10.1016/j.pain.2012.01.023.

78. Jeff Bryson W, Read JB, Bush JP, Edwards CL. The need for an integrated cognitive-behavioral model for co-occurring chronic pain and insomnia. *J Ration-Emotive Cogn-Behav Ther.* 2015;33(3):239–257. Available from: https://doi.org/10.1007/s10942-015-0213-z.

79. Koffel E, Krebs EE, Arbisi PA, Erbes CR, Polusny MA. The unhappy triad: pain, sleep complaints, and internalizing symptoms. *Clin Psychol Sci.* 2016;4(1):96–106. Available from: https://doi.org/10.1177/2167702615579342.

80. O'Brien EM, Waxenberg LB, Atchison JW, et al. Negative mood mediates the effect of poor sleep on pain among chronic pain patients. *Clin J Pain.* 2010;26(4):310–319. Available from: https://doi.org/10.1097/AJP.0b013e3181c328e9.

81. Valrie CR, Gil KM, Redding-Lallinger R, Daeschner C. Brief report: daily mood as a mediator or moderator of the pain–sleep relationship in children with sickle cell disease. *J Pediatr Psychol.* 2007;33(3):317–322. Available from: https://doi.org/10.1093/jpepsy/jsm058.

82. Owens JA, Spirito A, McGuinn M. The Children's Sleep Habits Questionnaire (CSHQ): psychometric properties of a survey instrument for school-aged children. *Sleep.* 2000;23(8):1043–1051. < https://pdfs.semanticscholar.org/b706/777c818bc92d554dc27d8f419e45f3f9929b.pdf > Accessed September 28, 2017.

83. LeBourgeois MK. The relationship between reported sleep quality and sleep hygiene in Italian and American adolescents. *Pediatrics.* 2005;115(1):257–265. Available from: https://doi.org/10.1542/peds.2004-0815H.

84. Essner B, Noel M, Myrvik M, Palermo T. Examination of the factor structure of the adolescent sleep–wake scale (ASWS. *Behav Sleep Med.* 2015;13(4):296–307. Available from: https://doi.org/10.1080/15402002.2014.896253.

85. de la Vega R, Miró J. The assessment of sleep in pediatric chronic pain sufferers. *Sleep Med Rev.* 2013;17(3):185–192. Available from: https://doi.org/10.1016/j.smrv.2012.04.002.

86. Buysse DJ, Reynolds CF, Monk TH, Berman SR, Kupfer DJ. The Pittsburgh Sleep Quality Index: a new instrument for psychiatric practice and research. *Psychiatry Res.* 1989;28(2):193–213. < http://www.ncbi.nlm.nih.gov/pubmed/2748771 > Accessed September 28, 2017.

87. Buysse DJ, Yu L, Moul DE, et al. Development and validation of patient-reported outcome measures for sleep disturbance and sleep-related impairments. *Sleep.* 2010;33(6):781–792. < http://www.ncbi.nlm.nih.gov/pubmed/20550019 > Accessed September 28, 2017.

88. Bastien CH, Vallières A, Morin CM. Validation of the Insomnia Severity Index as an outcome measure for insomnia research. *Sleep Med.* 2001;2(4):297–307. < https://www.researchgate.net/profile/Charles_Morin/publication/11903319_Validation_of_the_Insomnia_Severity_Index_ISI_as_an_outcome_measure_for_insomnia_research/links/02e7e52c0f301f16e3000000/Validation-of-the-Insomnia-Severity-Index-ISI-as-an-outcome-me > Accessed September 28, 2017.

89. Morin C. *Insomnia: Psychological Assessment and Management.* New York: Guilford Press; 1993. < http://psycnet.apa.org/record/1993-98362-000 > Accessed May 6, 2018.

90. Alfano CA, Pina AA, Zerr AA, Villalta IK. Pre-sleep arousal and sleep problems of anxiety-disordered youth. *Child Psychiatry Hum Dev.* 2010;41(2):156–167. Available from: https://doi.org/10.1007/s10578-009-0158-5.

91. Nicassio PM, Mendlowitz DR, Fussell JJ, Petras L. The phenomenology of the pre-sleep state: the development of the pre-sleep arousal scale. *Behav Res Ther.* 1985;23(3):263–271. Available from: https://doi.org/10.1016/0005-7967(85)90004-X.

92. Morin CM. Dysfunctional beliefs and attitudes about sleep: preliminary scale development and description. *Behav Ther.* 1994;(Summer)163–164.

93. Espie CA, Inglis SJ, Harvey L, Tessier S. Insomniacs' attributions. psychometric properties of the Dysfunctional Beliefs and Attitudes about Sleep Scale and the Sleep Disturbance Questionnaire. *J Psychosom Res.* 2000;48(2):141–148. < http://www.ncbi.nlm.nih.gov/pubmed/10719130 > Accessed September 28, 2017.

94. Gregory AM, Cox J, Crawford MR, et al. Dysfunctional beliefs and attitudes about sleep in children. *J Sleep Res.* 2009;18(4):422–426. Available from: https://doi.org/10.1111/j.1365-2869.2009.00747.x.

95. Horne R, Biggs S. Actigraphy and sleep/wake diaries. In: Wolfson A, Montgomery-Downs H, eds. *The Oxford Handbook of Infant, Child, and Adolescent Sleep and Behavior.* New York: Oxford University Press; 2013:189–203.

96. Buysse D, Ancoli-Israel S, Edinger J, Lichstein K, Morin C. Recommendations for a standard research assessment of insomnia. *Sleep.* 2006;29(9):1155–1173. Available from: https://doi.org/10.1093/sleep/29.9.1155.

97. Marshansky S, Mayer P, Rizzo D, Baltzan M, Denis R, Lavigne GJ. Sleep, chronic pain, and opioid risk for apnea. *Prog Neuro-Psychopharmacology Biol Psychiatry.* July 2017. Available from: https://doi.org/10.1016/j.pnpbp.2017.07.014.

98. Cheatle MD, Foster S, Pinkett A, Lesneski M, Qu D, Dhingra L. Assessing and managing sleep disturbance in patients with chronic pain. *Anesthesiol Clin.* 2016;34(2):379–393. Available from: https://doi.org/10.1016/j.anclin.2016.01.007.

99. Van Ryswyk E, Antic NA. Opioids and sleep-disordered breathing. *Chest.* 2016;150(4):934–944. Available from: https://doi.org/10.1016/j.chest.2016.05.022.

100. Kashikar-Zuck S, Swain NF, Jones BA, Graham TB. Efficacy of cognitive-behavioral intervention for juvenile primary fibromyalgia syndrome. *J Rheumatol.* 2005;32(8):1594–1602. < http://www.ncbi.nlm.nih.gov/pubmed/16078340 > Accessed September 28, 2017.

101. Palermo TM, Law EF, Fales J, Bromberg MH, Jessen-Fiddick T, Tai G. Internet-delivered cognitive-behavioral treatment for adolescents with chronic pain and their parents: a randomized controlled multicenter trial. *Pain.* 2016;157(1):174–185. Available from: https://doi.org/10.1097/j.pain.0000000000000348.

102. Wu JQ, Appleman ER, Salazar RD, Ong JC. Cognitive behavioral therapy for insomnia comorbid with psychiatric and medical conditions. *JAMA Intern Med.* 2015;175(9):1461. Available from: https://doi.org/10.1001/jamainternmed.2015.3006.

REFERENCES

103. Finan PH, Buenaver LF, Runko VT, Smith MT. Cognitive-behavioral therapy for comorbid insomnia and chronic pain. *Sleep Med Clin*. 2014;9(2):261−274. Available from: https://doi.org/10.1016/j.jsmc.2014.02.007.

104. Edinger JD, Wohlgemuth WK, Krystal AD, Rice JR. Behavioral insomnia therapy for fibromyalgia patients. *Arch Intern Med*. 2005;165(21):2527. Available from: https://doi.org/10.1001/archinte.165.21.2527.

105. Smith MT, Finan PH, Buenaver LF, et al. Cognitive-behavioral therapy for insomnia in knee osteoarthritis: a randomized, double-blind, active placebo-controlled clinical trial. *Arthritis Rheumatol*. 2015;67(5):1221−1233. Available from: https://doi.org/10.1002/art.39048.

106. Palermo TM, Beals-Erickson S, Bromberg M, Law E, Chen M. A single arm pilot trial of brief cognitive behavioral therapy for insomnia in adolescents with physical and psychiatric comorbidities. *J Clin Sleep Med*. 2017;13(3):401−410. Available from: https://doi.org/10.5664/jcsm.6490.

107. Blom K, Tarkian Tillgren H, Wiklund T, et al. Internet-vs. group-delivered cognitive behavior therapy for insomnia: a randomized controlled non-inferiority trial. *Behav Res Ther*. 2015;70:47−55. Available from: https://doi.org/10.1016/j.brat.2015.05.002.

II. SLEEP DISORDERS IN SPECIFIC MEDICAL CONDITIONS

CHAPTER 10

Traumatic Brain Injury

Marie-Christine Ouellet[1,2], Simon Beaulieu-Bonneau[1,2,3] and Charles M. Morin[2,3]

[1]Center for Interdisciplinary Research in Rehabilitation and Social Integration (CIRRIS), Québec, QC, Canada [2]School of Psychology, Université Laval, Québec, QC, Canada [3]CERVO Brain Research Centre, Québec, QC, Canada

OUTLINE

Introduction	222	*Differentiating Sleepiness from Fatigue*	230
Brief Overview of Traumatic Brain Injury (Prevalence, Etiology, Treatment)	223	**Insomnia**	**231**
		Risk Factors	*231*
		Assessment	*232*
		Pharmacological Interventions	*233*
Sleep—Wake Alterations in the Acute Phase After Traumatic Brain Injury and Impacts on Recovery	224	*Nonpharmacological Interventions*	*234*
		Clinical Vignette	**237**
Sleep Architecture in the Chronic Phase After Traumatic Brain Injury	225	**Excessive Daytime Sleepiness, Posttraumatic Hypersomnia, Sleep-Disordered Breathing, and Increased Need for Sleep**	**238**
Pathophysiology of Sleep—Wake Disturbances After Traumatic Brain Injury	226	*Assessment*	*240*
		Pharmacological Treatment	*240*
Potential Impacts of Sleep—Wake Disturbances on the Evolution of the Condition After Traumatic Brain Injury	228	*Nonpharmacological Treatment*	*241*
Cognitive Functioning	*228*	**Circadian Rhythm Sleep—Wake Disorders**	**242**
Functioning in the Chronic Phase	*229*	*Assessment of Circadian Rhythm Disorders*	*242*
Interaction With Pain	*229*		
Interaction With Psychopathology	*229*		

Handbook of Sleep Disorders in Medical Conditions
DOI: https://doi.org/10.1016/B978-0-12-813014-8.00010-X

© 2019 Elsevier Inc. All rights reserved.

Pharmacological and Nonpharmacological Treatment of Circadian Disturbances	242	**Conclusion**	243
		References	244
Areas for Future Research	243		

INTRODUCTION

Traumatic brain injury (TBI) is recognized as a leading cause of mortality and disability worldwide. Approximately 1.6 million individuals suffer a TBI in the United States each year and it is estimated that between 2.5 and 6.5 million people live with the repercussions of TBI every year.[1] The severity, duration, and persistence of TBI sequelae depend on the severity of the injury and on the interplay of various preinjury characteristics and postinjury factors. Persisting consequences of TBI may be felt at the physical, cognitive, emotional, and social levels[2] with impacts not only on the injured persons' lives but also on their families and communities in terms of social participation. Sleep disturbances are very frequent after TBI and may persist months or years postinjury. These take various forms including difficulties initiating or maintaining sleep, excessive sleepiness or fatigue during the day, changes in the sleep–wake rhythms, or an increased need for sleep. Sleep disturbances can emerge at different times in the post-TBI trajectory, as early as during the acute phase, or during hospitalization, but also in later phases such as during inpatient rehabilitation care or when the injured person reintegrates the community and attempts to resume preinjury social roles.[3–5] In their meta-analysis, Mathias et al.[6] found that regardless of injury severity, time since injury, or definition used, 50% of individuals with TBI have some form of sleep disturbance, and between 25% and 29% have a sleep disorder per se, the most frequent being insomnia and hypersomnia, followed by sleep apnea. The emergence of sleep–wake disturbances following TBI is explained by complex interactions between neurobiological, psychological, behavioral, and environmental factors, some of these arising after the TBI while others were premorbid. Alterations in sleep–wake patterns can exacerbate the consequences of TBI, including cognitive deficits, emotional problems, physical functioning, pain, fatigue, and, ultimately, functional independence, social participation, and quality of life.

In the past 20 years, a large literature has flourished describing the undeniably significant prevalence of sleep disorders after TBI. Despite this data, it remains unclear whether sleep–wake disturbances receive enough clinical attention. Moreover, research is still lacking in certain important areas, such as the impacts of sleep–wake impairments on the rehabilitation process, their interaction with functional recovery, and the adaptation of pharmacological and nonpharmacological interventions for sleep–wake disorders to the particular challenges of the TBI population. In Canada, recently published clinical guidelines give data-driven recommendations for the assessment and management of sleep–wake disturbances after mild TBI[7] and moderate–severe TBI,[8] attesting to an increasing attention to this field.

This chapter will present a brief overview of TBI, followed by sections on sleep–wake disturbances in the acute and chronic phase, pathophysiology, and potential impact of sleep–wake disturbances on TBI recovery. The three main types of sleep–wake disturbances will then be discussed, in terms of epidemiology, assessment, and treatment: insomnia, excessive daytime sleepiness (EDS) and associated disorders, and circadian sleep–wake rhythm disorders. Other sleep disorders such as parasomnias or restless leg syndrome have only been reported anecdotally post-TBI.

BRIEF OVERVIEW OF TRAUMATIC BRAIN INJURY (PREVALENCE, ETIOLOGY, TREATMENT)

TBI is defined as "an alteration of brain function (e.g., loss of or decreased level of consciousness, loss of memory for events immediately before – retrograde amnesia – or after – posttraumatic amnesia – the injury), or other evidence of brain pathology (i.e., visual, neuroradiologic, or laboratory confirmation of damage to the brain), caused by an external force (e.g., acceleration/deceleration movement of the brain, head being struck by or striking an object, foreign body penetrating the brain)."[9] The majority of TBIs are caused by blunt injuries (as opposed to penetrating injuries) and lead to multiple pathophysiological events including primary damage (i.e., focal lesions including contusions, lacerations, hematomas, or diffuse axonal injury) and secondary damage which may involve various pathophysiological cascades (e.g., ischemia, elevated intracranial pressure, or inflammatory response, abnormal neurotransmitter release, hormonal changes). Neurodegeneration processes such as cerebral atrophy have been observed to continue up to a year postinjury,[10] suggesting that the brain continues to reorganize over many months postinjury.

The two main causes of TBI are motor vehicle/traffic accidents and falls. Age is a major risk factor, with higher incidence of TBI in young children (0–4 years; falls), adolescents and young adults (15–24 years; motor vehicle accidents), and older adults (>75 years; falls).[11] Because of the aging of the population, TBI in the elderly is rapidly becoming a major public health issue, yet research concerning various post-TBI health outcomes (besides mortality) in this age group is still alarmingly limited. Men are twice more likely than women to sustain a TBI. Other risk factors include lower education, lower socioeconomic status, unemployment, and alcohol or drug abuse.[12,13]

TBIs are generally categorized into three levels of severity: mild, moderate, and severe. Several criteria can be used to assess TBI severity, including the Glasgow Coma Scale (GCS)[14] score (13–15: mild; 9–12: moderate; 8 or less: severe), the duration of loss of consciousness (0–30 minutes: mild; 30 minutes–24 hours: moderate; >24 hours: severe), the duration of posttraumatic amnesia (<1 day: mild; 1–7 days: moderate; >7 days: severe), and neurological and neuroimaging results. The vast majority of TBIs, between 70% and 90%, are classified as mild, while 5%–10% are moderate, and 5%–10% are severe.[15,16]

TBI is often referred to as a silent epidemic because, while it can significantly impact the lives of the injured persons, their families, and society in general, many cases of TBI are not identified or treated, and several of the sequelae are not readily visible (e.g., cognitive deficits, personality changes).[2] Following a mild TBI (or concussion), most individuals are asymptomatic by 3–12 months postinjury, but a minority of people (approximately

15%) continues to suffer from persistent symptoms.[17] Many persons sustaining a mild TBI receive little or even no clinical attention from a healthcare professional. When they decide to consult, some individuals will wait several days before visiting their primary care clinic. Of those who visit the emergency department (ED) after a head injury, several will be discharged after a short stay, hopefully with some recommendations on what to expect, recommendations for gradual return to activities, and signs necessitating further evaluation, although it is quite unclear how many patients receive adequate information after an ED visit for a concussion. A portion will receive rehabilitation services such as consultations in neuropsychology, occupational therapy, or vocational rehabilitation services. Following moderate-to-severe TBI, consciousness is altered for a greater period of time, and the duration of posttraumatic amnesia, in particular, is strongly associated with long-term functional consequences such as neuropsychological functioning or return to work. Compared to those with mild TBI, recovery takes longer and permanent sequelae are more common. In fact TBI is now widely admitted to be a chronic medical condition with many individuals living with evolving sequelae for the rest of their life.[18] The continuum of care following moderate or severe TBI is clearer, usually encompassing several steps that may include prehospital care, acute care [intensive care unit (ICU) management and initial hospitalization], inpatient rehabilitation, outpatient rehabilitation, and community reintegration services.[19]

TBI can affect multiple domains of functioning, including physical health (e.g., motor limitations, pain, sensory deficits), psychological health (e.g., depression, anxiety, substance use), cognition (e.g., impairments in attention, memory, and executive functions), and sociooccupational life (e.g., unemployment, unfitness to drive, marital difficulties, reduction of social network). Sleep—wake disturbances are among the most common consequences of TBI, and they interact with other conditions (e.g., psychopathology, pain, chronic fatigue), thereby potentially affecting rehabilitation, recovery, and return to premorbid functioning.

SLEEP—WAKE ALTERATIONS IN THE ACUTE PHASE AFTER TRAUMATIC BRAIN INJURY AND IMPACTS ON RECOVERY

Although research describing sleep in the acute phase post-TBI is still limited, it clearly indicates that sleep—wake patterns are disrupted in the first few days after a moderate or severe TBI. Patients treated in the ICU, regardless of diagnosis, have significantly disturbed sleep such as frequent awakenings, prolonged sleep-onset latency, and significant decreases and sometimes eradication of deep or slow-wave sleep and rapid eye movement (REM) sleep.[20,21] Patients are also often hypersomnolent in the first few days after the injury.[22–24] The earliest studies examining sleep after TBI[25,26] indicated that the absence of spontaneous sleep activity in the first few days postaccident was associated with a poor prognosis such as a vegetative state, death, or subsequent severe cognitive deficits. A more recent literature has emerged[27–31] clearly demonstrating that consciousness and cognitive recovery improve in a trajectory parallel to the evolution in sleep quality during the acute phase. Whether sleep quality contributes directly to cerebral recovery or represents mostly a marker of recovery is still unclear, and these hypothesized processes are probably not mutually exclusive.

Duclos et al.[27] used actigraphy to examine rest–activity cycle consolidation in patients upon their admission to the ICU and up to 10 days after moderate/severe TBI. In the first days following injury in this bed-ridden population, they observed activity bouts largely dispersed throughout the 24 hours of the day, suggesting that rest periods were highly fragmented. Half of the sample gradually improved by showing more and more consolidation of rest and activity periods over the 10 days of recording, and most of these individuals were no longer in posttraumatic amnesia. The other half however did not present such improvement, with only brief consolidated rest periods. The subgroup with evidence of improved rest–activity cycles clearly presented better outcomes such as a shorter stay in the ICU and hospital, and most importantly, lower disability at discharge as measured with the Disability Rating Scale.

In their study, Makley et al.[32] examined sleep quality in the first week after patients arrived in the inpatient rehabilitation ward by asking nursing staff to conduct hourly observations. They found that 68% of patients had disturbed sleep and that this subgroup was more functionally impaired. The same research team[30] observed that patients who were out of posttraumatic amnesia when admitted to the inpatient rehabilitation ward had significantly better sleep efficiency compared to those who continued to suffer from posttraumatic amnesia. Nakase-Richardson et al.[33] prospectively followed a large sample of consecutively hospitalized patients after moderate/severe TBI using the Galveston Orientation and Amnesia Test and one item of the Delirium Rating Scale-Revised evaluating sleep disturbance (based on nursing staff and other professionals' notes and family reports). The percentage of individuals with severely disorganized sleep decreased over the first month postinjury, but 84% of the sample still had sleep–wake disturbances upon admission to rehabilitation care after acute hospitalization. At 1-month post-TBI, 67% showed significant sleep abnormality, among which 39% had moderate-to-severe disorganization of sleep. When controlling for injury severity (GCS score), age, and time since injury, they found that patients exhibiting moderate or severe sleep disturbances at one month post-TBI had longer length of stay in acute care and rehabilitation, as well as longer posttraumatic amnesia. Taken together, these few studies clearly demonstrate the link between the normalization of the sleep–wake cycle and recuperation post-TBI.

As in other inpatient contexts, environmental factors in the ICU or during hospitalization may also contribute to sleep–wake disturbances in TBI patients, for example, noise, bright light, and frequent and painful interventions (e.g., administration of medication) during the night, and lack of routine.[21,34,35] Circadian rhythms can be affected by the absence of *zeitgebers* (e.g., light/dark cycles, meal timing, and social routines). However, the hospitalization environment alone cannot account for the whole picture of sleep disturbances observed in the acute phase postinjury.[29]

SLEEP ARCHITECTURE IN THE CHRONIC PHASE AFTER TRAUMATIC BRAIN INJURY

Studies using polysomnography (PSG) in the postacute phase after TBI corroborate subjective complaints by objectively measuring markers such as reduced total sleep time, increased wake time after sleep-onset and longer sleep-onset latencies.[36] When examining the sleep architecture, mixed results have been found, with some studies suggesting an increase in slow-wave sleep,[24,37] an increased proportion of non-REM (NREM) sleep stages, N1[4,38] and N2,[4,39] a decreased time spent in REM sleep,[4,39] and shorter latencies to REM sleep.[4,40,41] A recent meta-analysis focused specifically on alterations in sleep in the chronic phase[42] by including 14 studies conducted at least 6 months postinjury and including all levels of injury severity. The only clear difference that emerged in the TBI population compared to healthy controls was an increase in slow-wave sleep. However, this difference seemed to be attributable to a subgroup of patients with more severe injuries. Indeed, a closer look at the effect of injury severity revealed that only persons with moderate—severe TBI differed significantly from uninjured controls on a few sleep parameters: increased proportion of slow-wave sleep, reduced proportion of N2 sleep, and reduced sleep efficiency. An increase in slow-wave sleep (N3 stage) persisting in the chronic phase after TBI is emerging as a consistent result both in human and animal models of TBI.[43,44] In their meta-analysis, Mantua et al.[42] hypothesize that this increase in slow-wave sleep may be linked to ongoing neuroplasticity even several months postinjury, a result corroborated by a recent study in children and adolescents.[45] Indeed, slow-wave sleep is known to promote synaptic remodeling and axonal sprouting. They also suggest that the increase in slow-wave sleep may be at the expense of N2 sleep, which is known to have a role in memory consolidation, thus potentially having a detrimental effect on cognition. These hypotheses remain to be further investigated.

There are few studies of sleep architecture in mild TBI. Mollayeva et al. observed longer time spent awake compared to normative data, as well as less time spent in consolidated N2 and REM sleep.[46] Although alterations in the macrostructure of sleep do not seem to emerge as evidently in milder TBI, studies using spectral analysis reveal subtle alterations in electroencephalography (EEG) patterns, and some researchers have even suggested that these alterations in the microstructure of sleep may potentially serve as an objective marker of mild TBI.[47–49] For instance, Khoury et al.[50] observed lower delta and higher beta and gamma power (associated with arousal) in individuals with mild TBI compared to controls. In a group of athletes with mild TBI, Gosselin et al.[49] found that concussed athletes had increased delta activity and decreased alpha activity during wakefulness. They interpreted this result as suggesting a slower dissipation of sleep inertia, which may at least partially explain frequent subjective reports of fatigue and difficulties in daytime functioning after mild TBI.

PATHOPHYSIOLOGY OF SLEEP—WAKE DISTURBANCES AFTER TRAUMATIC BRAIN INJURY

There is a growing literature on various pathophysiological factors which may explain alterations in sleep—wake functions after TBI. TBI may induce structural lesions leading to various neurochemical and neuroelectric changes, which may explain the emergence of sleep—wake disturbances after TBI. Early studies clearly showed that the brainstem and the reticular formation were particularly vulnerable during TBI because of converging forces.[51–53] One study identified a link between the presence of sleep–wake disturbances and white matter abnormalities located in the parahippocampal region.[54] The same research team found that persons having sustained mild TBI and suffering from sleep disturbances had a greater tentorial length and narrower tentorial angle as measured with MRI data.[55] Damage to the tentorium was hypothesized to induce changes in the nearby-located pineal gland's capacity to produce melatonin. Melatonin secretion and its circadian regulation have indeed been found to be reduced both acutely and chronically post-TBI.[34,37,56] In a recent study, Grima et al.[57] examined the timing of melatonin secretion in nine individuals with severe TBI compared to nine healthy controls. In the TBI group, they observed a 1.5-hour delay of salivary dim light melatonin onset and a 42% decrease of overnight melatonin secretion compared to the control group.

Several studies indicate that TBI is associated with abnormal hypothalamic function, which could explain disturbances in the sleep–wake cycle, at least early following the injury. Postmortem examination of the brains of persons having succumbed to TBI is characterized by a loss in histamine and hypocretin wake-promoting neurons.[58,59] Baumann's team has described a decrease in secretion of hypocretin-I in the acute phase,[59–61] a characteristic also found in narcolepsy, which could thus be linked to the EDS frequently seen after TBI. Recent animal studies support these findings[62–65] with experimentally induced TBI leading to decreased numbers of hypocretin neurons in the lateral hypothalamus, difficulty for animals to sustain wakefulness, fragmented activity patterns during dark phases, increased transitions from wake to REM or NREM sleep,[65] cognitive impairments, and depression-like symptoms.[62] In a transcranial magnetic stimulation study in humans with TBI suffering from EDS, Nardone et al.[66] observed hypoexcitability of the cerebral cortex which they hypothesized was reflecting a dysregulation in the excitatory hypocretin-neurotransmitter system. Other studies have reported narcoleptic-like phenomena after TBI, for example, shortened latency to REM sleep[4,38,41,67] and sleep-onset REM periods (SOREMPs; i.e., sleep onset directly into REM sleep during daytime PSG).[4,39,61,68,69] Damage to the ascending reticular activating system may also aggravate EDS.[70] Recently, Valko et al.[71] found that TBI also damages the arousal-promoting neuronal circuiteries (serotonergic neurons of the raphe nucleus and noradrenergic neurons of the locus coeruleus) in the mesopontine tegmentum. These authors thus suggest that it is an aggregation of different injuries which lead to arousal problems after TBI.

The research on pathophysiological factors has mainly elucidated the etiology of EDS and probably pertains more specifically to more severe injuries. These pathophysiological mechanisms do not however fully account for the development of insomnia symptoms in a large proportion of patients up to months and even years postinjury. It is becoming increasingly evident that stress-related factors (including physiological, behavioral, emotional, and cognitive markers of stress), pain, fatigue, psychopathology, and environmental factors are also at play. Interestingly, the latter factors can more readily be modified through intervention.

POTENTIAL IMPACTS OF SLEEP–WAKE DISTURBANCES ON THE EVOLUTION OF THE CONDITION AFTER TRAUMATIC BRAIN INJURY

Cognitive Functioning

As mentioned earlier, a growing body of studies[27–31] conducted in the first few days or weeks post-TBI clearly indicate that the normalization of sleep patterns predicts or at least follows the trajectory of emergence from posttraumatic amnesia and cognitive recovery. In an EEG and magnetoencephalography study conducted at an average of 80 days after injury, Urakami et al.[72] found parallel improvement trajectories between sleep spindle abnormalities, consciousness, and cognitive functioning after moderate/severe TBI. Using actigraphy on a neurosurgical ward, Chiu et al.[73] found that total sleep time mediated the relationship between TBI severity and cognitive recovery as evaluated with the Rancho Los Amigos scale. Taken together, the findings linking sleep quality to recovery of consciousness in the acute phase point to the importance of giving more attention to sleep early in the rehabilitation process to optimize recovery.

In the longer term post-TBI, evidence is emerging showing that sleep–wake disturbances may exacerbate cognitive deficits. In 262 veterans with mild TBI,[74] Waldron-Perrine et al. noticed that a worse subjective insomnia severity and a shorter self-reported total sleep time the night prior to testing were correlated with poorer memory performance but not with performance on attention tasks or self-reported cognitive problems. When controlling for depression and anxiety, only total sleep time the night before testing continued to predict memory performance. The relationship between sleep and cognition is thus probably influenced by post-TBI levels of depression and anxiety.

In persons having suffered mild TBI but presenting with persistent symptoms, Dean and Sterr[75] also found a correlation between cognitive performance and sleep quality. In the same line, Theadom et al.[76] found that poorer sleep quality, along with other factors such as anxiety and depression, was linked to the presence of persisting cognitive complaints and lower social participation four years after mild TBI.

Using computerized sustained attention tasks, Bloomfield et al.[77] found that poor sleepers with TBI made significantly more commission errors than good sleepers. Those suffering from EDS also performed significantly worse than TBI survivors without EDS.[78] The same group found that TBI patients with obstructive sleep apnea (OSA) performed significantly worse on measures of attention and memory compared to patients without OSA.[79] In children, one study found that sleep problems predicted worse executive functioning measured with neuropsychological tests.[80] Similarly, daytime sleepiness as reported by adolescents was found to be associated with greater parent- and adolescent-reported executive function deficits.[81]

Still very limited research has explored whether treating sleep problems can improve cognition after TBI. In veterans with blast-induced mild TBI, one team observed improvement on the Montreal Cognitive Assessment after 9 weeks of combined behavioral counseling and prazosin.[82] Wiseman-Hakes et al.[83,84] also observed significant improvements on measures of attention, memory, processing speed, and language following individualized treatment for sleep disorders.

Functioning in the Chronic Phase

Sleep–wake disturbances after TBI can affect participation in rehabilitation therapy. Worthington and Melia[85] used rehabilitation staff reports to assess the nature and impact of sleep/arousal disturbances during delivery of rehabilitation treatments in the chronic phase of severe TBI. They observed that 47.4% of their sample ($N = 135$) had disturbed sleep or arousal and in 65.6% of these, the sleep–wake impairments were severe enough to interfere with rehabilitation interventions and patients' daily activities (e.g., patient unable to stay awake during an intervention or activity; patient missing appointments or opportunities due to late awakening time). Silva et al.[86] found that hypersomnolence was among the factors predicting posttraumatic confusion and suboptimal cooperation during inpatient rehabilitation.

In a recent chart review study of 417 adolescents aged 13–18 years having sustained mild TBI, Bramley et al.[87] found that sleep disturbances was associated with at least a threefold increase in recovery time. Sleep–wake disturbances also potentially affect adaptation to new limitations, for example, by making it more arduous for a person to cope with pain or cognitive deficits. Chaput et al.[88] showed that sleep–wake disturbances actually contribute to the chronicity of mood alterations and pain. In research with children having sustained TBI, sleep–wake disturbances predict problems with communication, self-care, and activity levels.[89]

Potential impacts of post-TBI sleep–wake disturbances on other medical issues have not yet been adequately documented, but a recent study revealed a 2.28-fold increased risk of suffering a stroke at a subsequent time when insomnia was observed in comorbidity with TBI.[90]

Interaction With Pain

Pain, which is particularly common after TBI (e.g., chronic headaches or pain related to orthopedic or other injuries sustained during the accident), is indeed clearly linked to sleep–wake disturbances.[67] For example, it has been shown to increase the risk for insomnia.[67,91] Pain can cause microarousals and wakefulness intrusions during NREM sleep, thus potentially leading to nonrestorative sleep or aggravating sleep–wake disturbances.[89,92] But sleep difficulties can also influence pain. For example, individuals with subjective sleep complaints are three times more likely to have headaches during the first 6 weeks after TBI.[88] Furthermore, when mild TBI is accompanied by pain, patients exhibit significantly more anxiety, depression, and pain catastrophizing, as well as alterations to the microstructure of sleep such as a greater increase in rapid EEG frequency bands (beta band), compared to mild TBI without pain.[50] This could reflect that brain injured patients with pain experience some level of arousal even when they are asleep, thereby potentially interfering with the restorative functions of sleep.[50,93]

Interaction With Psychopathology

Psychopathology is very prevalent after TBI. Between 31% and 65% of adults with TBI meet criteria for at least one psychiatric disorder over the first year postinjury, with major depression, anxiety disorders (e.g., generalized anxiety disorder), and posttraumatic stress disorder (PTSD) being the most frequent.[94–99] Sleep–wake disturbances, including insomnia, hypersomnia, or daytime sleepiness, are often a hallmark of mental health disorders and are thus clearly linked with post-TBI symptoms of depression or anxiety.[38,100,101] The relationship between post-TBI anxiety and sleep disturbances seems bidirectional with one exacerbating the other.[3,102] TBI is characterized by a myriad of psychosocial stressors with the many challenges that arise as injured persons attempt to adjust to new physical or cognitive limitations and make efforts to reintegrate different social roles. TBI survivors may face a variety of problems (e.g., family reorganization, loss of work capacity, financial difficulties, legal issues, stigma or incomprehension, relational strain), which can bring about significant worries, anxiety, rumination, and physical tension. The latter can in turn induce increased cognitive or emotional arousal at bedtime or during nighttime awakenings, thus contributing to insomnia.[103]

In children, one study found that psychosocial problems increase the risk for sleep disturbances, but sleep problems also predicted increased emotional or behavioral problems.[89] There also seems to be a link between post-TBI sleep disturbances and internalizing problems[104] and other psychiatric issues in children.[105] In a military sample, sleep quality was found to mediate the relationship between mild TBI and the presence of suicidal ideations in Iraq/Afghanistan veterans, suggesting that sleep disturbances potentiate the negative impacts of mild TBI on mood.[106]

There is a growing literature describing the joint incidence of TBI, PTSD, and pain in military personnel returning from combat zones, a constellation referred to as the polytrauma triad.[107] In a study of 116 individuals having sustained TBI during military deployment, Collen et al.[108] observed that 97% had sleep complaints, 91% fulfilled the criteria for at least one psychiatric condition (depression, PTSD, or anxiety disorders), and 94% had a prescription for a psychoactive medication. Those who had all three of these characteristics had significantly more severe sleep–wake problems compared to those who only displayed one or two. Other studies also support the finding that comorbidity between PTSD and TBI predicts more severe sleep–wake disturbances.[109–111] PTSD is known to cause nightmares and alterations of sleep structure such as reduced slow-wave sleep, increased stage N1, and increased REM density.[112]

Differentiating Sleepiness from Fatigue

Significant chronic fatigue is extremely common following TBI. This state is easily confused with sleepiness by both patients and health professionals. Post-TBI fatigue is measured predominantly with self-reported tools as it is a subjective feeling of weariness or lack of energy,[38,113,114] as opposed to sleepiness which is measurable by objective markers of drowsiness or reduced alertness (yawning, eyelids drooping).[115] Importantly, post-TBI fatigue can be present independently from sleep–wake disturbances and EDS,[114,116] but it can of course be associated with sleep disorders. Our team showed that, among TBI patients who did not have any symptoms of insomnia, 60% still reported suffering significantly greater fatigue than before the accident.[67,114]

In sum, literature is emerging showing that sleep—wake disturbances exacerbate, or at least clearly interact with other consequences of the TBI such as cognitive deficits, pain, psychopathology, or fatigue. Furthermore, the rehabilitation process seems to be hindered by the presence of sleep—wake disturbances, calling for increased attention to these issues.

INSOMNIA

Insomnia is the most frequently observed sleep disorder in TBI.[6] Symptoms of insomnia are very frequent, with studies using heterogeneous methodologies and definitions indicating a prevalence rate of 30%—70% at different time points postaccident.[5] Two early studies in the field documented the presence of an insomnia disorder using standardized criteria. At 4-month postinjury, Fichtenberg et al.[117] found that 30% or their sample of 50 patients consecutively admitted to a rehabilitation hospital suffered from an insomnia syndrome. A few years later, our team conducted a survey of 452 individuals who were on average 8 years post-TBI (mild to severe).[67] We found a very similar prevalence rate, with 29% fulfilling the criteria for an insomnia syndrome as defined by a combination of the DSM-IV and the International Classification of of Sleep Disorders criteria, suggesting that insomnia may adopt a chronic course or at least that TBI increases the lifetime prevalence of the disorder. Importantly 60% of these persons suffering from insomnia reported that they were not being treated for this disorder.

Patients who fulfilled the criteria for an insomnia disorder reported that it had significant impacts on their day-to-day functioning. In the same study, 60% indicated that insomnia affected their mood, 69% their cognition (e.g., attention, concentration, memory), and 57 % their social, leisure, or productive activities such as work.[67]

Risk Factors

Injury Severity

Although some studies do not establish a link between insomnia and injury severity,[118] a few studies suggest that insomnia complaints are associated with milder TBIs.[67,91,101,119—121] Several authors hypothesize that self-awareness explains these results: severe TBI can lead to significant *anosognosia*, whereas patients with mild TBI survivors are usually quite aware of their limitations and symptoms, and can therefore more readily be distressed by ongoing difficulties. Additional factors that potentially increase the likelihood of insomnia following TBI include anxiety and depressive symptoms, post-TBI fatigue, alcohol misuse, pain (in particular headaches), and dizziness.[67,91,118—120,122]

Medications

Some medications administered to patients with TBI can influence the quality or organization of sleep and, in some cases, contribute to insomnia, depending on the pharmacokinetics, dosage, and timing of drug administration. For example, analgesics, sedatives, corticosteroids, myorelaxants, anticonvulsants, and antidepressants are known to alter sleep architecture. Certain serotonin selective reuptake inhibitors decrease sleep efficiency and affect REM sleep onset and length[123] and carbamazepine can have detrimental impacts on REM sleep.[124]

Behavioral Factors

To compensate for insomnia, persons may engage in a variety of behaviors which can in fact contribute to crystallize the insomnia problem: irregular sleep—wake schedules, consumption of caffeine, alcohol or drugs, and napping during the day. TBI-related fatigue is accompanied by "recovery" behaviors, that is, taking naps or rest periods (i.e., lying down without sleeping) during the day, or spending more time in bed. These compensatory behaviors can desynchronize the sleep—wake cycle and potentially contribute to sleep disturbances.[125,126] Because of inability to work, some individuals may lack a routine of regular activities, such as getting up at the same time each morning, getting ready for work, eating at regular times, a routine that can help regulate the sleep—wake cycle.

Assessment

Assessment of sleep disorders in patients with TBI is globally similar to that in other populations with use of tools such as interviews, self-report questionnaires, and objective sleep measures (PSG, actigraphy). A variety of studies have used questionnaires such as the Insomnia Severity Index[125] or Pittsburgh Sleep Quality Index[127] with TBI samples attesting to their feasibility to assess insomnia symptoms in this population. Some teams have done validation studies of these instruments specifically in the TBI population.[128–130] Even if cognitive deficits are present, in most cases these to not preclude the use self-reported instruments. Only with a minority of patients for whom severe cognitive or language issues are present or who have significantly impaired self-awareness will alternative methods be necessary, such as observation of sleep patterns by family members or health professionals.

It is useful to conduct a comprehensive clinical interview documenting preinjury and postinjury sleep—wake patterns and habits including typical sleep—wake schedule, potential disturbances in daytime functioning (e.g., fatigue, vigilance, attention), and sleep-related behaviors and cognitions, including patients' potential worries and perceptions about the effects of sleep disturbances on their cerebral and functional recovery. The assessment should also cover the nature, severity, duration and evolution of insomnia, current and premorbid treatments for sleep difficulties, potential preinjury or postinjury psychopathology, and use of prescribed and over-the-counter medications, caffeine, energy drinks, alcohol, and drugs. Use of soft drugs is common after TBI[131] and should be well assessed. Indeed, our team found that 21% of insomnia sufferers admitted taking either alcohol or soft drugs such as marijuana specifically to help them sleep.[67] Key questions in the interview can also screen for sleep-related breathing disorders, periodic leg movements (PLMS), restless leg syndrome, and other sleep disorders.

Sleep diaries are very practical to monitor sleep habits and patterns, sleep duration and efficiency, quality of sleep, and the frequency and severity of insomnia symptoms such as prolonged sleep-onset latency or nighttime awakenings. Some patients may need more time and repetition to learn to use the sleep diary effectively. A close monitoring of its use in the first few days is helpful to ensure the patient is at ease with the tool. Adaptations to the sleep diary can be made, for example, by providing a simpler version with less questions and grosser estimates of time spent awake.[132] When severe cognitive deficits preclude any self-monitoring, actigraphy could complement the evaluation and be an interesting option.

When other sleep disorders such as OSA, restless legs syndrome, or PLMS need to be ruled out, PSG may be warranted, but such objective measurement is not required to pose an insomnia diagnosis, due to its subjective nature.

Pharmacological Interventions

In the general population, many pharmacological agents are prescribed for chronic insomnia, including benzodiazepine receptor agonists (BzRAs) and several drugs are used off-label (e.g., antidepressants such as trazodone or doxepin, antipsychotics such as gabapentin or tiagabine). BzRAs, which include benzodiazepines and more recent z-drugs (i.e., zaleplon, zopiclone, zolpidem), are the only pharmacological agents with sufficient empirical evidence for their recommendation as a treatment for insomnia, although they should be used for short-term rather than long-term management (i.e., daily usage no longer than 4–6 weeks) or on an occasional basis.[133,134]

In the TBI population specifically, data are scarce on the efficacy and safety of pharmacological agents for insomnia, and they should be prescribed with great caution. For instance, the use of hypnotic medications in itself, or due to pharmacological interactions with other products, may exacerbate post-TBI symptoms such as cognitive deficits, dizziness, motor impairments, and fatigue. Some agents may also lower the epileptic threshold, and the potential development of a substance use disorder (with abuse and dependence) may be higher in the context of TBI. Some authors suggest to completely avoid benzodiazepines after TBI, because of their potential negative consequences on memory and cognition and high risk of abuse.[135] However, benzodiazepine use is still common, as shown by a study reporting that 20% of participants were using benzodiazepines or other hypnotic drugs, on average 9 years after TBI.[85] In all cases, the pharmacological agent had been used for a longer period that what is recommended in UK clinical guidelines (a few months at most). Although newer agents such as z-drugs may be preferred to manage insomnia after TBI, they may still have some short- and long-term cognitive adverse effects, and their outcomes may vary depending on the subpopulation.[136]

Regarding the use of non-BzRA medications for insomnia, one study compared melatonin to amitriptyline, a tricyclic antidepressant, in a preliminary trial with seven TBI patients. Amitriptyline improved sleep duration and reduced sleep-onset latency while melatonin improved daytime alertness.[137] Finally, the combination of prazosin, often used for nightmares, and behavioral sleep counseling was associated with some successful outcomes in a veteran sample.[138]

Nonpharmacological Interventions

Cognitive-behavioral therapy (CBT) is a first-line treatment for persistent insomnia and has been shown to produce reliable and sustainable improvements in a wide array of clinical populations.[133,139] Very few studies have examined the efficacy of CBT for insomnia in the TBI population however. Our team was the first to study the potential efficacy of CBT in TBI patients. We made simple adaptations to the original CBT protocol developed by Morin[125] for example by including strategies to encourage attention (i.e., shorter sessions, breaks) and to optimize encoding of therapy recommendations, for example, by providing simplified written material to patients and using repetition. We also closely monitored sleep diary completion in the first few days to ensure comprehension and effective use of the diary. The cognitive therapy component was also simplified. Rather than using a socratic questioning approach, a more directive or educational approach was used: the Dysfunctional Beliefs and Attitudes about Sleep[140] scale was used to identify beliefs and attitudes to work on (e.g., "I absolutely need 8 hours of sleep to be functional during the day"), and to direct cognitive therapy component and offer alternative interpretations when the patient had difficulty coming up with these. We also paid particular attention to distress-provoking thoughts more specific to TBI, such as "If I don't sleep well, my brain will not recover from the injury," "If I cannot sleep well, I will never be able to return to work." We also involved significant others to promote adherence, motivation, and encoding of therapy recommendations, and added a fatigue management component which included pacing of activities, planning of activities into manageable chunks, incorporating rest periods, and exploring different ways to rest (besides going to bed). This protocol was evaluated in a first case study[141] and in a subsequent case series of 11 patients.[142] We obtained success rates comparable to those documented with samples of patients suffering from either primary insomnia or insomnia comorbid to medical disorders with 73% of patients showing significant reductions in total wake time and an increase in sleep efficiency, and these improvements were well maintained at a 3-month follow-up. In parallel to treatment effects on sleep, we observed a reduction in general or physical fatigue. Unfortunately, there was no significant improvement in cognitive fatigue. Post-TBI fatigue can thus not be expected to fully resolve after treatment for sleep disturbances and merits specific clinical attention. It is important to explain to TBI patients who receive CBT for insomnia that improving the quality or quantity of their sleep may not result in a complete alleviation of their TBI-related fatigue. Patients would undoubtedly benefit from a treatment focused specifically on fatigue; unfortunately, research on the interventions for post-TBI fatigue specifically is still very limited.

We found CBT for insomnia to be feasible with all participants, despite the fact that all had cognitive deficits documented in their medical files. Yet, none had any particular difficulty understanding the rationale of the recommendations or to complete self-monitoring. Results of this study suggested that CBT is feasible even in the presence of cognitive, emotional, or behavioral particularities of TBI patients. In Table 10.1, we present a series of recommendations on how to adapt CBT with TBI patients.

Recently, Nguyen et al.[143] conducted the first randomized controlled trial evaluating the efficacy of CBT for insomnia and fatigue in a sample of mild to severe patients with

INSOMNIA

235

TABLE 10.1 Suggestions for Adaptations to Cognitive-Behavioral Therapy for Insomnia in the Context of Traumatic Brain Injury

Cognitive deficits
- Build in repetition to ensure encoding of rationale of treatment strategies and nature of recommendations
- Closely monitor the efficacy of self-monitoring (e.g., sleep diary) to ensure comprehension and effective use. Use shorter, simplified versions of self-monitoring tools (e.g., sleep diaries) and other instruments if necessary
- Prefer shorter sessions (e.g., 30-40 min instead of 50−60 min) to avoid fatigue and adjust the timing/rhythm of session to the patient
- Prioritize behavioral strategies and introduce them gradually (e.g., restriction of time in bed alone initially, then gradually implement stimulus control)
- If necessary, simplify or avoid more complex or introspective strategies such as cognitive restructuring (e.g., use the Dysfunctional Beliefs and Attitudes about Sleep[140] to the direct cognitive therapy component, offer alternative interpretations)
- Provide simplified written and visual summaries of each cognitive-behavioral therapy-I strategy. Use metaphors, pictograms, or cue cards to help encoding and retention of treatment material
- Use portable electronic devices (e.g., smartphones) to foster compliance (e.g., set reminders to complete sleep diaries or to apply sleep window, refer to one of many available sleep apps)
- Consider the involvement of a significant other to whom the rationale and application of each strategy is described to promote compliance and support

Fatigue
- Consider multiple factors contributing to fatigue, some of which may not be easily modified (e.g., fatigue due to the traumatic brain injury itself) while others can be targeted (e.g., lack of activities, lack of exercise, lack of exposure to bright light)
- Evaluate the frequency, duration, timing, and conditions of daytime naps. Remain flexible in permitting short, carefully planned naps while avoiding those that are more likely to affect nighttime sleep
- Propose strategies other than napping or resting to increase the energy level (e.g., physical exercise, walking, pleasant activities, social activities)
- Schedule cognitive-behavioral therapy for insomnia sessions at a time when the energy level is optimal

Physical limitations
- Evaluate physical condition, pain, mobility issues, and motor impairments that may interfere with the implementation of cognitive-behavioral therapy for insomnia strategies. Contact and stay in touch with treating physician or other healthcare professionals about physical health particularities and medication use
- Discuss with the patient his/her perceptions and beliefs about the interrelationships between sleep, pain, fatigue, and physical functioning
- For patients with limited mobility, consider a modified version of stimulus control instructions (e.g., sitting up in bed when unable to fall asleep instead of getting out of bed and going to another room)
- When inactivity or physical deconditioning is an issue, gradually reintegrate simple, varied, and adapted activities

Behavioral particularities
- Provide structured, targeted session agendas to avoid distractions
- Immediately address disinhibition or inappropriate behavior to reiterate the purpose of the sessions
- Provide reinforcement and encouragement throughout the treatment process, especially with patients with apathy or lack of initiative
- Remain flexible (e.g., to missed appointments, lack of compliance) and involve patients in problem solving.
- As alcohol and recreational drug use is common, provide information on their impact on sleep and monitor their use during treatment

Anxiety and depression symptoms
- Carefully assess the presence of anxiety and depression symptoms, discuss these, and inform patients on their interaction with sleep−wake difficulties

(Continued)

TABLE 10.1 (Continued)

- Promote balance between activities (e.g., behavioral activation techniques to encourage pleasant, rewarding, useful, social, and physical activities)
- Be flexible when implementing behavioral strategies when significant anxiety symptoms are present (e.g., gradually decreasing time spent in bed rather than drastically decreasing it initially and then gradually increasing it)
- Encourage patients to seek support from family and friends

TBI. Their treatment protocol involved psychoeducation, reorganizing daily schedules (e.g., pacing tasks, planning breaks, promoting sleep hygiene), graded activity including physical activity, cognitive restructuring, core components of CBT for insomnia such as stimulus control and sleep restriction, and strategies for the management of physical and mental fatigue (e.g., dealing with time pressure, modifying tasks, or the environment). This comprehensive intervention yielded very promising results with 70% of the CBT group reaching clinically significant improvement in sleep (reduction of insomnia severity and overall improvement of sleep quality), and 55% on fatigue, compared to 27% and 30%, respectively, in the treatment as usual control group. These gains in sleep and fatigue were also accompanied by a more pronounced decrease in depressive symptoms. Results were also maintained two months after the end of treatment, which again attests to the efficacy of CBT in this patient population. The research team also examined which patient characteristics predicted positive treatment responses on sleep quality.[144] Interestingly, factors such as premorbid intelligence, time since injury, baseline severity of anxiety symptoms, and education did not predict treatment outcomes. However, younger persons, those with more severe depressive symptoms at baseline, and individuals with better scores on verbal memory measures were those who benefitted the most from therapy. This last finding suggests that neuropsychological function, in particular memory, does influence the efficacy of CBT for insomnia after TBI. This does not mean however that CBT should not be attempted when memory is affected. Indeed, Nguyen et al.[144] suggest that further adaptations to therapy may be required with these patients, for example, closer monitoring of skill practice during therapy or adjustment of the intensity or dosage of therapy (e.g., reducing the content to fewer CBT skills and repeating these more intensely, planning shorter and more frequent sessions).

The accessibility of CBT for insomnia is an issue regardless of the condition it may be associated with. Theadom et al.[145] recently made efforts in this direction by documenting the potential efficacy of online interventions to improve sleep after mild or moderate TBI. They conducted a pilot randomized controlled trial with 24 individuals with mild or moderate TBI, allocated to either an online CBT protocol or an online education condition. The CBT intervention included psychoeducation about sleep and how the environment can influence sleep, relaxation and mindfulness training, and restriction of time in bed. The control group obtained information about brain injury, sleep, and fatigue postinjury, including the effects of exercise, environmental factors, diet and substance use, and received some guidance on how to establishing a routine of periods of activity and rest. This quite active control group probably explains why the team did not observe significant between-group differences on objective sleep parameters (measured with actigraphy), quality of life, and cognitive functioning. Two patients with visual disturbances also had difficulty using the online program. They did however find a significant reduction in sleep disturbance as measured with the Pittsburgh Sleep Quality Index (PSQI) in the CBT group compared to the control group. These results support the feasibility of an online administration of CBT to improve sleep in adults with mild to moderate brain injury.

Behavioral interventions for sleep are also pertinent in the acute phase after TBI. De La Rue-Evans et al.[146] have put together a behavioral program including systematic routines, environmental changes, sleep hygiene education, and stimulus control, which may be implemented by nurses during rehabilitation. The program includes education for nurses and patients about sleep hygiene and various changes such as strategies for nurses to provide an environment more conducive to sleep (e.g., diminishing noise, dimming lights at night, opening blinds predawn, offering only decaffeinated drinks in the afternoon/evening), providing patients with a visible list of sleep-hygiene recommendations (e.g., keep room free of distractions), and establishing routines to keep patients active and awake during the day (waking patient at a regular time, minimizing naps). A pilot study suggests encouraging results for this intervention in terms of feasibility and clinical benefits.

CLINICAL VIGNETTE

François is a 42-year-old man who sustained a mild TBI and facial contusions in a bicycle accident. He is married, has two young children, and is working full time as a high school teacher. After his accident, he went to the nearest emergency department, underwent some tests, and was discharged the same day with verbal and written recommendations regarding concussion. In the first few days after his accident, he was sleepy during the day, slept much longer than usual at night, and experienced headaches, dizziness, and difficulty concentrating. He mostly stayed home and avoided physical or intellectual activities. Most of his symptoms disappeared within 3–4 weeks, although significant fatigue was still present on most days. To alleviate his fatigue, François napped once or twice per day, usually at the end of the afternoon before his children came back from school. He often felt exhausted early in the evening and went to bed around 8 or 9 p.m. on those nights. Although his fatigue and concentration difficulties gradually decreased, they still remained problematic and concerning for François, especially when he started reintegrating his leisure and family activities. Three months after his accident, he returned to work, on a progressive reintegration schedule. As he was getting closer to returning to full-time work, his sleep got progressively worse: he was able to fall asleep rapidly at night, but usually woke up around 1 or 2 a.m. and stayed awake for 30–120 min. He woke up much earlier than his alarm clock and was unable to sleep again. He was very worried that something was wrong with his brain, that he could not "shut it down" to sleep like before his TBI. He often anticipated that he would be exhausted the next day and that it would affect his concentration, his mood, and, worst of all, his teaching. He was impatient with his wife and children, and he reduced his leisure activities (e.g., physical exercise, family gatherings, social commitments) to keep all his energy for work. François felt that the situation was out of control, as he never experienced severe insomnia before his accident, other than a few nights in stressful periods where he had trouble falling asleep. He felt less and less confident in his ability to maintain his job and feared that he would damage his relationship with his family. He tried to take melatonin and other over-the-counter products for a few weeks, but these did not help much. Before going to his doctor to look for a stronger medication, he decided to try consulting a sleep specialist. He met with a psychologist with an expertise in sleep for eight sessions. He gradually realized that his insomnia and fatigue were not only related to his brain injury, but to multiple factors, some of which could be changed. He significantly reduced his napping, keeping short morning naps about twice per week. He changed his sleep schedule, going to bed later and at a regular time. He diversified his evening activities, reintegrated some physical exercise and short social events, and started using a relaxation technique to ease the transition between the evening and the night. When unable to fall back to sleep at night, he went out of bed and engaged in a quiet activity such as drawing or writing. He noted his anxiety-provoking thoughts during the day or evening and tried to analyze all angles of situations. He discussed his apprehensions with his wife more openly. It took several weeks of efforts and discipline, but his sleep started to improve. It did not become perfect, he still felt fatigued and overwhelmed at times, but François was much less worried about his insomnia and fatigue, and felt that he had tools to manage his difficulties

EXCESSIVE DAYTIME SLEEPINESS, POSTTRAUMATIC HYPERSOMNIA, SLEEP-DISORDERED BREATHING, AND INCREASED NEED FOR SLEEP

Several sleep disorders are associated with EDS, including sleep-related breathing disorders, narcolepsy, and posttraumatic hypersomnia. In the general population, subjective sleepiness, measured by self-reported questionnaires or visual analogs scales, is not always correlated with objective sleepiness, assessed with physiological measures such as PSG. This suggests that subjective and objective sleepiness measures may capture different aspects of this phenomenon. It may be even more so in the context of TBI, because fatigue, a prominent symptom, may be confounded with subjective sleepiness by patients. Between 14% and 57% of individuals with TBI report EDS,[4,35,61,91,102,147,148] but it is estimated that only 11%−25% of patients presenting with sleep−wake disturbances meet criteria for objective EDS measured by PSG.[61,78,149] A study by Verma et al. found that only 53% of patients with subjective sleepiness (Epworth Sleepiness Scale score >11) met the criterion for objective EDS on the Multiple Sleep Latency Test (MSLT; mean sleep-onset latency <5 minutes).[4] In another investigation on patients with head−neck trauma with subjective sleepiness, 28% met criteria for objective sleepiness.[68]

Research on factors associated with post-TBI EDS has been inconclusive. A positive relationship between a greater TBI severity and higher sleepiness has been found in at least two studies,[61,147] but others failed to replicate this finding.[78,148] There are some studies showing increased prevalence of EDS with time elapsed since the injury.[35,102] For a certain proportion of patients, EDS may have been present before the injury, possibly due to an untreated sleep disorder. In some cases, it may even have contributed to the TBI itself, for instance by triggering a motor vehicle accident, although this assumption is difficult to confirm. An Australian study found that subjective EDS is associated with increased anxiety and daytime napping.[102]

The presence of a sleep-related breathing disorder can be the main cause of EDS. In one study of persons with TBI complaining of EDS, 30%–40% were diagnosed with a sleep-related breathing disorder.[68,150] A study conducted in the early stages after TBI (on average 35 days postinjury), 30% of the sample had an abnormal apnea/hypopnea index (AHI; $\geq$5 events/hour) and 11% had a several abnormal AHI ($\geq$5 events/hour).[151] Sleep apnea is also frequent in the longer term, with two investigations reporting 30%[4] and 6%[152] of their participants having a severely abnormal AHI, 3–24 months, and 38$\pm$60 months postinjury, respectively. The majority of respiratory events during sleep are obstructive, representing 74% of all events in one study.[4]

Narcoleptic features have been reported in some TBI patients.[4,39,59] A review of 22 cases of posttraumatic narcolepsy highlights the heterogeneity of the clinical presentation.[153] In fact, the presenting symptom can be EDS, cataplexy, or both, symptom onset varying between a few hours to 18 months postinjury, and the condition is not associated with injury characteristics such as TBI severity, loss of consciousness, or neuroimaging findings. In most cases, however, posttraumatic narcolepsy is progressive. Verma et al.[4] reported that 32% of their sample with TBI who reported subjective sleepiness was diagnosed with narcolepsy on the basis of the MSLT.

According to the International Classification of Sleep Disorders, 3rd edition, a diagnosis of posttraumatic hypersomnia can be made (part of the broader diagnostic category of hypersomnia due to a medical condition) when: (1) the patient has an increased need to sleep or daytime lapses into sleep for at least three months; (2) the daytime sleepiness occurs as a consequence of TBI; (3) when the MSLT is used, mean sleep-onset latency is of 8 minutes or less, with fewer than two SOREMPs; and (4) the symptoms are not better explained by another sleep disorder, psychiatric disorder, or the effect of a substance.[154] Some authors suggest that 10%–30% of adults with TBI would meet criteria for posstraumatic hypersomnia.[61,78,152,155,156] In addition to EDS, patients with posttraumatic hypersomnia may also present fatigue, headaches, and cognitive impairment.[147]

In the past few years, one team[24,157] has proposed the term *posttraumatic pleiosomnia* to describe an increased need for sleep after TBI which can be independent from excessive daytime sleepiness. In a well-controlled prospective study, Imbach et al.[157] observed that compared to a control group, injured patients slept 1.5 hours more per 24 hour period, as measured with actigraphy 6 months post-TBI. They hypothesized that this increased need for sleep may be linked to an ongoing need for neuroplasticity, a hypothesis partly supported by their findings showing somewhat more consolidated NREM sleep in the TBI group. Pleiosomnia was however strongly linked with more severe injuries (lower Glasgow Coma Scale scores and presence of hemorrhage on CT scans performed in the acute phase), a result consistent with previous research[61,68,73,118] and underlining a differential trajectory or severity of sleep–wake disturbances in mild vs more severe brain injuries. In the same study, the authors also observed that decreased morning cortisol levels, as measured in the acute phase postinjury (first 2 weeks), was associated with objective EDS measured 6 months post-TBI. Also, 57% of the TBI group showed signs of objective sleepiness on the MSLT compared to 19% in the control group. Interestingly however, these patients tended to underestimate their EDS when assessed with a subjective measure, in this case the Epworth Sleepiness Scale.

Suzuki et al.[93] recently studied 56 individuals with mild TBI with actigraphy for about 7 days postinjury and at 1-month and 1-year postinjury. Compared to those without pain, they found that TBI patients with unrelieved pain had an increased need for sleep (required more than 8 hours per night) and took more naps during the day at 1-month postinjury, even when controlling for depression or age. Pain and increased sleep need was found to coexist in 29% of their sample.

Assessment

Based on their results suggesting that persons with TBI display a strong sleep state misperception with regard to their sleepiness level, Imbach et al.[157] suggest that self-reported screening instruments for sleepiness (e.g., the Epworth Sleepiness Scale) or for hypersomnia may not detect disorders related to EDS. Rather they suggest that patients should undergo PSG studies, or at least MSLT or actigraphic measures. The MSLT may be particularly useful to measure post-TBI EDS and to differentiate it from fatigue.[158] The Maintenance of Wakefulness Test, where the person is asked to try to stay awake for 40-minute periods with limited stimulation, can also be an interesting alternative in the TBI population. It is especially useful to evaluate how the patient is able to stay awake, which can inform daytime functioning linked to cognition or vigilance.[159,160] As described above in several studies, actigraphy is quite feasible after TBI and represents an interesting alternative when PSG is not available or is deemed too costly. It can also be used when PSG is more difficult to implement, for example with children, persons with cognitive impairment or behavioral issues such as apathy.[161] Zollman et al.[162] however suggest caution when using actigraphy with persons with motor impairment such as spasticity or paresis, or behavioral disturbances such as agitation or impulsivity. Placing the device on the least affected limb or using it in patient with a Rancho Los Amigos cognitive level of II or above are some recommendations they propose.

Pharmacological Treatment

Psychostimulants are often used in the treatment of EDS, but results from the available literature are mixed regarding their efficacy in the context of TBI. For instance, modafinil, which is effective to alleviate daytime sleepiness in narcolepsy, sleep-related breathing disorder, and sleep—wake disturbances related to shift work, has yielded contradictory results in TBI samples. First, a randomized controlled trial found that modafinil was not different from placebo on subjective fatigue and sleepiness measures, and participants using modafinil were more likely to present insomnia as a side effect.[163] A subsequent open-label trial reported slightly better results in patients with posttraumatic hypersomnia or narcolepsy using modafinil.[160] Another investigation showed benefits of modafinil in reducing sleepiness but not fatigue.[164] Finally, a recent 12-week randomized controlled trial compared three doses of modafinil (50, 150, 250 mg daily) and found that only the highest dose significantly improved objective (i.e., PSG-measured) daytime sleepiness compared to placebo, while none of the doses improved subjective sleepiness compared to placebo.[165] Similarly to modafinil, methylphenidate was shown to be effective in one study,[166] but did not reduce daytime sleep duration in another investigation.[167]

Nonpharmacological Treatment

Continuous Positive Airway Pressure for Sleep Apnea

The first-line treatment for OSA is continuous positive airway pressure (CPAP),[168] although its use in the context of TBI has seldom been investigated. One study observed significant improvements in the number of respiratory events during sleep (decreased AHI) after 3 months of CPAP in 13 patients with TBI, but subjective (measured by the Epworth Sleepiness Scale) and objective (assessed by the MSLT) daytime sleepiness were not improved.[160] One major issue with the treatment of sleep apnea with CPAP is compliance, as many patients are unable to tolerate the device throughout the night or for most nights. This may be even more problematic in patients with TBI, who can have cognitive and behavioral problems further impacting their compliance with CPAP.

Bright Light Therapy

Exposure to bright light is a promising avenue to alleviate daytime sleepiness, promote nighttime sleep, and regulate the sleep–wake cycle after TBI. Sinclair et al.[169] conducted a randomized controlled trial with 30 TBI participants complaining of fatigue or sleep–wake disturbances, comparing exposure to blue light (short wavelength), exposure to placebo yellow light, and a no-treatment condition. Light exposure lasted 45 minutes each morning for 4 weeks. Blue light therapy was associated with reduced fatigue and daytime sleepiness during treatment, which was not the case for the other two treatment conditions.

Naps

Daytime napping is often used by patients with TBI and may be recommended by healthcare professionals to manage fatigue, low energy, and daytime sleepiness. This behavior is not only common in the acute phase after the injury, but also in the long run. Indeed, Ouellet and Morin observed that 56% of their participants with TBI, on average 8 years post-TBI, were napping between three and seven times per week.[114] However, the beneficial and detrimental effects of daytime naps, and their optimal characteristics (frequency, duration, timing) have not yet been appropriately studied in the TBI population. On the one hand, daytime napping has the potential of having a positive impact on arousal and cognitive performance. On the other hand, negative outcomes may occur including sleep inertia (i.e., a prolonged period of reduced alertness upon awakening from a longer nap) and disturbed nighttime sleep.[170] The need for daytime napping should be evaluated on an individual basis, depending on the frequency and severity of daytime sleepiness, but not necessarily on fatigue alone. Napping should not be the only strategy to manage daytime fatigue, especially months and years after the injury, and alternatives should be found to manage energy (e.g., gradual implementation of physical exercise, alternating between tasks and leisure activities, planning social activities, adapting activities to reduce fatigue—for instance, by reducing duration or intensity). The timing and

242

10. TRAUMATIC BRAIN INJURY

duration of naps should also be revised particularly carefully, especially for individuals with insomnia. Ideally, naps should be shorter than 30 minutes (to avoid longer naps the use of an alarm clock is suggested) and be taken before 3:00 p.m.[125]

CIRCADIAN RHYTHM SLEEP–WAKE DISORDERS

Circadian rhythm sleep–wake disorders (CRSWD) are caused by a disrupted circadian time-keeping cerebral system, or by a misalignment of the circadian rhythm and the external environment.[154] CRSWDs include, among others, delayed sleep–wake phase disorder, advanced sleep–wake phase disorder, and irregular sleep–wake rhythm disorder. These disorders can cause difficulty initiating and maintaining sleep, and excessive sleepiness. For this reason, CRSWDs can be easily confounded with other sleep–wake disorders such as insomnia.[171] Following TBI, cases of CRSWDs have been described, most often delayed sleep–wake phase disorder.[172–175] In a study by Ayalon et al.[176], 15 individuals with mild TBI were diagnosed with a CRSWD (out of 42 complaining of insomnia), including eight with a delayed sleep–wake phase disorder (i.e., significant delay, usually of more than 2 hours, between sleep–wake schedule and conventional times) and seven with an irregular sleep–wake rhythm disorder (i.e., high day-to-day variability in sleep onset and awakening times). Melatonin secretion and body temperature rhythms were also disrupted in these individuals compared to controls. As most reports of CRSWDs after TBI are case studies, additional research is needed to gain better knowledge on the prevalence, correlates, and course of these disorders.

Assessment of Circadian Rhythm Disorders

Actigraphy can be useful to document shifts in sleep–wake patterns. Two questionnaires can also be helpful to document sleep–wake schedules: the Morningness/Eveningness Questionnaire[177] assesses preferences in engaging in certain activities in the morning or in the evening, classifying respondents into morning types (higher scores) or evening types (lower scores). The items of this instrument refer to hypothetical situations and as such can be more difficult to complete by TBI patients experiencing significant cognitive or self-awareness deficits. The Sleep Timing Questionnaire[178] is another potentially useful tool asking respondents to report on their usual sleep habits (e.g., usual bedtime and arising time on work days and off days).

Pharmacological and Nonpharmacological Treatment of Circadian Disturbances

Empirical evidence is even more limited for the treatment of CRSWD post-TBI. A case study showed that a 15-year-old girl with TBI with a delayed sleep phase disorder (i.e., her sleep–wake cycle was delayed almost 12 hours compared to conventional times) was successfully treated with melatonin.[173] One case study reported on a successful treatment of delayed sleep–wake phase disorder with chronotherapy in a 48-year-old individual.[175] The objective of chronotherapy is to gradually shift the sleep–wake schedule (bedtime

II. SLEEP DISORDERS IN SPECIFIC MEDICAL CONDITIONS

and arising time) toward more conventional times. Compliance is an issue, which may compromise outcome. Bright light therapy is another intervention that could be effective in treating CRSWDs. Patients have to be exposed to a source of bright light (typically 10,000 lx) early in the day (when dealing with delayed sleep—wake rhythm) or late in the day (when dealing with advanced sleep—wake rhythm).

AREAS FOR FUTURE RESEARCH

The literature in adults with TBI has significantly grown in the past decades, and it now counts several meta-analyses or systematic reviews;[6,36,42,179] It is becoming very clear that children and adolescents encounter significant sleep disturbances after TBI as well which require more scientific and clinical attention.[180–182] Research on sleep in older adults with TBI[183] is also still very limited, probably because of the many confounding factors at play. Neuroimaging studies are needed to elucidate the neurophysiology of sleep—wake disturbances, for example, microlesions in the brainstem.[157] There is also a clear need to evaluate the safety and efficacy of different treatment options including compliance and interactions with neurological recovery. Positive improvements in cognition, mood, or pain should accompany sleep-related treatment gains, but these remain to be specifically documented.

Differences between mild and more severe injuries and between acute and chronic sleep—wake disturbances need to be further investigated. It seems that sleep is affected differentially according to time since injury and injury severity. Indeed, some studies suggest that hypersomnia and increased need for sleep are linked to more severe injuries in the acute phase, whereas the literature pertaining to insomnia generally refers to a more chronic phase and does not indicate clear differences between severity groups or points toward more sleep complaints in milder injuries. It is possible that severe brain injury leads to a variety of pathophysiological processes which may explain sleepiness or pleiosomnia, whereas in the longer term psychological, behavioral and environmental factors may play a role in the appearance of insomnia. This is our interpretation of the current literature however, and much more research is needed to confirm this pattern.

CONCLUSION

Sleep—wake disturbances are unfortunately widespread sequelae of TBI, yet they are probably still often overlooked. Recent clinical guidelines[7,8] call for systematic attention to sleep—wake disturbances through screening and appropriate assessment. They also recommend using behavioral methods as a first choice to try to address sleep and fatigue issues and exert clinicians to take caution when administering medications. Improving sleep quality after TBI has the potential to promote recovery, facilitate the rehabilitation process, and ultimately enhance social participation for persons having sustained brain injury. As such, we hope that systematic attention is given to sleep—wake disturbances at any time point postinjury to ensure patients are adequately cared for.

KEY PRACTICE POINTS

- Complaints of poor sleep and sleepiness are extremely common in patients with TBI, regardless of injury severity and regardless of time since injury. The most frequent sleep–wake disorders observed after TBI are insomnia, hypersomnia, and sleep-related breathing disorders.
- Sleep disorders after TBI may hamper participation in rehabilitation and can exacerbate cognitive deficits brought about by TBI.
- Because sleep disorders can worsen the effects of the injury by further affecting quality of life, mental and physical health, individuals with TBI should be systematically investigated for the presence of underlying sleep disorders.
- Daytime sleepiness can be a symptom of a sleep–wake disorder (e.g., sleep-related breathing disorder, CRSWD, posttraumatic hypersomnia) and can be confounded with fatigue, one of the most common consequences of TBI. Fatigue is a subjective physical or mental state of low energy that is not necessarily related to poor sleep, while sleepiness is a physiological sleep drive.
- A comprehensive assessment is essential to document changes from preinjury sleep–wake patterns and the nature, frequency, severity of postinjury sleep–wake disturbances, comorbid medical and psychological symptoms or disorders, use of medications and substances, and activities.
- For assessment, we recommend using a variety of tools (e.g., clinical interview, objective measures, self-reported measures) and other sources of information. In some cases, it is useful to obtain information from a reliable informant (e.g., significant other, health professional), especially when the patient has significant cognitive limitations or impaired self-awareness. In most cases however, TBI does not preclude the use of self-report tools or clinical interviews.
- While PSG is the gold standard measure for several sleep–wake disorders such as sleep-related breathing disorders, clinical interviews and sleep diaries are the most commonly used methods to assess insomnia.
- There is a lack of empirical data on the efficacy and safety of pharmacological agents for sleep–wake disorders in the TBI population. Caution is warranted when prescribing a medication, especially when other medications are used, and psychiatric or medical comorbidities are present.
- CBT for insomnia is recommended as the first-line intervention for adults with insomnia. Some empirical evidence supports the use of CBT for insomnia in the context of TBI. Adaptations may be required in order to optimize the efficacy of treatment and compliance to behavioral and cognitive strategies.

References

1. Ragnarsson K, Clarke W, Daling J, et al. Rehabilitation of persons with traumatic brain injury. *J Am Med Assoc.* 1999;282(10):974–983.
2. Silver JM, McAllister TW, Yudofsky SC, eds. *Textbook of Traumatic Brain Injury*. Washington, DC: American Psychiatric Publishing; 2005.
3. Rao V, Spiro J, Vaishnavi S, et al. Prevalence and types of sleep disturbances acutely after traumatic brain injury. *Brain Injury*. 2008;22(5):381–386.

REFERENCES

245

4. Verma A, Anand V, Verma NP. Sleep disorders in chronic traumatic brain injury. *J Clin Sleep Med*. 2007;3 (4):357–362.

5. Ouellet M-C, Savard J, Morin CM. Insomnia following traumatic brain injury: a review. *Neurorehabil Neural Repair*. 2004;18(4):187–198.

6. Mathias JL, Alvaro PK. Prevalence of sleep disturbances, disorders, and problems following traumatic brain injury: a meta-analysis. *Sleep Med*. 2012;13(7):898–905.

7. Ontario Neurotrauma Foundation. *Section 7: Sleep-wake disturbances. Guideline for Concussion/Mild Traumatic Brain Injury & Persistent Symptoms*. 3rd ed Toronto, ON: Ontario Neurotrauma Foundation; 2018:38–42.

8. Institut national d'excellence en santé et en services sociaux, Ontario Neurotrauma Foundation. Clinical practice guideline for the rehabilitation of adults with moderate to severe traumatic brain injury. <https://www.inesss.qc.ca/fileadmin/doc/INESSS/Rapports/Traumatologie/Fundamental_and_priority_REC__ANG_final.pdf>; 2016.

9. Menon DK, Schwab K, Wright DW, Maas AI. Position statement: definition of traumatic brain injury. *Arch Phys Med Rehabil*. 2010;91(11):1637–1640.

10. Sidaros A, Skimminge A, Liptrot MG, et al. Long-term global and regional brain volume changes following severe traumatic brain injury: a longitudinal study with clinical correlates. *NeuroImage*. 2009;44(1):1–8.

11. Faul M, Xu L, Wald MM, Coronado VG. *Traumatic Brain Injury in the United States: Emergency Department Visits, Hospitalizations and Deaths 2002–2006*. Atlanta, GA: National Center for Injury Prevention and Control, Centers for Disease Control and Prevention. 2010.

12. Kraus JF, Chu LD. Epidemiology. In: Silver JM, McAllister TW, Yudofsky SC, eds. *Textbook of Traumatic Brain Injury*. Washington, DC: American Psychiatric Publishing; 2005:3–26.

13. Bruns Jr J, Hauser WA. The epidemiology of traumatic brain injury: a review. *Epilepsia*. 2003;44(Suppl 10):2–10.

14. Teasdale G, Jennett B. Assessment of coma and impaired consciousness. A practical scale. *Lancet*. 1974;304 (7872):81–84.

15. Cassidy JD, Carroll LJ, Peloso PM, et al. Incidence, risk factors and prevention of mild traumatic brain injury: results of the WHO Collaborating Centre Task Force on Mild Traumatic Brain Injury. *J Rehabil Med*. 2004;(43 Suppl)28–60.

16. Tagliaferri F, Compagnone C, Korsic M, Servadei F, Kraus J. A systematic review of brain injury epidemiology in Europe. *Acta Neurochir*. 2006;148(3):255–268.

17. Carroll LJ, Cassidy JD, Peloso PM, et al. Prognosis for mild traumatic brain injury: results of the WHO Collaborating Centre Task Force on Mild Traumatic Brain Injury. *J Rehabil Med*. 2004;(43 Suppl)84–105.

18. Corrigan JD, Hammond FM. Traumatic brain injury as a chronic health condition. *Arch Phys Med Rehabil*. 2013;94(6):1199–1201.

19. Cullen N, Meyer MJ, Aubut J-A, Bayley M, Teasell R. *Evidence-Based Review of Moderate to Severe Acquired Brain Injury: Module 3 – Efficacy and Models of Care Following Acquired Brain Injury*. In: (ABIEBR) ABIE-BR, ed2013: <http://www.abiebr.com/>.

20. Gabor JY, Cooper AB, Hanly PJ. Sleep disruption in the intensive care unit. *Curr Opin Crit Care*. 2001;7 (1):21–27.

21. Friese RS, Diaz-Arrastia R, McBride D, Frankel H, Gentilello LM. Quantity and quality of sleep in the surgical intensive care unit: are our patients sleeping? *J Trauma*. 2007;63(6):1210–1214.

22. Chiu HY, Chen PY, Chen NH, Chuang LP, Tsai PS. Trajectories of sleep changes during the acute phase of traumatic brain injury: a 7-day actigraphy study. *J Formos Med Assoc*. 2013;112(9):545–553.

23. Billiard M, Podesta C. Recurrent hypersomnia following traumatic brain injury. *Sleep Med*. 2013;14 (5):462–465.

24. Sommerauer M, Valko PO, Werth E, Baumann CR. Excessive sleep need following traumatic brain injury: a case-control study of 36 patients. *J Sleep Res*. 2013;22(6):634–639.

25. Evans BM, Bartlett JR. Prediction of outcome in severe head injury based on recognition of sleep related activity in the polygraphic electroencephalogram. *J Neurol Neurosurg Psychiatry*. 1995;59(1):17–25.

26. Ron S, Algom D, Hary D, Cohen M. Time-related changes in the distribution of sleep stages in brain injured patients. *Electroencephalogr Clin Neurophysiol*. 1980;48(4):432–441.

27. Duclos C, Dumont M, Blais H, et al. Rest–activity cycle disturbances in the acute phase of moderate to severe traumatic brain injury. *Neurorehabil Neural Repair*. 2014;28(5):472–482. Available from: http://dx.doi.org/10.1177/1545968313517756.

II. SLEEP DISORDERS IN SPECIFIC MEDICAL CONDITIONS

28. Duclos C, Dumont M, Arbour C, et al. Parallel recovery of consciousness and sleep in acute traumatic brain injury. *Neurology*. 2017;88(3):268−275.
29. Duclos C, Dumont M, Potvin MJ, et al. Evolution of severe sleep-wake cycle disturbances following traumatic brain injury: a case study in both acute and subacute phases post-injury. *BMC Neurol*. 2016;16(1):186.
30. Makley MJ, Johnson-Greene L, Tarwater PM, et al. Return of memory and sleep efficiency following moderate to severe closed head injury. *Neurorehabil Neural Repair*. 2009;23(4):320−326.
31. Sherer M, Yablon SA, Nakase-Richardson R. Patterns of recovery of posttraumatic confusional state in neurorehabilitation admissions after traumatic brain injury. *Arch Phys Med Rehabil*. 2009;90 (10):1749−1754.
32. Makley MJ, English JB, Drubach DA, Kreuz AJ, Celnik PA, Tarwater PM. Prevalence of sleep disturbance in closed head injury patients in a rehabilitation unit. *Neurorehabil Neural Repair*. 2008;22(4):341−347.
33. Nakase-Richardson R, Sherer M, Barnett SD, et al. Prospective evaluation of the nature, course, and impact of acute sleep abnormality after traumatic brain injury. *Arch Phys Med Rehabil*. 2013;94(5):875−882.
34. Paparrigopoulos T, Melissaki A, Tsekou H, et al. Melatonin secretion after head injury: a pilot study. *Brain Inj*. 2006;20(8):873−878.
35. Cohen M, Oksenberg A, Snir D, Stern MJ, Groswasser Z. Temporally related changes of sleep complaints in traumatic brain injured patients. *J Neurol Neurosurg Psychiatry*. 1992;55(4):313−315.
36. Grima N, Ponsford J, Rajaratnam SM, Mansfield D, Pase MP. Sleep disturbances in traumatic brain injury: a meta-analysis. *J Clin Sleep Med*. 2016;12(3):419−428.
37. Shekleton JA, Parcell DL, Redman JR, Phipps-Nelson J, Ponsford JL, Rajaratnam SM. Sleep disturbance and melatonin levels following traumatic brain injury. *Neurology*. 2010;74(21):1732−1738.
38. Ouellet M-C, Morin CM. Subjective and objective measures of insomnia in the context of traumatic brain injury: a preliminary study. *Sleep Med*. 2006;7(6):486−497.
39. Schreiber S, Barkai G, Gur-Hartman T, et al. Long-lasting sleep patterns of adult patients with minor traumatic brain injury (mTBI) and non-mTBI subjects. *Sleep Med*. 2008;9(5):481−487.
40. Stone JJ, Childs S, Smith LE, Battin M, Papadakos PJ, Huang JH. Hourly neurologic assessments for traumatic brain injury in the ICU. *Neurol Res*. 2014;36(2):164−169.
41. Williams BR, Lazic SE, Ogilvie RD. Polysomnographic and quantitative EEG analysis of subjects with long-term insomnia complaints associated with mild traumatic brain injury. *Clin Neurophysiol*. 2008;119(2):429−438.
42. Mantua J, Grillakis A, Mahfouz SH, et al. A systematic review and meta-analysis of sleep architecture and chronic traumatic brain injury. *Sleep Med Rev*. 2018;375(1):74−83.
43. Modarres M, Kuzma NN, Kretzmer T, Pack AI, Lim MM. EEG slow waves in traumatic brain injury: convergent findings in mouse and man. *Neurobiol Sleep Circadian Rhythms*. 2016;1. pii: S2451994416300025.
44. Sandsmark DK, Elliott JE, Lim MM, et al. Sleep−wake disturbances after traumatic brain injury: synthesis of human and animal studies. *Sleep*. 2017;40(5):101−107.
45. Mouthon AL, Meyer-Heim A, Kurth S, et al. High-density electroencephalographic recordings during sleep in children and adolescents with acquired brain injury. *Neurorehabil Neural Repair*. 2017;31(5):462−474.
46. Mollayeva T, Colantonio A, Cassidy JD, Vernich L, Moineddin R, Shapiro CM. Sleep stage distribution in persons with mild traumatic brain injury: a polysomnographic study according to American Academy of Sleep Medicine standards. *Sleep Med*. 2017;34:179−192.
47. Rowe RK, Harrison JL, O'Hara BF, Lifshitz J. Diffuse brain injury does not affect chronic sleep patterns in the mouse. *Brain Inj*. 2014;28(4):504−510.
48. Rao V, Bergey A, Hill H, Efron D, McCann U. Sleep disturbance after mild traumatic brain injury: indicator of injury? *J Neuropsychiatry Clin Neurosci*. 2011;23(2):201−205.
49. Gosselin N, Lassonde M, Petit D, et al. Sleep following sport-related concussions. *Sleep Med*. 2008;10:35−46.
50. Khoury S, Chouchou F, Amzica F, et al. Rapid EEG activity during sleep dominates in mild traumatic brain injury patients with acute pain. *J Neurotrauma*. 2013;30(8):633−641. Available from: https://doi.org/10.1089/neu.2012.2519. Epub 2013 Apr 1018.
51. Foltz EL, Schmidt RP. The role of the reticular formation in the coma of head injury. *J Neurosurg*. 1956;13 (2):145−154.
52. Denny-Brown DE, Russell WR. Experimental concussion: (section of neurology). *Proc R Soc Med*. 1941;34 (11):691−692.
53. Ward AA. Physiological basis of concussion. *J Neurosurg*. 1958;15(2):129−134.

REFERENCES

54. Fakhran S, Yaeger K, Alhilali L. Symptomatic white matter changes in mild traumatic brain injury resemble pathologic features of early Alzheimer dementia. *Radiology*. 2013;269(1):249–257.
55. Yaeger K, Alhilali L, Fakhran S. Evaluation of tentorial length and angle in sleep–wake disturbances after mild traumatic brain injury. *AJR Am J Roentgenol*. 2014;202(3):614–618.
56. Osier ND, Pham L, Pugh BJ, et al. Brain injury results in lower levels of melatonin receptors subtypes MT1 and MT2. *Neurosci Lett*. 2017;650:18–24.
57. Grima NA, Ponsford JL, St, Hilaire MA, et al. Circadian melatonin rhythm following traumatic brain injury. *Neurorehabil Neural Repair*. 2016;30(10):972–977.
58. Valko PO, Gavrilov YV, Yamamoto M, et al. Damage to histaminergic tuberomammillary neurons and other hypothalamic neurons with traumatic brain injury. *Ann Neurol*. 2015;77(1):177–182.
59. Baumann CR, Bassetti CL, Valko PO, et al. Loss of hypocretin (orexin) neurons with traumatic brain injury. *Ann Neurol*. 2009;66(4):555–559.
60. Baumann CR, Stocker R, Imhof HG, et al. Hypocretin-1 (orexin A) deficiency in acute traumatic brain injury. *Neurology*. 2005;65(1):147–149.
61. Baumann CR, Werth E, Stocker R, Ludwig S, Bassetti CL. Sleep–wake disturbances 6 months after traumatic brain injury: a prospective study. *Brain*. 2007;130(Pt 7):1873–1883.
62. Sabir M, Gaudreault PO, Freyburger M, et al. Impact of traumatic brain injury on sleep structure, electrocorticographic activity and transcriptome in mice. *Brain Behav Immun*. 2015;47:118–130.
63. Willie JT, Lim MM, Bennett RE, Azarion AA, Schwetye KE, Brody DL. Controlled cortical impact traumatic brain injury acutely disrupts wakefulness and extracellular orexin dynamics as determined by intracerebral microdialysis in mice. *J Neurotrauma*. 2012;29(10):1908–1921.
64. Skopin MD, Kabadi SV, Viechweg SS, Mong JA, Faden AI. Chronic decrease in wakefulness and disruption of sleep–wake behavior after experimental traumatic brain injury. *J Neurotrauma*. 2014;32:289–296.
65. Lim MM, Elkind J, Xiong G, et al. Dietary therapy mitigates persistent wake deficits caused by mild traumatic brain injury. *Sci Transl Med*. 2013;5(215):215ra173.
66. Nardone R, Bergmann J, Kunz A, et al. Cortical excitability changes in patients with sleep–wake disturbances after traumatic brain injury. *J Neurotrauma*. 2011;28:1165–1171.
67. Ouellet MC, Beaulieu-Bonneau S, Morin CM. Insomnia in patients with traumatic brain injury: frequency, characteristics, and risk factors. *J Head Trauma Rehabil*. 2006;21(3):199–212.
68. Guilleminault C, Yuen KM, Gulevich MG, Karadeniz D, Leger D, Philip P. Hypersomnia after head-neck trauma: a medicolegal dilemma. *Neurology*. 2000;54(3):653–659.
69. Beaulieu-Bonneau S, Morin CM. Sleepiness and fatigue following traumatic brain injury. *Sleep Med*. 2012;13(6):598–605.
70. Jang SH, Kwon HG. Injury of the ascending reticular activating system in patients with fatigue and hypersomnia following mild traumatic brain injury: two case reports. *Medicine*. 2016;95(6):e2628.
71. Valko PO, Gavrilov YV, Yamamoto M, et al. Damage to arousal-promoting brainstem neurons with traumatic brain injury. *Sleep*. 2016;39(6):1249–1252.
72. Urakami Y. Relationship between, sleep spindles and clinical recovery in patients with traumatic brain injury: a simultaneous EEG and MEG study. *Clin EEG Neurosci*. 2012;43(1):39–47.
73. Chiu HY, Lo WC, Chiang YH, Tsai PS. The effects of sleep on the relationship between brain injury severity and recovery of cognitive function: a prospective study. *Int J Nurs Stud*. 2014;51(6):892–899.
74. Waldron-Perrine B, McGuire AP, Spencer RJ, Drag LL, Pangilinan PH, Bieliauskas LA. The influence of sleep and mood on cognitive functioning among veterans being evaluated for mild traumatic brain injury. *Military Med*. 2012;177(11):1293–1301.
75. Dean PJ, Sterr A. Long-term effects of mild traumatic brain injury on cognitive performance. *Front Hum Neurosci*. 2013;7:30.
76. Theadom A, Starkey N, Barker-Collo S, Jones K, Ameratunga S, Feigin V. Population-based cohort study of the impacts of mild traumatic brain injury in adults four years post-injury. *PLoS One*. 2018;13(1):e0191655.
77. Bloomfield IL, Espie CA, Evans JJ. Do sleep difficulties exacerbate deficits in sustained attention following traumatic brain injury? *J Int Neuropsychol Soc*. 2010;16(1):17–25.
78. Castriotta RJ, Wilde MC, Lai JM, Atanasov S, Masel BE, Kuna ST. Prevalence and consequences of sleep disorders in traumatic brain injury. *J Clin Sleep Med*. 2007;3(4):349–356.

II. SLEEP DISORDERS IN SPECIFIC MEDICAL CONDITIONS

79. Wilde MC, Castriotta RJ, Lai JM, Atanasov S, Masel BE, Kuna ST. Cognitive impairment in patients with traumatic brain injury and obstructive sleep apnea. *Arch Phys Med Rehabil*. 2007;88(10):1284−1288.
80. Shay N, Yeates KO, Walz NC, et al. Sleep problems and their relationship to cognitive and behavioral outcomes in young children with traumatic brain injury. *J Neurotrauma*. 2014;31:1305−1312.
81. Osorio MB, Kurowski BG, Beebe D, et al. Association of daytime somnolence with executive functioning in the first 6 months after adolescent traumatic brain injury. *PM R*. 2013;5(7):554−562.
82. Ruff RL, Ruff SS, Wang XF. Improving sleep: initial headache treatment in OIF/OEF veterans with blast-induced mild traumatic brain injury. *J Rehabil Res Dev*. 2009;46(9):1071−1084.
83. Wiseman-Hakes C, Murray B, Moineddin R, et al. Evaluating the impact of treatment for sleep/wake disorders on recovery of cognition and communication in adults with chronic TBI. *Brain Inj*. 2013;27(12):1364−1376.
84. Wiseman-Hakes C, Victor JC, Brandys C, Murray BJ. Impact of post-traumatic hypersomnia on functional recovery of cognition and communication. *Brain Inj*. 2011;25(12):1256−1265.
85. Worthington AD, Melia Y. Rehabilitation is compromised by arousal and sleep disorders: results of a survey of rehabilitation centres. *Brain Inj*. 2006;20(3):327−332.
86. Silva MA, Nakase-Richardson R, Sherer M, Barnett SD, Evans CC, Yablon SA. Posttraumatic confusion predicts patient cooperation during traumatic brain injury rehabilitation. *Am J Phys Med Rehabil*. 2012;91(10):890−893.
87. Bramley H, Henson A, Lewis MM, Kong L, Stetter C, Silvis M. Sleep disturbance following concussion is a risk factor for a prolonged recovery. *Clin Pediatr (Phila)*. 2017;56(14):1280−1285.
88. Chaput G, Giguere JF, Chauny JM, Denis R, Lavigne G. Relationship among subjective sleep complaints, headaches, and mood alterations following a mild traumatic brain injury. *Sleep Med*. 2009;10(7):713−716.
89. Tham SW, Palermo TM, Vavilala MS, et al. The longitudinal course, risk factors, and impact of sleep disturbances in children with traumatic brain injury. *J Neurotrauma*. 2012;29(1):154−161.
90. Ao KH, Ho CH, Wang CC, Wang JJ, Chio CC, Kuo JR. The increased risk of stroke in early insomnia following traumatic brain injury: a population-based cohort study. *Sleep Med*. 2017;37:187−192.
91. Beetar JT, Guilmette TJ, Sparadeo FR. Sleep and pain complaints in symptomatic traumatic brain injury and neurologic populations. *Arch Phys Med Rehabil*. 1996;77(12):1298−1302.
92. Lew HL, Pogoda TK, Hsu PT, et al. Impact of the "polytrauma clinical triad" on sleep disturbance in a department of veterans affairs outpatient rehabilitation setting. *Am J Phys Med Rehabil*. 2010;89(6):437−445.
93. Suzuki Y, Khoury S, El-Khatib H, et al. Individuals with pain need more sleep in the early stage of mild traumatic brain injury. *Sleep Med*. 2017;33:36−42.
94. Bryant RA, O'Donnell ML, Creamer M, McFarlane AC, Clark CR, Silove D. The psychiatric sequelae of traumatic injury. *Am J Psychiatry*. 2010;167(3):312−320.
95. Gould KR, Ponsford JL, Johnston L, Schonberger M. The nature, frequency and course of psychiatric disorders in the first year after traumatic brain injury: a prospective study. *Psychol Med*. 2011;41(10):2099−2109.
96. Koponen S, Taiminen T, Hiekkanen H, Tenovuo O. Axis I and II psychiatric disorders in patients with traumatic brain injury: a 12-month follow-up study. *Brain Inj*. 2011;25(11):1029−1034.
97. Whelan-Goodinson R, Ponsford JL, Schonberger M, Johnston L. Predictors of psychiatric disorders following traumatic brain injury. *J Head Trauma Rehabil*. 2010;25(5):320−329.
98. Diaz AP, Schwarzbold ML, Thais ME, et al. Psychiatric disorders and health-related quality of life after severe traumatic brain injury: a prospective study. *J Neurotrauma*. 2012;29(6):1029−1037.
99. Hesdorffer DC, Rauch SL, Tamminga CA. Long-term psychiatric outcomes following traumatic brain injury: a review of the literature. *J Head Trauma Rehabil*. 2009;24(6):452−459.
100. Kempf J, Werth E, Kaiser PR, Bassetti CL, Baumann CR. Sleep−wake disturbances 3 years after traumatic brain injury. *J Neurol Neurosurg Psychiatry*. 2010;81(12):1402−1405.
101. Parcell DL, Ponsford JL, Redman JR, Rajaratnam SM. Poor sleep quality and changes in objectively recorded sleep after traumatic brain injury: a preliminary study. *Arch Phys Med Rehabil*. 2008;89(5):843−850.
102. Parcell DL, Ponsford JL, Rajaratnam SM, Redman JR. Self-reported changes to nighttime sleep after traumatic brain injury. *Arch Phys Med Rehabil*. 2006;87(2):278−285.
103. Morin C. *Insomnia: Psychological Assessment and Management*. New York: Guilford Press; 1993.

REFERENCES

104. Fischer JT, Hannay HJ, Alfano CA, Swank PR, Ewing-Cobbs L. Sleep disturbances and internalizing behavior problems following pediatric traumatic injury. *Neuropsychology*. 2018;32(2):161–175.
105. Morse AM, Garner DR. Traumatic brain injury, sleep disorders, and psychiatric disorders: an underrecognized relationship. *Med Sci (Basel)*. 2018;6(1). pii: E15.
106. DeBeer BB, Kimbrel NA, Mendoza C, et al. Traumatic brain injury, sleep quality, and suicidal ideation in Iraq/Afghanistan era veterans. *J Nerv Mental Dis*. 2017;205(7):512–516.
107. Lew HL, Otis JD, Tun C, Kerns RD, Clark ME, Cifu DX. Prevalence of chronic pain, posttraumatic stress disorder, and persistent postconcussive symptoms in OIF/OEF veterans: polytrauma clinical triad. *J Rehabil Res Dev*. 2009;46(6):697–702.
108. Collen J, Orr N, Lettieri CJ, Carter K, Holley AB. Sleep disturbances among soldiers with combat-related traumatic brain injury. *Chest*. 2012;142(3):622–630.
109. Macera CA, Aralis HJ, Rauh MJ, Macgregor AJ. Do sleep problems mediate the relationship between traumatic brain injury and development of mental health symptoms after deployment? *Sleep*. 2013;36(1):83–90.
110. Farrell-Carnahan L, Franke L, Graham C, McNamee S. Subjective sleep disturbance in veterans receiving care in the Veterans Affairs Polytrauma System following blast-related mild traumatic brain injury. *Mil Med*. 2013;178(9):951–956. Available from: https://doi.org/10.7205/MILMED-D-7213-00037.
111. Wallace DM, Shafazand S, Ramos AR, et al. Insomnia characteristics and clinical correlates in Operation Enduring Freedom/Operation Iraqi Freedom veterans with post-traumatic stress disorder and mild traumatic brain injury: an exploratory study. *Sleep Med*. 2011;12(9):850–859.
112. Kobayashi I, Boarts JM, Delahanty DL. Polysomnographically measured sleep abnormalities in PTSD: a meta-analytic review. *Psychophysiology*. 2007;44(4):660–669.
113. Elovic EP, Dobrovic NM, Fellus JL. Fatigue after traumatic brain injury. In: DeLuca J, ed. *Fatigue as a Window to the Brain*. London: MIT Press; 2005:89–105.
114. Ouellet M-C, Morin CM. Fatigue following traumatic brain injury: frequency, characteristics, and associated factors. *Rehabil Psychol*. 2006;51(2):140–149.
115. Pigeon WR, Sateia MJ, Ferguson RJ. Distinguishing between excessive daytime sleepiness and fatigue: Toward improved detection and treatment. *J Psychosom Res*. 2003;54:61–69.
116. Cantor JB, Ashman T, Gordon W, et al. Fatigue after traumatic brain injury and its impact on participation and quality of life. *J Head Trauma Rehabil*. 2008;23(1):41–51.
117. Fichtenberg NL, Zafonte RD, Putnam S, Mann NR, Millard AE. Insomnia in a post-acute brain injury sample. *Brain Inj*. 2002;16(3):197–206.
118. Hou L, Han X, Sheng P, et al. Risk factors associated with sleep disturbance following traumatic brain injury: clinical findings and questionnaire based study. *PLoS One*. 2013;8(10):e76087.
119. Clinchot DM, Bogner J, Mysiw WJ, Fugate L, Corrigan J. Defining sleep disturbance after brain injury. *Am J Phys Med Rehabil*. 1998;77(4):291–295.
120. Fichtenberg NL, Millis SR, Mann NR, Zafonte RD, Millard AE. Factors associated with insomnia among post-acute traumatic brain injury survivors. *Brain Inj*. 2000;14(7):659–667.
121. Mahmood O, Rapport LJ, Hanks RA, Fichtenberg NL. Neuropsychological performance and sleep disturbance following traumatic brain injury. *J Head Trauma Rehabil*. 2004;19(5):378–390.
122. Corrigan JD, Bogner JA, Mysiw WJ, Clinchot D, Fugate L. Systematic bias in outcome studies of persons with traumatic brain injury. *Arch Phys Med Rehabil*. 1997;78(2):132–137.
123. Ferguson JM. SSRI antidepressant medications: adverse effects and tolerability. *Prim Care Companion J Clin Psychiatry*. 2001;3(1):22–27.
124. Zafonte R, Mann N, Fitchenberg N. Sleep disturbance in traumatic brain injury: pharmacological options. *NeuroRehabilitation*. 1996;7:189–195.
125. Morin CM. *Insomnia: Psychological Assessment and Management*. New York: Guilford; 1993.
126. Webb WB, Agnew Jr. HW. Sleep efficiency for sleep–wake cycles of varied length. *Psychophysiology*. 1975;12 (6):637–641.
127. Buysse DJ, Reynolds 3rd CF, Monk TH, Berman SR, Kupfer DJ. The Pittsburgh Sleep Quality Index: a new instrument for psychiatric practice and research. *Psychiatry Res*. 1989;28(2):193–213.
128. Kaufmann CN, Orff HJ, Moore RC, Delano-Wood L, Depp CA, Schiehser DM. Psychometric characteristics of the insomnia severity index in veterans with history of traumatic brain injury. *Behav Sleep Med*. 2017;1–9. Available from: http://dx.doi.org/10.1080/15402002.2016.1266490.

II. SLEEP DISORDERS IN SPECIFIC MEDICAL CONDITIONS

129. Lequerica A, Chiaravalloti N, Cantor J, et al. The factor structure of the Pittsburgh Sleep Quality Index in persons with traumatic brain injury. A NIDRR TBI model systems module study. *NeuroRehabilitation*. 2014;35(3):485–492.

130. Fichtenberg NL, Putnam SH, Mann NR, Zafonte RD, Millard AE. Insomnia screening in postacute traumatic brain injury: utility and validity of the Pittsburgh Sleep Quality Index. *Am J Phys Med Rehabil*. 2001;80(5):339–345.

131. Beaulieu-Bonneau S, St-Onge F, Blackburn M-C, Banville A, Paradis-Giroux A-A, Ouellet M-C. Alcohol and drug use before and during the first year after traumatic brain injury. J Head Trauma Rehabil. 2018;33(3): E51-E60. Available from: http://dx.doi.org/10.1097/HTR.0000000000000341.

132. Ouellet M, Beaulieu-Bonneau S, Savard J, Morin C. *Manuel d'évaluation et d'intervention pour l'insomnie et la fatigue après un traumatisme craniocérébral*. Montreal: Centre de Liaison et d'interventions psychosociales; 2015. <http://www.cirris.ulaval.ca/fr/insomnie-et-fatigue-apres-un-traumatisme-craniocerebral>.

133. National Institutes of Health. National Institutes of Health State of the Science Conference statement on Manifestations and Management of Chronic Insomnia in Adults, June 13–15, 2005. *Sleep*. 2005;28(9):1049–1057.

134. Wilson SJ, Nutt DJ, Alford C, et al. British Association for Psychopharmacology consensus statement on evidence-based treatment of insomnia, parasomnias and circadian rhythm disorders. *J Psychopharmacol*. 2010;24(11):1577–1601.

135. Flanagan SR, Greenwald B, Wieber S. Pharmacological treatment of insomnia for individuals with brain injury. *J Head Trauma Rehabil*. 2007;22(1):67–70.

136. Larson EB, Zollman FS. The effect of sleep medications on cognitive recovery from traumatic brain injury. *J Head Trauma Rehabil*. 2010;25(1):61–67.

137. Kemp S, Biswas R, Neumann V, Coughlan A. The value of melatonin for sleep disorders occurring post-head injury: a pilot RCT. *Brain Inj*. 2004;18(9):911–919.

138. Ruff RL, Riechers 2nd RG, Wang XF, Piero T, Ruff SS. For veterans with mild traumatic brain injury, improved posttraumatic stress disorder severity and sleep correlated with symptomatic improvement. *J Rehabil Res Dev*. 2012;49(9):1305–1320.

139. Morin CM, Bootzin RR, Buysse DJ, Edinger JD, Espie CA, Lichstein KL. Psychological and behavioral treatment of insomnia: update of the recent evidence (1998–2004). *Sleep*. 2006;29(11):1398–1414.

140. Morin CM, Vallieres A, Ivers H. Dysfunctional beliefs and attitudes about sleep (DBAS): validation of a brief version (DBAS-16). *Sleep*. 2007;30(11):1547–1554.

141. Ouellet MC, Morin CM. Cognitive behavioral therapy for insomnia associated with traumatic brain injury: a single-case study. *Arch Phys Med Rehabil*. 2004;85(8):1298–1302.

142. Ouellet M-C, Morin CM. Efficacy of cognitive-behavioral therapy for insomnia associated with traumatic brain injury: a single-case experimental design. *Arch Phys Med Rehabil*. 2007;88(12):1581–1592.

143. Nguyen S, McKay A, Wong D, et al. Cognitive behavior therapy to treat sleep disturbance and fatigue after traumatic brain injury: a pilot randomized controlled trial. *Arch Phys Med Rehabil*. 2017;98(8):1508–1517. e1502.

144. Nguyen S, McKenzie D, McKay A, et al. Exploring predictors of treatment outcome in cognitive behavior therapy for sleep disturbance following acquired brain injury. *Disabil Rehabil*. 2018;40(16):1906–1913.

145. Theadom A, Barker-Collo S, Jones K, et al. A pilot randomized controlled trial of on-line interventions to improve sleep quality in adults after mild or moderate traumatic brain injury. *Clin Rehabil*. 2018;32(5):619–629.

146. De La Rue-Evans L, Nesbitt K, Oka RK. Sleep hygiene program implementation in patients with traumatic brain injury. *Rehabil Nurs*. 2013;38(1):2–10. Available from: https://doi.org/10.1002/rnj.1066.

147. Watson NF, Dikmen S, Machamer J, Doherty M, Temkin N. Hypersomnia following traumatic brain injury. *J Clin Sleep Med*. 2007;3(4):363–368.

148. Kempf J, Werth E, Kaiser PR, Bassetti CL, Baumann CR. Sleep–wake disturbances 3 years after traumatic brain injury. *J Neurol Neurosurg Psychiatry*. 2010;81(12):1402–1405.

149. Hartvigsen J, Boyle E, Cassidy JD, Carroll LJ. Mild traumatic brain injury after motor vehicle collisions: what are the symptoms and who treats them? A population-based 1-year inception cohort study. *Arch Phys Med Rehabil*. 2014;95(3 Suppl):S286–S294.

REFERENCES

150. Guilleminault C, Faull KF, Miles L, van den Hoed J. Posttraumatic excessive daytime sleepiness: a review of 20 patients. *Neurology*. 1983;33(12):1584–1589.
151. Webster JB, Bell KR, Hussey JD, Natale TK, Lakshminarayan S. Sleep apnea in adults with traumatic brain injury: a preliminary investigation. *Arch Phys Med Rehabil*. 2001;82(3):316–321.
152. Masel BE, Scheibel RS, Kimbark T, Kuna ST. Excessive daytime sleepiness in adults with brain injuries. *Arch Phys Med Rehabil*. 2001;82(11):1526–1532.
153. Ebrahim IO, Peacock KW, Williams AJ. Posttraumatic narcolepsy—two case reports and a mini review. *J Clin Sleep Med*. 2005;1(2):153–156.
154. American Academy of Sleep Medicine. *International Classification of Sleep Disorders*. 3rd ed Darien, IL: American Academy of Sleep Medicine; 2014.
155. Castriotta RJ, Lai JM. Sleep disorders associated with traumatic brain injury. *Arch Phys Med Rehabil*. 2001;82 (10):1403–1406.
156. Kemp S, Agostinis A, House A, Coughlan AK. Analgesia and other causes of amnesia that mimic post-traumatic amnesia (PTA): a cohort study. *J Neuropsychol*. 2010;4(Pt 2):231-236. Available from: http://dx.doi.org/10.1348/174866409X482614.
157. Imbach LL, Valko PO, Li T, et al. Increased sleep need and daytime sleepiness 6 months after traumatic brain injury: a prospective controlled clinical trial. *Brain*. 2015;15.
158. Mahowald M, Mahowald M. Sleep disorders. In: Rizzo M, Tranel D, eds. *Head Injury and Postconcussive Syndrome*. New York: Churchill Livingstone; 1996:285–304.
159. Arand D, Bonnet M, Hurwitz T, Mitler M, Rosa R, Sangal RB. The clinical use of the MSLT and MWT. *Sleep*. 2005;28(1):123–144.
160. Castriotta RJ, Atanasov S, Wilde MC, Masel BE, Lai JM, Kuna ST. Treatment of sleep disorders after traumatic brain injury. *J Clin Sleep Med*. 2009;5(2):137–144.
161. Muller U, Czymmek J, Thone-Otto A, Von Cramon DY. Reduced daytime activity in patients with acquired brain damage and apathy: a study with ambulatory actigraphy. *Brain Inj*. 2006;20(2):157–160.
162. Zollman FS, Cyborski C, Duraski SA. Actigraphy for assessment of sleep in traumatic brain injury: case series, review of the literature and proposed criteria for use. *Brain Inj*. 2010;24(5):748-754.
163. Jha A, Weintraub A, Allshouse A, et al. A randomized trial of modafinil for the treatment of fatigue and excessive daytime sleepiness in individuals with chronic traumatic brain injury. *J Head Trauma Rehabil*. 2008;23(1):52–63.
164. Kaiser PR, Valko PO, Werth E, et al. Modafinil ameliorates excessive daytime sleepiness after traumatic brain injury. *Neurology*. 2010;75(20):1780–1785.
165. Menn SJ, Yang R, Lankford A. Armodafinil for the treatment of excessive sleepiness associated with mild or moderate closed traumatic brain injury: a 12-week, randomized, double-blind study followed by a 12-month open-label extension. *J Clin Sleep Med*. 2014;10(11):1181–1191.
166. Lee H, Kim SW, Kim JM, Shin IS, Yang SJ, Yoon JS. Comparing effects of methylphenidate, sertraline and placebo on neuropsychiatric sequelae in patients with traumatic brain injury. *Hum Psychopharmacol*. 2005;20 (2):97–104.
167. Al-Adawi S, Burke DT, Dorvlo AS. The effect of methylphenidate on the sleep–wake cycle of brain-injured patients undergoing rehabilitation. *Sleep Med*. 2006;7(3):287–291.
168. Kushida CA, Littner MR, Hirshkowitz M, et al. Practice parameters for the use of continuous and bilevel positive airway pressure devices to treat adult patients with sleep-related breathing disorders. *Sleep*. 2006;29 (3):375–380.
169. Sinclair KL, Ponsford JL, Taffe J, Lockley SW, Rajaratnam SM. Randomized controlled trial of light therapy for fatigue following traumatic brain injury. *Neurorehabil Neural Repair*. 2014;28(4):303–313. Available from: https://doi.org/10.1177/1545968313508472. Epub 1545968313502013 Nov1545968313508478.
170. Dhand R, Sohal H. Good sleep, bad sleep! The role of daytime naps in healthy adults. *Curr Opin Pulm Med*. 2006;12(6):379–382.
171. Orff HJ, Ayalon L, Drummond SP. Traumatic brain injury and sleep disturbance: a review of current research. *J Head Trauma Rehabil*. 2009;24(3):155–165.
172. Boivin DB, Caliyurt O, James FO, Chalk C. Association between delayed sleep phase and hypernyctohemeral syndromes: a case study. *Sleep*. 2004;27(3):417–421.

II. SLEEP DISORDERS IN SPECIFIC MEDICAL CONDITIONS

173. Nagtegaal JE, Kerkhof GA, Smits MG, Swart AC, van der Meer YG. Traumatic brain injury-associated delayed sleep phase syndrome. *Funct Neurol*. 1997;12(6):345−348.
174. Patten SB, Lauderdale WM. Delayed sleep phase disorder after traumatic brain injury. *J Am Acad Child Adolesc Psychiatry*. 1992;31(1):100−102.
175. Quinto C, Gellido C, Chokroverty S, Masdeu J. Posttraumatic delayed sleep phase syndrome. *Neurology*. 2000;54(1):250−252.
176. Ayalon L, Borodkin K, Dishon L, Kanety H, Dagan Y. Circadian rhythm sleep disorders following mild traumatic brain injury. *Neurology*. 2007;68(14):1136−1140.
177. Horne JA, Ostberg O. A self-assessment questionnaire to determine morningness-eveningness in human circadian rhythms. *Int J Chronobiol*. 1976;4(2):97−110.
178. Monk TH, Buysse DJ, Kennedy KS, Pods JM, DeGrazia JM, Miewald JM. Measuring sleep habits without using a diary: the sleep timing questionnaire. *Sleep*. 2003;26(2):208−212.
179. Wiseman-Hakes C, Colantonio A, Gargaro J. Sleep and wake disorders following traumatic brain injury: a systematic review of the literature. *Crit Rev Phys Rehabil Med*. 2009;21(3-4):317−374.
180. Gagner C, Landry-Roy C, Laine F, Beauchamp MH. Sleep−wake disturbances and fatigue after pediatric traumatic brain injury: a systematic review of the literature. *J Neurotrauma*. 2015;32(20):1539−1552.
181. Theadom A, Starkey N, Jones K, et al. Sleep difficulties and their impact on recovery following mild traumatic brain injury in children. *Brain Inj*. 2016;30(10):1243−1248.
182. Singh K, Morse AM, Tkachenko N, et al. Sleep disorders associated with traumatic brain injury—a review. *Pediatric Neurol*. 2016;60(1):30−36.
183. Breed ST, Flanagan SR, Watson KR. The relationship between age and the self-report of health symptoms in persons with traumatic brain injury. *Arch Phys Med Rehabil*. 2004;85(4Suppl 2):S61−S67.

CHAPTER

11

Mild Cognitive Impairment and Dementia

Chenlu Gao[1], Michael K. Scullin[1] and Donald L. Bliwise[2]

[1]Department of Psychology and Neuroscience, Baylor University, Waco, TX, United States
[2]Department of Neurology, Emory University, Atlanta, GA, United States

OUTLINE

Introduction	254
Cognitive Aging	**254**
The Normal Cognitive Aging Process	254
Prevalence and Symptoms of Mild Cognitive Impairment and Dementia	254
Genetic, Physiological, and Behavioral Risk Factors of Mild Cognitive Impairment and Dementia	257
Sleep and Circadian Rhythm Changes With Normal Aging, Mild Cognitive Impairment, and Dementia	**258**
Characteristics and Structure of Sleep Predict Cognitive Function in Older People	260
Mechanisms Underlying Sleep and Cognition	260
Mild Cognitive Impairment/Dementia Is Associated With Aggravated Sleep Disturbances in Older Adults	261

Sleep Disordered Breathing in Mild Cognitive Impairment/Dementia	**262**
Epidemiology	262
Mechanisms	263
Sleep Disordered Breathing in Dementia Patients: Diagnostic Considerations	263
Treatment of Sleep Disorders in Mild Cognitive Impairment and Dementia	**264**
Interventions for Sleep Apnea	264
Behavioral Interventions for Disturbed Sleep in Dementia	264
Pharmacological Interventions	265
Areas for Future Research	**267**
Conclusion	**268**
References	**269**

Handbook of Sleep Disorders in Medical Conditions
DOI: https://doi.org/10.1016/B978-0-12-813014-8.00011-1

© 2019 Elsevier Inc. All rights reserved.

INTRODUCTION

As people age, cognitive function and sleep quality decline.[1–3] One view is that cognitive decline and poor sleep quality are normal by-products of aging. An alternative view is that sleep disturbances across the lifespan may contribute to pathological cognitive decline (including dementia). Sleep disorders are common in individuals with mild cognitive impairment (MCI) or dementia, and the severity of sleep disturbances often relates to the severity of dementia. Though scientists are still disentangling the causal direction for cognitive decline and sleep quality, in the present chapter, we cover the recent literature with attention given to potential mechanisms by which poor sleep could increase risk for MCI and dementia.

COGNITIVE AGING

The Normal Cognitive Aging Process

Many cognitive abilities naturally decline during maturation: executive function, memory, processing speed, language, and attention. For example, people at the age of 20 are at their peak in fluid intelligence, which is defined as the ability to think abstractly, reason, and identify relationships and patterns.[4] Young people excel at fluid intelligence tests and scored high on reasoning, speed, and memory tasks.[5] But, performance on fluid intelligence and many other cognitive tasks declines after age 30, with no amelioration. Though the causes of age-related cognitive decline are widely debated, age-related neuronal loss, synaptic connectivity reduction, and white matter integrity reduction are hypothesized to contribute to cognitive decline.[6,7] However, age-related changes in brain structure map imperfectly onto age-related changes in cognition.[8] Despite the neural changes associated with aging, general world knowledge and vocabulary increase with age[9]: In other words, the oldest man in the village is the wisest man in the village (see, e.g., Jung[10] senex archetype; King Nestor in Homer's *Odyssey*).

Prevalence and Symptoms of Mild Cognitive Impairment and Dementia

Neurocognitive disorders interfere with the "wise old man" archetype. The major types of neurocognitive disorders and corresponding features are listed in Table 11.1.[11] This chapter will focus on MCI and Alzheimer's disease (AD) in relation to sleep. MCI is a prodromal stage of dementia, in which there are noticeable cognitive impairments beyond effects due to normal aging (termed mild neurocognitive disorder in the Diagnostic and Statistical Manual of Mental Disorders 5th Edition). About 30% of older adults who have subjective memory complaints meet the criteria of MCI[12] and approximately half of MCI patients progress to dementia within 5 years.[13,14]

Dementia is characterized by even more severe cognitive disturbances and difficulty with independent living. Late-stage dementia commonly renders the individual bedbound, necessitating full-time caregiving. The average life expectancy of individuals with dementia is approximately 10 years.[11] The most common cause of dementia is AD. However, the

TABLE 11.1 Major Types of Neurocognitive Disorders (NCD)[11]

	Etiology and Features	Onset Age	Prevalence	Functional Features of Mild NCD	Functional Features of Major NCD
Alzheimer's disease	People with Alzheimer's disease gene (APOE4) are at elevated risk. Markers include cortical atrophy, amyloid-predominant neuritic plaques, and tau-predominant neurofibrillary tangles	Early onset form 40 and 50 s	5%–10% of people in 60 s	Decline in memory and learning, and other cognitive functions are prominent	Visuoconstructional/ Perceptual—motor ability and language functions are impaired
		Late onset form 70 and 80 s	60%–90% of all dementia cases	Executive functions are sometimes impaired	Social cognition is affected during late course
Frontotemporal	Neuronal death in frontal and anterior temporal lobes are common	20–80 s	Population prevalence 2–10 in 100,000	Symptoms include behavioral disinhibition, delusions, apathy or inertia, loss of sympathy or empathy; perseverative, stereotyped, or compulsive/ritualistic behavior; hyperorality and dietary changes	
	Genetic mutations associated with protein tau, the granule gene, and the C9ORF72 gene are usually present	Common in 50 s	5% of all dementia cases	Social cognition, executive functions, and language abilities are often impaired, but learning and memory functions are spared	
NCD with Lewy bodies	Synucleinopathy due to alpha-synuclein misfolding and aggregation is usually the cause. Low striatal dopamine transporter uptake is characteristic	50–80 s	0.1%–5% of the elderly population	Attention and alertness are commonly affected, sometimes accompanied by visual hallucinations, features of Parkinsonism, disordered rapid eye movement sleep behavior, and neuroleptic sensitivity	
		Common in 70 s	20%–35% of all dementia cases		
			Male-to-female: 1.5:1		
Vascular	Vascular NCD is a result of cerebrovascular diseases (e.g., infarcts, strokes, microvascular disease)	Any age	0.2% in 65–70 years old, 16% in people >80	Common features include impaired complex attention and frontal-executive function. Other functional impairments may occur depending on the location and severity of cerebrovascular disease	
		Common after 65	20–30% diagnosed within 3 months of stroke		
Traumatic brain injury	NCD occurs following traumatic brain injury	Traumatic brain injury rates are high in people under 4 years old, older adolescents, and people above 65	1.7 million TBIs occur in the US every year	Symptoms depend on the area and severity of injury. Common deficits occur in complex attention, executive function, learning and memory, information processing speed, and social cognition	
				In severe cases, aphasia, neglect, and constructional dyspraxia may occur	

(Continued)

TABLE 11.1 (Continued)

	Etiology and Features	Onset Age	Prevalence	Functional Features of Mild NCD	Functional Features of Major NCD
Human immunodeficiency virus infection	NCD occurs in individuals with human immunodeficiency virus infection. Etiologies include reduction in brain and white matter volume, and white matter hyperintensities	In developed countries, human immunodeficiency virus infection occurs primarily in adults	Mild NCD in 25% of people with human immunodeficiency virus Major NCD in fewer than 5% of people with human immunodeficiency virus	Patients often present impaired executive function, processing speed, abilities to complete demanding tasks. They have difficulty learning new material but are able to recall learned materials	Impairment in processing speed is prominent. Reduction in language fluency is common. Other symptoms may occur depending on the brain region affected
Parkinson's disease	NCD occurs in Parkinson's disease patients	50−80 s Common in early 60 s Mild NCD occurs early during the course of Parkinson's disease, but major NCD occurs relatively late	Parkinson's disease affects 0.5% people at 65−69, and 3% in people >85 75% Parkinson's disease patients will develop NCD during the course of Parkinson's disease	Deficits appear gradually. Apathy, depressed mood, anxiety, hallucinations, delusions, personality change, disordered rapid eye movement sleep behavior, and excessive daytime sleepiness are key characteristics of impairments	
Substance/ Medication-induced NCD	Neurocognitive deficits sustained beyond the common duration of intoxication and withdrawal	Any age Persistent use of substance/ medication past 50 years old increases the risk of developing NCD	Mild NCD in 30%−40% alcohol abusers. Prevalence for mild NCD is about one-third for other substance/ medication abuse, but major NCD is rare	Mild NCD often presents reduction in cognitive efficiency and attention. Motor symptoms can occur in both mild and major NCD	

hallmarks of AD, including amyloid plaques and neurofibrillary tangles, may only be confirmed by postmortem autopsy, which leads to 12%–23% misdiagnosis rate.[11,15] For example, individuals with vascular dementia (i.e., cognitive impairments due to stroke or other cerebrovascular events) may be misdiagnosed as having AD which may be relevant for treatment considerations.[16]

The primary risk factor for dementia is chronological age.[17] According to the Alzheimer's Association, in the United States, only 4% of people younger than 65 are diagnosed with dementia; 15% of people in the age group 65–74 develop dementia; 44% of people in the 75–84 age group develop dementia; after which, the diagnosis rate plateaus or slightly improves (e.g., due to survival bias).[18] As people's life expectancy increases in the United States and globally, we may see a growing rate of dementia patients in the elderly population.

Age is not the only demographic factor to increase risk for dementia. Caucasians have a lower prevalence of dementia than African-Americans and Hispanics.[19] For example, a study of 3608 participants showed that African-Americans were 4.4 times more likely to meet the criteria of MCI compared to Caucasian individuals.[20–23] Moreover, men have a lower prevalence of dementia than women at all age levels.[18] In addition, psychiatric symptoms such as depression increase dementia risk by 1.5 times.

Genetic, Physiological, and Behavioral Risk Factors of Mild Cognitive Impairment and Dementia

Genetic, health, and behavioral factors all contribute to the incidence of MCI and dementia. First, the Apolipoprotein E (ApoE) gene that works to transport lipid to repair neurons in the brain is the major genetic contributor to cognitive decline over age, and incidence of MCI and Alzheimer's dementia.[24] Every person has two alleles, out of ApoE ε2, ε3, and ε4 alleles. The presence of ApoE ε4 allele increases risk for cognitive impairments, whereas ε2 reduces risk. ApoE ε4 allele exists in 31% of healthy people but is present in 80% and 64% of familial and sporadic (no specific familial link) AD patients, respectively.[25] Individuals with one ApoE ε4 allele have a fivefold increased risk for AD, whereas individuals with two ApoE ε4 alleles have nearly a 20-fold increased risk.[26]

Cerebral health also relates to the prevalence of MCI. Magnetic resonance imaging (MRI) data show that individuals with more white matter lesions, larger ventricular volumes, more cortical atrophy, or more cerebral infarcts are more frequently diagnosed with MCI.[20,21] Although biomarker confirmation of AD [e.g., neuropathological findings or abnormal proteins in cerebrospinal fluid (CSF)] is desirable, neuroimaging can rule out causes of cognitive impairment, such as trauma or neoplasm. Neuroimaging techniques also play a critical role in assessing the progression of the disease and brain areas affected. Furthermore, treatments for presumed AD may be associated with changes in neuroimaging outcomes. For example, aerobic fitness training has been shown to increase hippocampal volume[27–29] and functional connectivity of frontal, posterior, and temporal cortices.[30]

The diagnoses of other health conditions, such as diabetes mellitus,[31] hypertension, or heart disease substantially increase the risk for MCI and dementia.[20,21] For instance, an observational study on 1969 older adults showed an odds ratio of 1.93 for nonamnestic

MCI (affecting cognitive domains other than memory) in people with a history of coronary heart disease compared with individuals free of heart disease.[32] Another large-scale epidemiological study demonstrated elevated risk for AD and vascular dementia in patients with Type 2 diabetes. The relationship between diabetes and dementia was strongest in ApoE ε4 allele carriers.[33]

Other than physiological factors, psychosocial and behavioral factors are also important in identifying people at risk for dementia. Depression, for example, is associated with the development of AD, even in cases where the symptoms of depression occurred 25 years prior to the onset of AD.[34] Moreover, in patients with MCI, those who met the criteria of major depressive disorder were 2.6 times more likely to progress into dementia 3 years later.[35] Conversely, factors that work in opposition to depression, such as being physically and cognitively active, reduce risk for developing dementia.[36,37]

Interestingly, all of the aforementioned risk factors are associated with sleep difficulties, and sleep difficulties have long been hypothesized to be connected to MCI and dementia.[38] Sleep disorders, including rapid eye movement (REM) sleep behavior disorders, sleep disordered breathing (SDB), and insomnia, are more commonly seen in individuals with MCI and dementia than in healthy control groups.[39] Nonetheless, other characteristics of sleep, including sleep duration,[40,41] sleep quality,[42] self-report sleep disturbances,[43] and consistency of sleep schedule,[44] are also prospectively related to the incidence of MCI and dementia.

Because cognitive capacities and sleep quality both decrease as people age, several theories have emerged to explain how sleep, cognition, and aging relate. One view is that the decline in sleep quality as people age causes a decrease in cognition, a second view is that the decrease in sleep quality is a consequence of neural aging, and a third view is that disturbed sleep quality and declining cognition are both by-products of normal aging. Understanding the relationship between sleep, cognition, and aging is crucial, as it may guide us in seeking treatments for sleep disorders and dementia.

SLEEP AND CIRCADIAN RHYTHM CHANGES WITH NORMAL AGING, MILD COGNITIVE IMPAIRMENT, AND DEMENTIA

Sleep patterns, duration, and quality gradually change with increasing age. When older adults self-report their sleep habits/quality, they often indicate difficulty maintaining sleep or waking too early,[45] and not feeling rested after sleep.[46] Similar results are observed when objectively measuring sleep. Consider, for example, a metaanalysis of 65 studies that objectively measured sleep. As Fig. 11.1 illustrates, nighttime total sleep duration gradually decreased from 450 minutes in 20-year olds to 350 minutes in 80-year olds, primarily because wake after sleep onset increased nearly fourfold.[47] A caveat to conclude that older adults need less sleep (or simply, sleep less overall) is that most of the studies included in the metaanalysis neglected to account for daytime napping. Older adults are more likely to nap during the day, and when their daytime naps are added to their nighttime sleep, there may be no age differences in total sleep duration.[48,49]

Though total sleep time may not differ across young and older adults, the quality/depth of sleep does. As shown in Fig. 11.2, proportion of time in stages 1 and 2 ("light

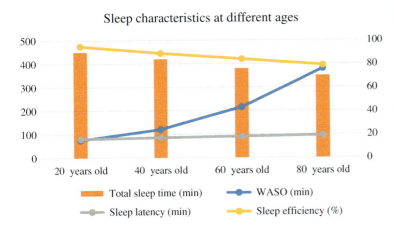

FIGURE 11.1 Sleep characteristics at different ages. Modified from Ohayon MM, Carskadon MA, Guilleminault C, et al. Meta-analysis of quantitative sleep parameters from childhood to old age in healthy individuals: developing normative sleep values across the human lifespan. Sleep. 2004;27(7):1255–1273.

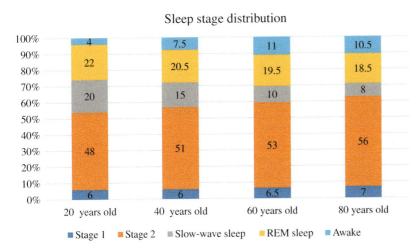

FIGURE 11.2 Sleep stage distribution as people age. Modified from Ohayon, Carskadon MA, Guilleminault C, et al. Meta-analysis of quantitative sleep parameters from childhood to old age in healthy individuals: developing normative sleep values across the human lifespan. Sleep. 2004;27(7):1255–1273.

sleep," "transitional sleep") is greater in older adults than in young adults. By contrast, slow-wave sleep (SWS) (now often referred to as N3 sleep) decreases by over 50%, and REM sleep decreases by 10% from 20- to 80-year olds. Thus, sleep is less deep in older adults than in young adults, which may be the cause or consequence of the increased frequency of nighttime awakenings in older adults. While reductions in sleep depth do not necessarily mean that older people *need* less sleep, they do represent a decline in the *ability* to sleep in older individuals.[50] The fact that women of all ages typically obtain more deep sleep relative to men of comparable age[51] also challenges the interpretation of these age-related changes.

There are intrinsic and extrinsic causes for the decline in sleep quality. For example, older people are more likely to be on medications or to have medical conditions that affect sleep.[52] Medications that affect the central nervous system can induce insomnia.[53] Medications that affect various neurotransmitter systems, along with some antibiotics, antihypertensives, and antidepressants, can also lead to insomnia. In addition, pain associated with various illnesses during the night constitutes one of the major sleep complaints in older adults,[54] as does nocturia.[55]

A major intrinsic cause of poor sleep in older adults may involve alterations in circadian rhythms. They become sleepy earlier in the evening and they wake earlier in the morning than young adults. This phenomenon is often cited as evidence of a "phase advance" of the sleep–wake cycle as a function of age in humans, although when sleep/wake rhythms are referenced to other chronobiologic markers, this may not always be the case.[39] Regardless of the apparent change in endogenous rhythmicity, the ability of the brain to process illumination, the major "zeitgeber" (i.e., time cue) that regulates rhythms in humans, is almost certainly impaired in old age. Light impacts the central circadian pacemaker, the suprachiasmatic nucleus (SCN) via photoreceptive ganglion cells, and 95-year-olds' photoreception abilities are only 10% of 10-year-olds'. Therefore, 10 times brighter light sources are needed in very elderly people to impact circadian rhythms as in young people.[56] Light exposure therapies can improve circadian rhythms and sleep in older adults,[57–59] although older people sometimes report dislike of the "glare" that occurs with such higher intensity light.

Characteristics and Structure of Sleep Predict Cognitive Function in Older People

The connection between sleep stages and specific cognitive abilities is unclear.[60] In young adults, small-scale experimental studies have shown a decline in spatial learning and motor/procedural memories following REM sleep restriction.[61–63] When SWS is experimentally fragmented, older individuals have difficulty encoding or retaining facts or experiences (declarative memory).[64,65] In the largest longitudinal study of sleep stages and cognition ($N = 2909$), community dwelling older adults who had less proportion of REM sleep had worse subsequent cognition.[66] Older adults with more stage 1 sleep also showed impaired subsequent cognition. They took more time to complete the Trails B and Digit Vigilance tests and had lower modified mini–mental state examination (MMSE) scores. Some of the decline might be due to sleep apnea: Individuals with greater hypoxemia during sleep performed worse than people who showed less or no hypoxemia.

Other sleep characteristics can also be associated with abnormal cognition. For instance, polysomnography (PSG) stage 2 sleep spindles are considered as a marker of efficient cortical–subcortical connectivity.[67] A few studies looking at whether increased amount of sleep spindles predict better cognitive performance showed positive results.[67,68] Fewer sleep spindles and K-complexes have also been identified as characteristics of AD patients.[69,70]

Mechanisms Underlying Sleep and Cognition

Sleep subserves numerous functions, including maintenance of immunologic, thermoregulatory, endocrine, and autonomic functions. Sleep promotes energy conservation, protein synthesis, and brain maturation. Sleep might also foster production of new neurons (neurogenesis).[71] Neurogenesis is an indicator of brain resilience and is particularly important to human psychosocial functions and cognition. Interestingly, sleep deprivation[72] and sleep fragmentation[73] hinder neurogenesis possibly by altering hypothalamic–pituitary–adrenal axis functions[74] and compromising neurotransmitter systems and neuroendocrine reactivity.[75] There is evidence that prolonged sleep restriction impairs neurotransmitter receptor systems such as serotonin,[76] dopamine,[77] norepinephrine,[78] and acetylcholine,[79] which may explain why sleep loss reduces neurogenesis because these neurotransmitters normally support cell proliferation.[80–83] In addition, the effect of sleep disturbances on neurogenesis may be mediated by the inflammatory response. Levels of interleukin 6, a marker of inflammation that is detrimental to neurogenesis, increase following sleep deprivation.[84–86]

The glymphatic system may also modify its operations during sleep. Animal research has suggested that during SWS, the amount of interstitial space fluid increases by 60%.[87] One consequence of such increased flow may be to clear waste elements (e.g., amyloid and tau) from the brain over a span of minutes rather than days, as would be predicted by simple diffusion alone. Although an intriguing and provocative finding, especially with implications for cognition, similar results were noted under anesthesia, and it remains unclear how this so-called glymphatic (i.e., glia-based) system operates in REM sleep. In addition, the directionality of such active clearance may simply be more variable when viewed with a higher density visual resolution than the unilateral flow shown by Xie et al.[87] Nonetheless, these findings are among the most exciting functions of sleep that potentially link a normal brain function associated with at least one component of sleep to elimination of proteins that, when abnormal, are associated with cognitive impairments.

Mild Cognitive Impairment/Dementia Is Associated With Aggravated Sleep Disturbances in Older Adults

Many of the sleep factors associated with age-related cognitive decline are also associated with MCI and dementia. Signs of sleep disorders such as REM sleep behavior disorder,[88] insomnia complaints,[89] and fragmented sleep[90] are more prevalent in MCI and dementia patients than in healthy controls. In addition, among individuals diagnosed with dementia, lower sleep efficiency and greater sleep fragmentation correlate with severity of dementia.[91] As dementia progresses, sleep problems become more severe. Decrements in SWS have sometimes been observed in dementia patients,[70] but decrements in REM sleep are consistently observed.[92,93] The neurotransmitter system that links REM sleep to AD may be acetylcholinesterase. Individuals with AD typically exhibit lowered choline acetyltransferase, and REM sleep is regulated by brainstem cholinergic neurons.[94,95]

Circadian rhythms may also be abnormal in AD patients. For example, some studies show that severely demented patients show a significant delay in acrophases (peak of circadian rhythm) and go to sleep later and wake up later than healthy controls,[96] whereas other studies suggest that they go to bed and wake up earlier than age-matched controls, essentially representing what appears to be an exaggeration of the pattern seen in normal older subjects.[97] In addition, many individuals with AD demonstrate sundowning. Sundowning is a state of confusion or agitation during late afternoon and early evening, which can be disturbing to both the patient and the caregiver.[38] The specific mechanisms behind sundowning are not fully understood, but most researchers agree that the circadian system plays a role. For example, fluctuations in body temperature acrophase (a common measure of circadian rhythm peak) correspond to severity of sundowning.[98] Some dementia symptoms may therefore be explained by the decline in SCN's ability to synchronize internal body rhythms with environmental cues. One potential solution is for patients with dementia to receive more daytime light, which they may not typically do for social, mobility, and other reasons.[96]

SLEEP DISORDERED BREATHING IN MILD COGNITIVE IMPAIRMENT/DEMENTIA

Epidemiology

There is increasing interest in the role of SDB in dementia.[99] SDB involves abnormal respiratory patterns during sleep, causing intermittent hypoxemia and/or recurrent arousals that burdens cerebral,[100,101] cardiovascular,[102] and metabolic systems.[103,104] The most common type of SDB is obstructive sleep apnea (OSA)/hypopnea, which is caused by complete or partial blockage of the airway during sleep (e.g., collapse of soft palate). An apnea is defined as a cessation in breathing for at least 10 seconds, whereas a hypopnea refers to an attenuation in breathing for 10 seconds or longer with an oxygen saturation level drop of $\geq 4\%$.[105,106] People with more than five apneic events per hour meet the clinical criteria for a sleep apnea diagnosis per US Center for Medicare Services criteria; however, defining the disease threshold may require age-dependent considerations such as a higher threshold (e.g., 30 events/h) among older populations (for a more comprehensive discussion of this issue, see Ref. [39]).

SDB is more common in overweight and male populations and also older adults of both sexes. For example, Mehra et al.[107] collected PSG data from 2911 older men (mean age = 76.38) and discovered that 26% of participants met the criteria for SDB, which is much greater than the general middle-aged population prevalence of 0.3%–5.0%.[108] In middle-aged individuals, greater body mass index, neck circumference, waist circumference, and hip circumference are all associated with higher likelihood of SDB, but there is evidence that these factors become less salient as predictors of SDB in older age groups. For example, in older adults, frailty and concurrent loss of weight can also be associated with SDB.[109,110] Existing literature conflicts as to whether older and middle-aged groups show the same number of medical comorbidities such as diabetes mellitus, hypertension, cardiovascular disease, and heart failure in relation to SDB.[39]

Malhotra et al.[111] proposed that the changes in upper airway anatomy and decreases in pharyngeal dilator muscle control are responsible for the increase in SDB risk in older adults (see also Schellenberg et al.[112]). MRI data show that, as people age, the bony shape surrounding the pharynx alters, the lateral skeleton grows, the soft palate becomes longer, and the size of parapharyngeal fat pads increases. Because of a more crowded upper airway, OSA occurs more easily when the pharyngeal muscles relax during sleep.

Pertinent to our focus on neurocognitive disorders, observational population-based studies have shown that SDB may elevate risk for MCI and dementia.[113–115] Women with

SDB were 1.85 times more likely to develop MCI or dementia prospectively over 4 years.[114] Those with particularly severe oxygen desaturation had 2.04 times increased risk of developing MCI or dementia. These findings imply that lack of oxygen contributes to the deterioration of cognitive functions, although they are in seeming contradiction with earlier studies of well-defined AD patients who showed no higher risk for SDB than age-matched controls.[116]

Mechanisms

According to Beebe and Gozal,[117] oxygen desaturation occurring with SDB hinders the chemical and cellular restorative processes during sleep, which eventually leads to cognitive dysfunction. There are now several studies in humans to support their model. For example, Carlson et al.[118] studied cerebral oxygenation during sleep in relation to cognitive functions in healthy older adults without sleep apnea. They showed that oxyhemoglobin saturation while participants slept was associated with daytime memory function. Compatible with these findings are SDB case–control neuroimaging studies that have shown that SDB is associated with changes in white matter[101,119] and the hippocampus.[120]

SDB-induced hypoxia may also hasten amyloid beta (Aβ) generation and tau phosphorylation regulation, which are commonly associated with the onset and progression of AD.[120] Osorio et al.[121] recruited 95 cognitively healthy elderly participants and tested phosphorylated-tau (p-tau), total-tau (t-tau), Aβ-42, and ApoE allele status from CSF samples. Using a home monitoring system, they also assessed SDB severity and oxygen saturation level during the night. In individuals with ApoE ε3 positive alleles, greater SDB severity was associated with greater p-tau, t-tau, and Aβ-42. By contrast, in individuals with ApoE ε2 positive alleles (known to protect against Alzheimer's risk), greater SDB severity was associated with lower levels of Aβ-42. Thus, there is a connection between SDB severity and AD biomarkers, but that connection is influenced by genetic factors.

It is highly likely that SDB might relate to other forms of impaired cognition (e.g., vascular dementia) via other health conditions such as chronic hypertension, cardiac disease, and vascular insufficiency. Observed associations might thus reflect whatever more general mechanisms (e.g., altered cerebral hemodynamics) mediate these relationships. For example, a large community-based study of 6132 middle-aged and older adults showed associations between severity of SDB and high blood pressure that exist in all ethnic groups and in both sexes.[122] Because hypertension has long been recognized to be associated with cognitive impairment and increase the likelihood of stroke, patients with SDB are at a higher risk for both major and minor brain infarcts and local ischemic and infarcts in the brain.[16,123–125]

Sleep Disordered Breathing in Dementia Patients: Diagnostic Considerations

There are several challenges to diagnosing SDB in patients with dementia. Patients with moderate and severe dementia exhibit limited capacity to consent to evaluations; patients with MCI typically exhibit relatively preserved capacity to consent.[126] Self-reported or caregiver-reported sleep characteristics (e.g., screening questionnaires) may be helpful for identifying the hallmark symptoms of sleep apnea (e.g., waking gasping for breath), but such questionnaires are prone to subjective biases. Actigraphy can be useful for diagnosing some sleep disorders (e.g., circadian disorders, insomnia), but actigraphy lacks the sensitivity/specificity for diagnosing SDB. In-laboratory PSG remains the gold standard method for diagnosing and evaluating SDB,[127] but now at-home PSG-based Levels III and IV screening devices can be used in some situations.[128] Caregiver observations cannot reliably replace objective assessments but can provide supplementary information regarding sleep disorders.

TREATMENT OF SLEEP DISORDERS IN MILD COGNITIVE IMPAIRMENT AND DEMENTIA

Interventions for Sleep Apnea

Continuous positive airway pressure (CPAP) is the standard treatment for OSA.[129] During sleep, pressurized room air (c.f., supplementary oxygen) is delivered through a nasal or face mask to keep the patient's airway open. The ventilation process is noninvasive and painless and usually does not cause serious complications or side effects.

CPAP use may partially ameliorate cognitive impairment in healthy individuals and patients with dementia.[130,131] For example, 12 months of CPAP use in dementia patients was associated with an improvement in general cognition (MMSE) and the ability to manage daily complicated tasks.[132] Apart from slowing progression of cognitive decline, Cooke et al.[133] also showed that dementia patients who sustained 13.3 months of CPAP use reduced depressive symptoms, decreased daytime sleepiness, and improved subjective sleep quality. Other benefits of CPAP use include improved cardiovascular and metabolic health.[134,135] Although most studies emphasized the importance of sustained and continuous use of CPAP, in practice, adherence to CPAP is moderate, especially among patients with depression symptoms.[136]

In nondementia populations, weight loss can be a useful treatment for sleep apnea; however, as mentioned above, many older patients are frail and do not incur high levels of obesity. Other avenues of treatment, such as oral appliance therapy (OAT), may be considered. Worn during sleep, OAT repositions the mandible (lower jaw) slightly forward and creates a more favorable upper airway caliber. If initial fitting can be tolerated by the patient, this may represent a viable option for CPAP. In nondementia patients, physical exercise (aerobic activity specifically) has been shown to reduce the severity of sleep apnea.[137] Demented nursing home patients who underwent strength training showed some improvement in their SDB in one study.[138]

Behavioral Interventions for Disturbed Sleep in Dementia

Poor sleep in dementia may improve with behavioral strategies. The three cardinal approaches involve (1) increased exposure to bright light, (2) avoidance of daytime napping, and (3) increased daytime activity. Of these approaches, enhanced illumination appears to have undergone the most extensive testing. Artificial bright light can be used in the morning to treat sleep disturbances.[139] In dementia patients, bright light therapy improves circadian rhythmicity[96] and cognition.[57–59] Moreover, bright light therapy may alleviate depressive symptoms,[140,141] mood disturbances,[142] and agitation.[143] Some studies reported no serious side effects of bright light treatment, even when utilizing 10,000 lx light for 30 minutes/day[144] or 1250 hours over 5 years[145,146]; however, other studies reported that some participants experienced jumpiness (8.8%), headaches (8.4%), and nausea (15.9%).[147] Moreover, older adults tend to nap more frequently and for longer duration during the day to compensate for disrupted nocturnal sleep, which may result in disruptions of the sleep–wake cycle.[148] Avoiding napping (especially after late afternoon) or limiting naps to less than 30 minutes has been recommended to alleviate insomnia.[148,149] Other studies have shown that exercise and social activity are effective in increasing sleep efficiency, Non-REM (NREM) sleep time, total sleep time, and subjective sleep quality in older adults, while decreasing sleep onset latency.[150–152]

Cognitive-behavioral therapy, and techniques such as self-relaxation and mindfulness, can effectively treat insomnia in older persons with intact cognition.[153,154] Zhang et al.[155] used sitting meditation and mindful yoga to treat chronic insomnia in healthy adults older than 75. Their 8-week intervention was successful in improving sleep quality and reducing depressive symptoms compared with the control group that only received standard care. Relaxation techniques may be beneficial to MCI and mild dementia patients but can be difficult to implement with moderate and severe dementia patients, because they may fail to comprehend or follow training instructions.[126] Other behavioral components for insomnia treatment include regularity of bedtime and wake up time, maintenance of quiet sleeping environment, avoidance of caffeine and alcohol in the evening, and avoidance of blue wavelength light at bedtime.

There may be value not only to educating MCI and dementia patients on sleep hygiene practices but also educating the caregivers of dementia patients. Caregivers can help patients to maintain a consistent bedtime routine (aiding consistency of sleep schedule), gain more morning light, reduce napping (aiding regulation of time in bed), and establish daily walking or other exercise routines.[156]

Pharmacological Interventions

Among nonprescription agents with the potential for improving sleep in demented patients, perhaps none has been as extensively researched as melatonin. Melatonin is an endogenous hormone produced by the pineal gland that serves to regulate the timing of sleep and wakefulness. Production of melatonin rises after sunset and then reaches its peak in the middle of the night, followed by a gradual decline after sunrise.[157] Disruption in melatonin secretion causes insomnia and sleep disturbance.[158] Exogenous melatonin used at supraphysiologic levels has been a popular over-the-counter evening supplement for individuals with insomnia complaints and circadian disturbances. Dosage ranging from 0.5 to 5 mg daily has been confirmed to be relatively safe,[159,160] but the safety and therapeutic benefits of routine use beyond 5 mg have not been fully investigated.[161]

Several melatonin studies in dementia patients have shown melatonin administration to help mitigate sundowning syndrome.[162–164] Whether melatonin can improve sleep quality in dementia patients is more equivocal. In AD patients, 2.5 and 10 mg daily melatonin use for 2 months did not influence objective, actigraphic sleep measures.[165] Moreover, caregiver-rated sleep quality improved in the 2.5 mg group, but not in the 10 mg group, suggesting that higher dosage does not mean better sleep quantity or quality. A metaanalysis of seven studies on melatonin use in dementia patients indicated that melatonin can increase total sleep time but that it does not consistently improve cognitive function.[166] Even though melatonin's known side effects occur at a low rate, the side effects of depression, sleepiness, dizziness, stomach cramps, and irritability indicate caution in recommending melatonin to patients with severe dementia or patients who are already taking multiple medications. Because melatonin is considered a dietary supplement and not a prescription medication, the quality assurance over its preparation, potency, and safety when purchased remains open to question.[167]

Given the modest effects of over-the-counter sleep medications, prescription hypnotic medications might deserve some consideration. Interestingly, epidemiological studies have reported that dementia patients use hypnotic medications more than age-matched cognitively healthy individuals. In a retrospective analysis of 271,365 participants over age 65, the rate of hypnotic use was 2.2 times more common in the dementia group.[168] In addition, among AD patients, those who were on donepezil (a cholinesterase inhibitor that treats AD symptoms) were about 2.5 times more likely to take hypnotics than those who were not on donepezil.[169] Most current FDA-approved medications with sedative/hypnotic indication can be categorized into two types: traditional benzodiazepines and nonbenzodiazepines which are site-specific gamma-aminobutyric acid (GABA) receptor agonists.

Benzodiazepines that have sedative/hypnotic indications (e.g., temazepam, triazolam, flurazepam, estazolam, and quazepam) aim to promote sensitivity of multiple GABA receptor subunits in the central nervous system and create hypnotic effects.[170] In the short term, such medications relieve insomnia but can cause dizziness and next day drowsiness, as well as increase risk for falls. They effectively shorten sleep latency and increase sleep duration but are also associated with an increased risk of hip fracture.[171] Triazolam was documented to produce amnesia and rebound insomnia, which occurs on its discontinuation,[172] which is a particularly detrimental feature of this short half-life drug. Long-term (3 months or more) nightly use of benzodiazepines may cause medication dependence; when benzodiazepine use is discontinued, decreased sensitivity of GABA receptors leads to withdrawal symptoms such as anxiety and insomnia.[173] A potential role for sleep medications hastening dementia remains a controversial topic with inconsistent evidence on this issue.[174–176] On the one hand, Pariente et al. argued that (1) benzodiazepines slow down $A\beta$ oligomer accumulation through reduced beta-secretase 1 and γ-secretase activity; (2) astrocytes at amyloid plague sites may have GABA-secreting activity, so predementia lesions would amplify benzodiazepine's negative effect on cognition; (3) benzodiazepines can limit the brain's neural compensation mechanism by lowering the brain activation level. On the other hand, the field lacks conclusive evidence of length of exposure, dose effects, and drug class (older benzodiazepines vs relatively newer site-specific agonists). Nevertheless, overall, benzodiazepine hypnotics should be used with caution, particularly in patients who have comorbid conditions or are on other medications.[177]

Nonbenzodiazepines (e.g., zolpidem, zaleplon, eszopiclone), collectively referred to as Z-drugs, are safer and often thought to have fewer side effects than benzodiazepines.[178] In older adults, Z-drugs improve sleep latency and subjective sleep quality but not necessarily total sleep time.[179] Z-drugs are generally well tolerated but can cause falls, memory loss, psychomotor problems, hallucinations, and sleep walking.[179,180] There are few well-done clinical trials of their utility in dementia. Epidemiological studies report that use of hypnotic medications (both benzodiazepines and Z-drugs) is associated with mortality risk,[181] though mortality is rare in clinical trials. Though the hypnotic—mortality association has been widely debated, there is little debate that the side effects of Z-drugs can be serious in older adults (with or without dementia). Newer drugs, such as the orexin antagonist, suvorexant, have yet to be tested in dementia patients.

AREAS FOR FUTURE RESEARCH

Despite evidence that SDB may be associated with impaired cognition, many issues remain uncertain. For example, SDB could reflect abnormalities in respiratory control that occur with neurodegeneration rather than a cause of the degeneration itself. It is also uncertain whether, given the large number of medical comorbidities associated with both SDB and dementia, SDB could be a unique cause of the decline or whether its effects might be mediated by other more widely recognized factors, such as cerebrovascular disease. Even if SDB was a unique cause for impaired cognition, it remains unclear which sleep-based mechanism(s) is(are) responsible for this association. For example, hypoxia and arousals both result from sleep apneic events, but we do not know whether the lack of oxygen or the increased sleep fragmentation and its effects at the level of neurons or glia may be responsible for the decline in cognition.[182] In addition, genetic factors may moderate the negative impact of SDB on dementia progression or onset. Finally, there remains a dearth of controlled clinical trials, which represent the standard for assigning cause and effect for a medical condition such as SDB causing dementia.

If poor sleep quality per se causes dementia, then treating sleep disturbances should reduce dementia risk. Again, there remains a lack of studies experimentally testing whether improving sleep can reduce the risk or delay the onset of dementia.[183] One approach would be to focus on sleep interventions in MCI patients; another approach might be to recruit preclinical individuals who have a high genetic risk of developing dementia (e.g., ApoE ε4 carriers or members of the Dominantly Inherited Alzheimer's Network). Given that most sleep medications have side effects, especially used in combination with antidementia drugs, nonpharmacological interventions may be preferable (e.g., cognitive-behavioral therapy), although newer sleep medications may have advantages and a better safety profile compared to other medications. Even if sleep interventions do not ultimately delay risk for dementia, the benefits of improving sleep include improved mood, daytime alertness, physical health, and quality of life.[184]

While the prevalence of sleep disorders is high in dementia patients, family caregivers of dementia patients are also concomitantly experiencing sleep disturbances.[185,186] The caregiving experience is time- and energy-consuming. Caregivers who are close family members of the patients also go through emotional distress. Thus, factors such as depression,[187] anxiety,[188] and stress[189] potentially mediate the relationship between caregiver burden and sleep disturbances. In addition, caregivers wake up at night to assist patients' medical needs and bathroom trips, which interferes with caregivers' sleep.[190] The severity of caregivers' sleep disturbances is related to care recipients' sleep, suggesting that treating caregivers' sleep problems may benefit the patients.[191] Specifically, better sleep quality in caregivers can alleviate depressive thoughts and perceived stress and improve emotions and memory for daily needs. Psychosocially and cognitively healthy caregivers provide better caregiving and emotional support to the patients, which is crucial to the well-being of the patients. Yet, more studies need to be conducted to clarify the direction of the association between caregivers' and patients' sleep disturbances. And, future studies should examine whether treating for caregivers' sleep disturbances enhances patients' quality of life.

Moreover, because AD is the leading cause of cognitive deficits and dementia, most research on sleep is limited in other types of dementia. Future research exploring sleep in relation to frontotemporal dementia, vascular dementia, and dementia caused by human immunodeficiency virus infection will help inform the mechanistic pathways by which poor sleep increases risk for cognitive decline.

Last, clinicians, researchers, and advocacy groups need to collaborate to improve dissemination of the importance of sleep to the general public. One recent review[115] claimed that as much as 15% of dementia could be accounted for by sleep issues. Whether future research can sustain this claim will remain to be seen. However, given the diverse (but heterogeneous) evidence in this area,[60] awareness of a potential role for sleep issues in cognition should be fostered rather than relegated as unimportant or merely a "nuisance." As but one example, sleep screening questions should be incorporated into a diversity of clinical settings, and educational programs should expand to raise awareness of sleep and cognitive health. Just as public programs have combated disease by raising awareness of smoking, transaturated fats, and sugar, so might we potentially enhance brain health by promoting better sleep habits in the general public.

CONCLUSION

Sleep and circadian rhythms change as people age, and these changes are exacerbated in individuals with MCI or dementia. Some research suggests that sleep quality or sleep physiology can predict cognitive function in older age. Potential mechanisms linking sleep quality to cognitive decline include neurogenesis (impaired by sleep deprivation) and glymphatic system function (clears beta amyloid and other metabolites during sleep). Longitudinal intervention studies are necessary to confirm or refute the hypothesis that maintaining 7–9 hours of sleep, and going to bed and waking up at the same time each day, decreases risk for pathological neuronal changes and subsequent dementia. For patients with dementia or MCI, keeping good sleep practices may improve quality of life, possibly also slowing the progression of cognitive impairments.

KEY PRACTICE POINTS

- Sleep disturbances contribute to cognitive symptoms and should be assessed and treated.
- CPAP is recommended for treating SDB, and adherence is key.
- Physical exercise and weight loss can be prescribed to individuals who are physically fit to treat mild SDB.
- Effective behavioral interventions for sleep disturbances include increased bright light exposure, increased daytime activity, and reduced daytime napping.
- Cognitive-behavioral therapy and relaxation techniques can be used to treat insomnia in high-functioning older adults.
- Educating caregivers of patients with cognitive disorders on sleep hygiene practices may lead to improved sleep in both caregivers and patients.
- Melatonin can treat insomnia or circadian disturbances and mitigate sundowning syndrome in dementia patients. Daily dosages from 0.5 to 5 mg are considered safe.
- Benzodiazepines relieve insomnia but have a range of potentially serious side effects.
- Nonbenzodiazepines (Z-drugs) can improve sleep in older adults, with fewer and less severe side effects than benzodiazepines, but caution is still advised.

References

1. Ortman JM, Velkoff VA, Hogan H. *An Aging Nation: The Older Population in the United States.* United States Census Bureau, Economics and Statistics Administration, US Department of Commerce; 2014.
2. Wimo A, Jönsson L, Bond J, et al. The worldwide economic impact of dementia 2010. *Alzheimers Dement.* 2013;9(1):1−11.
3. World Health Organization. *Dementia: A Public Health Priority.* World Health Organization; 2012.
4. Cattell RB. Theory of fluid and crystallized intelligence: a critical experiment. *J Educ Psychol.* 1963;54(1):1−22.
5. Salthouse TA. What and when of cognitive aging. *Curr Dir Psychol Sci.* 2004;13(4):140−144.
6. Gunning-Dixon FM, Brickman AM, Cheng JC, et al. Aging of cerebral white matter: a review of MRI findings. *Int J Geriatr Psychiatry.* 2009;24(2):109−117.
7. Bishop NA, Lu T, Yankner BA. Neural mechanisms of ageing and cognitive decline. *Nature.* 2010;464 (7288):529−535.
8. Park DC, Reuter-Lorenz P. The adaptive brain: aging and neurocognitive scaffolding. *Annu Rev Psychol.* 2009;60:173−196.
9. Harada CN, Love MCN, Triebel KL. Normal cognitive aging. *Clin Geriatr Med.* 2013;29(4):737−752.
10. Jung CG. *The Archetypes and the Collective Unconscious.* New York, NY: Pantheon Books; 1959.
11. American Psychiatric Association. *Diagnostic and Statistical Manual of Mental Disorders.* 5th ed Arlington, VA: American Psychiatric Publishing; 2013.
12. Jacinto AF, Brucki SMD, Porto CS, et al. Subjective memory complaints in the elderly: a sign of cognitive impairment? *Clinics.* 2014;69(3):194−197.
13. Petersen RC, Smith GE, Waring SC, et al. Mild cognitive impairment: clinical characterization and outcome. *Arch Neurol.* 1999;56(3):303−308.
14. Solfrizzi V, Panza F, Colacicco AM, et al. Vascular risk factors, incidence of MCI, and rates of progression to dementia. *Neurology.* 2004;63(10):1882−1891.
15. Gaugler JE, Svanum HA, Roth DL, et al. Characteristics of patients misdiagnosed with Alzheimer's disease and their medication use: an analysis of the NACC-UDS database. *BMC Geriatr.* 2013;13(1):137−146.
16. Scullin MK, Le D, Shelton JT. Healthy heart, healthy brain: hypertension affects cognitive functioning in older age. *Transl Issues Psychol Sci.* 2018. Available from: https://doi.org/10.1037/tps0000131.

17. Morris JC. Is Alzheimer's disease inevitable with age? *J Clin Invest*. 1999;104(9):1171−1173.
18. Alzheimer's Association. 2014 Alzheimer's disease facts and figures. *Alzheimers Dement*. 2014;10(2):e47−e92.
19. Tang MX, Cross P, Andrews H, et al. Incidence of AD in African-Americans, Caribbean Hispanics, and Caucasians in northern Manhattan. *Neurology*. 2001;56(1):49−56.
20. Lopez OL, Jagust WJ, DeKosky ST, et al. Prevalence and classification of mild cognitive impairment in the Cardiovascular Health Study Cognition Study: Part 1. *Arch Neurol*. 2003;60(10):1385−1389.
21. Lopez OL, Jagust WJ, Dulberg C, et al. Risk factors for mild cognitive impairment in the Cardiovascular Health Study Cognition Study: Part 2. *Arch Neurol*. 2003;60(10):1394−1399.
22. Petersen RC, Roberts RO, Knopman DS, et al. Prevalence of mild cognitive impairment is higher in men the mayo clinic study of aging. *Neurology*. 2010;75(10):889−897.
23. Roberts R, Knopman DS. Classification and epidemiology of MCI. *Clin Geriatr Med*. 2013;29(4):753−772.
24. Mahley RW, Rall SC. Apolipoprotein E: far more than a lipid transport protein. *Annu Rev Genomics Hum Genet*. 2000;1(1):507−537.
25. Corder EH, Saunders AM, Strittmatter WJ, et al. Gene dose of apolipoprotein E type 4 allele and the risk of Alzheimer's disease in late onset families. *Science*. 1993;261(5123):921−923.
26. Strittmatter WJ. Old drug, new hope for Alzheimer's disease. *Science*. 2012;335(6075):1447−1448.
27. Erickson KI, Voss MW, Prakash RS, et al. Exercise training increases size of hippocampus and improves memory. *Proc Natl Acad Sci U S A*. 2011;108(7):3017−3022.
28. Killgore WD, Olson EA, Weber M. Physical exercise habits correlate with gray matter volume of the hippocampus in healthy adult humans. *Sci Rep*. 2013;3:3457.
29. Ten Brinke LF, Bolandzadeh N, Nagamatsu LS, et al. Aerobic exercise increases hippocampal volume in older women with probable mild cognitive impairment: a 6-month randomised controlled trial. *Br J Sports Med*. 2015;49:248−254.
30. Voss MW, Prakash RS, Erickson KI, et al. Plasticity of brain networks in a randomized intervention trial of exercise training in older adults. *Front Aging Neurosci*. 2010;2(32):1−17.
31. Cheng G, Huang C, Deng H, et al. Diabetes as a risk factor for dementia and mild cognitive impairment: a meta-analysis of longitudinal studies. *Intern Med J*. 2012;42(5):484−491.
32. Roberts RO, Knopman DS, Geda YE, et al. Coronary heart disease is associated with non-amnestic mild cognitive impairment. *Neurobiol Aging*. 2010;31:1894−1902.
33. Peila R, Rodriguez BL, Launer LJ. Type 2 diabetes, ApoE gene, and the risk for dementia and related pathologies. *Diabetes*. 2002;51(4):1256−1262.
34. Green RC, Cupples A, Kurz A, et al. Depression as a risk factor for Alzheimer disease: the MIRAGE study. *Arch Neurol*. 2003;60(5):753−759.
35. Modrego PJ, Ferrández J. Depression in patients with mild cognitive impairment increases the risk of developing dementia of Alzheimer type: a prospective cohort study. *Arch Neurol*. 2004;61(8):1290−1293.
36. Valenzuela MJ. Brain reserve and the prevention of dementia. *Curr Opin Psychiatry*. 2008;21(3):296−302.
37. Hamer M, Chida Y. Physical activity and risk of neurodegenerative disease: a systematic review of prospective evidence. *Psychol Med*. 2009;39:3−11.
38. Bliwise DL, Carroll JS, Lee KA, et al. Sleep and "sundowning" in nursing home patients with dementia. *Psychiatry Res*. 1993;48(3):277−292.
39. Bliwise DL, Scullin MK. Normal aging. In: Kryger M, Roth T, Dement WC, eds. *Principles and Practice of Sleep Medicine*. 6th ed. Elsevier.
40. Keage HAD, Banks S, Yang KL, et al. What sleep characteristics predict cognitive decline in the elderly? *Sleep Med*. 2012;13(7):886−892.
41. Chen J, Espeland MA, Brunner RL, et al. Sleep duration, cognitive decline, and dementia risk in older women. *Alzheimers Dement*. 2016;12(1):21−33.
42. Potvin O, Lorrain D, Forget H, et al. Sleep quality and 1-year incident cognitive impairment in community-dwelling older adults. *Sleep*. 2012;35(4):491−499.
43. Sterniczuk R, Theou O, Rusak B, et al. Sleep disturbance is associated with incident dementia and mortality. *Curr Alzheimer Res*. 2013;10(7):767−775.
44. Tranah GJ, Blackwell T, Stone KL, et al. Circadian activity rhythms and risk of incident dementia and MCI in older women. *Ann Neurol*. 2011;70(5):722−732.
45. Mai E, Buysse DJ. Insomnia: prevalence, impact, pathogenesis, differential diagnosis, and evaluation. *Sleep Med Clin*. 2008;3(2):167−174.

REFERENCES

46. Foley DJ, Monjan AA, Brown SL, et al. Sleep complaints among elderly persons: an epidemiologic study of three communities. *Sleep*. 1995;18(6):425–432.
47. Ohayon MM, Carskadon MA, Guilleminault C, et al. Meta-analysis of quantitative sleep parameters from childhood to old age in healthy individuals: developing normative sleep values across the human lifespan. *Sleep*. 2004;27(7):1255–1273.
48. Buysse DJ, Browman BA, Monk TH, et al. Napping and 24-hour sleep/wake patterns in healthy elderly and young adults. *J Am Geriatr Soc*. 1992;40(8):779–786.
49. Hoch CC, Dew MA, Reynolds CF, et al. Longitudinal changes in diary- and laboratory-based sleep measures in healthy "old old" and "young old" subjects: a three-year follow-up. *Sleep*. 1997;20(3):192–202.
50. Scullin MK. Do older adults need sleep? A review of neuroimaging, sleep, and aging studies. *Curr Sleep Med Rep*. 2017;1–11.
51. Redline S, Kirchner HL, Quan SF, et al. The effects of age, sex, ethnicity, and sleep-disordered breathing on sleep architecture. *Arch Intern Med*. 2004;164:406–418.
52. Neikrug AB, Ancoli-Israel S. Sleep disorders in the older adult—a mini review. *Gerontology*. 2010;56(2):181–189.
53. Pagel JF, Parnes BL. Medications for the treatment of sleep disorders: an overview. *Prim Care Companion J Clin Psychiatry*. 2001;3(3):118–125.
54. Foley D, Ancoli-Israel A, Britz P, et al. Sleep disturbances and chronic disease in older adults: results of the 2003 National Sleep Foundation Sleep in America Survey. *J Psychosom Res*. 2004;56(5):497–502.
55. Bliwise DL, Foley DJ, Vitiello MV, et al. Nocturia and sleep disturbance in the elderly. *Sleep Med*. 2009;10:540–548.
56. Turner PL, Van Someren EJW, Mainster MA. The role of environmental light in sleep and health: effects of ocular aging and cataract surgery. *Sleep Med Rev*. 2010;14(4):269–280.
57. Graf A, Wallner C, Schubert V, et al. The effects of light therapy on mini–mental state examination scores in demented patients. *Biol Psychiatry*. 2001;50(9):725–727.
58. Riemersma-Van Der Lek RF, Swaab DF, Twisk J, et al. Effect of bright light and melatonin on cognitive and noncognitive function in elderly residents of group care facilities: a randomized controlled trial. *JAMA*. 2008;299(22):2642–2655.
59. Yamadera H, Ito T, Suzuki H, et al. Effects of bright light on cognitive and sleep–wake (circadian) rhythm disturbances in Alzheimer-type dementia. *Psychiatry Clin Neurosci*. 2000;54(3):352–353.
60. Scullin MK, Bliwise DL. Sleep, cognition, and normal aging: integrating a half century of multidisciplinary research. *Perspect Psychol Sci*. 2015;10(1):97–137.
61. Smith CT, Conway JM, Rose GM. Brief paradoxical sleep deprivation impairs reference, but not working, memory in the radial arm maze task. *Neurobiol Learn Mem*. 1998;69(2):211–217.
62. Bjorness TE, Riley BT, Tysor MK, et al. REM restriction persistently alters strategy used to solve a spatial task. *Learn Mem*. 2005;12(3):352–359.
63. Fu J, Li P, Ouyang X, et al. Rapid eye movement sleep deprivation selectively impairs recall of fear extinction in hippocampus-independent tasks in rats. *Neuroscience*. 2007;144(4):1186–1192.
64. Plihal W, Born J. Effects of early and late nocturnal sleep on declarative and procedural memory. *J Cogn Neurosci*. 1997;9(4):534–547.
65. Van Der Werf YD, Altena E, Schoonheim MM, et al. Sleep benefits subsequent hippocampal functioning. *Nature Neurosci*. 2009;12:122–123.
66. Blackwell T, Yaffe K, Ancoli-Israel S, et al. Associations between sleep architecture and sleep-disordered breathing and cognition in older community-dwelling men: the Osteoporotic Fractures in Men Sleep Study. *J Am Geriatr Soc*. 2011;59(12):2217–2225.
67. Schabus M, Hödlmoser K, Gruber G, et al. Sleep spindle-related activity in the human EEG and its relation to general cognitive and learning abilities. *Eur J Neurosci*. 2006;23(7):1738–1746.
68. Tamminen J, Payne JD, Stickgold R, et al. Sleep spindle activity is associated with the integration of new memories and existing knowledge. *J Neurosci*. 2010;30(43):14356–14360.
69. Montplaisir J, Petit D, Lorrain D, et al. Sleep in Alzheimer's disease: further considerations on the role of brainstem and forebrain cholinergic populations in sleep–wake mechanisms. *Sleep*. 1995;18(3):145–148.
70. Westerberg CE, Mander BA, Florczak SM, et al. Concurrent impairments in sleep and memory in amnestic mild cognitive impairment. *J Int Neuropsychol Soc*. 2012;18(3):490–500.

II. SLEEP DISORDERS IN SPECIFIC MEDICAL CONDITIONS

272   11. MILD COGNITIVE IMPAIRMENT AND DEMENTIA

71. Meerlo P, Mistlberger RE, Jacobs BL, et al. New neurons in the adult brain: the role of sleep and consequences of sleep loss. *Physiol Rev*. 2009;13(3):187−194.

72. Hairston IS, Little MT, Scanlon MD, et al. Sleep restriction suppresses neurogenesis induced by hippocampus-dependent learning. *J Neurophysiol*. 2005;94(6):4224−4233.

73. Guzman-Marin R, Bashir T, Suntsova N, et al. Hippocampal neurogenesis is reduced by sleep fragmentation in the adult rat. *Neuroscience*. 2007;148(1):325−333.

74. Mirescu C, Peters JD, Noiman L, et al. Sleep deprivation inhibits adult neurogenesis in the hippocampus by elevating glucocorticoids. *Proc Natl Acad Sci U S A*. 2006;103:19170−19175.

75. Novati A, Roman V, Cetin T, et al. Chronically restricted sleep leads to depression-like changes in neurotransmitter receptor sensitivity and neuroendocrine stress reactivity in rats. *Sleep*. 2008;31(11):1579−1585.

76. Roman V, Walstra I, Luiten PG, et al. Too little sleep gradually desensitizes the serotonin 1A receptor system. *Sleep*. 2005;28:1505−1510.

77. Nunes GP, Tufik S, Nobrega JN. Autoradiographic analysis of D_1 and D_2 dopaminergic receptors in rat brain after paradoxical sleep deprivation. *Brain Res Bull*. 1994;34(5):453−456.

78. Hipólide DC, Moreira KM, Barlow KBL, et al. Distinct effects of sleep deprivation on binding to norepinephrine and serotonin transporters in rat brain. *Biol Psychiatry*. 2005;29(2):297−303.

79. Bowers Jr MB, Hartmann EL, Freedman DX. Sleep deprivation and brain acetylcholine. *Science*. 1966;153 (3742):1416−1417.

80. Banasr M, Henry M, Printemps R, et al. Serotonin-induced increases in adult cell proliferation and neurogenesis are mediated through different and common 5-HT receptor subtypes in the dentate gyrus and the subventricular zone. *Neuropsychopharmacology*. 2004;29:450−460.

81. Borta A, Höglinger GU. Dopamine and adult neurogenesis. *J Neurochem*. 2006;100(3):587−595.

82. Dranovsky A, Hen R. Hippocampal neurogenesis: regulation by stress and antidepressants. *Biol Psychiatry*. 2006;59(12):1136−1143.

83. Mohapel P, Leanza G, Kokaia M, et al. Forebrain acetylcholine regulates adult hippocampal neurogenesis and learning. *Neurobiol Aging*. 2005;26(6):939−946.

84. Irwin MR, Wang M, Campomayor CO, et al. Sleep deprivation and activation of morning levels of cellular and genomic markers of inflammation. *Arch Intern Med*. 2006;166:1756−1762.

85. Haack M, Sanchez E, Mullington JM. Elevated inflammatory markers in response to prolonged sleep restriction are associated with increased pain experience in healthy volunteers. *Sleep*. 2007;30:1145−1152.

86. Monje ML, Toda H, Palmer TD. Inflammatory blockade restores adult hippocampal neurogenesis. *Science*. 2003;302(5651):1760−1765.

87. Xie L, Kang H, Xu Q, et al. Sleep drives metabolite clearance from the adult brain. *Science*. 2013;342 (6156):373−377.

88. Rongve A, Boeve BF, Aarsland D. Frequency and correlates of caregiver-reported sleep disturbances in a sample of persons with early dementia. *J Am Geriatr Soc*. 2010;58(3):480−486.

89. Merlino G, Piani A, Gigli GL, et al. Daytime sleepiness is associated with dementia and cognitive decline in older Italian adults: a population-based study. *Sleep Med*. 2010;11(4):372−377.

90. Lim ASP, Kowgier M, Yu L, et al. Sleep fragmentation and the risk of incident Alzheimer's disease and cognitive decline in older persons. *Sleep*. 2013;36(7):1027−1032.

91. Bliwise DL, Hughes M, McMahon PM, et al. Observed sleep/wakefulness and severity of dementia in an Alzheimer's disease special care unit. *J Gerontol A Biol Sci Med Sci*. 1995;50(6):303−306.

92. Bliwise DL. Sleep disorders in Alzheimer's disease and other dementias. *Clin Cornerstone*. 2004;6(1):16−28.

93. Pase MP, Himali JJ, Grima NA, et al. Sleep architecture and the risk of incident dementia in the community. *Neurology*. 2017;10−1212.

94. Bird TD, Stranahan S, Sumi SM, et al. Alzheimer's disease: choline acetyltransferase activity in brain tissue from clinical and pathological subgroups. *Ann Neurol*. 1983;14(3):284−293.

95. Van Dort CJ, Zachs DP, Kenny JD, et al. Optogenetic activation of cholinergic neurons in the PPT or LDT induces REM sleep. *Proc Natl Acad Sci U S A*. 2015;112(2):584−589.

96. Ancoli-Israel S, Martin JL, Kripke DF, et al. Effect of light treatment on sleep and circadian rhythms in demented nursing home patients. *J Am Geriatr Soc*. 2002;50(2):282−289.

97. Bliwise DL, Tinklenberg JR, Yesavage JA. Timing of sleep and wakefulness in Alzheimer's disease patients residing at home. *Biol Psychiatry*. 1992;31:1163−1165.

II. SLEEP DISORDERS IN SPECIFIC MEDICAL CONDITIONS

REFERENCES

273

98. Volicer L, Harper DG, Manning BC, et al. Sundowning and circadian rhythms in Alzheimer's disease. *Am J Psychiatry*. 2001;158(5):704–711.
99. Leng Y, McEvoy CT, Allen IE, et al. Association of sleep-disordered breathing with cognitive function and risk of cognitive impairment. *JAMA Neurol*. 2017;74(10):1237–1245.
100. Redline S, Yenokyan G, Gottlieb DJ, et al. Obstructive sleep apnea-hypopnea and incident stroke: the Sleep Heart Health Study. *Am J Respir Crit Care Med*. 2010;182(2):269–277.
101. Kim H, Yun CH, Thomas RJ, et al. Obstructive sleep apnea as a risk factor for cerebral white matter change in middle-aged and older general population. *Sleep*. 2013;36(5):709–715.
102. Drager LF, Polotsky VY, Lorenzi-Filho G. Obstructive sleep apnea: an emerging risk factor for atherosclerosis. *Chest*. 2011;140(2):534–542.
103. Kono M, Tatsumi K, Saibara T, et al. Obstructive sleep apnea syndrome is associated with some components of metabolic syndrome. *Chest*. 2007;131(5):1387–1392.
104. Tasali E, Ip MSM. Obstructive sleep apnea and metabolic syndrome alterations in glucose metabolism and inflammation. *Proc Am Thorac Soc*. 2008;5(2):207–217.
105. Berry RB, Brooks R, Gamaldo CE, et al. *The AASM Manual for the Scoring of Sleep and Associated Events: Rules, Terminology and Technical Specifications*. Darien, IL: American Academy of Sleep Medicine; 2012.
106. Kapur VK, Auckley DH, Chowdhuri S, et al. Clinical practice guideline for diagnostic testing for adult obstructive sleep apnea: an American Academy of Sleep Medicine clinical practice guideline. *J Clin Sleep Med*. 2017;13(3):479–504.
107. Mehra R, Stone KL, Blackwell T, et al. Prevalence and correlates of sleep-disordered breathing in older men: Osteoporotic Fractures in Men Sleep Study. *J Am Geriatr Soc*. 2007;55(9):1356–1364.
108. Young T, Peppard PE, Gottlieb DJ. Epidemiology of obstructive sleep apnea: a population health perspective. *Am J Respir Crit Care Med*. 2002;165(9):1217–1239.
109. Endeshaw YE, Unruh ML, Kutner M, et al. Sleep-disordered breathing and frailty in the Cardiovascular Health Study Cohort. *Am J of Epidemiol*. 2009;170:193–202.
110. Bliwise DL, Colrain IM, Swan GE, et al. Incident sleep disordered breathing in old age. *J Gerontol A Biol Sci Med Sci*. 2010;65:997–1003.
111. Malhotra A, Huang Y, Fogel R, et al. Aging influences on pharyngeal anatomy and physiology: the predisposition to pharyngeal collapse. *Am J Med*. 2006;119(1):72.e9–72.e14.
112. Schellenberg JB, Maislin G, Schwab RJ. Physical findings and the risk for obstructive sleep apnea: the importance of oropharyngeal structures. *Am J Respir Crit Care Med*. 2000;162(2):740–748.
113. Spira AP, Blackwell T, Stone KL, et al. Sleep-disordered breathing and cognition in older women. *J Am Geriatr Soc*. 2008;56(1):45–50.
114. Yaffe K, Laffan AM, Harrison SL, et al. Sleep-disordered breathing, hypoxia, and risk of mild cognitive impairment and dementia in older women. *JAMA*. 2011;306(6):613–619.
115. Bubu OM, Brannick M, Mortimer J, et al. Sleep, cognitive impairment, and Alzheimer's disease: a systematic review and meta-analysis. *Sleep*. 2017;40(1):1–18.
116. Bliwise DL, Yesavage JA, Tinklenberg J, et al. Sleep apnea in Alzheimer's Disease. *Neurobiol Aging*. 1989;10 (4):343–346.
117. Beebe DW, Gozal D. Obstructive sleep apnea and the prefrontal cortex: towards a comprehensive model linking nocturnal upper airway obstruction to daytime cognitive and behavioral deficits. *J Sleep Res*. 2002;11 (1):1–16.
118. Carlson BW, Neelon VJ, Carlson JR, et al. Cerebral oxygenation in wake and during sleep and its relationship to cognitive function in community-dwelling older adults without sleep disordered breathing. *J Gerontol A Biol Sci Med Sci*. 2011;66A(1):150–156.
119. Kamba M, Inoue Y, Higami S, et al. Cerebral metabolic impairment in patients with obstructive sleep apnoea: an independent association of obstructive sleep apnoea with white matter change. *J Neuro, Neurosurg Psychiatry*. 2001;71(3):334–339.
120. Daulatzai MA. Evidence of neurodegeneration in obstructive sleep apnea: relationship between obstructive sleep apnea and cognitive dysfunction in the elderly. *J Neurosci Res*. 2015;93(12):1778–1794.
121. Osorio RS, Ayappa I, Mantua J, et al. The interaction between sleep-disordered breathing and apolipoprotein E genotype on cerebrospinal fluid biomarkers for Alzheimer's disease in cognitively normal elderly individuals. *Neurobiol Aging*. 2014;35(6):1318–1324.

II. SLEEP DISORDERS IN SPECIFIC MEDICAL CONDITIONS

122. Nieto FJ, Young TB, Lind BK, et al. Association of sleep-disordered breathing, sleep apnea, and hypertension in a large community-based study. *JAMA*. 2000;283(14):1829–1836.

123. Munoz R, Duran-Cantolla J, Martínez-Villa E, et al. Severe sleep apnea and risk of ischemic stroke in the elderly. *Stroke*. 2006;37(9):2317–2321.

124. Young T, Finn L, Peppard PE, et al. Sleep disordered breathing and mortality: eighteen-year follow-up of the Wisconsin sleep cohort. *Sleep*. 2008;31:1071–1078.

125. Brown DL, McDermott M, Mowla A, et al. Brainstem infarction and sleep-disordered breathing in the BASIC sleep apnea study. *Sleep Med*. 2014;15(8):887–891.

126. Palmer BW, Harmell AL, Pinto LL, et al. Determinants of capacity to consent to research on Alzheimer's disease. *Clin Gerontol*. 2017;40(1):24–34.

127. dos Santos Silva M, Bazzana CM, de Souza AL, et al. Relationship between perceived sleep and polysomnography in older adults patients. *Sleep Sci*. 2015;8(2):75–81.

128. Collop NA, Anderson WM, Boehlecke B, et al. Portable monitoring task force of the American Academy of Sleep Medicine. *J Clin Sleep Med*. 2007;3:737–747.

129. Kushida CA, Littner MR, Hirshkowitz M, et al. Practice parameters for the use of the continuous and bilevel positive airway pressure devices to treat adult patients with sleep-related breathing disorders. *Sleep*. 2006;29(3):375–380.

130. Ancoli-Israel S, Palmer BW, Cooke JR, et al. Cognitive effects of treating obstructive sleep apnea in Alzheimer's disease: a randomized controlled study. *J Am Geriatr Soc*. 2008;56(11):2076–2081.

131. Sánchez AI, Martínez P, Miró E. CPAP and behavioral therapies in patients with obstructive sleep apnea: effects on daytime sleepiness, mood, and cognitive function. *Sleep Med Rev*. 2009;13:223–233.

132. Jung Y, Silber MH, Tippmann-Peikert M, et al. The effects of CPAP on cognitive and functional measures in patients with mild cognitive impairment and Alzheimer's disease. *Sleep*. 2017;40(suppl_1):A430–A431.

133. Cooke JR, Ayalon L, Palmer B, et al. Sustained use of CPAP slows deterioration of cognition, sleep, and mood in patients with Alzheimer's disease and obstructive sleep apnea: a preliminary study. *J Clin Sleep Med*. 2009;5(4):305–309.

134. Becker HF, Jerrentrup A, Ploch T, et al. Effect of nasal continuous positive airway pressure treatment on blood pressure in patients with obstructive sleep apnea. *Circulation*. 2003;107(1):68–73.

135. Kostopoulos K, Alhanatis E, Pampoukas K, et al. CPAP therapy induces favorable short-term changes in epicardial fat thickness and vascular and metabolic markers in apparently healthy subjects with obstructive sleep apnea–hypopnea syndrome (OSAHS). *Sleep Breath*. 2016;20(2):483–493.

136. Ayalon L, Ancoli-Israel S, Stepnowsky C, et al. Adherence to continuous positive airway pressure treatment in patients with Alzheimer disease and obstructive sleep apnea. *Am J Geriatr Psychiatry*. 2006;14(2):176–180.

137. Kline CE, Crowley EP, Ewing GB, et al. The effect of exercise training on obstructive sleep apnea and sleep quality: a randomized controlled trial. *Sleep*. 2011;34(12):1631–1640.

138. Herrick JE, Bliwise DL, Puri S, et al. Strength training and light physical activity reduces the apnea-hypopnea index in institutionalized older adults. *J Am Med Dir Assoc*. 2014;15:844–846.

139. Mishima K, Okawa M, Hishikawa Y, et al. Morning bright light therapy for sleep and behavior disorders in elderly patients with dementia. *Acta Psychiatr Scand*. 1994;89(1):1–7.

140. Hickman SE, Barrick AL, Williams CS, et al. The effect of ambient bright light therapy on depressive symptoms in persons with dementia. *J Am Geriatr Soc*. 2007;55(11):1817–1824.

141. Golden RN, Gaynes BN, Ekstrom RD, et al. The efficacy of light therapy in the treatment of mood disorders: a review and meta-analysis of the evidence. *Am J Psychiatry*. 2005;162(4):656–662.

142. Forbes D, Morgan DG, Bangma J, et al. Light therapy for managing sleep, behaviour, and mood disturbances in dementia. *Cochrane Database Syst Rev*. 2004;2:CD003946.

143. Lyketsos CG, Lindell Veiel L, Baker A, et al. A randomized, controlled trial of bright light therapy for agitated behaviors in dementia patients residing in long-term care. *Int J Geriatr Psychiatry*. 1999;14(7):520–525.

144. Kogan AO, Guilford PM. Side effects for short-term 10,000-lux light therapy. *Am J Psychiatry*. 1998;155(2):293–294.

145. Gallin PF, Terman M, Remé CE. Ophthalmologic examination of patients with seasonal affective disorder, before and after bright light therapy. *Am J Ophthalamol*. 1995;199:202–210.

146. Chesson AL, Littner M, Davilla D, et al. Practice parameters for the use of light therapy in the treatment of sleep disorders. *Sleep*. 1999;22(5):641–660.

REFERENCES

147. Terman M, Terman JS. Bright light therapy: side effects and benefits across the symptom spectrum. *J Clin Psychiatry*. 1999;60(11):799–808.
148. Kamel NS, Gammack JK. Insomnia in the elderly: cause, approach, and treatment. *Am J Med*. 2006;119 (6):463–469.
149. Krishnan P, Hawranik P. Diagnosis and management of geriatric insomnia: a guide for nurse practitioners. *J Am Assoc Nurse Pract*. 2008;20(12):590–599.
150. Richards KC, Lambert C, Beck CK, et al. Strength training, walking and social activity improve sleep in nursing home and assisted living residents: randomized controlled trial. *J Am Geriatr Soc*. 2011;59:214–223.
151. King AC, Oman RF, Brassington GS. Moderate-intensity exercise and self-rated quality of sleep in older adults. *JAMA*. 1997;277(1):32–37.
152. Alessi CA, Yoon EJ, Schnelle JF, et al. A randomized trial of a combined physical activity and environmental intervention in nursing home residents: do sleep and agitation improve? *J Am Geriatr Soc*. 1999;47 (7):784–791.
153. Black DS, O'Reilly GA, Olmstead R, et al. Mindfulness meditation and improvement in sleep quality and daytime impairment among older adults with sleep disturbances: a randomized clinical trial. *JAMA Intern Med*. 2015;175(4):494–501.
154. Sun J, Kang J, Wang P, et al. Self-relaxation training can improve sleep quality and cognitive functions in the older: a one-year randomised controlled trial. *J Clin Nurs*. 2013;22(9-10):1270–1280.
155. Zhang JX, Liu XH, Xie XH, et al. Mindfulness-based stress reduction for chronic insomnia in adults older than 75 years: a randomized, controlled, single-blind clinical trial. *Explore*. 2015;11(3):180–185.
156. McCurry SM, Gibbons LE, Logsdon RG, et al. Training caregivers to change the sleep hygiene practices of patients with dementia: the NITE-AD Project. *J Am Geriatr Soc*. 2003;51(10):1455–1460.
157. Liu X, Uchiyama M, Shibui K, et al. Diurnal preference, sleep habits, circadian sleep propensity and melatonin rhythm in healthy human subjects. *Neurosci Lett*. 2000;280(3):199–202.
158. Rikkert MO, Rigaud AS. Melatonin in elderly patients with insomnia—a systematic review. *Z Gerontol Geriatr*. 2001;34(6):491–497.
159. Suhner A, Schlagenhauf P, Johnson R, et al. Comparative study to determine the optimal melatonin dosage form for the alleviation of jet lag. *Chronobiol Int*. 1998;15(6):655–666.
160. Peck JS, LeGoff DB, Ahmed I, et al. Cognitive effects of exogenous melatonin administration in elderly persons: a pilot study. *Am J Geriatr Psychiatry*. 2004;12(4):432–436.
161. de Jonghe A, Korevaar JC, van Munster BC, et al. Effectiveness of melatonin treatment on circadian rhythm disturbances in dementia. Are there implications for delirium? A systematic review. *Int J Geriatr Psychiatry*. 2010;25(12):1201–1208.
162. Brusco LI, Márquez M, Cardinali DP. Melatonin treatment stabilizes chronobiologic and cognitive symptoms in Alzheimer's disease. *Neuro Endocrinol Lett*. 2000;21(1):39–42.
163. Cohen-Mansfiled J, Garfinkel D, Lipson S. Melatonin for treatment of sundowning in elderly persons with dementia—a preliminary study. *Arch Gerontol Geriatr*. 2000;31:65–76.
164. Cardinali DP, Brusco LI, Liberczuk C. The use of melatonin in Alzheimer's disease. *Neuro Endorinol Lett*. 2002;23:20–23.
165. Singer C, Tractenberg RE, Kaye J, et al. A multicenter, placebo-controlled trial of melatonin for sleep disturbance in Alzheimer's disease. *Sleep*. 2003;26(7):893–901.
166. Xu J, Wang LL, Dammer EB, et al. Melatonin for sleep disorders and cognition in dementia: a meta-analysis of randomized controlled trials. *Am J Alzheimers Dis Other Demen*. 2015;30(5):439–447.
167. Earland LA, Saxena PK. Melatonin natural health products and supplements: presence of serotonin and significant variability of melatonin content. *J Clin Sleep Med*. 2017;13:275–281.
168. Guthrie B, Clark SA, McCowan C. The burden of psychotropic drug prescribing in people with dementia: a population database study. *Age Ageing*. 2010;39(5):637–642.
169. Stahl SM, Markowitz JS, Gutterman EM, et al. Co-use of donepezil and hypnotics among Alzheimer's disease patients living in the community. *J Clin Psychiatry*. 2003;64(4):466–472.
170. Longo LP, Johnson B. Addition: Part I. Benzodiazepines-side effects, abuse risk and alternatives. *Am Fam Physician*. 2000;61(7):2121–2128.
171. Grad RM. Benzodiazepines for insomnia in community-dwelling elderly: a review of benefit and risk. *J Fam Pract*. 1995;41(5):473–482.

II. SLEEP DISORDERS IN SPECIFIC MEDICAL CONDITIONS

172. Greenblatt DJ, Harmatz JS, Zinny MA. Effect of gradual withdrawal on the rebound sleep disorder after discontinuation of triazolam. *N Eng J Med*. 1987;317(12):722–728.
173. de Gage SB, Moride Y, Ducruet T, et al. Benzodiazepine use and risk of Alzheimer's disease: case–control study. *BMJ*. 2014;349:g5205.
174. Gallacher J, Elwood P, Pickering J, et al. Benzodiazepine use and risk of dementia: evidence from the Caerphilly Prospective Study (CaPS). *J Epidemiol Community Health*. 2011;66:869–873.
175. Gray SL, Dublin S, Yu O, et al. Benzodiazepine use and risk of incident dementia or cognitive decline: prospective population based study. *BMJ*. 2016;352:i90.
176. Wu CS, Wang SC, Chang IS, et al. The association between dementia and long-term use of benzodiazepine in the elderly: nested case–control study using claims data. *Am J Geriatr Psychiatry*. 2009;17(7):614–620.
177. Pariente A, de Gage SB, Moore N, et al. The benzodiazepine-dementia disorders link: current state of knowledge. *CNS Drugs*. 2016;30(1):1–7.
178. Huedo-Medina TB, Kirsch I, Middlemass J, et al. Effectiveness of non-benzodiazepine hypnotics in treatment of adult insomnia: meta-analysis of data submitted to the Food and Drug Administration. *BMJ*. 2012;345: e8343.
179. Dolder C, Nelson M, McKinsey J. Use of non-benzodiazepine hypnotics in the elderly. *CNS Drugs*. 2007;21 (5):389–405.
180. Gunja N. In the Zzz zone: the effects of Z-drugs on human performance and driving. *J Med Toxicol*. 2013;9 (2):163–171.
181. Weich S, Pearce HL, Croft P, et al. Effect of anxiolytic and hypnotic drug prescriptions on mortality hazards: retrospective cohort study. *BMJ*. 2014;348:g1996.
182. Mander BA, Winer JR, Jagust WJ, et al. Sleep: a novel mechanistic pathway, biomarker, and treatment target in the pathology of Alzheimer's disease? *Trends Neurosci*. 2016;39(8):552–566.
183. Ju YS, Lucey BP, Holtzman DM. Sleep and Alzheimer disease pathology—a bidirectional relationship. *Nat Rev Neurol*. 2014;10(2):115–119.
184. McEvoy RD, Antic NA, Heeley E, et al. CPAP for prevention of cardiovascular events in obstructive sleep apnea. *N Engl J Med*. 2016;375(10):919–931.
185. Beaudreau SA, Spira AP, Gray HL, et al. The relationship between objectively measured sleep disturbance and dementia family caregiver distress and burden. *J Geriatr Psychiatry Neurol*. 2008;21(3):159–165.
186. Creese J, Bédard M, Brazil K, et al. Sleep disturbances in spousal caregivers of individuals with Alzheimer's disease. *Int Psychogeriatr*. 2008;20(1):149–161.
187. Carter PA. Family caregivers' sleep loss and depression over time. *Cancer Nurs*. 2003;26(4):253–259.
188. Liu S, Li C, Shi Z, et al. Caregiver burden and prevalence of depression, anxiety and sleep disturbances in Alzheimer's disease caregivers in China. *J Clin Nurs*. 2017;26:9–10.
189. Chiu YC, Lee YN, Wang PC, et al. Family caregivers' sleep disturbance and its associations with multilevel stressors when caring for patients with dementia. *Aging Mental Health*. 2014;18(1):92–101.
190. Gibson RH, Gander PH, Jones LM. Understanding the sleep problems of people with dementia and their family caregivers. *Dementia*. 2014;13(3):350–365.
191. Kotronoulas G, Wengstrom Y, Kearney N. Sleep and sleep–wake disturbances in care recipient–caregiver dyads in the context of a chronic illness: a critical review of the literature. *J Pain Symptom Manage*. 2013;45 (3):579–594.

II. SLEEP DISORDERS IN SPECIFIC MEDICAL CONDITIONS

CHAPTER

12

Stroke

Annette Sterr and James Ebajemito

School of Psychology, University of Surrey, Guildford, United Kingdom

OUTLINE

Introduction	277	Insomnia in the Chronic Phase of Stroke	284
Stroke: Brief Overview	278		
Incidence and Prevalence of Stroke	278	Sleep Interventions	285
Clinical Features	278	Light Therapy	285
Risk Factors	279	Cognitive Behavioral Therapy for Insomnia	285
Sleep Disorders in Stroke	279		
Sleep Disordered Breathing	279	Clinical Practice Considerations	286
Sleep Disorders	280	Areas for Future Research	286
Sleep-Stroke Interaction	281	Conclusion	287
Sleep-Recovery Interaction Model	283		
Physiological Changes	283	References	287
Psychological Changes	284		

INTRODUCTION

The World Health Organization defines stroke as a cardiovascular disease caused by an interruption of blood supply to the brain which cuts nutrient and oxygen supply leading to a rapid deterioration of neurological function of the affected brain region.[1] Stroke is a leading cause of chronic disability, affecting roughly 7 million individuals in the United States,[2] and more than 100,000 stroke cases in the United Kingdom each year.[3] Typically, a

Handbook of Sleep Disorders in Medical Conditions
DOI: https://doi.org/10.1016/B978-0-12-813014-8.00012-3

© 2019 Elsevier Inc. All rights reserved.

stroke begins as an acute medical condition which may affect motor, language, vision, and other cognitive processes. In severe cases, where critical brain areas are affected, a stroke may lead to coma or death. Although 10%–25% of stroke patients recover fully or with minor impairments,[4] a vast majority of stroke survivors sustain long-term difficulties that may affect their daily life. The recovery process from stroke is modulated by several comorbid and preexisting conditions, such as heart disease,[2,5] diabetes,[6] age,[7] or indeed sleep.[8] Sleep is hereby a particularly interesting factor in the trajectory of stroke recovery because of its relationship with brain plasticity and learning, physical and mental health, as well as daytime performance. The role of sleep in stroke recovery, rehabilitation, and long-term care is discussed in this chapter.

STROKE: BRIEF OVERVIEW

Incidence and Prevalence of Stroke

Stroke is a major public health problem. It is the third most common cause of death in the world and a leading cause of long-term disability in adults (www.strokecenter.org). Data from the Office of National Statistics report the estimated cost of stroke in the United Kingdom to be approximately £4 billion or 5.5% of expenditure on health care.[9] In the United States, every 40 seconds someone has a stroke, and every 4 minutes someone dies from stroke; the estimated cost of stroke in the United States is $33 billion.[10] With increasing affluence around the globe, stroke is becoming an increasing problem in the world's most populous nations. For example, stroke is now the most common cause of disability amongst young adults in India,[11] and one of the leading cause of death in China.[12] The incidence rates of strokes are rising in developing and threshold countries, such as Brazil[13,14] and Africa's most populous nation, Nigeria.[15] The economical strain and human hardship caused by stroke in these countries are even greater than the industrial nations due to underdeveloped public health services more generally, and little or no funding for poststroke rehabilitation in particular.[15–17]

The prevalence of stroke depends on the age and sex distribution of the population. In older individuals, the prevalence of stroke seems to be higher for both men and women, respectively.[18] Recent data indicate that in the industrialized western world, the incidence and mortality of stroke have declined over the last 20 years, due to major public health interventions regarding early recognition of warning signs and symptoms, and risk factor-related health behaviors,[19–21] as well as more effective treatment of causal risk factors, such as high blood pressure and high cholesterol levels.[20,22] In developing countries, however, stroke incidents are rising. Therefore, there is still a pressing need to reduce the disease burden from a social, societal, as well as economical perspective.

Clinical Features

In principle, there are two types of stroke, ischemic or hemorrhagic. Ischemic strokes are caused by a blockage of blood flow to specific areas of the brain (infarcts), for example, through deposits within the brain's blood vessels; they represent 67%−80% of all stroke cases.[23] Hemorrhagic stroke is caused by a rupture of blood vessels in the brain.

Risk Factors

Preexisting conditions, such as heart disease, diabetes, hypertension, and high cholesterol levels, may increase an individual's predisposition to stroke.[24] Moreover, the likelihood of stroke is greater in males and increases with age.[24] Risk factors for first ever as well as repeated strokes can further be aggravated by lifestyle choices, health habits, and environmental factors, such as stress, smoking, obesity, and exercise.[23] Notably, sleep apnea is the most prominent risk predictor of stroke even in the absence of other preexisting conditions.[25−29]

SLEEP DISORDERS IN STROKE

Sleep disorders are frequent after stroke, with 30% of adults suffering from single or multiple symptoms of insomnia.[30] Sleep disorders may occur as a result of poor health, environmental factors, or lifestyle choices.[31,32] With regards to stroke, sleep disorders may manifest as a direct consequence of the brain injury through damage to sleep regulating pathways in the brain, or indirectly through intrinsic factors, such as pain, stress, medication, and depression.[33,34] The most common forms of sleep disorders in stroke include sleep disordered breathing (SDB), sleep−wake disturbances, movement-related sleep disorders, and poststroke insomnia.

Sleep Disordered Breathing

Out of all the sleep disorders associated with stroke, SDB is the most pertinent factor because it can act as an independent trigger for a stroke event. SDB consists of a spectrum of sleep-related breathing problems, such as snoring, obstructive sleep apnea (OSA), and central sleep apnea. The most common form of SDB is OSA, which is the sudden cessation of breathing due to the collapse of the upper air passage. A metaanalysis investigating the frequency of sleep apnea across 29 studies, which included 1343 stroke patients, found that 72% of them had SDB. The severity of stroke is thereby associated with the specific form of the SDB, such as OSA.[35,36] Evidence further suggests that snoring is a major risk factor of stroke. For example, a study of 400 stroke patients and 400 controls found that the odds ratio for second stroke readmission was 3.2 for snorers compared to nonsnorers.[36] Evidence further indicates that snoring severity increases stroke mortality.[35−37]

SDBs can be treated through a number of interventions. The most effective intervention is the provision of continuous positive airway pressure which delivers a steady stream of air to the airways while the person is asleep. While very effective in treating the symptoms of SDB, compliance in patients is limited. Other approaches comprise oral appliances designed to prevent airways collapse and positional therapy promoting sleep in the best

Sleep Disorders

Sleep disorders are conditions which cause dissatisfaction with sleep quality, duration, or timing, such as narcolepsy, non−rapid eye movement (REM) sleep arousal disorders, nightmare disorder, REM sleep behavior disorder, and restless legs syndrome (RLS).[38,39] REM sleep behavior disorder occurs when patients act out vivid dreams, which may cause harm to themselves or their cosleeper. In stroke, this is particularly common in patients with brain stem infarcts.[40] Another condition is RLS, a sensorimotor disorder which causes periodic leg movements. It has been reported that acute stroke patients with RLS have worse clinical outcome[41] and this has been linked to a rise in blood pressure and sympathetic hyperactivity during RLS episodes.[26] Insomnia occurs when an individual does not feel restored after sleep due to the difficulty in initiating or maintaining sleep.[42]

Poststroke insomnia is the most frequent sleep disorder in the stroke population with an estimated incidence rate of 30%.[30] Insomnia causes physical and mental difficulties which includes fatigue, lethargy, depression, and impaired cognitive performance the following day.[43] In stroke, insomnia is highly prevalent in the acute and subacute phase, as evidenced by a range of methodologies, including the gold standard approach, polysomnography[44,45]. This is caused by a combination of the direct effects of stroke, such as loss of consciousness, pain, and the brain damage per se, as well as the treatment and medication thereof. Less research has been conducted with chronic stroke patients. However, there is converging evidence from subjective as well as objective measures that difficulties with sleeping remain a persistent problem in these patients.

For example, questionnaire- and actigraphy-based studies report chronic sleep difficulties in 34%−67% of patients,[46,47] with 48% satisfying the criteria for insomnia.[48] A recent meta-analysis of polysomnography studies comparing patients with stroke to controls corroborates the subjective patient reports with polysomnography evidence for reduced sleep efficiency and physiological markers of disturbed sleep initiation and maintenance.[45] In addition to sleep difficulties, stroke patients often report daytime fatigue and excessive daytime sleepiness as major problems in day-to-day life.[49] This is not only true for the early phases of stroke recovery but particularly so for the chronic phase, with an estimated 17%−34% of patients with stroke reporting disturbed daytime functioning due to excessive sleepiness[46,50] and fatigue (30%−69%)[49,51,52]. Daytime electroencephalogram measures in chronic motor stroke further indicate greater prevalence of slower frequencies in resting state wake electroencephalogram,[53] a phenomenon often interpreted as a marker of sleep deprivation.[54] Of course, daytime sleepiness can be a reflection of poor nocturnal sleep, and this has clearly been shown to be the case in patients with SDB. In this case, effective intervention for SDB also reduces excessive daytime sleepiness and fatigue. Besides the study of excessive daytime sleepiness and fatigue in the context of SDB, this issue is not extensively researched.[45,47] This is reflected by the fact that, at the point of writing this manuscript, there is no published study examining sleep propensity through the Multiple Sleep Latency Test, the physiological gold standard measure for daytime sleepiness.

Daytime sleepiness and fatigue represent a problem for patients.[55] Whether these symptoms are a direct or indirect consequence of the stroke is presently not clear. Either way, through stand-alone self-management approaches and/or in combination with primary interventions concerning sleep, patient benefit can be achieved. However, clinical practice in stroke rehabilitation and stroke care does not necessarily include measures for daytime sleepiness management.

Sleep-Stroke Interaction

As mentioned earlier, the SDB literature has shown a clear interaction between sleep and stroke occurrence. In addition to that, we argue that sleep is also relevant for stroke recovery, long-term outcome, and quality of life after stroke. This proposition is based on the notion that brain plasticity in terms of both, the *functional* and *structural* reorganization of neural circuits, is the key mechanism of poststroke recovery and, at the same time, is also a function of sleep. Specifically, it is now accepted that after stroke, the brain goes through stages of brain changes that, in part, represent a replication of the mechanisms at play during brain development.[56] In addition, functional and structural reorganization has been shown to be the key mechanism of effective practice-based interventions, such as constraint-induced movement therapy[57–59] for the treatment of hemiparesis, or constraint-induced aphasia therapy[60] for the treatment of language and communication deficits. At the same time there is a large literature in animals and humans demonstrating the importance of sleep for memory consolidation and learning.[61,62] Indeed, while the specific function of sleep is still not entirely clear, there is an emerging consensus that one of the main functions of sleep is the promotion of brain plasticity.[63] Sleep and stroke are therefore intrinsically linked via the mechanisms of brain plasticity. Initial evidence for this hypothesis comes from studies suggesting that poorer sleep might lead to poorer outcome.[64]

The relevance of sleep for stroke recovery, long-term outcome, and quality of life, however, goes beyond the principles of brain plasticity and also includes the role of physical and mental health. Again, both are strongly associated with sleep. For example, Alhola and Polo-Kantola[65] reported that poor sleep negatively impacts mental and physical health, day time function, learning, and memory. In the specific case of stroke, many patients suffer from depression and are often treated with antidepressants for many years. Sleep disorders, on the other hand, are strongly associated with poor mental health and a wide range of psychiatric disorders.[66] Depression in particular has thereby been identified as a main predictor of the transition from acute to chronic insomnia.[67,68] Together, the considerations outlined earlier have informed the sleep-stroke recovery model illustrated in Fig. 12.1. In essence, the model proposes that having a stroke induces physiological and psychological processes that negatively impact sleep. These processes interact with each other, and, importantly, their interaction changes over time. The psychological aspects of stroke thereby play a key role in the development and maintenance of sustained sleep disorders, in particular insomnia. This is conceptualized in the context of Spielman's 3P model of insomnia[69] as outlined below.

Spielman's 3P model of insomnia adapted for stroke: Spielman's 3P model of insomnia assumes that psychological and behavioral aspects can be clustered into predisposing, precipitating, and perpetuating factors involved in the development and maintenance of insomnia. *Predisposing* factors comprise characteristics that increase the vulnerability to sleep difficulties. *Precipitating* factors represent life events, medical and psychological strains that trigger insomnia. *Perpetuating* factors represent the psychological and behavioral elements that maintain sleep difficulties, such as changing bedtime or napping behavior, beliefs, and thoughts. This model has a clear application in the situation of stroke recovery as explained in detail in below.

Many of the risk factors for stroke, such as hypertension and obesity, also increase the vulnerability to sleep difficulties and hence represent predisposing factors. The stroke per se and events that take place following a stroke, such as being in a hospital

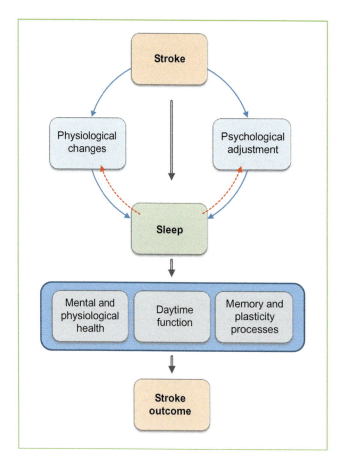

FIGURE 12.1 Sleep-stroke recovery model.[45] The graph illustrates the interplay of physiology and psychology in relation to sleep and stroke recovery.

II. SLEEP DISORDERS IN SPECIFIC MEDICAL CONDITIONS

environment, psychological impact of having had the stroke, pain, and physical problems, represent powerful precipitating factors. The cognitive strategies and behaviors patients often have to adopt in order to cope with their condition can act as perpetuating factors, e.g., napping during the day in response to feeling fatigued or depressed; going to bed early in response social isolation.his is particularly important in the context of the changing need for sleep and rest in the early phases post stroke, which is likely to require substantive time in bed during the day. The physical and cognitive deficits induced initially by the stroke are likely to cause excessive worrying and depression which in turn may affect nocturnal sleep. We therefore argue that patients who have suffered a stroke are likely to experience insomnia in the early phase of stroke and that the prevalence of potent perpetuating factors, such as medication, may increase the risk for the development of chronic insomnia through maladaptive psychological processes and behaviors. A wealth of literature suggests that cognitive behavioral therapy for insomnia (CBTi), a method specifically targeting the third P in Spielman's model, is an effective tool for treating insomnia[43]; it also represents a yet untapped opportunity for improving sleep and quality of life in those with stroke as described further below.

SLEEP-RECOVERY INTERACTION MODEL

As outlined earlier, the sleep recovery interaction model (Fig. 12.1) proclaims a reciprocal interaction between sleep and stroke. In essence it is assumed that the stroke causes physiological changes and psychological impacts that together affect sleep, while sleep is linked to a range of factors that are important for stroke recovery. The impact of the physiological and psychological factors and their potential influence of sleep evolves and changes throughout different phases of recovery. This is explained in detail in the following sections.

Physiological Changes

In most cases, stroke induces brain damage and is associated with an alteration in the state of consciousness, either caused directly as a result of the brain lesion or indirectly through symptom management (such as sedation or induced coma) and analgesic medication (anticoagulants, antihypertensive, antiplatelets).[70] Sleep can further be affected by the hemiparesis which may cause pain and difficulties with posture adjustments during sleep. Poor nocturnal sleep, fatigue, and excessive daytime sleepiness are therefore very common and well documented in the acute and postacute phase of stroke.[25,28] These sleep difficulties are likely to impact on stroke recovery in two ways. First, they may likely lead to a decrease in active participation in neurorehabilitation as well as societal participation and reintegration[47,71] and also increase the risk of accidents.[72,73] Moreover, as sleep plays an active role in neuroplasticity, memory consolidation, and learning, effective neurorehabilitation is likely to benefit from good sleep.[74−76]

Psychological Changes

Alongside and interlinked with the physiological impact of the stroke is the psychological impact of experiencing a stroke and the consequential psychological adjustments necessary for patients to adapt to their new life situation. These psychological aspects are the immediate consequences of the life event and the dramatic change it may bring to patients and their families, as well as long-term consequences, such as coming to terms with the sustained disability. For example, patients with severe injuries to the motor cortex sustain long-term disability and may be wheelchair bound for the rest of their lives. From a sleep-recovery perspective, these patients do not only have to overcome the initial trauma of the insult but go through weeks and month of rehabilitation and recovery with the hope for a better long-term outcome, that the pain would get better, that more function would come back, and that everyday activities would become less fatiguing. As time since stroke increases, and full recovery does materialize, the psychological burden of being a stroke survivor can indeed increase, which may lead to poor nocturnal sleep.[34,77] The latter is likely to aggravate the difficulties experienced in coping with everyday tasks and can feed a vicious circle that perpetuates the acute insomnia occurring in the early phase of stroke recovery.

Insomnia in the Chronic Phase of Stroke

Based on the above model, we conducted a series of studies in patients with moderate-to-severe hemiparesis at least 1-year poststroke using a mix of observational and descriptive studies. When asked "how is your sleep," 85% ($N = 61$) reported that their sleep had persistently changed since the stroke, with the majority feeling that their sleep difficulties started in hospital. Sixty-one percent napped more since their stroke and 25% felt persistently tired or fatigued. The Pittsburgh Sleep Quality Index identified 32% of the cohort as poor sleepers; according to the Insomnia Severity Index 50, 14% of patients had insomnia (compared to 7% in the general population). Further investigations showed that the experience of poor sleep poststroke was not a simple effect of the passage of time, that is, healthy persons of a similar age do not report to have experienced dramatic changes of their sleep in the previous few years. Similar findings were obtained in a sample of younger patients with chronic low-functioning hemiparesis.[47] A study using frequency analysis of electroencephalogram data acquired through the Karolinska Drowsiness Test[78] further showed a greater prevalence of slow frequency activity in those patients compared to age-matched controls.[53] This finding indirectly supports the idea of poorer sleep in chronic stroke, since electroencephalogram slowing is typically observed in situations of sleep deprivation. Together these findings expand the literature on sleep difficulties in the earlier phases poststroke and demonstrate specifically that insomnia can continue to be a clinically relevant problem in the chronic phase poststroke, when patients had the time to adjust to their disability and are living within the community. According to the sleep-stroke recovery model, at least in some of these patients, these sleep problems represent a preventable transition from *acute insomnia*, initiated by the stroke event, to a *chronic primary sleep disorder* if early intervention was sought. To what extent these sleep difficulties have also affected the efficacy of rehabilitation efforts and long-term recovery outcome needs to be examined through further research. A further issue to be considered is the impact of language and cognitive deficits can have on the assessment sleep difficulties.

SLEEP INTERVENTIONS

As mentioned earlier, the sleep-stroke recovery model proposes an interaction between sleep and stroke recovery such that (1) poorer sleep is detrimental to rehabilitation efficacy, long-term outcome, and quality of life and (2) sleep problems occurring in the early phase of recovery could be prevented from chronification through the provision of sleep interventions. Regarding the latter, there are three main categories: pharmacological interventions, light interventions, and CBTi. Pharmacological interventions represent the prescription of sleep promoting medications, such as benzodiazepines. However, drug-based intervention for sleep has problematic side effects and may interfere with neuroplasticity processes.[79] No pharmacological interventions, such as light therapy and CBTi, are therefore desirable.

Light Therapy

Light therapy is used to treat sleep difficulties arising from circadian alignment of the sleep–wake cycle, and/or depression.[80] Light therapy targets the body clock through bright light provision at certain times. It can be used to entrain the body clock and the sleep–wake pattern to the day–night cycle and the socially accepted sleep times.[81,82] Light therapy has thereby been successfully used to improve sleep in various neurological disorders, including Parkinson's disease[83] and cognitive impariment.[81] Moreover, recent studies have further shown that particular wavelength of light, such as blue light, can impact arousal, cognition, and brain function.[84] Light-therapy interventions, in particular blue light exposure applied to those with stroke, have yet to be trialed.

Cognitive Behavioral Therapy for Insomnia

CBTi is a psychological intervention designed to normalize sleep by untangling the predisposing, precipitating, and perpetuating factors specified in the 3P model of insomnia, and by specifically addressing the behaviors and emotions perpetuating physiological hyperarousal and insomnia. CBTi is by large accepted as a very effective and side-effect-free intervention.[43] In essence, it comprises four to six individual sessions where patients are first educated on how circadian rhythm and homeostatic sleep drive affect normal sleep, and how their own behaviors may contribute to their sleep difficulties. Patients are then helped to identify and develop healthy and more effective sleep behaviors. The typical CBTi protocol[85] comprises the following seven treatment sessions: (1) psychoeducation, (2) identifying barriers and developing a formulation, (3) sleep hygiene, stimulus control, and relaxation, (4) sleep scheduling, (5) managing emotions, (6) changing ways of thinking, and (7) relapse prevention.

The application of CBTi to patients with sustained disabilities, in particular those with motor deficits, needs to be adjusted to accommodate their needs and to make it safe. For example, sleep restriction can initially increase tiredness and hence an increased risk of falls. In a recent case study with five patients with chronic low-functioning motor stroke conducted by our group, we examined the feasibility and efficacy of stroke-adapted CBTi. To accommodate patient needs, psychoeducation components were adapted to include relevant information about stroke. When advising increased activity as part of sleep hygiene, consideration was given to the types of activity suitable for those with physical limitations. As part of stimulus control, the "15 minute" rule that requires a person to leave the bedroom should they not fall asleep within 15 minutes was felt to be too challenging for those with physical difficulties, and a relaxation routine was used instead. Sleep scheduling, which normally involves a retiming of bed/wake times and sleep restriction, was considered as an inappropriate risk to safety; therefore, "sleep compression" was used as a moderate alternative. A sleep hygiene rule of "healthy napping," which entailed napping only in bed, for no longer than 30 minutes, and not after 4 p.m., was used. The results suggest that the intervention protocol was feasible and allowed for high levels of standardization. Patients responded well to the intervention, although for some, the length of the session seemed to challenge their abilities, for example, becoming fatigued in a session. Further refinement and testing of the stroke-adapted CBTi are ongoing.

CLINICAL PRACTICE CONSIDERATIONS

At present, sleep is not typically considered in neurorehabilitation and/or the care of long-term stroke survivors. This is, for example, reflected in the United Kingdom's most recent stroke rehabilitation guidelines published by the National Institute for Health and Care Excellence,[86] which do not comment on sleep and give very limited attention to fatigue. In general clinical practice, many healthcare providers now recognize the importance of SDB as a major risk factor of stroke and henceforth include SDB diagnostics into their assessment pathway. Reflective of the operational and conceptual focus on SDB, these services often sit alongside divisions of pulmonary medicine. Based on the considerations outlined earlier, we suggest that these services need to be widened and integrated in a holistic model of stroke rehabilitation and stroke care.

AREAS FOR FUTURE RESEARCH

Research on the prevalence, evaluation, and treatment of insomnia in chronic stroke patients is still lacking. Furthermore, excessive daytime sleepiness and fatigue in the context of sleep disorder other than SDB also need to be more extensively researched. For example, future studies using the Multiple Sleep Latency Test could help to examine sleep propensity after stroke. Regarding intervention, research on how to prevent acute insomnia from transforming into a chronic primary sleep disorder is needed. The potential of light-therapy interventions specifically applied to persons with stroke is yet to be explored. Finally, there is a need to refine and thoroughly test the efficacy of stroke-adapted CBTi.

CONCLUSION

Sleep is an important risk factor for stroke occurrence as well as a likely modulator of neurorehabilitation efficacy, long-term recovery, and quality of life. In the early recovery phase of stroke, sleep problems are prevalent and may develop from a transient secondary symptom to a chronic primary disorder if left untreated. Conventional neurorehabilitation techniques as well as emerging techniques, such as neurostimulation, may be modulated by sleep.[71] Sleep should therefore be considered as a covariate in the assessment of neurorehabilitation efficacy.

KEY PRACTICE POINTS

- Improve sleep education amongst neurorehabilitation practitioners and those providing other forms of stroke care.
- Include a comprehensive insomnia assessment into assessment practice to identify patients having difficulties sleeping.
- Assessment may comprise the use of sleep logs, a comprehensive history, sleep questionnaires, and nocturnal polysomnography, particularly for SDB.
- Provide information materials on sleep difficulties for patients and their careers.
- Identify and strengthen the clinical pathways for access to sleep interventions.
- Build self-management tools for sleep, sleepiness, and fatigue into the rehabilitation program.
- Harness the power of sleep for the enhancement of neurorehabilitation through interventions combining sleep with practice-based treatments, as well as providing stand-alone sleep interventions.

References

1. WHO. Stroke, cerebrovascular accident. *Health Topics*. <http://www.who.int/topics/cerebrovascular_accident/en/>; 2017.
2. Roger VL, Go AS, Lloyd-Jones DM, et al. Heart disease and stroke statistics—2011 update: a report from the American Heart Association. *Circulation*. 2011;123(4):e18—e209. Available from: https://doi.org/10.1161/CIR.0b013e3182009701.
3. SSNAP. Clinical audit April—June 2015 public report. *Sentinel Stroke National Audit Programme*. <https://www.strokeaudit.org/Documents/Results/National/AprJun2015/AprJun2015-PublicReport.aspx>; 2015.
4. Lai SM, Studenski S, Duncan PW, Perera S. Persisting consequences of stroke measured by the Stroke Impact Scale. *Stroke*. 2002;33(7):1840—1844.
5. Writing Group M, Mozaffarian D, Benjamin EJ, et al. Heart disease and stroke statistics—2016 update: a report from the American Heart Association. *Circulation*. 2016;133(4):e38—e360. Available from: https://doi.org/10.1161/CIR.0000000000000350.

6. Chen R, Ovbiagele B, Feng W. Diabetes and stroke: epidemiology, pathophysiology, pharmaceuticals and outcomes. *Am J Med Sci.* 2016;351(4):380–386. Available from: https://doi.org/10.1016/j.amjms.2016.01.011.
7. Fonarow GC, Reeves MJ, Zhao X, et al. Age-related differences in characteristics, performance measures, treatment trends, and outcomes in patients with ischemic stroke. *Circulation.* 2010;121(7):879–891. Available from: https://doi.org/10.1161/Circulationaha.109.892497.
8. Bassetti CL. Sleep and stroke. *Semin Neurol.* 2005;25(1):19–32. Available from: https://doi.org/10.1055/s-2005-867073.
9. Saka O, McGuire A, Wolfe C. Cost of stroke in the United Kingdom. *Age Ageing.* 2009;38(1):27–32. Available from: https://doi.org/10.1093/ageing/afn281.
10. CDC. Stroke facts. Find Facts and Statistics about Stroke in the United States. <https://www.cdc.gov/stroke/facts.htm>; 2017.
11. Prabhakaran D, Jeemon P, Roy A. Cardiovascular diseases in India: current epidemiology and future directions. *Circulation.* 2016;133(16):1605–1620. Available from: https://doi.org/10.1161/CIRCULATIONAHA.114.008729.
12. Wang W, Jiang B, Sun H, et al. Prevalence, incidence, and mortality of stroke in China: results from a Nationwide Population-Based Survey of 480 687 adults. *Circulation.* 2017;135(8):759–771. Available from: https://doi.org/10.1161/CIRCULATIONAHA.116.025250.
13. de Carvalho JJ, Alves MB, Viana GA, et al. Stroke epidemiology, patterns of management, and outcomes in Fortaleza, Brazil: a hospital-based multicenter prospective study. *Stroke.* 2011;42(12):3341–3346. Available from: https://doi.org/10.1161/STROKEAHA.111.626523.
14. Lange MC, Cabral NL, Moro CH, et al. Incidence and mortality of ischemic stroke subtypes in Joinville, Brazil: a population-based study. *Arq Neuropsiquiatr.* 2015;73(8):648–654. Available from: https://doi.org/10.1590/0004-282X20150081.
15. Wahab KW, Okubadejo NU, Ojini FI, Danesi MA. Predictors of short-term intra-hospital case fatality following first-ever acute ischaemic stroke in Nigerians. *J Coll Physicians Surg Pak.* 2008;18(12):755–758. Available from: https://doi.org/12.2008/JCPSP.755758.
16. Adeloye D. An estimate of the incidence and prevalence of stroke in Africa: a systematic review and meta-analysis. *PLoS One.* 2014;9(6):e100724. Available from: https://doi.org/10.1371/journal.pone.0100724.
17. Martins SC, Pontes-Neto OM, Alves CV, et al. Past, present, and future of stroke in middle-income countries: the Brazilian experience. *Int J Stroke.* 2013;8(Suppl A100):106–111. Available from: https://doi.org/10.1111/ijs.12062.
18. Wyller TB, Bautzholter E, Holmen J. Prevalence of stroke and stroke-related disability in North Trondelag County, Norway. *Cerebrovasc Dis.* 1994;4(6):421–427.
19. Feigin VL, Lawes CMM, Bennett DA, Barker-Collo SI, Parag V. Worldwide stroke incidence and early case fatality reported in 56 population-based studies: a systematic review. *Lancet Neurol.* 2009;8(4):355–369. Available from: https://doi.org/10.1016/S1474-4422(09)70025-0.
20. Rothwell PM, Coull AJ, Giles MF, et al. Change in stroke incidence, mortality, case-fatality, severity, and risk factors in Oxfordshire, UK from 1981 to 2004 (Oxford Vascular Study). *Lancet.* 2004;363(9425):1925–1933. Available from: https://doi.org/10.1016/S0140-6736(04)16405-2.
21. Sarti C, Rastenyte D, Cepaitis Z, Tuomilehto J. International trends in mortality from stroke, 1968 to 1994. *Stroke.* 2000;31(7):1588–1601.
22. Elneihoum AM, Goransson M, Falke P, Janzon L. Three-year survival and recurrence after stroke in Malmo, Sweden: an analysis of stroke registry data. *Stroke.* 1998;29(10):2114–2117.
23. Leoo T, Lindgren A, Petersson J, von Arbin M. Risk factors and treatment at recurrent stroke onset: results from the Recurrent Stroke Quality and Epidemiology (RESQUE) study. *Cerebrovasc Dis.* 2008;25(3):254–260. Available from: https://doi.org/10.1159/000113864.
24. Goldstein LB, Amarenco P, Szarek M, et al. Hemorrhagic stroke in the Stroke Prevention by Aggressive Reduction in Cholesterol Levels study. *Neurology.* 2008;70(24 Pt 2):2364–2370. Available from: https://doi.org/10.1212/01.wnl.0000296277.63350.77.
25. Bassetti CL, Milanova M, Gugger M. Sleep-disordered breathing and acute ischemic stroke: diagnosis, risk factors, treatment, evolution, and long-term clinical outcome. *Stroke.* 2006;37(4):967–972. Available from: https://doi.org/10.1161/01.STR.0000208215.49243.c3.

REFERENCES

26. Hermann DM, Bassetti CL. Sleep-related breathing and sleep—wake disturbances in ischemic stroke. *Neurology.* 2009;73(16):1313—1322. Available from: https://doi.org/10.1212/WNL.0b013e3181bd137c.
27. Hermann DM, Bassetti CL. Role of sleep-disordered breathing and sleep—wake disturbances for stroke and stroke recovery. *Neurology.* 2016;87(13):1407—1416. Available from: https://doi.org/10.1212/WNL.0000000000003037.
28. Leng Y, Cappuccio FP, Wainwright NW, et al. Sleep duration and risk of fatal and nonfatal stroke: a prospective study and meta-analysis. *Neurology.* 2015;84(11):1072—1079. Available from: https://doi.org/10.1212/WNL.0000000000001371.
29. Qureshi AI, Giles WH, Croft JB, Bliwise DL. Habitual sleep patterns and risk for stroke and coronary heart disease: a 10-year follow-up from NHANES I. *Neurology.* 1997;48(4):904—911.
30. Roth T. Insomnia: definition, prevalence, etiology, and consequences. *J Clin Sleep Med.* 2007;3(5 Suppl):S7—S10.
31. Groeger JA, Zijlstra FR, Dijk DJ. Sleep quantity, sleep difficulties and their perceived consequences in a representative sample of some 2000 British adults. *J Sleep Res.* 2004;13(4):359—371. Available from: https://doi.org/10.1111/j.1365-2869.2004.00418.x.
32. Parish JM. Sleep-related problems in common medical conditions. *Chest.* 2009;135(2):563—572. Available from: https://doi.org/10.1378/chest.08-0934.
33. Bassetti CL. Sleep apnea and stroke. *Eur J Neurol.* 2005;12. 5-5.
34. Bassetti CL, Hermann DM. Sleep and stroke. *Handb Clin Neurol.* 2011;99:1051—1072. Available from: https://doi.org/10.1016/B978-0-444-52007-4.00021-7.
35. Kaneko Y, Hajek VE, Zivanovic V, Raboud J, Bradley TD. Relationship of sleep apnea to functional capacity and length of hospitalization following stroke. *Sleep.* 2003;26(3):293—297.
36. Spriggs DA, French JM, Murdy JM, Curless RH, Bates D, James OF. Snoring increases the risk of stroke and adversely affects prognosis. *Q J Med.* 1992;83(303):555—562.
37. Palomaki H, Partinen M, Juvela S, Kaste M. Snoring as a risk factor for sleep-related brain infarction. *Stroke.* 1989;20(10):1311—1315.
38. Abad VC, Guilleminault C. Diagnosis and treatment of sleep disorders: a brief review for clinicians. *Dialogues Clin Neurosci.* 2003;5(4):371—388.
39. Panossian LA, Avidan AY. Review of sleep disorders. *Med Clin North Am.* 2009;93(2):407—425. Available from: https://doi.org/10.1016/j.mcna.2008.09.001. ix.
40. Tang WK, Hermann DM, Chen YK, et al. Brainstem infarcts predict REM sleep behavior disorder in acute ischemic stroke. *BMC Neurol.* 2014;14:88. Available from: https://doi.org/10.1186/1471-2377-14-88.
41. Medeiros CA, de Bruin PF, Paiva TR, Coutinho WM, Ponte RP, de Bruin VM. Clinical outcome after acute ischaemic stroke: the influence of restless legs syndrome. *Eur J Neurol.* 2011;18(1):144—149. Available from: https://doi.org/10.1111/j.1468-1331.2010.03099.x.
42. NSF. Insomnia. *What is Insomnia?* <https://sleepfoundation.org/insomnia/content/what-is-insomnia>; 2017.
43. Reimann D, Baglioni C, Bassetti C, Bjorvatn B, Dolenc Groseli L, et al. European guideline for the diagnosis and treatment of insomnia. *J Sleep Res.* 2017;26(6):675—700. Available from: https://doi.org/10.1111/jsr.12594.
44. Kryger MH, Roth T, Dement WC. *Principles and Practice of Sleep Medicine.* Elsevier Science, Health Science Division; 2005.
45. Baglioni C, Nissen C, Schweinoch A, et al. Polysomnographic characteristics of sleep in stroke: a systematic review and meta-analysis. *PLoS One.* 2016;11(3):e0148496. Available from: https://doi.org/10.1371/journal.pone.0148496.
46. Schuiling WJ, Rinkel GJ, Walchenbach RW, De Weerd A. Disorders of sleep and wake in patients after subarachnoid hemorrhage. *Stroke.* 2005;36(3):578—582.
47. Sterr A, Herron K, Dijk D, Ellis J. Time to wake up: sleep problems and daytime sleepiness in long-term stroke. *Brain Inj.* 2008;22(7-8):575—579. Available from: https://doi.org/10.1080/02699050802189727.
48. Berg A, Palomäki H, Lehtihalmes M, Lönnqvist J, Kaste M. Poststroke depression: an 18-month follow-up. *Stroke.* 2003;34(1):138—143.
49. Ingles JL, Eskes GA, Phillips SJ. Fatigue after stroke. *Arch Phys Med Rehabil.* 1999;80(2):173—178.
50. Vock J, Achermann P, Bischof M, et al. Evolution of sleep and sleep EEG after hemispheric stroke. *J Sleep Res.* 2002;11(4):331—338. Available from: https://doi.org/10.1046/j.1365-2869.2002.00316.x.

II. SLEEP DISORDERS IN SPECIFIC MEDICAL CONDITIONS

51. Choi-Kwon S, Han SW, Kwon SW, Kim JS. Poststroke fatigue: characteristics and related factors. *Cerebrovasc Dis.* 2005;19(2):84−90. Available from: https://doi.org/10.1159/000082784.

52. Hinkle JL, Becker KJ, Kim JS, et al. Poststroke fatigue: emerging evidence and approaches to management: a scientific statement for healthcare professionals from the American Heart Association. *Stroke.* 2017;48(7): E159−E170. Available from: https://doi.org/10.1161/STR.0000000000000132.

53. Herron K, Dijk D, Dean P, Seiss E, Sterr A. Quantitative electroencephalography and behavioural correlates of daytime sleepiness in chronic stroke. *Biomed Res Int.* 2014;2014:794086. Available from: https://doi.org/10.1155/2014/794086.

54. Brunner D, Dijk D, Borbély A. Repeated partial sleep deprivation progressively changes in EEG during sleep and wakefulness. *Sleep.* 1993;16(2):100−113. Available from: https://doi.org/10.1093/sleep/16.2.100.

55. Sterr A, Furlan L. A case to be made: theoretical and empirical arguments for the need to reconsider fatigue in post-stroke motor rehabilitation. *Neural Regen Res.* 2015;10(8):1195−1197. Available from: https://doi.org/10.4103/1673-5374.162689.

56. Boyd L, Hayward K, Ward N, Stinear C, Rosso C, et al. Biomarkers of stroke recovery: consensus-based core recommendations from the stroke recovery and rehabilitation roundtable. *Neurorehabil Neural Repair.* 2017;31 (10-11):864−876. Available from: https://doi.org/10.1177/1545968317732680.

57. Gauthier L, Taub E, Perkins C, Ortmann M, Mark V, Uswatte G. Remodeling the brain—plastic structural brain changes produced by different motor therapies after stroke. *Stroke.* 2008;39(5):1520−1525. Available from: https://doi.org/10.1161/STROKEAHA.107.502229.

58. Kim H, Kim M, Koo Y, Lee H, Shin M, et al. Histological and functional assessment of the efficacy of constraint-induced movement therapy in rats following neonatal hypoxic−ischemic brain injury. *Exp Ther Med.* 2017;13(6):2775−2782. Available from: https://doi.org/10.3892/etm.2017.4371.

59. Murayama T, Numata K, Kawakami T, Tosaka T, Oga M, et al. Changes in the brain activation balance motor-related areas after constraint-induced movement therapy; a longitudinal fMRI study. *Brain Inj.* 2011;25 (11):1047−1057. Available from: https://doi.org/10.3109/02699052.2011.607785.

60. Zhang J, Yu J, Bao Y, et al. Constraint-induced aphasia therapy in post-stroke aphasia rehabilitation: a systematic review and meta-analysis of randomized controlled trials. *PLoS One.* 2017;12(8):E0183349. Available from: https://doi.org/10.1371/journal.pone.0183349.

61. Landmann K, Piosczyk F, Baglioni S, et al. The reorganisation of memory during sleep. *Sleep Med Rev.* 2014;18 (6):531−541. Available from: https://doi.org/10.1016/j.smrv.2014.03.005.

62. Landmann N, Kuhn M, Maier J-G, et al. REM sleep and memory reorganization: potential relevance for psychiatry and psychotherapy. *Neurobiol Learn Mem.* 2015;122(1):28−40. Available from: https://doi.org/10.1016/j.nlm.2015.01.004.

63. Kuhn M, Wolf E, Maier JG, et al. Sleep recalibrates homeostatic and associative synaptic plasticity in the human cortex. *Nat Commun.* 2016;7:12455. Available from: https://doi.org/10.1038/ncomms12455.

64. Siengsukon CF, Boyd LA. Sleep enhances implicit motor skill learning in individuals poststroke. *Top Stroke Rehabil.* 2008;15(1):1−12. Available from: https://doi.org/10.1310/tsr1501-1.

65. Alhola P, Polo-Kantola P. Sleep deprivation: impact on cognitive performance. *Neuropsychiatr Dis Treat.* 2007;3 (5):553−567.

66. Freeman D, Sheaves B, Goodwin GM, et al. The effects of improving sleep on mental health (OASIS): a randomised controlled trial with mediation analysis. *Lancet Psychiatry.* 2017;4(10):749−758. Available from: https://doi.org/10.1016/S2215-0366(17)30328-0.

67. Ellis J, Perlis M, Bastien C, Gardani M, Espie C. The natural history of insomnia: acute insomnia and first-onset depression. *Sleep.* 2014;37(1):97−106. Available from: https://doi.org/10.5665/sleep.3316.

68. Ellis J, Barclay N. Cognitive behaviour therapy for insomnia: state of the science or a stated science? *Sleep Med.* 2014;15(8):849−850. Available from: https://doi.org/10.1016/j.sleep.2014.04.008.

69. Spielman AJ, Caruso LS, Glovinsky PB. A behavioral perspective on insomnia treatment. *Psychiatr Clin North Am.* 1987;10(4):541−553.

70. Bourne RS, Mills GH. Sleep disruption in critically ill patients—pharmacological considerations. *Anaesthesia.* 2004;59(4):374−384. Available from: https://doi.org/10.1111/j.1365-2044.2004.03664.x.

71. Ebajemito JK, Furlan L, Nissen C, Sterr A. Application of transcranial direct current stimulation in neurorehabilitation: the modulatory effect of sleep. *Front Neurol.* 2016;7:54. Available from: https://doi.org/10.3389/fneur.2016.00054.

REFERENCES

72. Michael KM, Allen JK, Macko RF. Fatigue after stroke: relationship to mobility, fitness, ambulatory activity, social support, and falls efficacy. *Rehabil Nurs.* 2006;31(5):210–217.

73. Morley W, Jackson K, Mead GE. Post-stroke fatigue: an important yet neglected symptom. *Age Ageing.* 2005;34(3):313. Available from: https://doi.org/10.1093/ageing/afi082.

74. Beldarrain MG, Astorgano AG, Gonzalez AB, Garcia-Monco JC. Sleep improves sequential motor learning and performance in patients with prefrontal lobe lesions. *Clin Neurol Neurosurg.* 2008;110(3):245–252. Available from: https://doi.org/10.1016/j.clineuro.2007.11.004.

75. Siengsukon CF, Boyd LA. Sleep to learn after stroke: implicit and explicit off-line motor learning. *Neurosci Lett.* 2009;451(1):1–5. Available from: https://doi.org/10.1016/j.neulet.2008.12.040.

76. Walker MP, Brakefield T, Seidman J, Morgan A, Hobson JA, Stickgold R. Sleep and the time course of motor skill learning. *Learn Mem.* 2003;10(4):275–284. Available from: https://doi.org/10.1101/lm.58503.

77. Uhde TW, Cortese BM, Vedeniapin A. Anxiety and sleep problems: emerging concepts and theoretical treatment implications. *Curr Psychiatry Rep.* 2009;11(4):269–276.

78. Akerstedt T, Gillberg M. Subjective and objective sleepiness in the active individual. *Int J Neurosci.* 1990;52(1-2):29–37. Available from: https://doi.org/10.3109/00207459008994241.

79. Abad V, Guilleminault C. Pharmacological treatment of sleep disorders and its relationship with neuroplasticity. *Sleep Neuronal Plast Brain Funct.* 2015;25:503–553. Available from: https://doi.org/10.1007/7854_2014_365.

80. Sloane PD, Figueiro M, Cohen L. Light as therapy for sleep disorders and depression in older adults. *Clin Geriatr.* 2008;16(3):25–31.

81. Chui H, Chan P, Chu H, Hsiao S, Liu D, et al. Effectiveness of light therapy in cognitively impaired persons: a metaanalysis of randomized controlled trials. *J Am Geriatr Soc.* 2017;65(10):2227–2234. Available from: https://doi.org/10.1111/jgs.14990.

82. Souman J, Tinga A, te Pas S, van Ee R, Vlaskamp B. Acute alerting effects of light: a systematic literature review. *Behav Brain Res.* 2018;337(1):228–239. Available from: https://doi.org/10.1016/j.bbr.2017.09.016.

83. Amara AW, Chahine LM, Videnovic A. A systematic review of the literature on disorders of sleep and wakefulness in Parkinson's disease from 2005 to 2015. *Sleep Med Rev.* 2017;35(1):33–50. Available from: https://doi.org/10.1016/j.smrv.2016.08.001.

84. Alkozei A, Smith R, Dailey N, Bajaj S, Killgore W. Acute exposure to blue wavelength light during memory consolidation improves verbal memory performance. *PLoS One.* 2017;12(9):e0184884. Available from: https://doi.org/10.1371/journal.pone.0184884.

85. Perlis M, Benson-Jungquist C, Smith M, Posner D. *Cognitive behavioural treatment of insomnia: a session-by-session guide.* New York: Springer New York; 2005.

86. NICE. Stroke rehabilitation in adults. *Clinical Guideline [CG162].* <https://www.nice.org.uk/guidance/cg162>; 2013.

CHAPTER

13

Human Immunodeficiency Virus/AIDS

Kenneth D. Phillips[1], Robin F. Harris[2]
and Lisa M. Haddad[1]

[1]College of Nursing, East Tennessee State University, Johnson City, TN, United States
[2]College of Nursing, University of Tennessee, Knoxville, TN, United States

OUTLINE

Introduction	294	*Daytime Sleepiness*	299
HIV Disease	294	Symptom Clusters in HIV/AIDS	299
Sleep Disturbances and HIV Disease	294	Nonpharmacological Treatment for Insomnia	300
Prevalence	294	*Cognitive Behavioral Therapy*	300
Consequences	295	*Sleep Hygiene*	300
Insomnia	296	*Relaxation Training*	301
Overall Prevalence	296	*Stimulus Control Therapy*	302
Insomnia and Stage of HIV Infection	296	*Sleep Restriction*	302
Hypersomnia	297	Pharmacological Treatment for Insomnia	302
Obstructive Sleep Apnea	297	Areas for Future Research	303
Correlates of Sleep Quality in HIV/AIDS	298	Conclusion	303
Fatigue	298	References	303
Depression	298		

Handbook of Sleep Disorders in Medical Conditions
DOI: https://doi.org/10.1016/B978-0-12-813014-8.00013-5

© 2019 Elsevier Inc. All rights reserved.

INTRODUCTION

Despite considerable progress in detection and treatment, HIV infection remains highly prevalent worldwide. HIV-infected individuals being treated early and adhering to their antiretroviral medication have a life expectancy almost equivalent to that of their healthy counterparts. However, HIV infection is associated with many psychosocial and physical ordeals, including sleep difficulties. This chapter will overview what is known about sleep impairments in HIV-seropositive patients.

HIV DISEASE

HIV/AIDS remains a global pandemic that has claimed more than 35 million lives throughout the world since this infectious disease first appeared. In 2016, approximately 36.7 million people were living worldwide with HIV infection and it is estimated that 1 million persons died because of this disease. In that same year, 1.8 million people received a new diagnosis that they were infected with HIV.[1] Even though highly specific diagnostic and screening tests and antiretroviral medications have been developed since 1981, only 54% of the adults and 43% of the children are currently receiving antiretroviral therapy worldwide. The African continent remains the most affected region of the world with 25.6 million people living with HIV disease in 2016. Between 2000 and 2016, new HIV infections fell by 39% from 3 million infections to 1.8 million infections and HIV-related deaths declined by one-third.[1]

The decline in deaths related to HIV infections can be attributed to greater and earlier preventative education, highly specific diagnostic and screening tests, and the development of potent combinations of highly active antiretroviral therapy (HAART).[1] A young adult who is diagnosed with HIV infection today and begins antiretroviral therapy before the number of CD4 + T helper cells drops below 500 cell/mm^3 is likely to live for several decades. Avoiding substance abuse and being treated for physical health issues such as coinfection with tuberculosis and hepatitis C virus and receiving care for mental health issues such as depression and schizophrenia and faithful adherence to antiretroviral medications may live to approximately the same age as others his or her own age. Nearly one-third of all people living with HIV/AIDS (PLWHA) in the United States are greater than 50 years of age.[2]

SLEEP DISTURBANCES AND HIV DISEASE

Prevalence

A metaanalysis was conducted in 2015 to estimate the pooled prevalence of sleep disturbances in PLWHA.[3] The authors found 27 that met their inclusion criteria which were that the articles were published in English, contained original data, study participants tested positive or self-reported HIV infection, were 18 years of age or older, and provided frequency data for sleep disturbances. The original research studies were conducted in North America, South America, Africa, Europe, and Asia giving a worldwide representation. In the first stage of analysis, use of an innovative method of statistical analysis (Wilson's score method), the researchers calculated the prevalence of self-reported sleep disturbances. In the second stage of the metaanalysis, a random effect model was performed to estimate the effects of covariates that were included in the original studies. The researchers who conducted the metaanalysis concluded that the overall prevalence of self-reported sleep disturbances was 58.0% (95% CI = 49.6–66.6).

Consequences

Sleep is an indicator of a person's overall health status and general well-being. Physically, psychologically, and emotionally healthy people sleep better than those who have physical, psychological, or emotional health problems.[4] Protecting the quality of sleep and finding ways to restore sleep quality are important to maintaining health and well-being. People who have frequent sleep disorders often experience decreased energy, mood disorders, and increased levels of stress. Poor sleep quality is associated with poorer working memory, daytime irritability, poorer concentration, decreased attention spans, poorer job performance, greater anxiety, more accidents, and relationship difficulties.[5,6]

In PLWHA, sleep disturbances have been found to be associated with daytime sleepiness, fatigue, depression, cognitive impairment, neurobehavioral dysfunctions, and reduced quality of life. PLWHA with sleep disorders are more likely to be poorly adherent to HAART.[7,8] Sleep disturbance and poor adherence may be associated with decline in virologic control and numerous changes to the immune system.[9,10] It has been hypothesized, and in many studies demonstrated, that there is a bidirectional relationship between sleep and immunity. Sleep quality has important roles in modulating the immune response. Sleep impairment is associated with impaired proliferation of lymphocytes such as CD4 + and CD8 + lymphocytes which have very important roles in the quantity of HIV + which are produced and the number of CD4 + cells which are destroyed by HIV infection.[11]

Sleep disturbances are conditions that alter sleep patterns of persons. Sleep disorders are classified as insomnia, hypersomnia, circadian rhythm sleep disorders, sleep apnea, narcolepsy and cataplexy, parasomnias, and sleep-related movement disorders.[12] It is not altogether clear what types of sleep problems PLWHA experience or if the sleep disturbance is related to the infection or to the medications and their side effects.[13] Poor sleep experienced by PLWHA has been associated with disease progression, medication therapy, side effects of medications, employment status, and lack of knowledge about behaviors that promote good sleep. The three sleep disturbances that have been most often studied in relationship to HIV disease are insomnia, hypersomnia, and obstructive sleep apnea (OSA).[14–16]

INSOMNIA

Overall Prevalence

Insomnia and fatigue are among the earliest and most distressing complaints of PLWHA.[17] Insomnia is a common complaint that is often reported before diagnosis of HIV infection and continues throughout the illness. Insomnia is defined as difficulty falling asleep (sleep-onset insomnia), difficulty staying asleep (sleep-maintenance insomnia), awakening too early (late insomnia), or unrefreshing sleep in combination with at least one daytime symptom such as insomnia, fatigue, daytime sleepiness, or irritability.[18] Insomnia is characterized by getting too little sleep or simply having poor quality sleep.[19–21] The incidence and prevalence of insomnia in HIV-infected persons remain higher in PLWHA than in the general population.[22] Yet, sleep disturbances are typically underreported, underdiagnosed, and undertreated in HIV disease.[22]

Lee et al.[13] recruited a convenience sample of 292 adults living with HIV. Participants wore a wrist actigraph for three nights to measure sleep and activity and completed the Pittsburgh Sleep Quality Index to report subjective sleep quality. Difficulty falling asleep was measured using the sleep latency score obtained by wrist actigraphy. Sleep latency greater than 30 minutes was seen in 34% of the sample. Wake after sleep onset (WASO) was used to measure sleep maintenance, which is a percentage of time spent awake after the first episode of sleep. WASO could range of 0% of time to 100% of time. The mean time spent awake after sleep onset as measured by wrist actigraphy was 20.9% ($\pm$14.8%), which indicates severe fragmentation of the sleep. Total sleep time ranged from less than 2 hours to more than 11 hours and sleep efficiency (ratio of total sleep time on total time spent in bed) ranged from 18% to 98% in that study. A significant finding of that study was that participants who slept less than 6 hours had lower CD4 + T-cell counts and higher viral loads.[13]

Insomnia and Stage of HIV Infection

Acute HIV infection develops within 2–4 weeks after a person becomes infected with HIV. The symptoms of this stage of infection are flu-like and may include fever, chills, headache, rash, lymphadenopathy, and musculoskeletal pain. It is during the acute stage that HIV multiplies rapidly and attacks the CD4 cells which are normally the immune cells which fight off viral infections, like HIV. Due to the highly increased number of HIV-infected cells during this phase, the infected person is very contagious which increases the risk of HIV transmission. This stage of HIV infection often goes undiagnosed because the symptoms of this stage mimic the symptoms of influenza, mononucleosis, and other less life-threatening illnesses.[23]

Chronic HIV infection is also known as the asymptomatic or clinical latency period of HIV infection. During this stage, the replication of HIV decreases drastically. The person with HIV infection may or may not experience any symptoms normally associated with HIV infection but can still transmit HIV infection to others. Insomnia in early stages of HIV infection has been shown to be variable across studies. This variation is due to inconsistency in research trials regarding methodology, participant's self-report as a measure of insomnia, and small sample sizes.[24]

AIDS is the most severe stage of HIV infection. The viral load rises rapidly during this phase and the CD4 cells decline drastically. A CD4 count of less than 200 cells/mm^3 indicates that the person has progressed from HIV infection to AIDS. It is during this phase that a person's immune system has become so compromised that opportunistic infections and opportunistic malignancies can begin to appear. Insomnia is common in advanced or late stage HIV/AIDS and has been reported to be more prevalent in those with cognitive impairment.[24]

HYPERSOMNIA

Hypersomnia refers to sleeping too much and is usually found in the most severe stage of HIV infection (AIDS). Adults, as a general rule, need 7–9 hours of sleep per night. Hypersomnia in PLWHA is associated with greater fatigue and decreased cognitive function.[6] An interesting finding regarding total sleep time, insomnia, and poor sleep quality in data collected from 640 Parisian PLWHA who were still ambulatory was that insomnia and short sleep were greater in persons with a CD4 count >500 cells/mm^3, and hypersomnia (long sleepers) was greater in those with a CD4 cell count <500 cells/mm^3. This indicates that long sleep times are associated with greater severity of HIV infection.[25]

OBSTRUCTIVE SLEEP APNEA

Obstructive sleep apnea (OSA) is coming to greater attention. OSA is caused by a repeated closure of the upper airway during sleep which results in frequent arousal and sleep fragmentation.[26] These repeated closures of the upper airway produce periods of absent breathing and reduced blood oxygen levels. During these periods of absent breathing, the person awakens, choking and gasping for air and upper airway patency is restored.[27] Adenotonsillar hypertrophy is an early manifestation of HIV disease and also a significant predictor of OSA.[28] Additional clinical symptoms include excessive daytime sleepiness and reduced quality of life. OSA is a known risk factor for cardiovascular diseases (i.e., myocardial infarction, cardiac arrhythmias, hypertension, and stroke), metabolic disease, and cognitive impairment.[29]

As research increases regarding OSA and HIV disease, it is becoming apparent that OSA is common in PLWHA and that previously recognized risk factors associated with OSA, such as obesity and advanced age, are not as useful in predicting OSA in PLWHA. It is also acknowledged that many cases of OSA are undiagnosed and untreated. Treating OSA in PLWHA may decrease fatigue, improve quality of life, and lengthen life expectancy.[30] Diagnostic assessment for OSA should be performed particularly in patients with adenotonsillar hypertrophy and increased snoring.[28]

CORRELATES OF SLEEP QUALITY IN HIV/AIDS

Sleep disturbances are common in HIV/AIDS and contribute to use of sleep medications, poor work performance, depression, fatigue, anxiety, increased medical problems, impaired cognitive function, and decreased quality of life.[31] Correlates of sleep disturbance appear to be enmeshed due to overlap conceptually and in terms of measurement instruments.[24,32]

Fatigue

Fatigue is defined as lack of energy, sleepiness, exhaustion, and inability to get enough rest.[33] Fatigue occurs when a person's physical and psychological resources do not meet the demands for those resources. In HIV/AIDS, fatigue is very common and has debilitating effects.[34,35] In fact, the most frequent and troublesome complaint in HIV disease is fatigue,[36,37] with rates of 60%–88% of PLWHA experiencing symptoms of fatigue.[38–43] Justice et al.[44] reported that the most common functional status in HIV disease is fatigue and that greater fatigue is linked to lower survival rates.

Fatigue is a complex, multidimensional process and therefore is difficult to treat.[26] Some causes of fatigue that have been identified include lack of rest, poor exercise habits, improper or inadequate diet, psychological distress (depression and anxiety), use of recreational drugs, side effects of antiretroviral medications, sleep disturbances, fever, anemia, and low CD4 cell counts.[33,37,45] The relationship between fatigue and higher levels of psychological stress has been well documented.[37,45,46] Some scientists have speculated that HIV-related fatigue is a result of physiological factors such as low CD4 count and high HIV viral load,[39,47] while others attribute it to psychosocial variables such as depression.[38]

Fatigue is associated with work problems, decrease in social activities, and decrease in daily activities of living.[44] In HIV disease, fatigue is associated with poor antiretroviral adherence and immunity/virological failure.[48,49] HIV persons who are highly fatigued from sleep disturbances are more vulnerable to additional negative consequences of fatigue secondary to the increase in psychosocial stress they experience.[50]

Sleep quality is significantly related to fatigue. Robbins et al.[51] and Phillips et al.[41] reported that sleep quality predicted fatigue in PLWHA. It is difficult to determine from the literature if sleep quality causes fatigue or if the fatigue causes poor sleep quality. Byun et al.[6] looked at sleep, fatigue, and problems with cognitive function in adults living with HIV, finding that increase in reported fatigue was significantly associated with perception of cognitive problems in adults with HIV. Wilson et al.[52] studied the frequency and burden of HIV-associated symptoms experienced by HIV-infected individuals. The most frequent and burdensome symptoms were muscle aches/joint pain, fatigue, and poor sleep.

Depression

Depression may accelerate the progression of HIV disease even in those who are receiving optimum treatment for this condition. PLWHA report a higher level of depression (20%−30%) than the general population.[53−55] Multiple reasons are reported for this higher prevalence including high viral load, direct effects of the virus on the immune system, symptom burden, emotional reaction to the diagnosis, and low social acceptance.[53,56] Depression is linked to decreased quality of life, reduced medication adherence,[57] increased substance use, poor treatment outcomes, rapid decline in CD4 count, and increased mortality.[58−60] Aouizerat et al.[61] reported that depressive symptoms were strongly associated with sleep disturbances, fatigue, and lack of energy. The relationship between fatigue, depression, and sleep problems is well documented in HIV disease.[62]

Due to the increase prevalence of psychological symptoms in HIV/AIDS, there is an increased need for mental health services to treat these individuals; however, the supply of these specialty providers has not kept up with the demand.[55] For depression per se, data collected from a recent study suggest that only 18% of PLWHA with major depressive disorder are receiving any treatment, only 7% are receiving adequate care, and only 5% have achieved remission. This finding is very troubling in view of the fact that depression is extremely harmful whether a person is HIV-infected or not, but major depression disorder may affect a person's decision to remain in medical care and adherence to antiretroviral therapy.[63] Depression's strong relationship to sleep disturbances raises the value of research regarding effective treatments for sleep disturbances in reducing depression and improving life expectancy in HIV disease.

Daytime Sleepiness

Fatigue and excessive daytime sleepiness are the most common symptoms of OSA.[64] Daytime sleepiness and fatigue are caused by fragmented sleep. Lee et al.[13] report that daytime sleepiness and fatigue are common and can lead to difficulty concentrating, poor cognitive functioning, depression, and inability to be productive during the day.[13]

SYMPTOM CLUSTERS IN HIV/AIDS

Symptom management has been of great interest in caring for people with HIV/AIDS. In the past, there has been great attention focused on individual symptoms that are experienced by PLWHA. However, interest of symptom management has begun transitioning from single to multiple symptoms, because PLWHA commonly experience multiple concurrent symptoms and that the individual symptoms reported by PLWHA are additive in their impact on other concurrent symptoms. Dodd, Miaskowski, and Paul first introduced the idea of symptom clusters and studied their collective effects on functional status and quality of life in patients with cancer. Symptom clusters were defined by these researchers as "three or more concurrent symptoms that are related to each other but are not required to have the same etiology" (see Ref. [65], p. 468).

Four studies have examined possible symptom clusters in PLWHA.[42,66−68] To the best of our knowledge, no one has examined the possibility of sleep, fatigue, and depression as a symptom cluster in HIV although these individual symptoms frequently occur concurrently. As reported in the correlates of sleep disturbances presented above, there is moderate-to-strong correlation of these three symptoms in a majority of studies of PLWHA. It is altogether possible that these three and other symptoms are part of a symptom cluster. In studying symptom clusters, it is important to consider a wide range of variables that might contribute to these symptoms. These factors would include physiological, psychological, environmental, and social variables. Appropriate analytical approaches, such as cluster analysis and factor analysis, can be utilized to study clusters of symptoms in PLWHA.

NONPHARMACOLOGICAL TREATMENT FOR INSOMNIA

Various psychological interventions have been shown to be useful for treatment of insomnia as either first-line treatment or in conjunction with pharmacological therapies.[69,70] Cognitive behavioral therapy (CBT), sleep hygiene, relaxation training, progressive muscle relaxation, stimulus control therapy, and sleep restriction are psychological interventions used in treatment of insomnia and will be further discussed below.[69,70] However, the literature is sparse on the treatment of insomnia in PLWHA. Studies conducted in HIV-infected patients are needed to assess whether these treatments, of which efficacy is established in healthy individuals and patients with other medical conditions (e.g., cancer), are also useful in the context of HIV disease.

Cognitive Behavioral Therapy

CBT is a first-line treatment for insomnia that has been shown to be safe and effective.[71] CBT for insomnia is a multifaceted treatment typically composed of stimulus control, sleep restriction, cognitive restructuring, and sleep hygiene. Both CBT and cognitive behavioral therapy with relaxation are used as first-line treatments for insomnia. Modalities for CBT include face-to-face, telephone, or web-based therapy either individually or in group therapy sessions. Multiple studies have shown benefit with these treatments for measures of sleep including sleep onset latency and sleep quality, number and duration of awakenings, and total time of sleep. CBT has been shown to be effective in adults including younger, middle, and older adults.[72–76] Recently, a pilot study of 22 PLWHA supported the feasibility of a brief behavioral treatment for insomnia (sleep restriction and stimulus control) in this population and provided preliminary evidence of its efficacy.[77] These promising findings need to be replicated in large randomized-controlled trials.

Sleep Hygiene

Sleep hygiene behaviors facilitate effective sleep. This refers to those behaviors that surround activities prior to beginning sleep and activities to adjust environmental factors to promote sleep.[78,79] Sleep-related behaviors include a healthy diet, controlling caffeine intake, consistent bedtime, avoidance of strenuous activities, alcohol, and tobacco near bedtime, and limiting daytime napping. Environmental factors include maintaining temperature control in the sleeping room and avoiding use of electronic devices preceding bedtime.[80] In a survey of 328 working individuals, researchers found that participants who did not adhere to sleep hygiene principles prior to sleep were less engaged in work-related tasks while at work.[81]

Caffeine reduction to facilitate sleep has been studied in PLWHA in a randomized clinical trial.[14] The experimental group received instruction to gradually decrease caffeine intake leading to avoidance of caffeine for 30 days. The experimental group had 90% reduction of caffeine compared to 6% reduction in the control group. There was no difference in sleep quality reported from pretest to posttest between groups. When sleep quality was controlled for health status, researchers noted a 35% improvement in sleep quality in the experimental group.[14] Another randomized-controlled trial of 40 PLWHA evaluated the System CHANGE-HIV, which included a sleep hygiene component and behavioral modification, compared to control. In this study, Webel et al.[82] found that the intervention group had increased sleep duration (+10 min/night), improved sleep efficiency (+2.3%), and less sleep fragmentation than the control group (usual care) at posttreatment. Further research with larger samples and more consistency in sleep hygiene education principles is warranted to evaluate the efficacy of sleep hygiene education in PLWHA.[31,83] Sleep hygiene education is used commonly in clinical settings. Research to inform clinicians about best practices for sleep hygiene principles and patient education is needed.

Relaxation Training

Relaxation training to include progressive muscle relaxation and techniques to control negative thoughts at bedtime has been used to facilitate sleep.

Progressive muscle relaxation. Progressive muscle relaxation is a technique that has been shown to be effective in the treatment of chronic insomnia.[31] This technique involves alternate brief contraction and relaxation of different muscle groups. The individual becomes familiar with the sensations associated with muscle contraction and muscle relaxation. With continued practice over time, the individual can sense the difference between tensed muscle and relaxed muscle without having to consciously contract muscle groups to effectively manage chronic insomnia.[84,85] Similarly, autogenic training using autosuggestion facilitates muscle relaxation and recognition of relaxing respirations and slowing heart rate to promote sleep.[86,87]

Guided imagery and meditation. The Association of Sleep Medicine recommends the use of guided imagery to facilitate sleep.[31,88,89] In guided imagery technique, the individual imagines a safe and comfortable environment and is guided to both visualize and feel muscle relaxation. Meditation techniques can also be used to facilitate sleep. These techniques incorporate the use of words or movement over a set time period to promote relaxation,[90] typically 10–20 minutes, and involve refocusing the mind on the meditation if other thoughts occur during this time period.[31,89,91]

Stimulus Control Therapy

Stimulus control therapy has been shown to be an effective intervention for insomnia. This therapeutic technique has at its premise of the principles of classical conditioning. The goal of this strategy for individuals with chronic insomnia is to associate the bedroom and bed only with sleep and positive thoughts or emotions about sleep. Stimulus control therapy includes using the bedroom only for sleep or sexual activity, establishing set sleep and wake times avoiding daytime naps and avoid staying in bed when awake. The individual is instructed to move to a different room other than the bedroom for periods of wakefulness or when trying to relax to become sleepy. Stimulus control therapy can be beneficial for individuals with chronic insomnia but may take several consecutive weeks of therapy to be successful for the management of chronic insomnia.[31,92,93]

Sleep Restriction

Sleep restriction therapy has been shown to improve sleep with longer periods of sleep at a time and fewer awakenings during sleep.[94–96] This therapy requires the use of a sleep diary to document actual time slept. It is based on the natural sleep drive that the longer time an individual is awake, the sleep drive increases resulting in longer and higher quality sleep. At the beginning of therapy, the individual spends time in bed equal only to the amount of time recorded in the diary. As time progresses, the amount of time spent in bed increases or decreases based on the actual sleep time until the individual reaches the time of actual sleep. Individuals can experience longer and deeper sleep with this approach.[94–96]

PHARMACOLOGICAL TREATMENT FOR INSOMNIA

Treatment for sleep disorders in PLWHA should be initiated with nonpharmacological interventions. There is significant concern about drug interactions between antiretroviral medications and psychotropic medications due to similar metabolic pathways.[97] When sleep disturbances become so troublesome and nonpharmacological treatment are not available or are not effective, pharmacological treatments become necessary. Melatonin, a nonprescription medication, can be used before starting prescription medications. Four prescription medications (doxepin, mirtazapine, oxazepam, and zaleplon) are considered to be first-line medications for treating chronic insomnia in HIV/AIDS, but more evidence from randomized-controlled trials is needed in HIV/AIDS.[98] Ramelton, a melatonin agonist, may be used to improve sleep in HIV disease, but this has not been sufficiently tested. Coadministration of ramelton and fluvoxamine has been shown to increase ramelton levels in the serum from by 100 to 200-fold. Melatonin receptors are present on lymphocytes and may result in increases in the proinflammatory cytokine interleukin-2 which may help suppress HIV replication. This report was published in 2006 and clinical trials in HIV have not followed.[99] No randomized controlled trials with a placebo arm have been conducted to assess the efficacy and safety of hypnotic medications specifically in PLWHA. The above-described recommendations are based on reports from the literature. Selection of sedatives takes into consideration metabolism of antiretroviral medications and hepatic and renal failure.

AREAS FOR FUTURE RESEARCH

Randomized-controlled clinical trials to assess the efficacy and applicability of both non-pharmacological and pharmacological interventions for sleep disorders in PLWHA are needed in the future. In particular, clinical trials evaluating the usefulness of CBT are warranted. With regards to the pharmacological management, doxepin, mirtazapine, oxazepam, and zaleplon should be studied and compared with placebos. In addition to these sedatives, others should be investigated in separate studies to make the studies more generalizable to PLWHA who are dealing with sleep disturbances. In these studies, immune variables should be tested to determine the effects of the various sedatives on immunity.

CONCLUSION

Advances in antiretroviral therapy for treatment of HIV have contributed to increased life expectancy in PLWHA. Sleep disorders are common in PLWHA and are present in varying degrees in early to advanced stage HIV. Major types of sleep disorders in HIV include insomnia, hypersomnia, and OSAs which have been discussed in this chapter. Many of the symptoms of HIV infection overlap with symptoms of sleep disturbances that are experienced in HIV disease. The treatments for sleep disturbances are psychological, nonpharmacological, and pharmacological. Treatment of sleep disorders in this patient population can improve immune function, quality of life, and functional capacity and prevent conditions associated with chronic alterations in sleep patterns in PLWHA.

KEY PRACTICE POINTS

- Attention to sleep quality is important to optimize immune function in persons with HIV/AIDS.
- Sleep disturbances should be assessed and treated in HIV/AIDS as they are common and contribute to depression, fatigue, poor work performance, impaired cognition, increased medical problems, and decreased quality of life.
- The most common sleep disorders in HIV/AIDS are insomnia, hypersomnia, and OSA.
- Treatment of sleep disorders in this population can improve immune function, quality of life, and functional capacity.
- Both nonpharmacological (CBT, sleep hygiene education, relaxation, stimulus control, sleep restriction) and pharmacological treatments (e.g., Melatonin, Zaleplon) can be considered for insomnia starting with the least invasive approaches.

References

1. WHO. HIV/AIDS; 2017. <http://www.who.int/mediacentre/factsheets/fs360/en/>.
2. CDC. HIV Among People Aged 50 and Over; 2018. <https://www.cdc.gov/hiv/group/age/olderamericans/index.html> Accessed 06.09.18.

3. Wu J, Wu H, Lu C, Guo L, Li P. Self-reported sleep disturbances in HIV-infected people: a meta-analysis of prevalence and moderators. *Sleep Med.* 2015;16(8):901–907.
4. Lee KA, Gay C, Byun E, Lerdal A, Pullinger CR, Aouizerat BE. Circadian regulation gene polymorphisms are associated with sleep disruption and duration, and circadian phase and rhythm in adults with HIV. *Chronobiol Int.* 2015;32(9):1278–1293.
5. Ances BM, Ellis RJ. Dementia and neurocognitive disorders due to HIV-1 infection. *Semin Neurol.* 2007;27(1):86–92.
6. Byun E, Gay CL, Lee KA. Sleep, fatigue, and problems with cognitive function in adults living with HIV. *J Assoc Nurses AIDS Care.* 2016;27(1):5–16.
7. Omonuwa TS, Goforth HW, Preud'homme X, Krystal AD. The pharmacologic management of insomnia in patients with HIV. *J Clin Sleep Med.* 2009;5(3):251–262.
8. Phillips KD, Moneyham L, Murdaugh C, et al. Sleep disturbance and depression as barriers to adherence. *Clin Nurs Res.* 2005;14(3):273–293.
9. Irwin MR, Cole JC, Nicassio PM. Comparative meta-analysis of behavioral interventions for insomnia and their efficacy in middle-aged adults and in older adults 55 + years of age. *Health Psychol.* 2006;25(1):3–14.
10. Vgontzas AN, Zoumakis E, Bixler EO, et al. Adverse effects of modest sleep restriction on sleepiness, performance, and inflammatory cytokines. *J Clin Endocrinol Metab.* 2004;89(5):2119–2126.
11. Ibarra-Coronado EG, Pantaleón-Martínez AM, Velazquéz-Moctezuma J, et al. The bidirectional relationship between sleep and immunity against infections. *J Immunol Res.* 2015;2015.
12. AASM. *International Classification of Sleep Disorders*. 3rd ed. Darien, IL: American Academy of Sleep Medicine; 2014. <https://learn.aasm.org/Public/Catalog/Details.aspx?id = %2FgqQVDMQIT%2FEDy86PWgqgQ%3D%3D&returnurl = %2fUsers%2fUserOnlineCourse.aspx%3fLearningActivityID%3d%252fgqQVDMQIT%252fEDy86PWgqgQ%253d%253d> Accessed 07.03.18.
13. Lee KA, Gay C, Portillo CJ, et al. Types of sleep problems in adults living with HIV/AIDS. *J Clin Sleep Med.* 2012;8(1):67–75.
14. Dreher HM. The effect of caffeine reduction on sleep quality and well-being in persons with HIV. *J Psychosom Res.* 2003;54(3):191–198.
15. Koppel BS, Bharel C. Use of amitriptyline to offset sleep disturbances caused by efavirenz. *AIDS Patient Care STDS.* 2005;19(7):419–420.
16. Nunez M, Gonzalez de Requena D, Gallego L, Jimenez-Nacher I, Gonzalez-Lahoz J, Soriano V. Higher efavirenz plasma levels correlate with development of insomnia. *J Acquir Immune Defic Syndr.* 2001;28(4):399–400.
17. Lee KA, Portillo CJ, Miramontes H. The influence of sleep and activity patterns on fatigue in women with HIV/AIDS. *J Assoc Nurses AIDS Care.* 2001;12(suppl):19–27.
18. Rosenberg RS, Van Hout S. The American Academy of Sleep Medicine inter-scorer reliability program: respiratory events. *J Clin Sleep Med.* 2014;10(4):447–454.
19. Gamaldo CE, Spira AP, Hock RS, et al. Sleep, function and HIV: a multi-method assessment. *AIDS Behav.* 2013;17(8):2808–2815.
20. Phillips KD, Sowell RL, Boyd M, Dudgeon WD, Hand GA, Mind-Body Research G. Sleep quality and health-related quality of life in HIV-infected African-American women of childbearing age. *Qual Life Res.* 2005;14(4):959–970.
21. Taibi DM. Sleep disturbances in persons living with HIV. *J Assoc Nurses AIDS Care.* 2013;24(1).
22. Rubinstein ML, Selwyn PA. High prevalence of insomnia in an outpatient population with HIV infection. *J Acquir Immune Defic Syndr Hum Retrovirol.* 1998;19(3):260–265.
23. Cohen MS, Gay CL, Busch MP, Hecht FM. The detection of acute HIV infection. *J Infect Dis.* 2010;202(suppl 2(S2)):S270–S277.
24. Reid S, Dwyer J. Insomnia in HIV infection: a systematic review of prevalence, correlates, and management. *Psychosom Med.* 2005;67(2):260–269.
25. Faraut B, Malmartel A, Ghosn J, et al. Sleep disturbance and total sleep time in persons living with HIV: a cross-sectional study. *AIDS Behav.* 2018;1–11.
26. Goswami U, Baker JV, Wang Q, Khalil W, Kunisaki KM. Sleep apnea symptoms as a predictor of fatigue in an urban HIV clinic. *AIDS Patient Care STDS.* 2015;29(11):591–596.
27. Kunisaki KM, Akgun KM, Fiellin DA, et al. Prevalence and correlates of obstructive sleep apnoea among patients with and without HIV infection. *HIV Med.* 2015;16(2):105–113.

REFERENCES

305

28. Epstein LJ, Strollo PJ, Donegan RB, Delmar J, Hendrix C, Westbrook PR. Obstructive sleep apnea in patients with human immunodeficiency virus (HIV) disease. *Sleep.* 1995;18(5):368–376.

29. Jaffe LM, Kjekshus J, Gottlieb SS. Importance and management of chronic sleep apnoea in cardiology. *Eur Heart J.* 2013;34(11):809–815.

30. Owens RL, Hicks CB. A wake-up call for human immunodeficiency virus (HIV) providers: obstructive sleep apnea in people living with HIV. *Clin Infect Dis.* 2018;67(3):472–476.

31. Morgenthaler T, Kramer M, Alessi C, et al. Practice parameters for the psychological and behavioral treatment of insomnia: an update. An American Academy of Sleep Medicine report. *Sleep.* 2006;29(11):1415–1419.

32. Nokes KM, Kendrew J. Correlates of sleep quality in persons with HIV disease. *J Assoc Nurses AIDS Care.* 2001;12(1):17–22.

33. Adinolfi A. Assessment and treatment of HIV-related fatigue. *J Assoc Nurses AIDS Care.* 2001;12(suppl):29–34. quiz35-38.

34. Breitbart W, McDonald MV, Rosenfeld B, Monkman ND, Passik S. Fatigue in ambulatory AIDS patients. *J Pain Symptom Manage.* 1998;15(3):159–167.

35. Heaton RK, Clifford DB, Franklin DR, et al. HIV-associated neurocognitive disorders persist in the era of potent antiretroviral therapy. *Neurology.* 2010;75(23):2087–2096.

36. Barroso J, Pence BW, Salahuddin N, Harmon JL, Leserman J. Physiological correlates of HIV-related fatigue. *Clin Nurs Res.* 2008;17(1):5–19.

37. Salahuddin N, Barroso J, Leserman J, Harmon JL, Pence BW. Daytime sleepiness, nighttime sleep quality, stressful life events, and HIV-related fatigue. *J Assoc Nurses AIDS Care.* 2009;20(1):6–13.

38. Barroso J, Carlson JR, Meynell J. Physiological and psychological markers associated with HIV-related fatigue. *Clin Nurs Res.* 2003;12(1):49–68.

39. Henderson M, Safa F, Easterbrook P, Hotopf M. Fatigue among HIV-infected patients in the era of highly active antiretroviral therapy. *HIV Med.* 2005;6(5):347–352.

40. Millikin CP, Rourke SB, Halman MH, Power C. Fatigue in HIV/AIDS is associated with depression and subjective neurocognitive complaints but not neuropsychological functioning. *J Clin Exp Neuropsychol.* 2003;25(2):201–215.

41. Phillips KD, Sowell RL, Rojas M, Tavakoli A, Fulk LJ, Hand GA. Physiological and psychological correlates of fatigue in HIV disease. *Biol Res Nurs.* 2004;6(1):59–74.

42. Voss JG. Predictors and correlates of fatigue in HIV/AIDS. *J Pain Symptom Manage.* 2005;29(2):173–184.

43. Jong E, Oudhoff LA, Epskamp C, et al. Predictors and treatment strategies of HIV-related fatigue in the combined antiretroviral therapy era. *AIDS.* 2010;24(10):1387–1405.

44. Justice AC, Rabeneck L, Hays RD, Wu AW, Bozzette SA. Sensitivity, specificity, reliability, and clinical validity of provider-reported symptoms: a comparison with self-reported symptoms. Outcomes Committee of the AIDS Clinical Trials Group. *J Acquir Immune Defic Syndr.* 1999;21(2):126–133.

45. Leserman J, Barroso J, Pence BW, Salahuddin N, Harmon JL. Trauma, stressful life events and depression predict HIV-related fatigue. *AIDS Care.* 2008;20(10):1258–1265.

46. Pence BW, Barroso J, Leserman J, Harmon JL, Salahuddin N. Measuring fatigue in people living with HIV/AIDS: psychometric characteristics of the HIV-related fatigue scale. *AIDS Care.* 2008;20(7):829–837.

47. Simmonds MJ, Novy D, Sandoval R. The differential influence of pain and fatigue on physical performance and health status in ambulatory patients with human immunodeficiency virus. *Clin J Pain.* 2005;21(3):200–206.

48. Al-Dakkak I, Patel S, McCann E, Gadkari A, Prajapati G, Maiese EM. The impact of specific HIV treatment-related adverse events on adherence to antiretroviral therapy: a systematic review and meta-analysis. *AIDS Care.* 2013;25(4):400–414.

49. Marconi VC, Wu B, Hampton J, et al. Early warning indicators for first-line virologic failure independent of adherence measures in a South African urban clinic. *AIDS Patient Care and STDs.* 2013;27(12):657–668.

50. Marion I, Antoni M, Pereira D, et al. Distress, sleep difficulty, and fatigue in women co-infected with HIV and HPV. *Behav Sleep Med.* 2009;7(3):180–193.

51. Robbins JL, Phillips KD, Dudgeon WD, Hand GA. Physiological and psychological correlates of sleep in HIV infection. *Clin Nurs Res.* 2004;13(1):33–52.

52. Wilson NL, Azuero A, Vance DE, et al. Identifying symptom patterns in people living with HIV disease. *J Assoc Nurses AIDS Care.* 2016;27(2):121–132.

II. SLEEP DISORDERS IN SPECIFIC MEDICAL CONDITIONS

53. Bing EG, Burnam MA, Longshore D, et al. Psychiatric disorders and drug use among human immunodeficiency virus-infected adults in the United States. *Arch Gen Psychiatry.* 2001;58(8):721–728.
54. Ciesla JA, Roberts JE. Meta-analysis of the relationship between HIV infection and risk for depressive disorders. *Am J Psychiatry.* 2001;158(5):725–730.
55. Sowa NA, Bengtson A, Gaynes BN, Pence BW. Predictors of depression recovery in HIV-infected individuals managed through measurement-based care in infectious disease clinics. *J Affect Disord.* 2016;192:153–161.
56. Atkinson JH, Heaton RK, Patterson TL, et al. Two-year prospective study of major depressive disorder in HIV-infected men. *J Affect Disord.* 2008;108(3):225–234.
57. Dalmida SG, Holstad MM, Fox R, Delaney AM. Depressive symptoms and fatigue as mediators of relationship between poor sleep factors and medication adherence in HIV-positive women. *J Res Nurs.* 2015;20(6):499–514.
58. Ickovics JR, Hamburger ME, Vlahov D, et al. Mortality, CD4 cell count decline, and depressive symptoms among HIV-seropositive women: longitudinal analysis from the HIV Epidemiology Research Study. *JAMA.* 2001;285(11):1466–1474.
59. Kacanek D, Jacobson DL, Spiegelman D, Wanke C, Isaac R, Wilson IB. Incident depression symptoms are associated with poorer HAART adherence: a longitudinal analysis from the Nutrition for Healthy Living study. *J Acquir Immune Defic Syndr.* 2010;53(2):266–272.
60. Pence BW, Miller WC, Gaynes BN, Eron Jr. JJ. Psychiatric illness and virologic response in patients initiating highly active antiretroviral therapy. *J Acquir Immune Defic Syndr.* 2007;44(2):159–166.
61. Aouizerat BE, Gay CL, Lerdal A, Portillo CJ, Lee KA. Lack of energy: an important and distinct component of HIV-related fatigue and daytime function. *J Pain Symptom Manage.* 2013;45(2):191–201.
62. Lerdal A, Gay CL, Aouizerat BE, Portillo CJ, Lee KA. Patterns of morning and evening fatigue among adults with HIV/AIDS. *J Clin Nurs.* 2011;20(15–16):2204–2216.
63. Pence BW, O'Donnell JK, Gaynes BN. Falling through the cracks: the gaps between depression prevalence, diagnosis, treatment, and response in HIV care. *AIDS.* 2012;26(5):656–658.
64. Chervin RD. Sleepiness, fatigue, tiredness, and lack of energy in obstructive sleep apnea. *Chest.* 2000;118(2):372–379.
65. Dodd MJ, Miaskowski C, Paul SM. Symptom clusters and their effect on the functional status of patients with cancer. *Oncol Nurs Forum.* 2001;28(3):465–470.
66. Sousa KH, Tann SS, Kwok OM. Reconsidering the assessment of symptom status in HIV/AIDS care. *J Assoc Nurses AIDS Care.* 2006;17(2):36–46.
67. Cook PF, Sousa KH, Matthews EE, Meek PM, Kwong J. Patterns of change in symptom clusters with HIV disease progression. *J Pain Symptom Manage.* 2011;42(1):12–23.
68. Namisango E, Harding R, Katabira ET, et al. A novel symptom cluster analysis among ambulatory HIV/AIDS patients in Uganda. *AIDS Care.* 2015;27(8):954–963.
69. Narenda N, Walls K, Hill E. Insomnia in chronic low back pain: Nonpharmacological physiotherapy interventions. *Int J Ther Rehabil.* 2013;20(10):510–516.
70. Singh AN. Recent advances in pharmacotherapy of insomnia. *Int Med J.* 2016;23(6):602–604.
71. NIH. NIH state-of-the-science conference statement on manifestations and management of chronic insomnia in adults. *NIH Consens Sci Statements.* 2005;22(2):1–30.
72. Hauk L. Treatment of chronic insomnia in adults: ACP guideline. *Am Fam Physician.* 2017;95(10):669–670.
73. Morin CM, Benca R. Chronic insomnia. *Lancet.* 2012;379(9821):1129–1241.
74. Trauer JM, Qian MY, Doyle JS, Rajaratnam SM, Cunnington D. Cognitive behavioral therapy for chronic insomnia: a systematic review and meta-analysis. *Ann Intern Med.* 2015;163(3):191–204.
75. Morin CM, Culbert JP, Schwartz SM. Nonpharmacological interventions for insomnia: a meta-analysis of treatment efficacy. *Am J Psychiatry.* 1994;151(8):1172–1180.
76. Gooneratne NS, Vitiello MV. Sleep in older adults: normative changes, sleep disorders, and treatment options. *Clin Geriatr Med.* 2014;30(3):591–627.
77. Buchanan DT, McCurry SM, Eilers K, Applin S, Williams ET, Voss JG. Brief behavioral treatment for insomnia in persons living with HIV. *Behav Sleep Med.* 2018;16(3):244–258.
78. Homsey M, O'Connell K. Use and success of pharmacologic and nonpharmacologic strategies for sleep problems. *J Am Acad Nurse Pract.* 2012;24(10):612–623.

REFERENCES

79. Gigli GL, Valente M. Should the definition of "sleep hygiene" be antedated of a century? A historical note based on an old book by Paolo Mantegazza, rediscovered. To place in a new historical context the development of the concept of sleep hygiene. *Neurol Sci.* 2013;34(5):755–760.
80. AASM. *Sleep Hygiene: Behaviors that Help Promote Sound Sleep.* Westchester, IL: American Academy of Sleep Medicine; 2002.
81. Barber L, Grawitch MJ, Munz DC. Are better sleepers more engaged workers? A self-regulatory approach to sleep hygiene and work engagement. *Stress Health.* 2013;29(4):307–316.
82. Webel AR, Moore SM, Hanson JE, Patel SR, Schmotzer B, Salata RA. Improving sleep hygiene behavior in adults living with HIV/AIDS: a randomized control pilot study of the SystemCHANGE(TM)-HIV intervention. *Appl Nurs Res.* 2013;26(2):85–91.
83. Stepanski EJ, Wyatt JK. Use of sleep hygiene in the treatment of insomnia. *Sleep Med Rev.* 2003;7(3):215–225.
84. Jacobson E. *Progressive Relaxation.* 2nd ed. Oxford, England: University of Chicago Press; 1938.
85. Yilmaz CK, Kapucu S. The effect of progressive relaxation exercises on fatigue and sleep quality in individuals with COPD. *Holist Nurs Pract.* 2017;31:369–377.
86. Schultz JH, Luthe W. *Autogenic Training: A Psychophysiologic Approach in Psychotherapy.* vol. 91. Grune & Stratton; 1959.
87. Bowden A, Lorenc A, Robinson N. Autogenic training as a behavioural approach to insomnia: a prospective cohort study. *Prim Health Care Res Dev.* 2012;13(2):175–185.
88. Morin CM, Colecchi C, Stone J, Sood R, Brink D. Behavioral and pharmacological therapies for late-life insomnia: a randomized controlled trial. *JAMA.* 1999;281(11):991–999.
89. Schutte-Rodin S, Broch L, Buysse D, Dorsey C, Sateia M. Clinical guideline for the evaluation and management of chronic insomnia in adults. *J Clin Sleep Med.* 2008;4(5):487–504.
90. Benson H. *The Relaxation Response.* G. K. Hall; 1976.
91. Morin CM, Bootzin RR, Buysse DJ, Edinger JD, Espie CA, Lichstein KL. Psychological and behavioral treatment of insomnia: update of the recent evidence (1998–2004). *Sleep.* 2006;29(11):1398–1414.
92. Riedel BW, Lichstein KL, Peterson BA, Epperson MT, Means MK, Aguillard RN. A comparison of the efficacy of stimulus control for medicated and nonmedicated insomniacs. *Behav Modif.* 1998;22(1):3–28.
93. Espie CA, Lindsay WR, Brooks DN, Hood EM, Turvey T. A controlled comparative investigation of psychological treatments for chronic sleep-onset insomnia. *Behav Res Ther.* 1989;27(1):79–88.
94. Spielman AJ, Saskin P, Thorpy MJ. Treatment of chronic insomnia by restriction of time in bed. *Sleep.* 1987;10(1):45–56.
95. Friedman L, Bliwise DL, Yesavage JA, Salom SR. A preliminary study comparing sleep restriction and relaxation treatments for insomnia in older adults. *J Gerontol.* 1991;46(1):1–8.
96. Halson SL. Neurofeedback as a potential nonpharmacological treatment for insomnia. *Biofeedback.* 2017;45(1):19–20.
97. Phillips KD, Gunther M. *Sleep and HIV Disease.* Illustrated ed. New York: Springer; 2015.
98. Goodkin K, Kompella S, Kendell SF. End-of-life care and bereavement issues in human immunodeficiency virus-AIDS. *Nurs Clin North Am.* 2018;53(1):123–135.
99. Kast RE, Altschuler EL. Co-administration of ramelton and fluvoxamine to increase levels of interleukin-2. *Med Hypotheses.* 2006;67(6):1389–1390.

CHAPTER

14

Inflammatory Arthropathies

Regina M. Taylor-Gjevre[1] and John A. Gjevre[2]

[1]Division of Rheumatology, Department of Medicine, University of Saskatchewan, Saskatoon, SK, Canada [2]Division of Respiratory and Sleep Medicine, Department of Medicine, University of Saskatchewan, Saskatoon, SK, Canada

OUTLINE

Introduction	309	Biologic Therapies and Sleep in Inflammatory Arthritis	317
Nature and Prevalence of Sleep Disturbance/Disorders in Inflammatory Arthritis Patients	311	Effect of CPAP on Inflammation and Pain in Inflammatory Arthritis	318
Sleep Quality	312	Areas for Future Research	319
Insomnia and Rheumatic Diseases	313		
Obstructive Sleep Apnea	314	Conclusion	319
Restless Legs Syndrome	315	References	320
Polysomnographic Studies in Inflammatory Arthritis	316	Further Reading	324
Cytokines and Sleep	317		

INTRODUCTION

Sleep problems are common in people with arthritis and other rheumatologic conditions.[1] Many people with arthritis including inflammatory arthritis have a component of recurrent or chronic pain.[2] Pain by itself has been known to be associated with sleep issues. In the Sleep in America poll 2015, which focused on pain and sleep: it was found that patients with chronic

Handbook of Sleep Disorders in Medical Conditions
DOI: https://doi.org/10.1016/B978-0-12-813014-8.00014-7

© 2019 Elsevier Inc. All rights reserved.

pain reported lower sleep quality, more sleep problems, and greater sleep debt. Only 36% of respondents with chronic pain reported good or very good sleep. Difficulty sleeping was felt to interfere with work by 52% of people with chronic pain responding to this survey. Sleep difficulty was also considered to interfere with mood (55%), daily activities (46%), enjoyment of life (50%), and relationships (37%). Those experiencing pain noted greater sensitivity to environmental factors disturbing sleep.[3] An earlier Sleep in America poll in 2003 included a specific response assessment for people with self-identified diagnoses of arthritis. The poll reported that, in older adults diagnosed with "arthritis" compared to those without such a diagnosis, there are higher proportions of people reporting sleeping less than 6 hours/night (15%), having fair/poor sleep quality (29%), report insomnia or sleep fragmentation (56%), and report having a sleep problem (72%).[4]

There has been recognition of interaction between both inflammation and pain with aspects of sleep. Within this chapter on sleep disorders and inflammatory arthropathies, the focus will be primarily on the rheumatoid arthritis (RA) and the axial spondyloarthropathy (AxSpA) patient populations. These two diagnoses represent categories of inflammatory arthropathies that have relatively high prevalence within the general population and for which there has been a volume of research into sleep-related issues.

RA is the most common of the inflammatory polyarthropathies with an overall prevalence of approximately 0.5%–1% in general populations.[5] There is a female predominance with an approximately 3:1 ratio of women/men affected.[5] There are both late age of onset and younger age of onset categories recognized.[6] There have been some variations in prevalence observed in different geo-ethnic populations.[7] In Canadian Indigenous populations, for example, rates have been reported to be higher for RA compared to the general population.[8] Other associations with disease incidence include some genetic/major histocompatibility antigenic linkages or predispositions.[9] External or environmental exposures have been linked with development of RA. Most prominently of these is cigarette smoking, which has also been demonstrated to have a strong association with development of disease-specific autoantibody formation.[9] Other environmental associations include farming exposures, which have been associated with higher prevalence rates for RA.[10]

RA is a polyarthritis meaning that it involves more than three joints. It is a chronic arthritis that can be destructive and disabling with substantial human and socioeconomic cost.[11] Although most appendicular skeleton joints can be affected by RA, the classic pattern of involvement includes the proximal interphalangeal joints, the metacarpal–phalangeal joints, the wrists, and the metatarsal–phalangeal joints. The axial skeleton is generally spared with the exception of the cervical spine. Atlantoaxial involvement in this region is a particular concern for longer standing RA patients, with the consequent potential for cervical instability and neurological compromise.[12]

Approximately 70% of RA patients are "sero-positive" for disease-associated autoantibodies. These include the classic "rheumatoid factor" and the more recently recognized anticyclic citrullinated peptide autoantibody, which are widely available for diagnostic testing.[13] Sero-positivity is generally felt to herald more aggressive inflammatory disease and to be more likely to be associated with extraarticular manifestations of rheumatoid disease such as rheumatoid nodules, pulmonary manifestations (pleural effusions, interstitial lung disease, etc.), vasculitis, and inflammatory eye involvement.[14]

Pathologically, rheumatoid synovium is characterized by synovial cell proliferation and by an inflammatory cell infiltrate with pannus formation. Inflammatory cells include

T and B lymphocytes, plasma cells, dendritic cells, and natural killer cells. Pannus formation is associated with erosion of adjacent bone, joint effusions, and intraarticular cartilaginous damage.[15] These histological changes are often associated with an acute phase response in the serum as manifested by circulating inflammatory markers such as C-reactive protein (CRP) and inflammatory cytokine molecules. Elevated levels of various inflammatory cytokines, including tumor necrosis factor-alpha (TNF-α) and interleukin-6 (IL-6), are evident in active rheumatoid disease.[16] These observations have formed the basis for targeted biologic therapeutic strategies that have proven to be a major advance in treatment of RA. Specifically, anti-TNF-α antibodies or receptor strategies, and anti-IL-6 antibodies form part of the biologic therapy armamentarium commonly used in treatment of refractory RA.[17]

AxSpA are a group of inflammatory articular disorders that include ankylosing spondylitis (AS), axial spondyloarthropathic variants of psoriatic arthritis, inflammatory bowel disease-associated arthropathies, and reactive arthritis. There has historically been a sense that these disorders predominantly affect younger men; however, more recent research suggests that the incidence distribution may be less gender weighted than previously understood.[18] Newer criteria for these disorders accommodate greater diversity in presentation.[19] Currently, prevalence estimates for AS approximate 0.2%−0.5% in the North American population.[20] There has been some geo-ethnic variation in prevalence estimates with for example Canadian Haida populations reportedly having substantially higher prevalence rates.[21] Associations of AxSpA with other chronic inflammatory disorders including psoriasis and inflammatory bowel disease are well recognized.[22] A strong association between AS and selected HLA-B27 polymorphisms has been long acknowledged.[23]

AxSpA is dominantly an enthesopathic inflammatory disorder with a propensity toward new bone formation. The sacroiliac joints are nearly always involved and any region of the axial skeleton can be affected, although classically an ascending pattern is observed. Inflammatory disease in the appendicular skeleton, particularly large joints in the lower extremities, may also feature in the clinical presentation. Extraarticular features including ocular, cardiac, or pulmonary involvement may be seen.[22] The initial osseous inflammatory lesion often involving juxta-articular entheses may over time lead to ankylosis and fusion of involved joints. Histologically, there is evidence of increased vascularity, endothelial cell activation, and mononuclear cell infiltration.[24] The absence of circulating autoantibodies associated with this diagnosis led to the earlier term of "sero-negative spondyloarthropathies." However, there is often laboratory evidence of an inflammatory process with elevated circulating acute phase markers such as CRP, and evidence of increased tissue level and circulating inflammatory cytokines including TNF-α. These observations have led to the successful utilization of biologic strategies, particularly anti-TNF-α agents in the chronic treatment or management of AxSpA.[25]

NATURE AND PREVALENCE OF SLEEP DISTURBANCE/DISORDERS IN INFLAMMATORY ARTHRITIS PATIENTS

Sleep problems are a frequently voiced concern for patients with inflammatory arthritis. There has been greater recognition in recent years of the prevalence and impact of sleep dysfunction and primary sleep disorders in these patients with rheumatic diseases.[1]

312 14. INFLAMMATORY ARTHROPATHIES

Abnormalities of sleep reported in people with inflammatory arthritis vary from distinct sleep disorders such as obstructive sleep apnea (OSA) and restless legs syndrome (RLS) to the spectrum of diminished sleep quality, insomnia, and sleep fragmentation. These reports are based on research utilizing various questionnaire instruments and, in some settings, objective polysomnographic (PSG) or actigraphic technologically derived data.

A number of different published questionnaire instruments have been utilized to explore sleep health, hygiene, and quality. Questionnaires have been also developed to investigate potential primary sleep disorder symptomatology. These various instruments are more fully described in the chapter on assessment of sleep and sleep disorders. The Pittsburgh Sleep Quality Index (PSQI),[26] the Medical Outcome Study Sleep Scale,[27] the Berlin Questionnaire,[28] the Epworth Sleepiness Scale,[29] and the International Restless Legs Syndrome Study Group (IRLSSG) Criteria[30] are all amongst instruments that have been applied within the inflammatory arthropathy populations.

Results from questionnaire-based studies in the inflammatory arthritis populations can be separated into those which have a more general evaluation of sleep, sleepiness, or sleep quality and those specifically evaluating potential primary sleep disorders. The following paragraphs which address findings from questionnaire-based studies are structured to initially review sleep quality—focused results and subsequently to examine findings from more specific primary sleep disorder—oriented instruments in these populations.

Sleep Quality

The PSQI is a widely used questionnaire instrument for estimation of sleep quality. Abnormal sleep quality as determined by elevated PSQI scores has been found in a high proportion of patients with rheumatologic disorders. The PSQI instrument provides a weighted score based on questionnaire responses within seven separate domains. These seven domains include subjective assessment of sleep quality, sleep latency, sleep duration, sleep efficiency, sleep disturbances, daytime dysfunction, and pharmacologic sleep-aid usage.[26]

Overall sleep quality scores have been reported by several investigators to be abnormal in more than half of RA patient populations. In an examination of specific domains or subscores within the PSQI sleep quality instrument measurement in 145 RA patients, the following were found: fairly bad/bad subjective sleep quality (27.6% of respondents), sleep latency of more than 30 minutes (19.3%), sleep duration <6 hours (20.7%), habitual sleep efficiency <74% (25.5%), fairly bad/bad daytime dysfunction (18.6%), use of medication sleep-aids more than weekly (25.5%), and fairly bad/bad sleep disturbance scores (44.1%). Awakening at night three or more times per week was reported by 52.4%, with the majority of these patients attributing this to requiring use of the washroom (51%) and second being awoken by pain (33.1%).[31] Karatas et al. reported PSQI scores reflecting poor sleep quality in 60% of RA patients ($n = 35$).[32]

Similar findings have been reported in patients with other rheumatologic conditions including AS. Karatas et al. reported in a study of 34 AS patients that impaired sleep quality evaluated by PSQI was evident in 57.9% of the study group.[33] Batmaz et al. studied a

II. SLEEP DISORDERS IN SPECIFIC MEDICAL CONDITIONS

group of 80 AS patients compared to 52 controls and found PSQI scores to be significantly higher indicating worse sleep quality in the AS group.[34]

Using the medical outcomes study (MOS) sleep questionnaire in a RA population of 8676 patients in comparison to a group of control subjects, Wolfe et al. reported increased sleep disturbance in RA patients.[35] The sleep disturbance level reported in this study was somewhat higher in women and lessened with age. Austad et al. evaluated 986 patients from the Oslo RA Register using the MOS sleep instrument and found associations for sleep disturbance with pain, fatigue, and disease activity in these patients.[36]

The MOS-sleep scale (MOS-SS) has also been applied in the AS population in comparison to controls and also within therapeutic trials (as described in the section below on cytokines and sleep). Karadag et al. reported increased MOS-SS component scores in a study of 171 AS patients compared to 86 age/gender-matched controls.[37]

The Epworth Sleepiness Scale measures daytime sleepiness that is frequently associated with OSA but is not specific to that primary sleep disorder.[29] It is an easily applied questionnaire that has been utilized in the RA population. Fragiadaki found the mean Epworth Sleepiness Scale (ESS) score in 15 RA patients to be 6.9 (4.8) and subsequently improving with tocilizumab therapy.[38] Purabdollah et al. reported a mean ESS score of 13.1 (5.6) in a sample of 210 RA patients.[39] Abbasi et al. reported daytime sleepiness to be present in 25% of 100 RA patients compared to 15% of a matched control group.[40]

Erb et al. reported on a study of 17 AS patients, in which the mean ESS score was 9.7 (5.5).[41] In a mixed rheumatology clinic population ($n = 423$), which included RA (34.3%) and AS (22%), the mean ESS was 7.30 (4.98) with 25.7% of the population scoring in the abnormal range (>10). No significant differences in scores were observed between patients in diagnostic categories.[42]

Insomnia and Rheumatic Diseases

Insomnia is defined by the American Academy of Sleep Medicine as a "subjective report of difficulty with sleep initiation, duration, consolidation, or quality that occurs despite adequate opportunity for sleep, and that results in some form of daytime impairment."[43,44] Insomnia symptoms are fairly common with one consensus paper indicating that it occurs in approximately 30% of adults.[45] However, insomnia disorder with daytime impairment is less prevalent with estimates of 10%.[46] The etiology of insomnia in the overall population is likely multifactorial with potential contribution from demographics, familial factors, psychological aspects, and medical comorbidities.[47,48] Management of insomnia should be tailored to the individual patient but in general includes recommendations to improve sleep hygiene and to consider cognitive—behavioral therapy (CBT).[43] Pharmacological interventions may be utilized but the strength of the recommendations is overall weak.[49]

Disrupted sleep and insomnia are fairly common in patients with rheumatic diseases with 29%—56% of patients with arthritis affected.[1] Various studies have shown that the majority of RA patients have disrupted sleep or poor sleep quality.[50] In the large National Health Interview Survey,[51] patients with RA had insomnia significantly more often than controls (32.1% vs 13.9%). Other researchers have noted insomnia impacts at least 50% of

314

RA patients[52] while a recent RA sleep survey found self-reported insomnia in over 73% of patients.[39]

In the general population, the standard treatment for insomnia is CBT. However, it is less clear whether this would be as effective in RA patients since their sleep is often disrupted by pain and the ideal approach would be to optimize pain control. Nevertheless, CBT would be an area of future research in the RA patient population.

In part, these findings may be due to chronic pain causing arousals and lead to a vicious circle with sleep deprivation causing worse pain control. Interestingly, there seems to be a bidirectional component with insomnia itself being a significant risk factor for RA.[53] It has been long recognized that there is a reciprocal relationship between sleep quality and pain, including RA related.[54] A recent Belgian study found that poor RA control was associated with a reduction in sleep quality.[55] Other researchers have also noted poor sleep quality associated with RA disease activity.[56]

Certain pharmacological therapies for rheumatic diseases, such as corticosteroids, may impact sleep and contribute toward insomnia.[57] Other agents such as methotrexate and etanercept improve sleep by reducing RA inflammation.[58] Polysomnogram studies have confirmed that anti-TNF therapies increase sleep efficiency and decrease wake after sleep onset.[31] Finally, it should be noted that many RA patients experience fatigue which is only partially related to RA disease activity. Instead, fatigue in RA has strong correlations with sleep and other factors.[59]

Overall, there is a complicated relationship between insomnia and rheumatic diseases. Clearly, rheumatic disease patients have an increased prevalence of insomnia but the causative relationship remains unclear. Poor RA disease control is associated with worse insomnia symptoms. Controlling pain and improving overall RA disease control improve sleep. Further research is needed to broaden our understanding of insomnia in RA and other inflammatory rheumatic diseases.

Obstructive Sleep Apnea

OSA is a common primary sleep disorder for which the clinical presentation, pathophysiology, investigation, and treatment are described in a separate chapter. Most studies on OSA in the context of rheumatic diseases are smaller cohort studies; however using administrative health data, a recent population-based retrospective cohort study from Taiwan identified a 75% greater incidence rate of OSA in the RA cohort compared to controls.[60] A separate administrative health data study also using the Taiwan national database found rheumatic disease, including RA to be overrepresented in the OSA population compared to controls without OSA.[61]

In a general rheumatology clinic population study (34.3% RA, 22% AS) using the Berlin Questionnaire, which allows differentiation of scores to indicate high risk or low risk of OSA, 35.2% of responses from 423 participants were categorized as high risk for OSA.[42] The authors found no significant differences in Berlin categorization proportions between study patients in different arthritis diagnostic categories. In studies within the specific RA population, high risk for OSA by the Berlin Questionnaire scoring has ranged between 31% and 48%.[57,62] One study of 25 RA patients reported a correlation of high risk for OSA

II. SLEEP DISORDERS IN SPECIFIC MEDICAL CONDITIONS

Berlin questionnaire categorization with abnormal Epworth sleepiness score (coefficient of 0.6).[63] The Berlin Questionnaire is a well-validated tool for people in the community who are at risk for OSA.[28] However, it should be noted that the actual risk of OSA can vary based on the patient population and other factors.

Increased risk for OSA within these patient populations as with the general population is likely conferred by multiple factors including increased body mass index. Characteristics more specifically related to RA and its sequelae which enhance risk for OSA include temporomandibular joint pathology, cricoarytenoid joint involvement, retrognathia, and micrognathia.[64] Development of these complications from RA in particular may increase risk for OSA in a mechanism directly related to airway closure dynamics. Cervical spine involvement and resultant instability have been associated with a high prevalence of OSA. Shoda et al. reported a 79% prevalence of OSA in 29 RA patients with progressive myelopathy secondary to cervical spine pathology.[65]

Restless Legs Syndrome

RLS, also known as Willis–Ekbom disease, is a common primary sleep disorder that increases in prevalence with age and has been observed to be more prevalent in some chronic medical conditions. The clinical presentation, causal factors, spectrum of disease, investigations, diagnosis, and treatment of this entity are described more fully in a separate chapter.

RLS has been described as increased in frequency in the RA population. Reynolds et al. studied hospitalized RA patients in 1986 and found 30% of patients to have RLS.[66] Salih et al., using 1986 published guidelines for RLS diagnosis, found 25% of RA patients to meet RLS criteria.[67] Application of the 2003 IRLSSG criteria identified 27.7% of RA patients as meeting requirements for a diagnosis of RLS.[68] These patients indicated ability to distinguish between discomfort related to their arthritis and RLS symptoms. In support of clarity of these perceptions, an actigraphic study has demonstrated significantly higher frequencies of nocturnal periodic limb movements in RA patients meeting RLS criteria than in RA patients who did not have RLS.[69]

IRLSSG criteria defined RLS was present in 30.8% of 130 ankylosing patients compared to 13.2% of age/gender-matched controls in a study reported by Tekatas and Pamuk.[70] Similarly, Demirci et al. reported 36.4% of 108 AS patients met IRLSSG criteria compared to 14% of the 64 controls in this study. Demirci also reported significantly higher RLS severity scores and significantly poorer sleep quality scores in the AS patients than controls. RLS severity scores were independently associated with PSQI scores, AS quality of life questionnaire results, and insomnia severity index scores.[71]

The mechanism by which RLS may be increased in frequency in patients with inflammatory arthritis is not clear. With nonsteroidal antiinflammatory drug therapy being commonly employed by patients with inflammatory arthritis, it is possible that subclinical iron deficiency may be a contributing factor in the development of this syndrome.[72] An additional consideration may link to chronic pain through a dopaminergic neural pathway predisposing to development of RLS.[73]

POLYSOMNOGRAPHIC STUDIES IN INFLAMMATORY ARTHRITIS

Although questionnaire instruments have advantages in capacity for application to large numbers of people, the gold standard for evaluation of sleep and also for diagnosis of sleep disorders does involve quantitative measurement through sleep studies, although some sleep diagnoses (e.g. insomnia) may not require objective measurements. For example, clinical criteria are utilized to diagnose RLS. However, to obtain quantitative measurements, a sleep test procedure must be done. The quantitative sleep testing may include actigraphy, level 3 home-based sleep testing, level 1 in-lab sleep testing, and other measures. The most comprehensive test is a level 1 in-lab polysomnogram and the nature of PSG studies is described in detail in an earlier chapter.

Several investigators have utilized polysomnography in study of RA. Hirsch et al. studied 19 RA patients and 19 matched controls finding preserved sleep architecture but evidence of severe sleep fragmentation.[74] They also reported "enhanced presence of primary sleep disorders." Drewes et al. reported increased periodic leg movements and increased alpha-electroencephalogram (EEG) activity in RA patients ($n = 41$) versus controls.[75] Lavie studied PSG findings in 13 RA patients in conjunction with a tenoxicam therapeutic trial. Eight of the 13 patients studied were found to have evidence of sleep disorders. Four had periodic leg movements, three had sleep apneas, and one had both findings.[76] Mahowald et al. identified 16 chronic active RA patients with early onset of fatigue (less than 6 hours following morning awakening). Multiple sleep latency testing revealed 7 of 16 to be hypersomnolent. During PSG examination, all 16 exhibited frequent arousals and periodic limb movements. Two cases of sleep apnea were identified.[77] Zamarron studied PSG findings before and after first anti-TNF treatment in six RA patients, one of the six was identified as having OSA.[78] Alamoudi reported a series of 10 patients who had acquired retrognathia related to RA, who had undergone overnight PSG studies. Three of these patients had severe OSA, three had milder OSA, and four had apnea—hypopnea indexes of less than 10.[79] Mutoh et al. studied hospitalized RA patients with overnight PSG finding moderate-to-severe sleep apnea in 13 of 62 patients studied (20.9%). Temporomandibular joint abnormality severity and health assessment questionnaire disability index were identified as risk factors for a positive study in this inpatient population.[80] Bjurstrom et al. studied 24 RA patients and 48 controls by all-night PSG findings lower levels of sleep efficiency and a higher level of stage 3 sleep in the RA patients compared to controls.[81] Gjevre et al. reported a study of 25 RA patients who had undergone PSG in which 40% reported abnormal daytime sleepiness; however, 68% were found to have an abnormal PSG consistent with OSA.[63]

In AS patients, Solak et al. reported a prevalence of 22.6% for OSA based on PSG results in 31 AS patients.[82] Yamamoto reported a case study of a patient with severe OSA related to juxtaposition of the oropharynx during sleep with a markedly ossified and hypertrophied cervical anterior longitudinal ligament.[83]

These PSG studies generally examine relatively small patient numbers within a research setting making extrapolation of findings to the general RA or AS populations uncertain. PSG has also occasionally been used in clinical trials to evaluate change associated with specific therapeutic interventions in these populations.

CYTOKINES AND SLEEP

Cytokine mediators have been recognized as key players in the inflammatory disease processes in RA and other rheumatic diseases. Biologic therapeutic strategies using antibodies targeting-specific cytokines have proven efficacious in treatment of both RA and AS. The targeting of individual cytokines may have implications for sleep as well as inducing improvement in inflammatory activity. The role of various cytokines in mediating sleep regulation has been recognized with differential associations for promotion/inhibition of sleep phases. TNF-α and IL-6 have been observed to exhibit diurnal rhythms with peaks during sleep periods and trough levels during wake times.[84,85] TNF-α has been extensively studied in sleep disorders. A shift to higher daytime levels of TNF-α has been noted in people with insomnia experiencing increasing daytime fatigue.[86] A functional alteration in the TNF-α system has been observed in patients with narcolepsy.[87] An increase in TNF-α levels have been noted in patients with OSA. A TNF-α (-308A) gene polymorphism has been reported in OSA patients compared to population controls in a UK study.[88] Intermittent hypoxia has been reported to be the strongest predictor of elevated TNF-α levels in OSA patients.[89] Studies examining treatment with continuous positive airway pressure (CPAP) in the general OSA population have reported normalization of circulating TNF-α levels.[89] In terms of TNF receptors, Yue et al. observed an association between increased arousals on PSG and higher circulating levels of sTNF-R1.[90]

Within the RA population, in a study which included both PSG and cytokine measurements, Bjurstrom et al. examined 24 RA patients and 48 matched controls that underwent overnight PSG and diurnal assessment of TNF and IL-6 levels (spontaneous and Toll-like receptor-4-stimulated monocytic production). Compared to controls, RA patients showed lower sleep efficiency, higher percentage stage 3 sleep, and higher percentages of spontaneous and stimulated TNF and stimulated IL-6 expression. The authors concluded that their findings supported an interrelationship and hypothesized a feedback loop between sleep maintenance, slow-wave sleep, and cytokine-specific cellular inflammation.[81]

BIOLOGIC THERAPIES AND SLEEP IN INFLAMMATORY ARTHRITIS

In light of the hypothesized relationship between specific cytokines, sleep maintenance/regulation, and inflammatory disease, as well as the recognized impact of arthritis associated pain on sleep, there have been some clinical therapeutic trials that have included sleep measurements as an outcome measure.

Anti-TNF-α therapy (etanercept) has been used in nonrheumatologic OSA patients with resultant decrease in apnea—hypopnea index and daytime sleepiness. Within the RA population, Zamarron reported improvement in sleep latency and sleep efficiency after one infusion of anti-TNF-α therapy (infliximab).[78] Improvement in sleep efficiency and awakening after sleep onset time were also observed in 10 RA patients studied before and approximately 2 months after onset of anti-TNF therapy.[91] These findings were supported by Detert et al. in a prospective study of 36 RA patients. In that study, those treated with etanercept were observed to have significant improvement in sleep efficiency, total sleep time, and stage 2 sleep duration.[92] Conversely, Wolfe et al. did not observe any significance differences in MOS-sleep scores in a large RA population treated with anti-TNF therapy.[35]

In the AS population, MOS-SS has been used to monitor change with use of adalimumab (anti-TNF humanized antibody). In a trial of 1250 AS patients, it was observed that adalimumab use was associated with significant improvement in the MOS-SS domains.[93] Another trial of a different anti-TNF agent (certolizumab pegol) in an AS population also revealed significant improvement in MOS-SS components associated with use of the therapeutic agent.[94] Some improvement in MOS-sleep measures were also observed by Dougados et al. in early nonradiographic axial spondyloarthritis patients treated with etanercept (anti-TNF biologic agent).[95] Karadag reported association of anti-TNF therapy in a study of 171 AS patients, with better sleep parameters as measured by the MOS-sleep questionnaire.[37] PSG changes observed in AS patients revealed significantly greater sleep efficiency, total sleep time, sleep onset latency, and arousals in 28 patients treated with anti-TNF biologic agents compared to 31 patients on nonsteroidal antiinflammatory therapies alone. Significantly better scores on PSQI were also observed in those treated with anti-TNF agents.[96]

IL-6 has been noted to be elevated in people with sleep disturbances.[86] Fragiadaki et al. reported that tocilizumab (anti-IL-6 antibody) therapy was associated with significant improvement in Epworth sleepiness scale and PSQI sleep quality scores in a study of 15 treated RA patients.[38]

Other biologic agents undergoing therapeutic trials in RA patients have included abatacept (selective T-cell costimulation modulator), which was reported by Wells et al. to be associated with significant improvement in sleep disturbance and other sleep aspects using the MOS-sleep instruments in the ATTAIN and AIM studies.[97] A trial of tofacitinib (janus kinase inhibitor) in combination with conventional disease-modifying antirheumatic drugs in 795 patients reported statistically significant differences at 3 months in MOS-SS results.[98]

Overall, there is evidence to support the concept that sleep disturbances and sleep-related symptomatology could be impacted by cytokine-specific targeting strategies. To what extent this effect relates to feedback loop mediation versus indirect benefit through pain alleviation and reduced inflammatory disease activity is not fully understood.

EFFECT OF CPAP ON INFLAMMATION AND PAIN IN INFLAMMATORY ARTHRITIS

General population OSA patients have elevated levels of inflammation and CPAP therapy helps to reduce systemic inflammation. A recent metaanalysis of 35 studies investigating inflammatory markers (utilizing CRP, IL-6, IL-8, or TNF-α) revealed that CPAP therapy could partially suppress systemic inflammation in the general OSA population and adherence to therapy (>4 hours/night and >3 months duration) was associated with better results.[99] OSA has been shown to activate inflammatory pathways and the risk of cardiovascular events[100] while CPAP therapy may decrease those cardiovascular events.[101] While it is likely that CPAP therapy would reduce inflammation in patients with inflammatory arthropathies and OSA, there are no known studies explicitly evaluating that aspect. Likewise, while patients with inflammatory arthropathies often have pain disrupting their sleep, there is little data to evaluate the effect of CPAP on their pain or global health. It has been hypothesized that OSA causes increased sensitivity to pain due to fragmented sleep. An intriguing study evaluated 12 pain-free OSA patients and found that CPAP therapy reduced their pain sensitivity to external stimuli.[102]

AREAS FOR FUTURE RESEARCH

The exact inflammatory relationship between OSA and rheumatic disease remains unclear. How does OSA affect the response to anti-TNF-α therapy in patients with RA and AS? Are there RA or AS subgroups more at risk for poor disease control because of ongoing inflammatory stimulus from OSA? Also, what is the potential impact of CPAP therapy on inflammatory markers in this patient population? In addition, while CBT has long been the standard therapy for insomnia in the general population, the benefit of CBT for insomnia in RA patients is less established. There is evidence for the utility of CBT in treatment of RA-related fatigue[59]; however, comparable experience with CBT for treatment of insomnia in the RA population has not been reported. Finally, there are little data on the effect of CPAP on pain or global health measures in patients with these inflammatory arthropathies. The need for future research is significant given the prevalence of sleep disorders in this patient population.

CONCLUSION

Poorer sleep quality scores have been associated with higher mental and general fatigue levels, depression, inflammatory disease activity measures, and experienced pain levels. Decreases in quality of life and functional status measures have also been associated with poor sleep and sleep disorders in this patient population.[31,34−36,68,103] In keeping with observations in the general population, patients with inflammatory arthritis would also be susceptible to increased risk of cardiorespiratory complications related to sleep disorders, OSA in particular.

RA and AS represent two major categories of inflammatory arthritis. The high prevalence of poor sleep quality in patients with these inflammatory arthropathies is well recognized, as are the associated features of increased pain, depression, and disease activity measures. Results from several investigators also indicate an increased frequency of primary sleep disorders, specifically OSA and RLS in these patient populations. Disease activity in these inflammatory arthropathies is associated with increased proinflammatory cytokine profiles including IL-6 and TNF-α. The influence of cytokine mediators particularly TNF-α in sleep regulation may be particularly important in understanding sleep disturbances in this clinical setting. With initiation of biologic strategies targeting individual cytokine mediators in the RA and AS populations in particular, some investigators have reported improvement in sleep parameters for patients receiving these therapies.

KEY PRACTICE POINTS

- There is an increased prevalence of sleep disorders in patients with inflammatory arthropathies.
- Certain rheumatic disease sequelae such as retrognathia, temporomandibular joint (TMJ), and cricoarytenoid joint pathology may increase the risk of OSA.
- Sleep disorders may be linked to increased pain perception and sense of fatigue in rheumatic disease patients.
- Inflammatory cytokine levels are elevated in OSA. Treatment with anti-TNF-α agents may improve certain sleep parameters; however, the exact relationship between OSA and inflammatory arthropathies remains unclear.
- An awareness of the increased risk of sleep disorders in rheumatology clinic patients is important. Utilization of simple sleep screening questionnaires may be beneficial.

References

1. Abad VC, Sarinas PS, Guilleminault C. Sleep and rheumatologic disorders. *Sleep Med Rev.* 2008;12(3):211−228.
2. Barbour KE, Boring M, Hlemick CG, et al. Prevalence of severe joint pain among adults with doctor-diagnosed arthritis—United States, 2002−2014. *Centers Disease Control Prev: Morb Mortal Wkly Rep.* 2016;65 (39):1052−1056.
3. National Sleep Foundation. Sleep and pain poll. Summary of findings. <https://sleepfoundation.org/sleep-polls-data/sleep-in-america-poll/2015-sleep-and-pain>; 2015 Accessed 02.08.17.
4. National Sleep Foundation. Sleep and aging poll. Summary of findings. <https://sleepfoundation.org/sleep-polls-data/sleep-in-america-poll/2003-sleep-and-aging>; 2003 Accessed 02.08.17.
5. Widdifield J, Paterson JM, Bernatsky S, et al. The epidemiology of rheumatoid arthritis in Ontario, Canada. *Arthritis Rheumatol.* 2014;66(4):786−793.
6. Innala L, Berglin E, Moller B, et al. Age at onset determines severity and choice of treatment in early rheumatoid arthritis: a prospective study. *Arthritis Res Ther.* 2014;16:R94.
7. Tobon GJ, Youinou P, Saraux A. The environment, geo-epidemiology and autoimmune disease: rheumatoid arthritis. *J. Autoimmun.* 2010;35(1):10−14.
8. McDougall C, Hurd K, Barnabe C. Systemic review of rheumatic disease epidemiology in the indigenous populations of Canada, the United States, Australia, and New Zealand. *Semin Arthritis Rheum.* 2017;46(5):675−686.
9. Stolt P, Bengtsson C, Nordmark B, et al. Quantification of the influence of cigarette smoking on rheumatoid arthritis: results from a population based case-control study, using incident cases. *Ann Rheum Dis.* 2003;62 (9):835−841.
10. Taylor-Gjevre RM, Trask C, King N, et al. Prevalence and occupational impact of arthritis in Saskatchewan farmers. *J Agromedicine.* 2015;20(2):205−216.
11. Albers JM, Kuper HH, van Riel PL, et al. Socio-economic consequences of rheumatoid arthritis in the first years of the disease. *Rheumatology.* 1999;38(5):423−430.
12. Tehlirian CV, Bathon JM. Rheumatoid arthritis: A. Clinical and laboratory manifestations. In: Stone JH, Crofford LJ, White PH, eds. *Primer on the rheumatic diseases.* 13th ed. New York: Springer Science + Business Media; 2008:114−121.
13. Taylor P, Gartemann J, Hsieh J, et al. A systematic review of serum biomarkers anti-cyclic citrullinated peptide and rheumatoid factor as tests for rheumatoid arthritis. *Autoimmune Dis.* 2011;2011:815038.
14. Young A, Koduri G. Extra-articular manifestations and complications of rheumatoid arthritis. *Best Pract Res Clin Rheumatol.* 2007;21(5):907−927.
15. Waldenburger JM, Firestein GS. Rheumatoid arthritis B. Epidemiology, pathology and pathogenesis. In: Stone JH, Crofford LJ, White PH, eds. *Primer on the rheumatic diseases.* 13th ed. New York: Springer Science + Business Media; 2008:122−132.

REFERENCES

16. Noack M, Miossec P. Selected cytokine pathways in rheumatoid arthritis. *Semin Immunopathol.* 2017;39 (4):365–383.
17. Rein P, Mueller RB. Treatment with biologicals in rheumatoid arthritis: an overview. *Rheumatol Ther.* 2017;4 (2):247–261.
18. Haroon NN, Paterson JM, Li P, et al. Increasing proportion of female patients with ankylosing spondylitis: a population-based study of trends in the incidence and prevalence of AS. *BMJ Open.* 2014;4(12):e006634.
19. Rudwaleit M, van der Heijde D, Landewe R, et al. The development of Assessment of SpondyloArthritis International Society classification criteria for axial spondyloarthritis (part II): validation and final selection. *Ann Rheum Dis.* 2009;68(6):777–783.
20. Reveille JD. Epidemiology of spondyloarthritis in North America. *Am J Med Sci.* 2011;341(4):284–286.
21. Gofton JP, Chalmers A, Price GE, et al. HL-A 27 and ankylosing spondylitis in B.C. Indians. *J Rheumatol.* 1984;11(5):572–573.
22. van der Heijde D. Ankylosing spondylitis: A. Clinical features. In: Stone JH, Crofford LJ, White PH, eds. *Primer on the rheumatic diseases.* 13th ed. New York: Springer Science + Business Media; 2008:1193–1199.
23. Yang T, Duan Z, Wu S, et al. Association of HLA-B27 genetic polymorphisms with ankylosing spondylitis susceptibility worldwide: a meta-analysis. *Mod Rheumatol.* 2014;24(1):150–161.
24. Braun J. Ankylosing spondylitis B: pathology and pathogenesis. In: Stone JH, Crofford LJ, White PH, eds. *Primer on the rheumatic diseases.* 13th ed New York: Springer Science + Business Media; 2008:200–207.
25. Maxwell LJ, Zochling J, Boonen A, et al. TNF-alpha inhibitors for ankylosing spondylitis. *Cochrane Database Syst Rev.* 2015;18(4):CD005468.
26. Buysse DJ, Reynolds CFI, Monk TH, et al. The Pittsburgh Sleep Quality Index: a new instrument for psychiatric practice and research. *Psychiatry Res.* 1989;28:193–213.
27. Rand Health Medical Outcome Study: Sleep Scale Survey Instrument. <https://www.rand.org/health/surveys_tools/mos/sleep-scale.html> Accessed 02.08.17.
28. Netzer NC, Stoohs RA, Netzer CM, Clark K, Strohl KP. Using the Berlin Questionnaire to identify patients at risk for the sleep apnea syndrome. *Ann Intern Med.* 1999;131(7):485–491.
29. Johns MJ. A new method for measuring daytime sleepiness: the Epworth Sleepiness Scale. *Sleep.* 1991;14:540–545.
30. Allen RP, Picchietti D, Hening A, et al. Restless legs syndrome: diagnostic criteria, special considerations, and epidemiology. A report from the restless legs syndrome diagnosis and epidemiology workshop at the National Institutes of Health. *Sleep Med.* 2003;4:101–119.
31. Taylor-Gjevre RM, Gjevre JA, Nair B, et al. Components of sleep quality and sleep fragmentation in rheumatoid arthritis and osteoarthritis. *Musculoskel Care.* 2011;9(3):152–159.
32. Karatas G, Bal A, Yuceege M, et al. The evaluation of sleep quality and response to anti-tumor necrosis factor alpha therapy in rheumatoid arthritis patients. *Clin Rheumatol.* 2017;36(1):45–50.
33. Karatas G, Bal A, Yuceege M. Evaluation of sleep quality in patients with ankylosing spondylitis and efficacy of anti-TNF-alpha therapy on sleep problems: a polysomnographic study. *Int J Rheum Dis.* 2017;21 (6):1263–1269.
34. Batmaz I, Sariyildiz MA, Dilek B, et al. Sleep quality and associated factors in ankylosing spondylitis: relationship with disease parameters, psychological status and quality of life. *Rheumatol Int.* 2013;33(4):1039–1045.
35. Wolfe F, Michaud K, Li T. Sleep disturbance in patients with rheumatoid arthritis: evaluation by medical outcomes study and visual analogue sleep scales. *J Rheumatol.* 2006;33:1942–1951.
36. Austad C, Kvien TK, Olsen IC, et al. Sleep disturbance in patients with rheumatoid arthritis is related to fatigue, disease activity, and other patient-reported outcomes. *Scand J Rheumatol.* 2017;46(2):95–103.
37. Karadag O, Nakas D, Kalyoncu U, et al. Effect of anti-TNF treatment on sleep problems in ankylosing spondylitis. *Rheumatol Int.* 2012;32(7):1909–1913.
38. Fragiadaki K, Tektonidou MG, Konsta M, et al. Sleep disturbances and interleukin 6 receptor inhibition in rheumatoid arthritis. *J Rheumatol.* 2012;39(1):60–62.
39. Purabdollah M, Lakdizaji S, Rahmani A, et al. Relationship between sleep disorders, pain and quality of life in patients with rheumatoid arthritis. *J Caring Sci.* 2015;4(3):233–241.
40. Abbasi M, Yazdi Z, Rezaie N. Sleep disturbances in patients with rheumatoid arthritis. *Niger J Med.* 2013;22 (3):181–186.

41. Erb N, Karokis D, Delamere JP, et al. Obstructive sleep apnoea as a cause of fatigue in ankylosing spondylitis. *Ann Rheum Dis.* 2003;62(2):183–184.
42. Taylor-Gjevre RM, Gjevre JA, Nair B, et al. Hypersomnolence and sleep disorders in a rheumatic disease patient population. *J Clin Rheumatol.* 2010;16(6):255–261.
43. Schutte-Rodin S, Broch L, Buysse D, Dorsey C, Sateia M. Clinical Guideline for the Evaluation and Management of Chronic Insomnia in Adults. *J Clin Sleep Med.* 2008;4(5):487–504.
44. American Academy of Sleep Medicine. *International classification of sleep disorders.* 3rd ed. Darien, IL: American Academy of Sleep Medicine; 2014.
45. Ancoli-Israel S, Roth T. Characteristics of insomnia in the United States: results of the 1991 National Sleep Foundation Survey. *Sleep.* 1999;22(Suppl. 2):S347–S353.
46. National Institutes of Health. National Institutes of Health State of the Science Conference Statement on manifestations and management of chronic insomnia in adults. *Sleep.* 2005;28(9):1049–1057.
47. Jarrin DC, Morin CM, Rochefort A, et al. Familial aggregation of insomnia. *Sleep.* 2017;40(2). Available from: https://doi.org/10.1093/sleep/zsw053.
48. LeBlanc M, Mérette C, Savard J, Ivers H, Baillargeon L, Morin CM. Incidence and risk factors of insomnia in a population-based sample. *Sleep.* 2009;32(8):1027–1037.
49. Sateia MJ, Buysse DJ, Krystal AD, Neubauer DN, Heald JL. Clinical practice guideline for the pharmacologic treatment of chronic insomnia in adults: an American Academy of Sleep Medicine clinical practice guideline. *J Clin Sleep Med.* 2017;13(2):307–349.
50. Guo G, Fu T, Yin R, et al. Sleep quality in Chinese patients with rheumatoid arthritis: contributing factors and effects on health-related quality of life. *Health Qual Life Outcomes.* 2016;14:151. Available from: https://doi.org/10.1186/s12955-016-0550-3. PMCID: PMC5111274.
51. Louie GH, Tektonidou MG, Caban-Martinez AJ, Ward MM. Sleep disturbances in adults with arthritis: prevalence, mediators, and subgroups at greatest risk. Data From the 2007 National Health Interview Survey. *Arthritis Care Res (Hoboken).* 2011;63(2):247–260. Available from: https://doi.org/10.1002/acr.20362.
52. Buenaver LF, Smith MT. Sleep in rheumatic diseases and other painful conditions. *Curr Treat Options Neurol.* 2007;9(5):325–336.
53. Sivertsen B, Lallukka T, Salo P, et al. Insomnia as a risk factor for ill health: results from the large population-based prospective HUNT Study in Norway. *J Sleep Res.* 2014;23(2):124–132.
54. Moldofsky H. Sleep and pain. *Sleep Med Rev.* 2001;5(5):385–396.
55. Westhovens R, Van der Elst K, Matthys A, Tran M, Gilloteau I. Sleep problems in patients with rheumatoid arthritis. *J Rheumatol.* 2014;41(1):31–40. Available from: https://doi.org/10.3899/jrheum.130430. PMID: 24293569.
56. Sariyildiz MA, Batmaz I, Bozkurt M, et al. Sleep quality in rheumatoid arthritis: relationship between the disease severity, depression, functional status and the quality of life. *J Clin Med Res.* 2014;6(1):44–52. Available from: https://doi.org/10.4021/jocmr1648w.
57. Goes ACJ, Reis LAB, Silva MBG, et al. Rheumatoid arthritis and sleep quality. *Rev Bras Reumatol Engl Ed.* 2017;57(4):294–298.
58. Straub RH, Detert J, Dziurla R, et al. Inflammation is an important covariate for the crosstalk of sleep and the HPA axis in rheumatoid arthritis. *Neuroimmunomodulation.* 2017;24(1):11–20. Available from: https://doi.org/10.1159/000475714.
59. Katz P. Causes and consequences of fatigue in rheumatoid arthritis. *Curr Opin Rheumatol.* 2017;29(3):269–276. Available from: https://doi.org/10.1097/BOR.0000000000000376.
60. Shen TC, Hang LW, Liang SJ, et al. Risk of obstructive sleep apnoea in patients with rheumatoid arthritis: a nationwide population-based retrospective cohort study. *BMJ Open.* 2016;6(11):e013151.
61. Chen WS, Chang YS, Chang CC, et al. Management and risk reduction of rheumatoid arthritis in individuals with obstructive sleep apnea: a nationwide population-based study in Taiwan. *Sleep.* 2016;39(10):1883–1890.
62. Reading SR, Crowson CS, Rodeheffer RJ, et al. Do rheumatoid arthritis patients have a higher risk for sleep apnea? *J Rheumatol.* 2009;36(9):1869–1872.
63. Gjevre JA, Taylor-Gjevre RM, Nair BV, et al. Do sleepy rheumatoid arthritis patients have a sleep disorder? *Musculoskel Care.* 2012;10(4):187–195.
64. Taylor-Gjevre RM, Nair BV, Gjevre JA. Obstructive sleep apnea in relation to rheumatic disease. *Rheumatology.* 2013;52(1):15–21.

REFERENCES

65. Shoda N, Seichi A, Takeshita K, et al. Sleep apnea in rheumatoid arthritis patients with occipitocervical lesions: the prevalence and associated radiographic features. *Eur Spine J.* 2009;18:905–910.
66. Reynolds G, Blake DR, Pall HS, et al. Restless legs syndrome and rheumatoid arthritis. *BMJ.* 1986;292:659–660.
67. Salih AM, Gray RE, Mills KR, et al. A clinical, serological and neurophysiological study of restless legs syndrome in rheumatoid arthritis. *Br J Rheumatol.* 1994;33:60–63.
68. Taylor-Gjevre RM, Gjevre JA, Skomro R, et al. Restless legs syndrome in a rheumatoid arthritis patient cohort. *J Clin Rheumatol.* 2009;15(1):12–15.
69. Taylor-Gjevre RM, Gjevre JA, Nair BV. Increased nocturnal periodic limb movements in rheumatoid arthritis patients meeting questionnaire diagnostic criteria for restless legs syndrome. *BMC Musculoskelet Disord.* 2014;15:378.
70. Tekatas A, Pamuk ON. Increased frequency of restless leg syndrome in patients with ankylosing spondylitis. *Int J Rheum Dis.* 2015;18(1):58–62.
71. Demirci S, Demirci K, Dogru A, et al. Restless legs syndrome is associated with poor sleep quality and quality of life in patients with ankylosing spondylitis: a questionnaire-based study. *Acta Neurol Belg.* 2016;116 (3):329–336.
72. Allen RP. Controversies and challenges in defining the etiology and pathophysiology of restless legs syndrome. *Am J Med.* 2007;120:S13–S21.
73. Kumru H, Albu S, Vidal J, et al. Dopaminergic treatment of restless legs syndrome in spinal cord injury patients with neuropathic pain. *Spinal Cord Ser Cases.* 2016;18(2):16022.
74. Hirsch M, Carlander B, Verge M, et al. Objective and subjective sleep disturbances in patients with rheumatoid arthritis. *Arthritis Rheum.* 1994;37:41–49.
75. Drewes AM, Nielsen KD, Hansen B, et al. A longitudinal study of clinical symptoms and sleep parameters in rheumatoid arthritis. *Rheumatology.* 2000;39:1287–1289.
76. Lavie P, Nahir M, Lorber M, et al. Nonsteroidal anti-inflammatory drug therapy in rheumatoid arthritis patients: lack of association between clinical improvement and effects on sleep. *Arthritis Rheum.* 1991;34:655–659.
77. Mahowald MW, Mahowald ML, Bundle SR, et al. Sleep fragmentation in rheumatoid arthritis. *Arthritis Rheum.* 1989;32:974–983.
78. Zamarron C, Maceiras F, Mera A, et al. Effect of the first infliximab infusion on sleep and alertness in patients with active rheumatoid arthritis. *Ann Rheum Dis.* 2004;63:88–90.
79. Alamoudi OS. Sleep disordered breathing in patients with acquired retrognathia secondary to rheumatoid arthritis. *Med Sci Monit.* 2006;12(12):CR530–CR534.
80. Mutoh T, Okuda Y, Mokuda S, et al. Study on the frequency and risk factors of moderate-to-severe sleep apnea syndrome in rheumatoid arthritis. *Mod Rheumatol.* 2016;26(5):681–684.
81. Bjurstrom MF, Olmstead R, Irwin MR. Reciprocal relationship between sleep macrostructure and evening and morning cellular inflammation in rheumatoid arthritis. *Psychosom Med.* 2017;79(1):24–33.
82. Solak O, Fidan F, Dundar U, et al. The prevalence of obstructive sleep apnoea syndrome in ankylosing spondylitis patients. *Rheumatology.* 2009;48(4):433–435.
83. Yamamoto J, Okamoto Y, Shibuya E, et al. Obstructive sleep apnea syndrome induced by ossification of the anterior longitudinal ligament with ankylosing spondylitis. *Nihon Kokyuki Gakkai Zasshi.* 2000;38(5):413–416.
84. Petrovsky N, McNair P, Harrison LC. Diurnal rhythms of proinflammatory cytokines: regulation by plasma cortisol and therapeutic implications. *Cytokine.* 1998;10:307–312.
85. Krueger JM, Fang J, Taishi P, et al. Sleep, a physiologic role for IL-1 beta and TNF-alpha. *Ann N Y Acad Sci.* 1998;856:148–159.
86. Vgontzas AN, Zoumakis M, Papanicolaou DA, et al. Chronic insomnia is associated with a shift of interleukin-6 and tumor necrosis factor secretion from nighttime to daytime. *Metabolism.* 2002;51:887–892.
87. Himmerich H, Beitinger PA, Fulda S, et al. Plasma levels of tumor necrosis factor alpha and soluble tumor necrosis factor receptors in patients with narcolepsy. *Arch Intern Med.* 2006;166:1739–1743.
88. Riha RL, Brander P, Vennelle M, et al. Tumour necrosis factor-alpha (-308) gene polymorphism in obstructive sleep apnoea–hypopnoea syndrome. *Eur Respir J.* 2005;26:673–678.
89. Ryan S, Taylor CT, McNicholas WT. Predictors of elevated nuclear factor-kappa B dependent genes in obstructive sleep apnea syndrome. *Am J Respir Crit Care Med.* 2006;174:824–830.

II. SLEEP DISORDERS IN SPECIFIC MEDICAL CONDITIONS

90. Yue HJ, Mills PJ, Ancoli-Israel S, et al. The roles of TNF-alpha and the soluble TNF receptor I on sleep architecture in OSA. *Sleep Breath.* 2009;13(3):263–269.

91. Taylor-Gjevre RM, Gjevre JA, Nair BV, et al. Improved sleep efficiency after anti-tumor necrosis factor-alpha therapy in rheumatoid arthritis patients. *Ther Adv Musculoskel Dis.* 2011;3(5):227–233.

92. Detert J, Dziurla R, Hoff P, et al. Effects of treatment etanercept versus methotrexate on sleep quality, fatigue and selected immune parameters in patients with active rheumatoid arthritis. *Clin Exp Rheumatol.* 2016;34:848–856.

93. Rudwaleit M, Gooch K, Michel B, et al. Adalimumab improves sleep and sleep quality in patients with active ankylosing spondylitis. *J Rheumatol.* 2011;38(1):79–86.

94. Sieper J, Kivitz A, van Tubergen A, et al. Impact of certolizumab pegol on patient-reported outcomes in patients with axial spondyloarthritis. *Arthritis Care Res.* 2015;67(10):1475–1480.

95. Dougados M, Tsai WC, Saaibi DL, et al. Evaluation of health outcomes with etanercept treatment in patients with early nonradiographic axial spondyloarthritis. *J Rheumatol.* 2015;42(10):1835–1841.

96. In E, Turgut T, Gulkesen A, et al. Sleep quality is related to disease activity in patients with ankylosing spondylitis: a polysomnographic study. *J Clin Rheumatol.* 2016;22(5):248–252.

97. Wells G, Li T, Tugwell P. Investigation into the impact of abatacept on sleep quality in patients with rheumatoid arthritis and the validity of the MOS-Sleep questionnaire sleep disturbance scale. *Ann Rheum Dis.* 2010;69(10):1768–1773.

98. Strand V, Kremer JM, Gruben D, et al. Tofacitinib in combination with conventional disease-modifying antirheumatic drugs in patients with active rheumatoid arthritis: patient-reported outcomes from a phase III randomized controlled trial. *Arthritis Care Res.* 2017;69(4):592–598.

99. Xie X, Pan L, Ren D, Du C, Guo Y. Effects of continuous positive airway pressure therapy on systemic inflammation in obstructive sleep apnea: a meta-analysis. *Sleep Med.* 2013;14(11):1139–1150. Available from: https://doi.org/10.1016/j.sleep.2013.07.006.

100. Ryan S, Taylor CT, McNicholas WT. Selective activation of inflammatory pathways by intermittent hypoxia in obstructive sleep apnea syndrome. *Circulation.* 2005;112(17):2660–2667.

101. Martínez-García MA, Campos-Rodríguez F, Catalán-Serra P, et al. Cardiovascular mortality in obstructive sleep apnea in the elderly: role of long-term continuous positive airway pressure treatment: a prospective observational study. *Am J Respir Crit Care Med.* 2012;186(9):909–916. Available from: https://doi.org/10.1164/rccm.201203-0448OC.

102. Khalid I, Roehrs T, Hudgel D, Roth T. Continuous positive airway pressure in severe obstructive sleep apnea reduces pain sensitivity. *Sleep.* 2011;34(12):1687–1691. Available from: https://doi.org/10.5665/sleep.1436.

103. Loppenthin K, Esbensen BA, Jennum P, et al. Sleep quality and correlates of poor sleep in patients with rheumatoid arthritis. *Clin Rheumatol.* 2015;34(12):2029–2039.

Further Reading

Viatte S, Plant D, Raychaudhuri S. Genetics and epigenetics of rheumatoid arthritis. *Nat Rev Rheumatol.* 2013;9(3):141–153.

Ataka H, Tanno T, Miyashita T, et al. Occipitocervical fusion has potential to improve sleep apnea in patients with rheumatoid arthritis and upper cervical lesions. *Spine.* 2010;35:E971–E975.

CHAPTER

15

Chronic Fatigue Syndrome and Fibromyalgia

Fumiharu Togo[1], Akifumi Kishi[1] and Benjamin H. Natelson[2]

[1]Educational Physiology Laboratory, Graduate School of Education, The University of Tokyo, Tokyo, Japan [2]Department of Neurology, Icahn School of Medicine at Mount Sinai, New York, NY, United States

OUTLINE

Introduction	326	Consequences of Sleep Disorders and Abnormalities	332
Brief Overview of Chronic Fatigue Syndrome and Fibromyalgia	326	*Sleep-Related Symptoms*	332
Diagnosis	326	*Impacts on Pain and Fatigue*	332
Prevalence	327	Assessment	333
Etiology	327	Treatment	333
		Pharmacological Interventions	333
Sleep Disorders and Abnormalities in Chronic Fatigue Syndrome and Fibromyalgia	328	*Nonpharmacological Interventions*	334
		Other Treatments	336
Sleep Disorders	328	Areas for Future Research	337
Sleep Abnormalities in Patients Without Sleep Disorders	328	Conclusion	337
		References	338

Handbook of Sleep Disorders in Medical Conditions
DOI: https://doi.org/10.1016/B978-0-12-813014-8.00015-9

© 2019 Elsevier Inc. All rights reserved.

INTRODUCTION

Chronic fatigue syndrome (CFS) is a medically unexplained condition occurring mostly in women and is characterized by persistent or relapsing fatigue that lasts at least 6 months and substantially interferes with normal activities. In addition to severe fatigue, one of the symptoms used for diagnosing CFS is "unrefreshing sleep," and, in fact, this sleep-related problem is the most common complaint among patients with severe medically unexplained fatigue. An obvious possibility is that patients with this problem have an underlying sleep disorder or substantial amounts of interrupted sleep which may be responsible for the genesis of the illness.

Fibromyalgia (FM) is a medically unexplained illness characterized by four quadrant pain lasting at least 3 months and accompanied by multiple areas of tenderness on palpation of at specific points of the body using 4 kg force. FM occurs more often in women than men but is quite common in both sexes, occurring in approximately 3% of the population. As sleep difficulties are part of the 2010 standard diagnostic criteria, complaints of poor and nonrestorative sleep indicative of insomnia are common and have been associated with intense pain, fatigue, sleepiness, and cognitive difficulties in FM. FM frequently occurs in conjunction with CFS.

CFS and FM share considerable overlapping symptoms, including sleep-related complaints. However, differences between CFS and FM have been reported (e.g., substance P levels in cerebrospinal fluid[1,2] and plasma prolactin response to tryptophan in female patients[3]), and research focusing on uncovering differences between these medically unexplained illnesses is helpful to understand differences in their pathophysiologic processes, rather than focusing on their similarities.[4] In this chapter, we will review studies on sleep in CFS and FM patients in order to better understand their sleep problems, effects of treatment of their sleep problems, and differences between them.

BRIEF OVERVIEW OF CHRONIC FATIGUE SYNDROME AND FIBROMYALGIA

Diagnosis

The main feature in CFS is persistent or relapsing fatigue lasting for at least 6 months, severe enough to substantially interfere with the person's normal activities. Because it is diagnosed based on clinical criteria, it is largely a diagnosis of exclusion. The fatigue is unrelieved by rest and exacerbated by exercise. While depression may be associated with fatigue, CFS can occur in the absence of depressive disorders; for example, approximately 60% of patients studied in our team's research center were negative for depression (Natelson, personal communication). In addition to severe fatigue, diagnosis using the Fukuda Case Definition[5] requires the report of constant or recurring problems with four of the following symptoms: impaired memory or concentration, sore throat, tender cervical or axillary nodes, muscle pain, joint pain unaccompanied by redness or swelling, new headaches, unrefreshing sleep, and post exertion malaise. A new clinical case definition developed by the Institute of Medicine in 2015[6] presents simplified criteria: patients must have new onset of fatigue producing a substantial reduction in activity and accompanied by postexertional malaise, unrefreshing sleep, and at least one of the following manifestations: cognitive impairment or orthostatic intolerance.

According to the 1990 American College of Rheumatology (ACR) criteria, FM is a medically unexplained illness characterized by four quadrant pain lasting at least 3 months and accompanied by multiple areas of tenderness (i.e., at least 11 of the 18 designated tender points) on palpation of the body using 4 kg force. In 2010, the ACR proposed new diagnostic criteria which include a widespread pain index (WPI), a measure of the number of painful body regions of 19 areas of the body, and a symptom severity scale, a measure of the severity of symptoms such as fatigue, sleep problems, and cognitive problems. For the diagnosis of FM, a WPI score 7 or more with a symptom severity score of 5 or greater or a WPI score of 3—6 with a symptom severity score greater than or equal to 9 is required. Kaseeska et al. have reported that using this case definition blurs differences between CFS and FM, and so these updated criteria are not used by all researchers.

Prevalence

Studies on prevalence estimates in Brazil, Hong Kong, Japan, the United Kingdom, and the United States based on the Fukuda Case Definition[5] of CFS have reported 0.19%—6.4% in adults.[8] The illness affects women twice as often as men (0.52% vs 0.29%) in the United States[14] and is often disabling.

FM occurs in approximately 2%—3% of the population of countries in Europe (i.e., France, Germany, Italy, Portugal, and Spain)[9] and the United States.[10] The illness occurs more often in women than men. The prevalence of women to men ratio ranges from 40:1,[11] 6:1[10] to 2:1.[9]

Etiology

It is generally agreed that CFS is a heterogeneous disease and does not have a single cause. About a quarter of our study patients report an acute flu-like illness onset, suggestive of a viral infection, but a specific agent has not been consistently identified. However, many patients have abnormalities in central nervous system function. It has been proposed that some patients with CFS have a mild encephalopathy with varying adverse effects on different areas of the brain and different aspects of brain function producing with varying severity, abnormalities in information processing, sympathetic/parasympathetic nervous system balance, and/or endocrine function. The common complaint of unrefreshing sleep[12–16] seems to be caused by poor sleep efficiency[12,17–19] and multiple arousals[12,17,19] that lead to chronic sleep deprivation. Disturbed sleep in some CFS patients may be the result of an immune regulatory disorder or an imbalance in sleep-promoting and sleep-inhibiting cytokines.[20] Such an imbalance in addition to impaired regulation of other immune active substances within the brain and brain and spinal fluid abnormalities[21] may also reflect an encephalopathic process.

Although the etiology of FM is unclear, FM is characterized by widespread body pain, abnormal pain processing in the central nervous system, sleep disturbance, fatigue, and often psychological distress.[22,23] Patients with FM have hyperalgesia (increased sensitivity to painful stimuli) and/or allodynia (pain in response to normally nonpainful stimuli), suggesting that the patients have dysfunctional central pain processing rather than a pathological abnormality in the region of the body where the patients experiences pain. Sleep problems are common in patients with FM. Epidemiologic studies[24,25] reported that more than 90% of patients with FM complain about sleep problems such as difficulty falling asleep, difficulty falling back to sleep after waking up during nocturnal sleep, and unrefreshing sleep. An imbalance in sleep-promoting and sleep-inhibiting cytokines may cause disturbed sleep in some FM patients.[20,26]

SLEEP DISORDERS AND ABNORMALITIES IN CHRONIC FATIGUE SYNDROME AND FIBROMYALGIA

Sleep Disorders

Although untreated sleep apnea and narcolepsy which may explain the presence of chronic fatigue are exclusions for the diagnosis of CFS according to the 1994 Centers for Disease Control and Prevention (CDC) Case Definition,[5] polysomnography is not recommended in evaluating a patient with severe fatigue. As a result, several early studies suggested that as many as one-half of the individuals with CFS have mild sleep apnea syndrome (five or more episodes per hour of apnea/hypopnea), periodic leg movements, or the restless leg syndrome.[18,27,28] Other later studies with more stringent criteria for these disorders did not find this result.[29–31] In a population-based study of CFS in the United States,[30] 18.6% of CFS patients had sleep disorders. Among patients with sleep disordered breathing (SDB) and insomnia, a recent study reported that rates of CFS were 10% in SDB and 19% in insomnia.[32]

A study[33] reported a relatively low rate of SDB in patients with FM (i.e., 11%). However, later studies suggest that FM is often accompanied by a substantial amount of SDB (45%–96%),[34–37] except for our own work (i.e., 7% of CFS + FM patients).[29] The presence of obesity or overweight in women[38] may contribute to SDB, varying from frank sleep apnea to inspiratory airflow limitation with arousals.[34] Among patients with sleep apnea, one study found that patients with FM had the same frequency of sleep apnea as normal controls.[39] Another study reported that the estimated prevalence rate of FM patients was 6%.[40] One group suggested that as many as 33% of individuals with FM had the restless leg syndrome,[41] while the prevalence rate in CFS + FM patients in our own work was 7%.[29] A genetic study found common genetic characteristics between FM and narcolepsy.[42]

Sleep Abnormalities in Patients Without Sleep Disorders

Macrostructure of Sleep

Polysomnographic studies suggest that the sleep architecture is abnormal in CFS patients. The most consistent abnormality is significantly reduced sleep efficiency (i.e., the proportion of time spent sleeping relative to the time available for sleeping) when compared to controls[12,17–19]; the reported average values range from clearly abnormal (i.e., 76.5%)[17] to those within the normal range (i.e., 90%).[19] From one study providing data on individual patients, 75% of CFS patients had reduced sleep efficiency.[18] Sleep disturbance in CFS patients is obvious because they often show increases in time needed to fall asleep[12,19] and multiple periods of awakenings or arousals.[12,17,19] Decrease in total duration of Stage 4 sleep has also been reported.[17]

We have described the sleep architecture of a sample of female CFS or CFS + FM patients during a fixed period of their menstrual cycle and after excluding patients with diagnosable sleep disorders and coexisting major depressive disorder to reduce patient pool heterogeneity.[29] These patients differed significantly from matched controls in showing evidence of sleep disruption in the form of significantly reduced total sleep time, reduced sleep efficiency, and shorter bouts of sleep than healthy controls. In comparison with controls, a night of sleep in CFS or CFS + FM patients had little effect on both changes in self-rated sleepiness and fatigue before and after sleep. However, interestingly, for patients only, ratings of sleepiness and fatigue correlated well with total sleep duration and efficiency. For example, self-rated sleepiness and fatigue before sleep correlated positively with sleep efficiency and duration of Stage 4 sleep and negatively with sleep latency. Dichotomizing the patients into a group that felt sleepier after a night of sleep than before sleep (i.e., a.m. sleepier) and a group that felt less sleepy after a night of sleep (i.e., a.m. less sleepy) reduced the variability of the sleep structures considerably.[29] Those patients reporting less sleepiness after a night's sleep had sleep structures similar to those for healthy controls except for a shorter total sleep time and a commensurate reduction in Stage 2 sleep; moreover, they reported lower fatigue and pain following sleep. In contrast, patients in the a.m. sleepier group had the greatest abnormalities of sleep architecture, including poor sleep efficiency, longer sleep latency, and more disrupted sleep as manifested by a higher percentage of short-duration sleep runs, than either controls or patients in the a.m. less sleepy group.

As the time since awakening from sleep increases, sleep latency decreases,[43] and one early study of young adults reported an average sleep latency of 30 seconds after a night of sleep deprivation.[44] We have studied latency to fall asleep in patients with CFS and healthy controls, previously habituated to sleeping in a sleep lab, after such a night of sleep deprivation in our laboratory.[20] Nine out of twelve healthy subjects fell asleep within 6 minutes; however, three subjects took longer—falling asleep within 9 minutes. The CFS patients as a group showed a significantly longer latency to fall asleep after sleep deprivation, but the study population fell out into two groups with the largest group of 10 out of 15 patients falling asleep within 6 minutes. However, the remaining five patients remained awake for a longer period than any control, suggesting that they may have a disorder of arousal. Sleep latency following sleep deprivation correlated inversely with sleep efficiency for the patients with CFS. Our results indicate that some CFS patients may have a disorder of arousal which interferes with normal sleep and may, at least in part, be responsible for their disabling fatigue.

In many FM patients, a host of studies strongly suggest that the pattern of sleep is abnormal. The most consistent abnormality is significantly increased Stage 1 sleep compared to healthy controls.[39,45–50] Sleep disturbance in patients with FM is obvious because polysomnographic studies have shown longer sleep latencies,[47,48,51] more wakefulness,[47] reduced sleep efficiency,[47,48] reduced Stage 2 sleep,[48] and reduced Stage 4 sleep[45,52] in FM patients compared to healthy control subjects of similar age. Patients with FM awaken more easily[53] and compared to healthy controls have higher levels of physical activity during the night.[54,55]

Even when compared to control subjects with similar sleep efficiencies, FM patients showed more arousals[39,56] and increased Stage 1 sleep.[39] Molony et al. reported that patients with FM had three times more microarousals (brief sleep interruptions lasting 5–19 seconds) per hour than did healthy controls.[39] These results indicate that patients with FM have poor sleep quality with fragmented sleep.

Microstructure of Sleep

An alpha-electroencephalography (EEG) anomaly during nonrapid eye movement (non-REM) sleep has been considered a biologic correlate of chronic pain and a possible basis of nonrestorative sleep complaints in patients with FM.[57–61] The alpha-EEG anomaly is excessive alpha wave intrusion into slower frequency EEG activity which has been interpreted as a heightened arousal state during non-REM sleep.[60,62] However, this has not been found consistently across studies.[51] Alpha–delta sleep is an abnormal sleep EEG rhythm characterized by alpha activity that is superimposed on delta waves of Stages 3 and 4 sleep.[63] Horne and Shackell found that the mean alpha activity in Stages 2–4 sleep was greater for the patients with FM than for healthy controls.[51] Branco et al. studied alpha and delta activity and the alpha–delta ratio across sleep cycles in patients with FM and healthy controls.[57] The alpha–delta sleep anomaly occurred in almost all patients with fragmented sleep. Perlis et al. found that alpha-EEG sleep was associated with perception of shallow sleep and an increased tendency to display arousal in response to external auditory stimuli in patients with FM.[53] A pharmacological intervention study has found that sodium oxybate reduced alpha-EEG anomaly in patients with FM.[64]

Most studies on alpha-EEG anomaly in patients with FM have been based on visual and hence relatively subjective analysis of the EEG. Using spectral analysis, a quantitative measurement is provided not only for the alpha component of the EEG but also for other existing frequency components. Drewes et al. examined spectral EEG patterns and found that patients with FM showed more power in the alpha (higher frequency) band and a decrease in the lower frequency bands in Stages 2–4 sleep and all sleep cycles.[65] However, the alpha-EEG anomaly is not specific for patients with FM in which it also can be seen in some healthy individuals[50,51,62] and in patients with disorders such as rheumatoid arthritis and CFS.[66,67]

Symptoms of unrefreshing sleep are reported to be greater when the cyclic alternating pattern (CAP, periodic appearance of delta waves and K-complexes) of EEG occupies a greater percent of sleep.[68] Sforza et al. suggested that bursts of delta waves and K-complexes were expressions of subcortical arousals representing a real arousal response with tachycardia similar to that seen during cortical arousals.[69] Patients with FM have increased amounts of CAP—more so in the more severely symptomatic patients,[70] while there is no study on CAP in patients with CFS.

Sleep Stage Dynamics

In sleep stage dynamics, we have found robust differences between healthy controls and patients with CFS.[71] Although the distribution of duration of each sleep stage is not different between healthy controls and patients with CFS, probabilities of transition from both Stage 1 sleep and REM sleep to wake are significantly greater in patients with CFS than healthy controls, indicating that the influence of factors interfering with the continuation of Stage 1 sleep and REM sleep may be different between healthy controls and CFS patients, while the fundamental mechanisms determining durations of each sleep stage are similar. CFS patients might not have a dysfunction in systems maintaining each sleep stage, but they may have a disturbed switching mechanism governing sleep stage transitions. Our data suggest that the major complaint of CFS patients of "unrefreshing sleep" may be derived from this sudden arousal from both Stage 1 sleep and REM sleep.

One study[72] showed that sleep stage dynamics was different between patients with FM and healthy controls. Patients with FM showed a parameter that reflects shortened durations of Stage 2 sleep periods. Although shorter Stage 2 sleep durations did not predict daytime sleepiness, these did predict pain which is the main symptom in FM. Short Stage 2 sleep durations may be associated with sleep fragmentation or pressure for recovery sleep.

We have recently compared sleep stage dynamics between patients with CFS alone and CFS + FM and found specific differences between groups despite the fact that global sleep architecture did not differ significantly.[73] Patients with CFS alone have greater probabilities of transitions from REM sleep to wake than healthy controls. This result could be interpreted as a lower sleep pressure for patients with CFS alone. In contrast, patients with CFS + FM have greater probabilities of transitions from wake, Stage 1 sleep, and REM sleep to Stage 2 sleep and from Stage 2 sleep to slow-wave sleep than healthy controls, suggesting an increased sleep pressure in CFS + FM. Transitions from wake and REM sleep to Stage 2 sleep are unusual transitions in healthy adults.[71] CFS + FM also has greater probabilities of transitions from slow-wave sleep to wake and Stage 1 sleep, suggesting that this may be the specific sleep problem of CFS + FM.

There are reports of decreased level of central serotonin in FM patients.[74,75] In contrast, central serotonin responses have been reported to be upregulated in CFS patients.[3,76] We have recently reported that the administration of central monoaminergic (serotonergic and dopaminergic) antagonists alter dynamical sleep stage transitions from Stage 2 sleep to slow-wave sleep; probability of transition from Stage 2 sleep to slow-wave sleep was significantly increased when a central serotonergic and dopaminergic antagonist was administered.[77] Such monoaminergic systems are closely related with pain modulation.[78] Thus, the imbalance of central monoaminergic (serotonergic) systems in FM patients would lead to abnormalities of pain modulations and sleep regulations.

Autonomic Nervous System Activity During Sleep

One study reported an association between unrefreshing sleep and reduced heart rate variability (HRV) during sleep.[79] Specifically, the study showed that time- and frequency-domain HRV during sleep were significantly lower in patients with CFS compared to healthy controls and that HRV parameters could predict subjective sleepiness in CFS patients.[79] A population-based study also showed reduced HRV during sleep in patients with CFS compared to age, sex, and body mass index—matched healthy controls.[80]

We[81] have assessed short-term HRV during each sleep stage using a fractal scaling exponent ($\alpha 1$), analyzed by the detrended fluctuation analysis method, after stratifying patients into those who reported more or less sleepiness after a night of sleep (a.m. sleepier or a.m. less sleepy, respectively) and found that patients in the a.m. sleepier group showed higher fractal scaling index $\alpha 1$ during non-REM sleep than healthy controls, although standard polysomnographic measures did not differ between the groups. In addition, the fractal scaling index $\alpha 1$ during non-REM sleep was higher than that during awake periods after sleep onset for healthy controls and patients in the a.m. less sleepy group but did not differ between sleep stages for patients in the a.m. sleepier group. For patients, changes in self-reported sleepiness before and after the night correlated positively with the fractal scaling index $\alpha 1$ during non-REM sleep. These results suggest that R-wave to R-wave (RR) interval (i.e., the time interval between consecutive R peaks in the electrocardiogram) dynamics or autonomic nervous system activity during non-REM sleep might be associated with disrupted sleep in patients with CFS.

CONSEQUENCES OF SLEEP DISORDERS AND ABNORMALITIES

Sleep-Related Symptoms

In the 1994 CDC Case Definition of CFS,[5] unrefreshing sleep is one of the symptoms used for diagnosing CFS. In fact, this sleep-related problem is the most common complaint among patients with severe medically unexplained fatigue.[82] Sleep apnea and narcolepsy are sleep-related conditions which may explain the presence of chronic fatigue but are exclusions for the diagnosis of CFS. In the Canadian Consensus Criteria for Myalgic Encephalomyelitis/CFS,[83] sleep dysfunction such as unrefreshing sleep, poor sleep quality, and rhythm disturbances (i.e., early, middle or late insomnia, with reversed or irregular insomnia, hypersomnia, and abnormal diurnal variation of energy levels, including reversed or chaotic diurnal rest and sleep rhythms) is a required criterion for diagnosis of CFS. Treatable sleep disorders such as upper airway resistance syndrome, obstructive sleep apnea/central sleep apnea, and restless leg syndrome are exclusions.

Although sleep difficulties are not part of previous standard diagnostic criteria,[84] complaints of poor and nonrestorative sleep consistent with insomnia are common in patients with FM. An early study showed that 65.7% of patients with FM reported nonrestorative sleep.[85] Recently, two epidemiologic studies[24,25] reported that more than 90% of patients with FM complain about sleep problems such as difficulty falling asleep, difficulty falling back to sleep after waking up during nocturnal sleep, and unrefreshing sleep. Sleep is also one of the domains which associate most strongly with the patients' overall impression of improvement.[86]

Impacts on Pain and Fatigue

Sleep disruption and/or partial sleep deprivation in healthy people can produce the hallmark symptoms of CFS and FM—namely, marked daytime fatigue,[87] musculoskeletal achiness,[59,88,89] and cognitive problems.[87]

One study reported that sleep disturbances led to exacerbation of pain in patients with FM.[54] One recent study reported that negative mood (i.e., depression and anxiety), which is common among chronic pain patients or poor sleepers, almost fully mediated the relationship between sleep and pain in chronic pain patients.[90] A moderating impact of depressive symptoms on the relationship between sleep and pain was also reported in another study.[91]

ASSESSMENT

Sleep disturbances in CFS/FM are assessed similarly to that of any patient with a sleep complaint. The usual way this is done clinically is by taking a standard sleep history which involves questioning about time to bed (should be constant ±45 minutes maximum) and determining that the patient does not use the bed for any other function than sleep; other questions should include determination of sleep latency, frequency of arousals with ability or inability to fall back to sleep; total sleep time; and if a longer sleep duration is accompanied by report of better quality sleep. Specific questions related to detecting sleep pathology would inquire (1) about sudden onset of sleep episodes at unusual times such as while speaking or while driving as well as determining whether cataplexy exists and (2) whether the patient's bedmate notes breath arrest in the middle of or at the end of a bout of snoring.

In addition to collecting this clinical sleep history, the use of a questionnaire such as the Insomnia Severity Index[92] or the Epworth Sleepiness Scale can be useful. For example, Natelson et al.[32] studied 159 consecutive patients with complaints of disturbed sleep deemed serious enough to warrant an overnight sleep study. Twenty one (13%) had insomnia. In addition to these patients, 122 were found to have SDB. Of these, a strikingly large number, 13%, which is 30 times higher than the community prevalence, fulfilled the 1994 case definition for CFS; in contrast, 4% was found to fulfill the 1990 case definition for FM, a rate identical to that reported in community samples; these data support the possibility that CFS and FM may be caused by different underlying pathophysiological processes. Since SDB can produce the fatigue, pain, and unrefreshing sleep indicative of CFS, the treating physician needs to know which patients should undergo sleep studies. In that sense, the Epworth Sleepiness Scale score may be useful since scores higher than 15 were found to be indicative of a higher risk of having SDB.

TREATMENT

Pharmacological Interventions

A few studies have reported on the effects of pharmacological interventions on sleep disorders, sleep disturbances, or sleep quality in patients with CFS, with few benefits documented. Sleep disturbance was moderately improved in 70% patients with CFS after 6 weeks of treatment with nefazodone, an antidepressant agent.[93] Studies on the effects of other drugs did not report improvements in sleep problems in patients with CFS. Total and subscores of the Pittsburgh Sleep Quality Index (PSQI) were not changed by galantamine hydrobromide.[94] Sleep quality assessed by self-reported questionnaire was not improved by valganiciclovir.[95] Sleep latency during daytime was not changed by dextroamphetamine.[96] Insomnia score was not improved at week 8 and week 30 from the baseline by low-dose clonidine hydrochloride in adolescents with CFS.[97] In addition, low-dose clonidine had a concomitant negative effect on physical activity in adolescents with CFS; suggesting low-dose clonidine is not clinically useful in CFS.

Results in the field of FM are more promising. A review paper reported that antidepressants improve sleep in patients with FM.[98] Studies on the effects of amitriptyline,[99,100] cyclobenzaprine,[101] and sodium oxybate[102] also reported positive effects in sleep problems in patients with FM. In a double-blind trial of zolpidem, Moldofsky et al.[103] found that patients with FM reported increased total sleep time, fewer awakenings, and reduced sleep onset latency. In a double-blind study of zopiclone, slow-wave sleep was increased in patients with FM.[104]

Nonpharmacological Interventions

Exercise

Exertion is a particularly interesting avenue to study in CFS because a disabling and characteristic feature of CFS patients is that even minimal exertion produces a dramatic worsening of symptoms.[105] No such effect occurs in healthy controls, and, in fact, some reports suggest that acute exercise can actually improve sleep.[106] Using standard cardiac-type stress tests to probe effects of exertion on symptoms in CFS, Sisto et al. found that CFS patients reported more fatigue as much as 4 days after the exercise stress test.[107] Next, they used actigraphy to monitor activity before and after exercise and found that activity levels also fell significantly 4 days after the exercise stress test.[108] Yoshiuchi et al. recently replicated and extended these findings in real time and demonstrated that CFS symptoms worsen several days after maximal exercise but that neither mood nor cognitive function was affected.[109] They interpreted these changes in activity as evidence supporting the patients' complaint of worsening of symptoms induced by exercise or effort.

To our knowledge, only our team[110,111] has compared sleep in CFS patients before and after exercise. We recently investigated the influence of an acute bout of exercise on polysomnography and self-reported measures of sleep.[110] CFS patients as a group have disrupted sleep characterized by significantly poorer quality sleep than controls. However, the patients as a group showed evidence of improved sleep after exercise. The results were clearer after we used the same stratification strategy that we had used in our earlier work,[29] that is, splitting subjects into those who were either sleepier or less sleepy after a night of sleep. As expected, exercise improved the sleep quality of healthy controls who had reported decreased morning sleepiness after the baseline sleep night. Contrary to our expectation, it had the same result in less sleepy CFS patients who reported having decreased morning sleepiness. However, patients who reported increased morning sleepiness showed no improvement in sleep disruption, but exercise did not exacerbate their sleep disturbance. These patients also had the lowest average sleep efficiency of any of the groups studied. Because exercise did not produce a significant worsening of sleep morphology in CFS, the complaints of symptom worsening, which are reported to occur the next day after exertion, cannot be explained by disruption in sleep. After exercise, approximately half the patients actually slept better than on their baseline study night, whereas the rest simply did not improve.

We also calculated transition probabilities and rates between sleep stages and cumulative duration distributions (as a measure of continuity) of each sleep stage and sleep as a whole.[111] After exercise, healthy controls showed a significantly greater probability of transition from N1 to N2 and a lower rate of transition from N1 to wake than at baseline; patients with CFS showed a significantly greater probability of transition from N2 to N3 and a lower rate of transition from N2 to N1. These findings suggest improved quality of sleep after exercise. After exercise, controls had improved sleep continuity, whereas CFS had less continuous N1 and more continuous REM sleep. However, CFS had a significantly greater probability and rate of transition from REM to wake than controls. Probability of transition from REM to wake correlated significantly with increases in subjective fatigue, pain, and sleepiness overnight in CFS—suggesting these transitions may relate to patient complaints of unrefreshing sleep. Thus, exercise promoted transitions to deeper sleep stages and inhibited transitions to lighter sleep stages for controls and CFS, but CFS also reported increased fatigue and continued to have REM sleep disruption. This dissociation suggests possible mechanistic pathways for the underlying pathology of CFS.

Two randomized controlled studies on graded exercise therapy have found improvements in sleep in patients with CFS. Powell et al. reported that individual patient education to encourage graded exercise improved sleep problems as well as physical functions, although changes in the amount of exercise were not reported.[112] In addition, improvements in sleep problems were maintained at 2-year follow-up.[113] Another group[114] found that patients in the graded exercise therapy group had better scores on a sleep disturbance scale (i.e., Jenkins Sleep Scale) compared with patients in usual medical care alone group at 52 weeks. The medical care consisted of an explanation of CFS, generic advice, such as to avoid extremes of activity and rest, specific advice on self-help, and symptomatic pharmacotherapy. One study,[115] however, did not find improvements in PSQI score by a 12-week supervised graded aerobic exercise program, although fatigue and peak oxygen uptake were improved.

In patients with FM, Andrade et al. examined the effects of strength training on quality of sleep.[116] The training consisted 24 training sessions, three times per week for 8 weeks. After 8 weeks of the training, patients showed improvements in self-reported quality of sleep.

Cognitive Behavioral Therapy

Two studies in patients with CFS have reported improvements in sleep with cognitive behavioral therapy. One study[117] found that patients in a group-based 12-week cognitive behavioral stress-management intervention group showed improvements in unrefreshing sleep. In this intervention, patients were trained with specific relaxation techniques, including progressive muscle relaxation and visualization techniques, and were taught to better recognize how stress impacts them emotionally and physically, and the relationship between thoughts, feeling, and behaviors. Another study using a randomized control trial[114] indicated that cognitive behavioral therapy led to better scores on a sleep disturbance scale (i.e., Jenkins Sleep Scale) compared with patients in a specialist medical care alone group at 52 weeks. During cognitive behavioral therapy, patients addressed unhelpful cognitions, including fears about symptoms or activity by testing these in behavioral experiments. The experiments consisted of establishing a baseline of activity and rest and a regular sleep pattern, and then making planned gradual increases in both physical and mental activities. Furthermore, patients were helped to address social and emotional obstacles to improvement through problem-solving. A multi-convergent therapy which combined cognitive behavioral therapy and graded exercise therapy also showed improvements in sleep quality at posttherapy in patients with CFS.[118] The study found that after approximately 10 one-hour sessions, the rates of patients who reported improvement in sleep was 83.3% in the therapy group, compared to 26.6% in a relaxation group, and 11.1% in a control group. The improvement in the multi-convergent therapy group was maintained at 6-month follow-up. During cognitive behavior therapy, patients worked through the following stages: (1) exploration of the predisposing, precipitating, and perpetuating factors, (2) exploration of a model of illness as well as neurophysiological model of neuroplasticity, (3) introduction of behavior modification in relation to cognitions, (4) exploration of anxiety/depression, (5) identifying positive and negative patterns of behavior pertaining to fatigue etc., (6) coping strategies, (7) exploration of sleep problems and rectification where problems existed (sleep hygiene), and (8) application of techniques (meditation etc.) for behavior modification.

In patients with FM, a study has reported that cognitive behavioral therapy for insomnia is effective for sleep disturbance in the patients. Using sleep-logs, Edinger et al.[119] found a reduction of about 50% in nocturnal wake time in a group treated with cognitive behavioral therapy compared to a 20% reduction in a sleep hygiene therapy group, and 3.5% in a group receiving usual care during 6 weeks. In addition, 57% in the cognitive behavioral group improves in objective sleep (i.e., total sleep time, total wake time, and sleep efficiency) by the end of the intervention, while 17% and 0% in the sleep hygiene and control groups, respectively.

Other Treatments

Besides cognitive behavioral therapy for insomnia in patients with FM, there are no FDA-approved treatments for the sleep disturbance in CFS and/or FM; so the physician is often left to try treatments that have shown efficacy in other patient populations. Melatonin before sleep can sometimes be effective.[120] Valerian root is nontoxic and worthy of consideration.[121] While tricyclic antidepressants such as amitriptyline have sedating effects as well as effects in reducing both focal (headache)[122] and widespread pain,[123] the side effect profile of weight gain, dry mouth and postural instability limit their use. Low doses of doxepin (6 mg or less) have been shown to improve sleep quality[124] and a low-dose preparation of this medication has been recently approved by the FDA for insomnia. Patients often ask for treatment with standard sedatives such as zolpidem or eszopiclone (Natelson, personal communication). While these medications do induce sleep, patients with CFS or FM seldom report an improved quality of sleep and continue to report having substantial problems with their sleep being unrefreshing (Natelson, personal communication). The company manufacturing oxybate did a number of double-blind, placebo-controlled trials of this drug to relieve pain and the sleep complaints in FM; although these studies were positive[102] and encouraging, the company did not pursue getting an indication from the US FDA to treat FM—probably related to the abuse-potential of this class of drugs. Since use of this medication is heavily regulated within the United States, physicians wanting to prescribe this drug must interact with the drug's manufacturer.

AREAS FOR FUTURE RESEARCH

Limited research has been conducted on the effects of nonpharmacological interventions on sleep problems in patients with CFS or FM. Effects of exercise therapy and cognitive behavioral therapy on objective measures of sleep are promising but have only been examined in one study. Future studies in this area using objective measures of sleep using polysomnography and actigraphy are needed. Regarding protocols of exercise therapy, details of exercise on the effects of sleep problems on the patients are still unclear. The effects of intensity, duration, frequency, and mode of exercise on sleep problems have yet to be examined. These are important for establishing how to prescribe exercise therapy individually. In addition, considering that problems in sleep regulation may be different between patients with CFS and FM (e.g., more arousals during Stage 1 sleep and REM sleep in patients with CFS and greater probabilities of transitions from slow-wave sleep to wake and Stage 1 sleep in patients with FM), the most effective strategy in nonpharmacological interventions, as well as the most effective pharmacological agents, may be different between them.

CONCLUSION

Patients with CFS and FM often have sleep-related problems. Polysomnographic studies have shown sleep problems in CFS, that is, increased Stage 1 sleep, reduced slow-wave sleep, more arousals, prolonged sleep onset, reduced sleep efficiency, and microarousals or subcortical arousals during sleep. Although these problems are also shown in patients with FM, dynamic aspects of sleep show different patterns between CFS and FM patients. Patients with CFS + FM had greater probabilities of transitions from wake, Stage 1 sleep, and REM sleep to Stage 2 sleep and from Stage 2 sleep to slow-wave sleep than healthy controls, suggesting the increased sleep pressure in CFS + FM which is probably caused by disruption in slow-wave sleep. In contrast, CFS alone has greater probabilities of transitions from REM sleep to awake than healthy controls, suggesting the existence of lower sleep pressure in CFS alone. Finding such differences is support for the thesis that CFS is a different illness from FM, associated with different problems in sleep regulation. Studies on pharmacological and nonpharmacological interventions such as antidepressants, graded exercise, and cognitive behavior therapy have shown some improvements in sleep problems but remain limited. After a bout of acute exercise, our polysomnographic study showed that patients with CFS had better sleep or unchanged sleep compared with their

338

baseline sleep. However, it should be noted that, after an acute exercise stress test, CFS patients reported more fatigue for several days and their activity levels during daily life also fell, which should be considered when seeking the optimal strategy for exercise therapy in patients with CFS and/or FM.

KEY PRACTICE POINTS

- No significant difference in rates of severe sleep disorders between CFS and controls or between those with both CFS and FM and controls.
- FM patients may have increased rates of sleep disordered breathing (SDB).
- Patients thought to have CFS with Epworth Sleepiness Scores of 16 or higher have a high probability of having coexisting SDB and should undergo PSG testing to confirm and then possibly to treat.
- Patients with sleep disorders undergoing PSG were found to have greatly elevated rates of CFS compared to community controls but the same rates of FM as in community samples.
- However, whether treatment for SDB will relieve the symptoms of CFS and/or FM remains unclear although one study in CFS did not find improvement by continuous positive airway pressure (CPAP) intervention.
- The disturbed sleep reported by patients with FM can often be markedly helped by a program of cognitive behavioral therapy aimed at improving sleep quality. This therapeutic effect will probably carry over to those with CFS also.

References

1. Russell IJ, Orr MD, Littman B, et al. Elevated cerebrospinal fluid levels of substance P in patients with the fibromyalgia syndrome. *Arthritis Rheum.* 1994;37(11):1593−1601.
2. Evengard B, Nilsson CG, Lindh G, et al. Chronic fatigue syndrome differs from fibromyalgia. No evidence for elevated substance P levels in cerebrospinal fluid of patients with chronic fatigue syndrome. *Pain.* 1998;78 (2):153−155.
3. Weaver SA, Janal MN, Aktan N, Ottenweller JE, Natelson BH. Sex differences in plasma prolactin response to tryptophan in chronic fatigue syndrome patients with and without comorbid fibromyalgia. *J Womens Health (Larchmt).* 2010;19(5):951−958.
4. Lange G, Natelson BH. Chronic fatigue syndrome. In: Mayer EA, Bushnell MC, eds. *Functional Pain Syndromes: Presentation and Pathophysiology.* 1st ed. Seattle, WA: IASP Press; 2009:245−261.
5. Fukuda K, Straus SE, Hickie I, Sharpe MC, Dobbins JG, Komaroff A. The chronic fatigue syndrome: a comprehensive approach to its definition and study. International Chronic Fatigue Syndrome Study Group. *Ann Intern Med.* 1994;121(12):953−959.
6. Committee on the Diagnostic Criteria for Myalgic Encephalomyelitis/Chronic Fatigue Syndrome, Board on the Health of Select Populations, Institute of Medicine. *Beyond Myalgic Encephalomyelitis/Chronic Fatigue Syndrome: Redefining an Illness.* Washington, DC: National Academies Press; 2015.
7. Kaseeska K, Brown M, Jason LA. Comparing two fibromyalgia diagnostic criteria in a cohort of chronic fatigue syndrome patients. *Bull IACFS/ME.* 2011;19(1):47−57.
8. Johnston S, Brenu EW, Staines DR, Marshall-Gradisnik S. The adoption of chronic fatigue syndrome/myalgic encephalomyelitis case definitions to assess prevalence: a systematic review. *Ann Epidemiol.* 2013;23(6):371−376.
9. Branco JC, Bannwarth B, Failde I, et al. Prevalence of fibromyalgia: a survey in five European countries. *Semin Arthritis Rheum.* 2010;39(6):448−453.

REFERENCES

10. Wolfe F, Ross K, Anderson J, Russell IJ, Hebert L. The prevalence and characteristics of fibromyalgia in the general population. *Arthritis Rheum.* 1995;38(1):19–28.
11. Senna ER, De Barros AL, Silva EO, et al. Prevalence of rheumatic diseases in Brazil: a study using the COPCORD approach. *J Rheumatol.* 2004;31(3):594–597.
12. Sharpley A, Clements A, Hawton K, Sharpe M. Do patients with "pure" chronic fatigue syndrome (neurasthenia) have abnormal sleep? *Psychosom Med.* 1997;59(6):592–596.
13. Nisenbaum R, Jones JF, Unger ER, Reyes M, Reeves WC. A population-based study of the clinical course of chronic fatigue syndrome. *Health Qual Life Outcomes.* 2003;1:49.
14. Jason LA, Richman JA, Rademaker AW, et al. A community-based study of chronic fatigue syndrome. *Arch Intern Med.* 1999;159(18):2129–2137.
15. Nisenbaum R, Reyes M, Unger ER, Reeves WC. Factor analysis of symptoms among subjects with unexplained chronic fatigue – what can we learn about chronic fatigue syndrome? *J Psychosom Res.* 2004;56 (2):171–178.
16. Hamaguchi M, Kawahito Y, Takeda N, Kato T, Kojima T. Characteristics of chronic fatigue syndrome in a Japanese community population Chronic fatigue syndrome in Japan. *Clin Rheumatol.* 2011;30(7):895–906.
17. Fischler B, Le BO, Hoffmann G, Cluydts R, Kaufman L, De MK. Sleep anomalies in the chronic fatigue syndrome. A comorbidity study. *Neuropsychobiology.* 1997;35(3):115–122.
18. Krupp LB, Jandorf L, Coyle PK, Mendelson WB. Sleep disturbance in chronic fatigue syndrome. *J Psychosom Res.* 1993;37(4):325–331.
19. Morriss R, Sharpe M, Sharpley AL, Cowen PJ, Hawton K, Morris J. Abnormalities of sleep in patients with the chronic fatigue syndrome. *Br Med J.* 1993;306(6886):1161–1164.
20. Nakamura T, Togo F, Cherniack NS, Rapoport DM, Natelson BH. A subgroup of patients with chronic fatigue syndrome may have a disorder of arousal. *Open Sleep J.* 2010;3:6–11.
21. Natelson BH, Mao X, Stegner AJ, et al. Multimodal and simultaneous assessments of brain and spinal fluid abnormalities in chronic fatigue syndrome and the effects of psychiatric comorbidity. *J Neurol Sci.* 2017;375:411–416.
22. Centers for Disease Control and Prevention (CDC). Fibromyalgia. <http://www.cdc.gov/arthritis/basics/fibromyalgia.htm>; 2015 Accessed 01.08.18.
23. National Institute of Arthritis and Musculoskeletal and Skin Diseases (NIAMS). Fibromyalgia: questions and answers about fibromyalgia. <http://www.niams.nih.gov/health_Info/Fibromyalgia/default.asp>; 2014 Accessed 01.08.18.
24. Bigatti SM, Hernandez AM, Cronan TA, Rand KL. Sleep disturbances in fibromyalgia syndrome: relationship to pain and depression. *Arthritis Rheum.* 2008;59(7):961–967.
25. Theadom A, Cropley M, Humphrey KL. Exploring the role of sleep and coping in quality of life in fibromyalgia. *J Psychosom Res.* 2007;62(2):145–151.
26. Togo F, Natelson BH, Adler GK, et al. Plasma cytokine fluctuations over time in healthy controls and patients with fibromyalgia. *Exp Biol Med (Maywood).* 2009;234(2):232–240.
27. Buchwald D, Pascualy R, Bombardier C, Kith P. Sleep disorders in patients with chronic fatigue. *Clin Infect Dis.* 1994;18(Suppl. 1):S68–S72.
28. Le Bon O, Fischler B, Hoffmann G, et al. How significant are primary sleep disorders and sleepiness in the chronic fatigue syndrome? *Sleep Res Online.* 2000;3(2):43–48.
29. Togo F, Natelson BH, Cherniack NS, FitzGibbons J, Garcon C, Rapoport DM. Sleep structure and sleepiness in chronic fatigue syndrome with or without coexisting fibromyalgia. *Arthritis Res Ther.* 2008;10(3):R56.
30. Reeves WC, Heim C, Maloney EM, et al. Sleep characteristics of persons with chronic fatigue syndrome and non-fatigued controls: results from a population-based study. *BMC Neurol.* 2006;6:41.
31. Ball N, Buchwald DS, Schmidt D, Goldberg J, Ashton S, Armitage R. Monozygotic twins discordant for chronic fatigue syndrome: objective measures of sleep. *J Psychosom Res.* 2004;56(2):207–212.
32. Pejovic S, Natelson BH, Basta M, Fernandez-Mendoza J, Mahr F, Vgontzas AN. Chronic fatigue syndrome and fibromyalgia in diagnosed sleep disorders: a further test of the 'unitary' hypothesis. *BMC Neurol.* 2015;15:53.
33. May KP, West SG, Baker MR, Everett DW. Sleep-apnea in male-patients with the fibromyalgia syndrome. *Am J Med.* 1993;94(5):505–508.

II. SLEEP DISORDERS IN SPECIFIC MEDICAL CONDITIONS

34. Gold AR, Dipalo F, Gold MS, Broderick J. Inspiratory airflow dynamics during sleep in women with fibromyalgia. *Sleep*. 2004;27(3):459–466.
35. Shah MA, Feinberg S, Krishnan E. Sleep-disordered breathing among women with fibromyalgia syndrome. *J Clin Rheumatol*. 2006;12(6):277–281.
36. Germanowicz D, Lumertz MS, Martinez D, Margarites AF. Sleep disordered breathing concomitant with fibromyalgia syndrome. *J Bras Pneumol*. 2006;32(4):333–338.
37. Rosenfeld VW, Rutledge DN, Stern JM. Polysomnography with quantitative EEG in patients with and without fibromyalgia. *J Clin Neurophysiol*. 2015;32(2):164–170.
38. Moldofsky H. Management of sleep disorders in fibromyalgia. *Rheum Dis Clin North Am*. 2002;28(2):353–365.
39. Molony RR, MacPeek DM, Schiffman PL, et al. Sleep, sleep apnea and the fibromyalgia syndrome. *J Rheumatol*. 1986;13(4):797–800.
40. Lario BA, Teran J, Alonso JL, Alegre J, Arroyo I, Viejo JL. Lack of association between fibromyalgia and sleep-apnea syndrome. *Ann Rheum Dis*. 1992;51(1):108–111.
41. Viola-Saltzman M, Watson NF, Bogart A, Goldberg J, Buchwald D. High prevalence of restless legs syndrome among patients with fibromyalgia: a controlled cross-sectional study. *J Clin Sleep Med*. 2010;6(5):423–427.
42. Spitzer AR, Broadman M. A retrospective review of the sleep characteristics in patients with chronic fatigue syndrome and fibromyalgia. *Pain Pract*. 2010;10(4):294–300.
43. Devoto A, Lucidi F, Violani C, Bertini M. Effects of different sleep reductions on daytime sleepiness. *Sleep*. 1999;22(3):336–343.
44. Carskadon MA, Dement WC. Effects of total sleep loss on sleep tendency. *Percept Mot Skills*. 1979;48(2):495–506.
45. Anch AM, Lue FA, MacLean AW, Moldofsky H. Sleep physiology and psychological aspects of the fibrositis (fibromyalgia) syndrome. *Can J Psychol*. 1991;45(2):179–184.
46. Cote KA, Moldofsky H. Sleep, daytime symptoms, and cognitive performance in patients with fibromyalgia. *J Rheumatol*. 1997;24(10):2014–2023.
47. Drewes AM, Svendsen L, Nielsen KD, Taagholt SJ, Bjerregard K. Quantification of Alpha-EEG activity during sleep in fibromyalgia: a study based on ambulatory sleep monitoring. *J Musculoskelet Pain*. 1994;2(4):33–53.
48. Landis CA, Lentz MJ, Tsuji J, Buchwald D, Shaver JL. Pain, psychological variables, sleep quality, and natural killer cell activity in midlife women with and without fibromyalgia. *Brain Behav Immun*. 2004;18(4):304–313.
49. Leventhal L, Freundlich B, Lewis J, Gillen K, Henry J, Dinges D. Controlled study of sleep parameters in patients with fibromyalgia. *J Clin Rheumatol*. 1995;1(2):110–113.
50. Shaver JL, Lentz M, Landis CA, Heitkemper MM, Buchwald DS, Woods NF. Sleep, psychological distress, and stress arousal in women with fibromyalgia. *Res Nurs Health*. 1997;20(3):247–257.
51. Horne JA, Shackell BS. Alpha-like EEG activity in non-REM sleep and the fibromyalgia (fibrositis) syndrome. *Electroencephalogr Clin Neurophysiol*. 1991;79(4):271–276.
52. Lashley FR. A review of sleep in selected immune and autoimmune disorders. *Holist Nurs Pract*. 2003;17(2):65–80.
53. Perlis ML, Giles DE, Bootzin RR, et al. Alpha sleep and information processing, perception of sleep, pain, and arousability in fibromyalgia. *Int J Neurosci*. 1997;89(3–4):265–280.
54. Affleck G, Urrows S, Tennen H, Higgins P, Abeles M. Sequential daily relations of sleep, pain intensity, and attention to pain among women with fibromyalgia. *Pain*. 1996;68(2–3):363–368.
55. Korszun A, Young EA, Engleberg NC, Brucksch CB, Greden JF, Crofford LA. Use of actigraphy for monitoring sleep and activity levels in patients with fibromyalgia and depression. *J Psychosom Res*. 2002;52(6):439–443.
56. Jennum P, Drewes AM, Andreasen A, Nielsen KD. Sleep and other symptoms in primary fibromyalgia and in healthy controls. *J Rheumatol*. 1993;20(10):1756–1759.
57. Branco J, Atalaia A, Paiva T. Sleep cycles and alpha–delta sleep in fibromyalgia syndrome. *J Rheumatol*. 1994;21(6):1113–1117.
58. Moldofsky H, Scarisbrick P, England R, Smythe H. Musculosketal symptoms and non-REM sleep disturbance in patients with "fibrositis syndrome" and healthy subjects. *Psychosom Med*. 1975;37(4):341–351.
59. Moldofsky H, Scarisbrick P. Induction of neurasthenic musculoskeletal pain syndrome by selective sleep stage deprivation. *Psychosom Med*. 1976;38(1):35–44.
60. Moldofsky H. Sleep and fibrositis syndrome. *Rheum Dis Clin North Am*. 1989;15(1):91–103.

REFERENCES

61. Roizenblatt S, Moldofsky H, Benedito-Silva AA, Tufik S. Alpha sleep characteristics in fibromyalgia. *Arthritis Rheum.* 2001;44(1):222−230.
62. Scheuler W, Stinshoff D, Kubicki S. The alpha-sleep pattern. Differentiation from other sleep patterns and effect of hypnotics. *Neuropsychobiology.* 1983;10(2−3):183−189.
63. McNamara ME. Alpha sleep: a mini review and update. *Clin Electroencephalogr.* 1993;24(4):192−193.
64. Scharf MB, Baumann M, Berkowitz DV. The effects of sodium oxybate on clinical symptoms and sleep patterns in patients with fibromyalgia. *J Rheumatol.* 2003;30(5):1070−1074.
65. Drewes AM, Nielsen KD, Taagholt SJ, Bjerregard K, Svendsen L, Gade J. Sleep intensity in fibromyalgia: focus on the microstructure of the sleep process. *Br J Rheumatol.* 1995;34(7):629−635.
66. Moldofsky H, Lue FA, Smythe HA. Alpha EEG sleep and morning symptoms in rheumatoid arthritis. *J Rheumatol.* 1983;10(3):373−379.
67. Moldofsky H, Saskin P, Lue FA. Sleep and symptoms in fibrositis syndrome after a febrile illness. *J Rheumatol.* 1988;15(11):1701−1704.
68. Terzano MG, Parrino L. Origin and significance of the cyclic alternating pattern (CAP). Review article. *Sleep Med Rev.* 2000;4(1):101−123.
69. Sforza E, Jouny C, Ibanez V. Cardiac activation during arousal in humans: further evidence for hierarchy in the arousal response. *Clin Neurophysiol.* 2000;111(9):1611−1619.
70. Rizzi M, Sarzi-Puttini P, Atzeni F, et al. Cyclic alternating pattern: a new marker of sleep alteration in patients with fibromyalgia? *J Rheumatol.* 2004;31(6):1193−1199.
71. Kishi A, Struzik ZR, Natelson BH, Togo F, Yamamoto Y. Dynamics of sleep stage transitions in healthy humans and patients with chronic fatigue syndrome. *Am J Physiol Regul Integr Comp Physiol.* 2008;294(6): R1980−R1987.
72. Burns JW, Crofford LJ, Chervin RD. Sleep stage dynamics in fibromyalgia patients and controls. *Sleep Med.* 2008;9(6):689−696.
73. Kishi A, Natelson BH, Togo F, Struzik ZR, Rapoport DM, Yamamoto Y. Sleep-stage dynamics in patients with chronic fatigue syndrome with or without fibromyalgia. *Sleep.* 2011;34(11):1551−1560.
74. Juhl JH. Fibromyalgia and the serotonin pathway. *Altern Med Rev.* 1998;3(5):367−375.
75. Neeck G, Riedel W. Neuromediator and hormonal perturbations in fibromyalgia syndrome: results of chronic stress? *Baillieres Clin Rheumatol.* 1994;8(4):763−775.
76. Afari N, Buchwald D. Chronic fatigue syndrome: a review. *Am J Psychiatry.* 2003;160(2):221−236.
77. Kishi A, Yasuda H, Matsumoto T, et al. Sleep stage transitions in healthy humans altered by central monoaminergic antagonist. *Methods Inf Med.* 2010;49(5):458−461.
78. Bannister K, Bee LA, Dickenson AH. Preclinical and early clinical investigations related to monoaminergic pain modulation. *Neurotherapeutics.* 2009;6(4):703−712.
79. Burton AR, Rahman K, Kadota Y, Lloyd A, Vollmer-Conna U. Reduced heart rate variability predicts poor sleep quality in a case-control study of chronic fatigue syndrome. *Exp Brain Res.* 2010;204(1):71−78.
80. Boneva RS, Decker MJ, Maloney EM, et al. Higher heart rate and reduced heart rate variability persist during sleep in chronic fatigue syndrome: a population-based study. *Auton Neurosci.* 2007;137(1-2):94−101.
81. Togo F, Natelson BH. Heart rate variability during sleep and subsequent sleepiness in patients with chronic fatigue syndrome. *Auton Neurosci.* 2013;176(1-2):85−90.
82. Unger ER, Nisenbaum R, Moldofsky H, et al. Sleep assessment in a population-based study of chronic fatigue syndrome. *BMC Neurol.* 2004;4:6.
83. Carruthers BM, Jain AK, De Meirleir KL, et al. Myalgic encephalomyelitis/chronic fatigue syndrome. *J Chron Fatigue Syndr.* 2003;11(1):7−115.
84. Wolfe F, Smythe HA, Yunus MB, et al. The American College of Rheumatology 1990 criteria for the classification of fibromyalgia. Report of the Multicenter Criteria Committee. *Arthritis Rheum.* 1990;33 (2):160−172.
85. White KP, Speechley M, Harth M, Ostbye T. The London Fibromyalgia Epidemiology Study: comparing the demographic and clinical characteristics in 100 random community cases of fibromyalgia versus controls. *J Rheumatol.* 1999;26(7):1577−1585.
86. Arnold LM, Zlateva G, Sadosky A, Emir B, Whalen E. Correlations between fibromyalgia symptom and function domains and patient global impression of change: a pooled analysis of three randomized, placebo-controlled trials of pregabalin. *Pain Med.* 2011;12(2):260−267.

II. SLEEP DISORDERS IN SPECIFIC MEDICAL CONDITIONS

87. Martin SE, Engleman HM, Deary IJ, Douglas NJ. The effect of sleep fragmentation on daytime function. *Am J Respir Crit Care Med.* 1996;153(4 Pt 1):1328–1332.
88. Lentz MJ, Landis CA, Rothermel J, Shaver JL. Effects of selective slow wave sleep disruption on musculoskeletal pain and fatigue in middle aged women. *J Rheumatol.* 1999;26(7):1586–1592.
89. Onen SH, Alloui A, Gross A, Eschallier A, Dubray C. The effects of total sleep deprivation, selective sleep interruption and sleep recovery on pain tolerance thresholds in healthy subjects. *J Sleep Res.* 2001;10(1):35–42.
90. O'Brien EM, Waxenberg LB, Atchison JW, et al. Negative mood mediates the effect of poor sleep on pain among chronic pain patients. *Clin J Pain.* 2010;26(4):310–319.
91. O'Brien EM, Waxenberg LB, Atchison JW, et al. Intraindividual variability in daily sleep and pain ratings among chronic pain patients: bidirectional association and the role of negative mood. *Clin J Pain.* 2011;27(5):425–433.
92. Bastien CH, Vallieres A, Morin CM. Validation of the Insomnia Severity Index as an outcome measure for insomnia research. *Sleep Med.* 2001;2(4):297–307.
93. Hickie I. Nefazodone for patients with chronic fatigue syndrome. *Aust N Z J Psychiatry.* 1999;33(2):278–280.
94. Blacker CV, Greenwood DT, Wesnes KA, et al. Effect of galantamine hydrobromide in chronic fatigue syndrome: a randomized controlled trial. *JAMA.* 2004;292(10):1195–1204.
95. Montoya JG, Kogelnik AM, Bhangoo M, et al. Randomized clinical trial to evaluate the efficacy and safety of valganciclovir in a subset of patients with chronic fatigue syndrome. *J Med Virol.* 2013;85(12):2101–2109.
96. Olson LG, Ambrogetti A, Sutherland DC. A pilot randomized controlled trial of dexamphetamine in patients with chronic fatigue syndrome. *Psychosomatics.* 2003;44(1):38–43.
97. Sulheim D, Fagermoen E, Winger A, et al. Disease mechanisms and clonidine treatment in adolescent chronic fatigue syndrome: a combined cross-sectional and randomized clinical trial. *JAMA Pediatr.* 2014;168(4):351–360.
98. Goldenberg DL, Burckhardt C, Crofford L. Management of fibromyalgia syndrome. *JAMA.* 2004;292(19):2388–2395.
99. Carette S, Mccain GA, Bell DA, Fam AG. Evaluation of amitriptyline in primary fibrositis – a double-blind, placebo-controlled study. *Arthritis Rheum.* 1986;29(5):655–659.
100. Goldenberg DL, Felson DT, Dinerman H. A randomized, controlled trial of amitriptyline and naproxen in the treatment of patients with fibromyalgia. *Arthritis Rheum.* 1986;29(11):1371–1377.
101. Tofferi JK, Jackson JL, O'Malley PG. Treatment of fibromyalgia with cyclobenzaprine: a meta-analysis. *Arthritis Rheum.* 2004;51(1):9–13.
102. Spaeth M, Bennett RM, Benson BA, Wang YG, Lai C, Choy EH. Sodium oxybate therapy provides multidimensional improvement in fibromyalgia: results of an international phase 3 trial. *Ann Rheum Dis.* 2012;71(6):935–942.
103. Moldofsky H, Lue FA, Mously C, Roth-Schechter B, Reynolds WJ. The effect of zolpidem in patients with fibromyalgia: a dose ranging, double blind, placebo controlled, modified crossover study. *J Rheumatol.* 1996;23(3):529–533.
104. Drewes AM, Andreasen A, Jennum P, Nielsen KD. Zopiclone in the treatment of sleep abnormalities in fibromyalgia. *Scand J Rheumatol.* 1991;20(4):288–293.
105. Komaroff AL, Buchwald D. Symptoms and signs of chronic fatigue syndrome. *Rev Infect Dis.* 1991;13(Suppl. 1):S8–S11.
106. Youngstedt SD, O'Connor PJ, Dishman RK. The effects of acute exercise on sleep: a quantitative synthesis. *Sleep.* 1997;20(3):203–214.
107. Sisto SA, LaManca J, Cordero DL, et al. Metabolic and cardiovascular effects of a progressive exercise test in patients with chronic fatigue syndrome. *Am J Med.* 1996;100(6):634–640.
108. Sisto SA, Tapp WN, LaManca JJ, et al. Physical activity before and after exercise in women with chronic fatigue syndrome. *QJM.* 1998;91(7):465–473.
109. Yoshiuchi K, Cook DB, Ohashi K, et al. A real-time assessment of the effect of exercise in chronic fatigue syndrome. *Physiol Behav.* 2007;92(5):963–968.
110. Togo F, Natelson BH, Cherniack NS, Klapholz M, Rapoport DM, Cook DB. Sleep is not disrupted by exercise in patients with chronic fatigue syndromes. *Med Sci Sports Exerc.* 2010;42(1):16–22.
111. Kishi A, Togo F, Cook DB, et al. The effects of exercise on dynamic sleep morphology in healthy controls and patients with chronic fatigue syndrome. *Physiol Rep.* 2013;1(6):e00152.

REFERENCES

343

112. Powell P, Bentall RP, Nye FJ, Edwards RH. Randomised controlled trial of patient education to encourage graded exercise in chronic fatigue syndrome. *BMJ*. 2001;322(7283):387—390.

113. Powell P, Bentall RP, Nye FJ, Edwards RH. Patient education to encourage graded exercise in chronic fatigue syndrome. 2-year follow-up of randomised controlled trial. *Br J Psychiatry*. 2004;184:142—146.

114. White PD, Goldsmith KA, Johnson AL, et al. Comparison of adaptive pacing therapy, cognitive behaviour therapy, graded exercise therapy, and specialist medical care for chronic fatigue syndrome (PACE): a randomised trial. *Lancet*. 2011;377(9768):823—836.

115. Fulcher KY, White PD. Randomised controlled trial of graded exercise in patients with the chronic fatigue syndrome. *Br Med J*. 1997;314(7095):1647—1652.

116. Andrade A, Vilarino GT, Bevilacqua GG. What is the effect of strength training on pain and sleep in patients with fibromyalgia? *Am J Phys Med Rehabil*. 2017;96(12):889—893.

117. Lopez C, Antoni M, Penedo F, et al. A pilot study of cognitive behavioral stress management effects on stress, quality of life, and symptoms in persons with chronic fatigue syndrome. *J Psychosom Res*. 2011;70 (4):328—334.

118. Thomas M, Sadlier M, Smith A. The effect of Multi Convergent Therapy on the psychopathology, mood and performance of Chronic Fatigue Syndrome patients: a preliminary study. *Couns Psychother Res*. 2006;6 (2):91—99.

119. Edinger JD, Wohlgemuth WK, Krystal AD, Rice JR. Behavioral insomnia therapy for fibromyalgia patients — a randomized clinical trial. *Arch Intern Med*. 2005;165(21):2527—2535.

120. Danilov A, Kurganova J. Melatonin in chronic pain syndromes. *Pain Ther*. 2016;5(1):1—17.

121. Bent S, Padula A, Moore D, Patterson M, Mehling W. Valerian for sleep: a systematic review and meta-analysis. *Am J Med*. 2006;119(12):1005—1012.

122. Xu XM, Liu Y, Dong MX, Zou DZ, Wei YD. Tricyclic antidepressants for preventing migraine in adults. *Medicine*. 2017;96(22).

123. Lawson K. A brief review of the pharmacology of amitriptyline and clinical outcomes in treating fibromyalgia. *Biomedicines*. 2017;5(2):E24.

124. Katwala J, Kumar AK, Sejpal JJ, Terrence M, Mishra M. Therapeutic rationale for low dose doxepin in insomnia patients. *Asian Pac J Trop Dis*. 2013;3:331—336.

II. SLEEP DISORDERS IN SPECIFIC MEDICAL CONDITIONS

CHAPTER

16

Multiple Sclerosis

Christian Veauthier[1] and Friedemann Paul[2]

[1]Charité—Universitätsmedizin Berlin, Corporate Member of Freie Universität Berlin, Humboldt-Universität zu Berlin, and Berlin Institute of Health, Interdisciplinary Center for Sleep Medicine, Berlin, Germany [2]Charité—Universitätsmedizin Berlin, Corporate Member of Freie Universität Berlin, Humboldt-Universität zu Berlin, and Berlin Institute of Health, NeuroCure Clinical Research Center, Berlin, Germany

OUTLINE

Introduction	346	Impact and Quality of Life	353
Brief Overview of Multiple Sclerosis	347		
Sleep-Related Breathing Disorders	347	Hypnotic Use Due to Insomnia and Its Relationship With Fatigue	353
Insomnia	348	Fatigue and Employment Status	354
Restless Legs Syndrome and Periodic Limb Movement Disorder	350	Assessment of Sleep Disorders in Multiple Sclerosis	354
Narcolepsy and Hypersomnia	351	*Diagnostic Interview*	354
REM Sleep Behavior Disorder	352	*Questionnaires*	354
Relationship of Sleep Disorders With Fatigue, Depression, Disease Course, and Disability	352	*Polygraphic Recordings and Polysomnography*	354
		Differential Diagnosis	355

Handbook of Sleep Disorders in Medical Conditions
DOI: https://doi.org/10.1016/B978-0-12-813014-8.00016-0

© 2019 Elsevier Inc. All rights reserved.

Treatment of Sleep Disorders in		Sleep Disorders and Fatigue in	
Multiple Sclerosis	355	Neuromyelitis Optica Spectrum	
Pharmacological Interventions	355	Disorders	357
Empirical Evidence	356	Clinical Vignette	358
Nonpharmacological Interventions	357		
Treatment of Fatigue: The Berlin		Areas for Future Research and	
Treatment Algorithm	357	Conclusion	358
		References	359

INTRODUCTION

Nearly three quarters of multiple sclerosis (MS) patients suffer from sleep disorders, which are also relevant contributors to MS-related fatigue.[1] Despite 30 years of research and multiple randomized controlled trials (RCTs), no proven effective pharmacotherapy of MS-related fatigue exists (apart from perhaps vitamin D).[2–6] Against this background, the treatment of underlying sleep disorders as a relevant contributor to fatigue in MS becomes increasingly important to alleviate fatigue in MS patients.[7,8] MS patients suffering from sleep disorders show a reduced quality of life (QoL) compared to those without sleep disorders.[9]

The most common sleep medical condition in MS is chronic insomnia.[1,10] Self-medication with over-the-counter drugs can induce daytime fatigue, in addition to fatigue caused by sleep disruption.[11,12] The prevalence of restless legs syndrome (RLS) is up to four times higher in MS compared to the general population.[13–16] Compared to MS patients without RLS, MS patients with RLS have a significantly higher prevalence of spinal cord lesions.[17]

Conversely, no precise data about the prevalence of obstructive sleep apnea syndrome (OSAS) and central sleep apnea (CSA) in MS exists.[18] Inflammatory brain tissue damage and neuroaxonal loss due to MS (focal lesions, diffuse damage to the normal appearing white matter, brain atrophy) might conceivably cause CSA similar to that reported in stroke, in which CSA has been described as a consequence of brainstem infarction.[19–24] Single case reports have suggested that MS lesions in the brainstem might trigger REM sleep behavior disorders (RBDs) and that MS lesions in the hypothalamus might lead to secondary narcolepsy.[25] In narcolepsy type 1 (NT1) patients, a possible genetic predisposition for MS has long been debated due to the linkage of the human leucocyte antigen (HLA) allele DQB1*06:02 (present in >98% of NT1 patients) to the HLA-DRB1*15:01 allele, which itself is associated with MS.[25–28]

The aim of this chapter is to provide practical instructions for diagnosing and treating sleep disorders and fatigue in MS and the related condition neuromyelitis optica spectrum disorder (NMOSD) and to discuss the presumptive or known underlying pathophysiological mechanisms.

Sleep disorders in MS can lead to fatigue, and MS-related fatigue is a key factor in reduced QoL and participation in the workforce.[18,29] Lack of diagnosis and treatment of sleep disorders is a bottleneck in improving individualized treatment plans of MS patients. Furthermore, in MS patients, sleep disturbances do not seem to be associated with any measures of objective cognitive impairment, but correlate with perceived cognitive impairment and perception of altered executive functions (e.g., planning/organization) and prospective memory.[30]

BRIEF OVERVIEW OF MULTIPLE SCLEROSIS

MS is a chronic autoimmune demyelinating and neurodegenerative disease of the central nervous system (brain, optic nerve, and spinal cord) and predominantly affects young adults with a female-to-male ratio of approximately 3—4:1.[31–35] Although considered primarily an inflammatory condition, axonal loss, and neuronal demise in the brain, spinal cord and retina are prevalent from the earliest disease stages and become more prominent as the disease advances.[35–47] A combination of genetic and environmental factors (smoking, obesity, vitamin D deficiency, Epstein—Barr-virus, etc.) appear to determine the individual disease risk.[48–58] MS affects more than two million people worldwide, with the highest prevalence rates in North America and Europe, and the disease is one of the most common disabling nontraumatic neurological conditions in young adults.[59–61]

Any functional system of the central nervous system may be affected (vision, motor functions, coordination, gait, eye movements, sensory functions, vegetative functions); however, less "tangible" features such as depression, fatigue, sleep disorders, and cognitive impairment are at least as frequent and burdensome.[18,32,36,62–69] Although incurable, a number of so-called disease-modifying (or "immunomodulatory") drugs that reduce relapse rates, and in some cases slow disability progression, are available.

SLEEP-RELATED BREATHING DISORDERS

Although the exact prevalence of OSAS in MS patients is unknown, it appears to be a very frequent comorbidity.[25] Kaminska et al.[70] diagnosed OSAS using polysomnography (PSG) in more than half (58%) of MS patients ($n = 36$), but also in 49% of control persons ($n = 15$).

Another recent study using PSG found sleep apnea in 6 of 66 consecutive MS patients (12%).[1] Follow-up studies in these patients suggest that sleep medical treatment as continuous positive airway pressure (CPAP) therapy may reduce fatigue.[7,8,71] Kallweit et al., in a polygraphy (PG) study of 69 consecutive MS patients with severe fatigue, defined as a Fatigue Severity Scale (FSS) score equal or greater than 5.0, found OSAS in 28 patients (41%).[71,72] The majority of these patients (87%) suffered from secondary progressive MS, and 13% had a relapsing remitting form.[71] Chen et al.[73] investigated 21 consecutive MS patients and 11 control persons also using PSG and found OSAS in none of the patients or controls. These inconsistent findings, due to differences in sample size, patient selection, continents and regions (Europe, Asia, North America) and analytical approaches, underscore the necessity for large multicentric, preferably also multinational epidemiological studies in this field.[74,75]

One recent study discussed the potential relationship of CSA with brainstem lesions due to MS.[19] In a retrospective analysis of polysomnographic data from 48 MS patients and 48 healthy control subjects from the database of a sleep center, the authors found an increased frequency of central apnea in MS patients compared to non-MS controls, with brainstem involvement confirmed by magnetic resonance imaging (MRI), underscoring the role of the brainstem in regulation of breathing. Further studies are required to clarify this issue.

It is important to bear in mind that OSAS is not always accompanied by sleepiness and that many OSAS patients suffer only from fatigue and tiredness.[76] Therefore, it is important not to overlook OSAS in MS patients and mistakenly attribute possible symptoms (e.g., fatigue, tiredness, and sleepiness) to MS itself. MS patients should be investigated by PG or PSG with a low threshold of suspicion (e.g., one study recommended PG or PSG in the case of score greater than 34 on the Modified Fatigue Impact Scale—MFIS).[3,19,77−80]

INSOMNIA

The International Classification of Sleep Disorders, third edition (ICSD-3) defines insomnia as *a persistent difficulty with sleep initiation, duration, consolidation, or quality that occurs despite adequate opportunity and circumstances for sleep, and results in some form of daytime impairment.*[81] Furthermore, the ICSD-3 requires that *individuals who report these sleep related symptoms in the absence of daytime impairment are not regarded as having an insomnia disorder that warrants clinical attention other than education and reassurance.* Thus, in the case of reported sleep problems, daytime impairment is then the decisive diagnostic criterion for a diagnosis of insomnia, which is of crucial importance given the fact that the majority of MS patients (nearly 90%) complain of fatigue,[18,72,77,82] and it is sometimes impossible to determine whether fatigue results from sleep disturbances or is related to MS itself.

In MS patients suffering from comorbid insomnia, the chronological order of appearance of different symptoms and thorough diagnostic interviews are very important. Physicians should ask patients if insomnia (mainly difficulties falling asleep) occurred prior to a diagnosis of MS or afterwards.

Insomnia may present as initial, middle or terminal insomnia depending on whether it presents as difficulties in falling asleep (initiation), maintaining sleep, or waking up too early and is usually associated with decreased sleep efficacy.[81] In terminal insomnia, patients wake up earlier than desired which is often associated with depression.[83]

Insomnia comorbid to MS is associated with the same clinical features and the same known psychological factors perpetuating insomnia as in insomniac individuals free of neurological disease: increased levels of cognitive and somatic arousal, higher endorsement of dysfunctional beliefs about the consequences of insomnia on daytime functioning, and worry about insomnia and more frequent engagement in sleep-related safety behaviors.[84]

Nearly one quarter of MS patients are affected by insomnia. In a multicentre, hospital-based cross-sectional study (206 MS patients; 88% relapsing-remitting MS; RR-MS) chronic insomnia was diagnosed in 46 of 206 MS patients (22.3%) using a modified version of the brief insomnia questionnaire, whose scoring rules were adapted to comply with the ICSD-3.[10] Chronic insomnia was more frequent in women (87%) compared to men (13%) and was associated with lower QoL (assessed using the EQ-5D scale, the Portuguese version of the EuroQol health states).[85,86] Another study found chronic insomnia using the older ICSD-2 criteria in 17 of 66 consecutive MS patients (26%).[1,87]

No comparative study investigating the prevalence of insomnia in MS patients compared to the general population using the same diagnostic instruments has been performed.[18,88] Furthermore, the reported prevalence of insomnia in the general population depends on criteria used: although about one-third of the general population is present with insomnia symptoms, only 9%−15% report daytime consequences and 8%−18% sleep dissatisfaction, and only 6.6% fulfill the insomnia criteria of the Diagnostic and Statistical Manual fourth edition.[89] None of these data are based on the newer ICSD-3 criteria.[89] Insomnia is, as mentioned above, more frequent in women than in men and the prevalence of insomnia symptoms generally also increases with age.[89,90] In the general population, dysfunctional beliefs and attitudes about sleep (such as worries and concerns about consequences of disturbed sleep, feeling of helplessness and hopelessness and expectations about sleep medication) may be instrumental in perpetuating insomnia.[91] Two studies demonstrated that cognitive behavioral therapy (CBT) improved insomnia in MS patients, but despite overall improvement, insomnia persisted, at varying levels, in most participants. CBT may serve as an effective clinical intervention of insomnia in MS, but larger prospective randomized trails are needed, and further research is necessary.[92,93]

In MS patients suffering from insomnia with comorbid depression, telephone-administered CBT improved insomnia (in particular sleep onset insomnia), depression, and anxiety, but despite this improvement, nearly half of participants continued to report insomnia. Participants with residual insomnia were more likely to have major depressive disorder, greater MS severity, and higher levels of anxiety.[94]

With regard to the pharmacological treatment of insomnia, mostly with nonbenzodiazepine hypnotic agents or antidepressants, we refer to the recently published *AASM clinical practice guideline*.[95] This guideline uses two grades of recommendations: "strong"—a recommendation that clinicians should, under most circumstances follow, and "weak"—which does not indicate ineffectiveness but posits that the outcome of these recommendations is less certain and that the patient-care strategy might not be appropriate for all patients.[95] None of the reviewed drugs in the guideline fulfilled the criteria for a strong recommendation. Weakly recommended for the treatment of insomnia were suvorexant, eszopiclone, zaleplon, zolpidem, triazolam, temazepam, ramelteon, doxepin, trazodone, tiagabine, diphenhydramine, melatonin, tryptophan, and valerian.[95] Specifically in MS patients, Attarian et al.[96] found in a double-blind, placebo-controlled pilot trial over seven weeks an increase of the total sleep time under treatment with eszopiclone compared with placebo. Products that contain diphenhydramine can induce daytime fatigue in MS patients and should be prescribed with caution.[11] Mirtazapine can trigger RLS and given the high prevalence of RLS in MS, patients should be carefully monitored for RLS while treated with mirtazapine.[97] Melatonin ameliorates experimental autoimmune encephalitis and serum levels of melatonin are lower in MS patients compared to control persons.[98,99]

Melatonin plays an important role in pathogenesis, inflammatory response, and motor neuron degeneration and is a mediator of vitamin D neuro-immunomodulatory effects in MS.[100–103] High vitamin D serum levels seem to reduce frequency of MS relapses, but MS relapses occur more frequently in the springtime and summer, when the vitamin D levels are especially high.[104,105] Here, melatonin could be an additional regulator of the immune response, as melatonin production is stimulated by darkness and melatonin levels are higher in winter.[98,106] Some authors suggest that melatonin deficiency may also be among the factors involved in the occurrence of depression in MS patients.[107,108] Moreover, two large population-based case-control studies with together 6472 cases and 7409 controls demonstrated that subjects who have been exposed to shift work at various ages were suffering more often from MS compared to subjects who have not, and the authors discussed disturbed melatonin secretion and enhanced proinflammatory responses as potential underlying mechanism.[109] Despite numerous studies on the association of melatonin and MS, no study on treatment of insomnia with melatonin in patients with MS has thus far been published.

RESTLESS LEGS SYNDROME AND PERIODIC LIMB MOVEMENT DISORDER

RLS is a clinical diagnosis based on the International RLS Study Group (IRLSSG) guideline criteria.[110] The ICSD-3 defines RLS *as an urge to move the legs, usually accompanied by or thought to be caused by uncomfortable and unpleasant sensations in the legs which (i) begin or worsen during periods of rest or inactivity, (ii) are relieved by movement, and (iii) occur (exclusively or predominantly) in the evening or night.*[81]

The prevalence of RLS in MS is four times higher than in the general population. Manconi et al.[13] found RLS in 164 of 861 MS patients (19%) compared to 27 of 649 control persons (4.2%). The treatment of RLS in MS patients does not differ from the treatment of RLS in the general population. As iron deficiency plays a major role in the etiology of RLS,[111] treatment comprises iron substitution in the case of low ($<75\,\mu g/L$) or insufficient ($<20\,\mu g/L$) ferritin serum levels and dopaminergic therapy, for example, with ropinirole, pramipexole or rotigotine.[112,113] Furthermore opioids and alpha-2-delta-ligands ($\alpha 2\delta$-ligands; pregabalin or gabapentin) have also been reported to be effective.[112–114]

Every MS patient should be asked about the abovementioned RLS criteria (urge to move the legs, worsening during periods of rest and relieved by movement, predominantly in the evening or night). If it is difficult to distinguish between spasticity and pain and RLS, a probatory dopaminergic therapy might help to differentiate between them.

A MRI study using diffusion tensor imaging fractional anisotropy found that MS patients with comorbid RLS display a higher rate of cervical cord damage compared to MS patients without RLS.[16] Another recent study showed a higher REM-periodic limb movement (PLM)-index (number of PLMs per hour of REM sleep) in MS patients with a higher level of disability as quantified by the EDSS.[115] This suggests a causal relationship between spinal cord damage and RLS in MS.

In many cases, the etiology of RLS (idiopathic vs secondary due to MS) cannot be determined precisely. A family history argues for an idiopathic RLS, but even in these patients with a positive family history of RLS, secondary RLS due to MS cannot be ruled out. However, secondary RLS should be treated as idiopathic RLS and the recommended treatment is identical. Finally, all RLS patients should ideally fill in the International RLS Rating Scale in order to evaluate the course of the disease under treatment (before treatment and three months after treatment initiation).[116]

NARCOLEPSY AND HYPERSOMNIA

The clinical features of NT1 include excessive daytime sleepiness (EDS) and signs of REM sleep dissociation (cataplexy, hypnagogic hallucinations, and sleep paralysis).[81] In 1998, Sakurai et al.[117] and de Lecea et al.[118] described a neurotransmitter they named hypocretin and orexin, respectively. Lack of hypocretin resulting from cell loss in the lateral hypothalamus causes EDS and REM sleep dissociation.[117,118] The hypothalamus secretes two different peptides, hypocretin-1 and hypocretin-2.[117,118] Hypocretin-1 (HCRT-1) can be measured in the cerebrospinal fluid (CSF) and is reduced in NT1 (HCRT-2 can presumably be found as well in the CSF, but little is known about it due to the lack of a commercially available diagnostic kit for HCRT-2 measurements and HCRT-2 levels are normally not investigated in narcolepsy). Therefore, NT1 is also called hypocretin deficiency syndrome.[81] The diagnosis of narcolepsy type 2 also requires the presence of EDS for at least 3 months, reduced sleep latencies in the multiple sleep latency test (MSLT) ≤8 minutes and two or more sleep onset REM periods, but normal HCRT-1 values.

Over 98% of NT1 patients have HLA types DQB1*06:02 (or alternatively DQA1*01:02).[25] Lorenzoni et al.[27] found no significant association between narcolepsy, MS, and HLA-DQB1*06:02.

NT1 presumably results from a complex interaction of genetic disposition and environmental factors. Although an autoimmune etiology has been proposed, definite proof is lacking. HLA DRB1*15:01 is also frequently found in both NT1 and MS, and consequently a common etiology or shared pathophysiology has been hypothesized.[25,26] However, only 15 NT1 patients with a concomitant MS diagnosis have been reported in the literature to date.[25] For most of these patients, MRI scans showed bilateral hypothalamic lesions and CSF examination, where performed, identified decreased HCRT-1 (lower than 110 pg/mL; normal values greater than 200 pg/mL). In two of those patients, who were reexamined after i.v. corticosteroid therapy, a normalization of HCRT-1 CSF values and EDS was reported. Therefore, in the case of narcolepsy due to MS relapse with hypothalamic lesions corticosteroid pulse therapy should be given. Apart from the treatment of relapses, the management of narcolepsy and hypersomnia in patients with MS does not differ from that in the general population.

REM SLEEP BEHAVIOR DISORDER

RBD is characterized by dream-enacting episodes, which occasionally result in violence against the bed partner.[81] Brainstem lesions (especially in the dorsal pons) can cause RBD and may represent the initial signs of MS; however, only five cases of patients suffering from MS and comorbid MS have been published.[119–121]

Idiopathic RBD mostly occurs in men over 50 years of age.[81] Therefore, while MRI of the brain is recommended for all RBD patients, cranial MRI is mandatory in any young MS patients presenting with comorbid RBD.[120] RBD can improve after therapy with adrenocorticotropic hormone,[120] and consequently, not overlooking a possible diagnosis of RBD is very important. Antidepressants (in particular SSRI) can provoke RBD; such medication should be withdrawn slowly once RBD has been diagnosed.[25,81]

RELATIONSHIP OF SLEEP DISORDERS WITH FATIGUE, DEPRESSION, DISEASE COURSE, AND DISABILITY

As mentioned above, nearly 90% of MS patients suffer from fatigue.[18,72,77,82] MS patients with a secondary progressive disease course are more likely to have concomitant sleep disorders compared to patients with RR-MS.[1] Two cross-sectional studies have shown a significant relationship between sleep disorders and fatigue.[1,70] In one study, 66 consecutive MS patients were investigated with two home-based polysomnographies.[1] Notably, 96% of patients with fatigue suffered from a comorbid and treatable sleep disorder.[7] In a nonrandomized, open-label follow-up study,[7] patients were divided in three subgroups according to compliance with the recommended sleep medical treatment (mostly CPAP for OSAS, dopaminergic drugs for RLS, antidepressants or zolpidem and CBT for insomnia). Fatigue improved significantly in patients with good compliance, nonsignificantly in patients with moderate compliance and worsened in noncompliant patients.

In the second cross-sectional study, Kaminska et al.[70] found OSAS in 36 of 62 (58%) MS patients and in 15 of 32 controls (49%) and concluded that fatigue was associated with OSAS in MS. In a follow-up study, fatigue severity in patients with sleep disorders who had received specific sleep medical treatment improved significantly compared to untreated patients with sleep disorders—whereas fatigue worsened in MS patients without sleep disorders.[7] These studies demonstrate that a rigorous treatment of sleep disorders may reduce fatigue severity in MS.

Major depression is the most common comorbidity in MS.[3,122] Depression is an important confounder of studies investigating fatigue in MS as there is a substantial overlap between fatigue and depression.[64,123,124] CPAP may improve depression in the general population. No studies on the effect of CPAP on depression in MS patients suffering from OSAS have been performed to date.

IMPACT AND QUALITY OF LIFE

Health-related QoL (HRQoL) is decreased in MS compared to the general population.[125] The Nottingham Health Profile (NHP), a valid and reliable indicator of subjective health status in physical, social, and emotional areas, comprises six subscales (*sleep, physical mobility, energy, pain, emotional reactions*, and *social isolation*); the maximum of any subscale is 100 and as a result, the maximum of the NHP total score is 600 (the higher the NHP values, the lower the HRQoL).[126] In a recent study investigating the impact of sleep disorders on HRQoL in 66 MS patients using the NHP,[9] results showed a significantly lower HRQoL in MS patients suffering from comorbid sleep disorders ($n = 49$) (meaning increased NHP values) compared to MS patients without ($n = 17$); this is held true not only for the NHP total score and the category of "sleep" but also for the "energy" and "emotional" categories of the NHP. In this study, MS patients suffering from OSAS also showed increased NHP values in the "physical abilities" category. The low NHP values in MS patients without sleep disorders (mean-NHP 67.3), the moderately increased NHP values of MS patients suffering from RLS or PLM disorder (PLMD) (mean-NHP 119.9) and the clearly increased NHP total scores of MS insomnia patients (mean-NHP 220.3) and MS OSAS patients (mean-NHP 239.6) underline the impact of sleep disorders on HRQoL in MS. [For comparison only: mineworkers (equivalent with a high HRQoL) and fit elderly persons showed a very low global mean global NHP score of 8.8 and 12.4, respectively, pregnant women at 37 weeks, a NHP score of 127.0, fracture victims, 129.6 and chronically ill elderly patients, 156.4; especially high NHP values (271.3) were obtained in patients with osteoarthrosis].[9,126]

OSAS patients (without MS diagnosis) also showed increased global NHP median values (median NHP 218) with especially high NHP values in the "physical abilities" category, which is related to an objectively low level of physical activity, as measured by actigraphy.[127] Therefore, given that NHP values were similar across OSAS patients, regardless of whether they had MS or not, OSAS might play even the greater role in contributing to poor HRQoL.

HYPNOTIC USE DUE TO INSOMNIA AND ITS RELATIONSHIP WITH FATIGUE

In contrast to prescribed hypnotics, over-the-counter hypnotic use (diphenhydramine-containing products, as well as benzodiazepine-containing products) may lead to or aggravate fatigue in MS patients.[11] Antihistaminic drugs taken at night may exert carryover effects with impaired daytime functioning the following day. To date, epidemiological data on the frequency of hypnotic use in MS compared with the general population is lacking.

Research and treatment of MS-related fatigue remains a challenging task because the underlying mechanisms are widely unknown and evidence-based treatment options are sparse (in particular, neither the Food and Drug Administration nor the European Medicines Agency have yet approved a drug treatment for MS-related fatigue).[18]

FATIGUE AND EMPLOYMENT STATUS

Unemployment rates in MS range from 22% to 80%.[128,129] Among the strongest contributors to unemployment and premature retirement are cognitive dysfunction and fatigue.[3,128,130−134] This underscores the need to identify the underlying causes of fatigue and their possible treatment.

ASSESSMENT OF SLEEP DISORDERS IN MULTIPLE SCLEROSIS

Diagnostic Interview

In general, assessment of sleep disorders in MS patients and the general population does not differ. Due to the increased prevalence of RLS in MS (and the occasional difficulty of distinguishing between RLS and pain and spasticity), RLS symptoms should be addressed very carefully and with a low threshold of suspicion and in MS patients suffering from RLS dopaminergic treatment is recommended.

Questionnaires

Ideally, all fatigued MS patients should fill out the Epworth Sleepiness Scale (ESS) for measuring sleepiness, the Pittsburgh Sleep Quality Index (PSQI) or the Insomnia Severity Index (ISI) for insomnia, the STOPBang Questionnaire for measuring the risk of OSAS, and the MFIS for fatigue (or alternatively the FSS).[135−141] In addition, the Beck Depression Inventory (BDI-II) should be filled out to distinguish between fatigue and depression. But most fatigue questionnaires (FSS, MFIS, and others) show a substantial overlap between fatigue and depression.[18,64,142]

Polygraphic Recordings and Polysomnography

In general, PSQI values >5 indicate sleep disturbances. In one study investigating sleep disorders in MS patients using PSG, a global PSQI score >5 yielded a diagnostic sensitivity of 75% and a diagnostic specificity 64.7% (positive predictive value 85.7% and negative predictive value 47.8%).[7] In line with an earlier paper from Buysse et al.,[140] a threshold value of >5 in the PSQI appeared to be the best cutoff-point to predict sleep disturbances. A MFIS score of >34 yielded a diagnostic sensitivity of 71.4% and a specificity of 82.4% for predicting sleep disorders as detected by PSG (positive predictive value

92.1% and negative predictive value 50%). By jointly applying the MFIS cutoff of 34 and the PSQI cutoff of 5 (either MFIS > 34 or PSQI > 5), a higher sensitivity of 89.8% was achieved (specificity 58.8%, positive predictive value 86.3%, negative predictive value 66.7%). These data show that the combination of two straight-forward, self-rating questionnaires, which take just a few minutes to complete, are a good screening instrument for sleep disorders in MS-related fatigue and can predict sleep disorders with good sensitivity, good positive predictive value, and sufficient specificity.

Treatment of MS-related fatigue remains difficult, because no specific pharmacotherapy exists.[3,143] Cognitive behavioral therapy for fatigue and physical exercise may improve fatigue, while potential efficacious interventions such as deep transcranial magnetic stimulation require confirmation in larger studies.[3,144] As—in contrast to fatigue—specific and efficacious treatment options for the majority of sleep disorders are available, early diagnosis is the key. Therefore, a careful medical history, investigating the possibility of all sleep disturbances (including insomnia, RLS, nocturia and pain, sleep-related breathing disorders) should be taken from all fatigued MS patients. Not uncommonly, this step is unfortunately omitted in MS patients, as the treating neurologist may attribute daytime tiredness to MS-related fatigue and thus regard a PSG as unnecessary.[79] Other symptoms indicative of sleep disorders such as PLMDs, sleep-related breathing disorders, and certain other sleep disorders are often not noticed by either patients or their bed partners. Therefore, as sleep disorders are often not diagnosed in MS, all fatigued MS-patients with MFIS-values >34 or PSQI-values >5 should be referred to a sleep specialist and possibly undergo subsequent PSG.

Differential Diagnosis

For some MS patients suffering from fatigue, distinguishing between minor sleep problems that do not require a specific treatment and severe insomnias with daytime impairment is not possible by diagnostic interview alone, due to other MS-related symptoms, such as fatigue and depression. Depression, medication-related side effects, and reduced levels of physical activity at home and at work are important confounders of MS-related fatigue and must be taken into account. In unclear cases, further sleep medical investigation (e.g., admission to a sleep laboratory) is recommended. Neurogenic bladder affects up to 80% of patients with MS, and overactive bladder symptom scores are associated with reduced QoL, which increases with the number of walks to the toilet during the night and should be treated.[18,145,146]

TREATMENT OF SLEEP DISORDERS IN MULTIPLE SCLEROSIS

Pharmacological Interventions

Insomnia should be treated according to the recommendations of the *European guideline for the diagnosis and treatment of insomnia* and the *AASM Clinical Practice Guideline* with non-benzodiazepine hypnotic agents or antidepressants.[95,147] RLS treatment should primarily follow the recommendations of the combined task force of the *IRLSSG* in conjunction with the *European RLS Study Group (EURLSSG)* and the *RLS Foundation (RLSF)* with α2δ-ligands and dopaminergic drugs.[148]

Empirical Evidence

The following recommendations have not been specifically devised for MS patients but are general recommendations for the treating of sleep disorders proposed by the American Academy of Sleep Medicine (AASM) which are nonetheless applicable to MS patients suffering from sleep disorders.

A task force of the AASM reviewed RCTs of the treatment of chronic insomnia in adults.[95] The level of evidence in terms of quality of evidence, the balance of benefits and harms, and patient values and preferences was assessed using the Grading of Recommendations Assessment, Development, and Evaluation (GRADE) approach.[149] According to the GRADE system, a strong recommendation is one that clinicians should follow under most circumstances. While the *AASM's systematic review* of the RCTs for chronic insomnia did not identify any strong recommendations, several drug treatments were categorized as weak recommendations: suvorexant, eszopiclone, zaleplon, zolpidem, triazolam, temazepam, ramelteon, doxepin, trazodone, tiagabine, diphenhydramine, melatonin, tryptophan, and valerian.

The *European guideline* for the diagnosis and treatment of insomnia suggests that benzodiazepines and benzodiazepine receptor agonists (high-quality evidence according to GRADE) and sedative antidepressants (moderate-quality evidence) may be used in the short term if CBT is ineffective. Antihistamines and antipsychotics (strong recommendation—low- to very-low-quality evidence) as well as melatonin and phytotherapy (weak recommendation—low-quality evidence) are not recommended.[147]

The *IRLSSG*, the *EURLSSG*, and the *RLSF* recommended α2δ-ligands such as gabapentin and pregabalin and dopaminergic drugs for the treatment of RLS.[148] α2δ-ligands are effective and have little risk of augmentation (augmentation is defined as worsening of symptom severity manifested by an earlier onset of symptoms in the afternoon or evening compared with before treatment initiation).[148] Doses of dopaminergic drugs should be as low as possible in order to avoid augmentation. Patients with low ferritin levels (<75 μg/L) should be given appropriate iron supplementation (even if the values are still in the lowest normal region).

Narcolepsy patients with MS should be treated as narcolepsy patients without MS according to the current guidelines and recommendations.[150–153] Several studies showed a good efficacy of clonazepam 0.5−2 and/or melatonin 3−9 mg to treat RBD (with and without MS), but large multicenter, double-blind, placebo-controlled studies are lacking.[154–156]

Clinical Particularities

As MS relapse can trigger narcolepsy and RBD, MS patients presenting for the first time with symptoms of these sleep disorders should be carefully assessed and corticosteroid treatment administered if necessary. Of course, MS patients suffering from comorbid narcolepsy or RBD should be treated as narcolepsy or RBD patients without MS as recommended in the corresponding guidelines.[25,153,157]

Nonpharmacological Interventions

The standard treatment of OSAS is the CPAP therapy.[158] CPAP therapy in people with MS does not differ from CPAP therapy in the general population. There are three studies reporting an improvement of daytime fatigue after CPAP therapy in MS patients, while studies on the adherence to CPAP therapy over a longer period are lacking.[7,8,71] As mentioned above, insomnia should be treated primarily with CBT.[92,93,147,159] Group-based CBT, individual CBT, or internet-based CBT are effective.[92,93,160,161]

Treatment of Fatigue: The Berlin Treatment Algorithm

Identifying the underlying factor or, as is often the case, combination of factors contributing to MS fatigue remains a challenge, particularly in light of the many different guidelines and recommendations that only address single or subsets of possible causes. One stepwise clinical approach for diagnosing and treating MS-related fatigue is the Berlin Treatment Algorithm.[3] The diagnosis and treatment are divided into five phases (A–E). Phase A represents basic recommendations: diagnosis of fatigue and/or a sleep disorder using the MFIS and PSQI, and ruling out other causes, such as hypothyroidism, renal failure, hepatic insufficiency and anemia, chronic obstructive pulmonary disease (COPD), sedative drug intake, and smoking.

Phase B comprises a diagnostic interview focusing mainly on possible causes of fatigue such as nocturia, depression, and sleep disorders (OSA, insomnia, RLS). Physicians should also address sleep deprivation and ask about time spent in bed (currently and in the past, on workdays, and at weekends) and fatigue severity during workdays compared to rest days. Sleep diaries and actigraphies can be helpful in this issue. In the case of sleep deprivation, patients should be encouraged to go earlier to bed and to avoid working or watching TV late in the evening.

Phase C recommends supplementation of vitamin D to blood levels of 100–150 nmol/L (if blood calcium levels are not higher than 10.5 mg/dL) or drug therapy with fampridine in MS patients with walking impairment and EDSS values of 4–7. (Fampridine is prescribed to improve mobility but seems to be effective against fatigue as well).

In phase D, MS patients with consistently increased MFIS > 34 and PSQI values >5 should be referred to a sleep laboratory in order to diagnose or exclude other underlying sleep disorders according to the ICSD-3.

Phase E consists of rehabilitation (exercise or physical therapy, educational interventions such as mindfulness training, and behavioral interventions) and CBT corresponding to CBT for insomnia as well as more specific internet-based CBT self-management programs with and without the use of email support.[160–162]

SLEEP DISORDERS AND FATIGUE IN NEUROMYELITIS OPTICA SPECTRUM DISORDERS

NMOSDs are a group of rare CNS conditions characterized by severe inflammatory damage to the optic nerve, spinal cord, and brainstem.[163] Long considered a rare variant of MS, the detection of antibodies to a highly specific serum biomarker, astrocyte water channel aquaporin-4 (AQP4), identified NMOSD as an independent nosologic entity with an immunopathogenesis distinct from MS.[164–172] As in MS, fatigue is also an extremely common symptom, with prevalence rates of up to 70%,[173–179] and is also associated with poor HRQoL.

Literature on sleep disorders in NMOSD is emerging. A disrupted sleep architecture was recently reported in NMOSD patients, who displayed reduced sleep efficiency and a higher PLM index compared to healthy controls. Interestingly, a high proportion of the patients with PLM had infratentorial lesions.[180] Moreover, the frequency and severity of RLS has been reported as higher in NMOSD compared to healthy controls.[181] In NMOSD, RLS has been associated with a longer disease duration and higher neurological disability. In another study, poor sleep was associated with fatigue and reduced QoL.[182] Although no interventional studies have specifically investigated the impact of sleep medical treatment on fatigue and HRQoL in NMOSD, using the same fatigue and sleep questionnaires in NMOSD as in MS and performing PSG in the case of high fatigue scores or poor sleep as assessed using the PSQI is advisable. Another similarity to MS is that some NMOSD patients have been identified with bilateral hypothalamic lesions and subsequent HCRT-1 deficiency leading to NT1.[183] Recently, antibodies to myelin oligodendrocyte glycoprotein (MOG) have been reported in a subset of patients with an NMOSD phenotype seronegative for AQP4 antibodies and in few patients with MS[184–188]; however, it is a contentious issue as to whether these patients should be diagnosed with an NMOSD or with a distinct neuroimmunological disease entity.[189] Literature on sleep disorders in patients with MOG antibodies is absent.

CLINICAL VIGNETTE

A 56-year-old female with a diagnosis of secondary progressive MS, whose fatigue levels had already led to disability-related retirement 6 years earlier, presented at the NeuroCure Clinical Research Center (neuroimmunology outpatient clinic). Neurological examination yielded an EDSS of 4.5 (limited walking distance), and fatigue scores according to the MFIS of 58 (min.– max. = 0–84). PSG showed OSAS and medical history revealed RLS (the patient had reported unpleasant sensations in the legs previously, but these had been interpreted as spasticity and pain due to advanced neurological disability). MSLT showed no early REM, but sleep latency was shortened to 7.5 minutes. Dopaminergic treatment and CPAP therapy were initiated. Four months later, a control PSG identified significant continuing sleep fragmentation due to nocturia and pain in her left knee. Administration of an alpha receptor blocker and a knee cast led finally to a significant improvement of subjective fatigue and a decrease of MFIS scores to 23. A third PSG demonstrated a normal hypnogram. This case highlights that multiple, potentially treatable factors in the context of a neurological disease may cause sleep disruption and subsequent severe fatigue.

AREAS FOR FUTURE RESEARCH AND CONCLUSION

A recent case-control study has suggested that sleep disturbance might be a trigger for MS relapse.[190] This is an important point that should be further investigated in longitudinal studies using questionnaires and PSG data, and the impact of shift work on MS relapses should be investigated in these studies as well. As mentioned above, there is a huge number of studies investigating the role of melatonin on autoimmune and inflammatory processes in MS, but studies investigating the efficacy of melatonin in treating insomnia are lacking. Moreover, there is a single study on the relationship between brainstem lesions and CSA and prospective studies are needed in order to elucidate this relationship.[19] The relationship between sleep disorders and cognitive impairment should be investigated as well. Finally, the treatment of sleep disorders may improve fatigue.

KEY PRACTICE POINTS

- Sleep disorders should be diagnosed and treated as in patients without MS in the general population. Diagnosis and treatment of MS-related fatigue should routinely include sleep disorders, in addition to depression, drug side effects and low activity due to motor skill impairment as possible causes.
- In the case of severe fatigue or daytime impairment, polygraphic or polysomnographic investigations should be performed.
- MS patients presenting with narcolepsy or RBD symptoms appearing for the first time should be assessed for possible MS relapse and, if necessary, immune therapies should be administered.
- The Modified Fatigue Impact Scale or the Fatigue Severity Scale, the Epworth Sleepiness Scale, the Beck Depression Inventory (BDI-II) (or a similar depression scale), and the Pittsburgh Sleep Quality Index (or the Insomnia Severity Index) should be completed.
- Diagnosing any underlying treatable sleep disorder is paramount in MS as MS-related fatigue severity may be significantly reduced by successful treatment of a related sleep medical condition.

References

1. Veauthier C, Radbruch H, Gaede G, et al. Fatigue in multiple sclerosis is closely related to sleep disorders: a polysomnographic cross-sectional study. *Mult Scler Houndmills Basingstoke Engl.* 2011;17(5):613–622. Available from: https://doi.org/10.1177/1352458510393772.
2. Brañas P, Jordan R, Fry-Smith A, Burls A, Hyde C. Treatments for fatigue in multiple sclerosis: a rapid and systematic review. *Health Technol Assess Winch Engl.* 2000;4(27):1–61.
3. Veauthier C, Hasselmann H, Gold SM, Paul F. The Berlin Treatment Algorithm: recommendations for tailored innovative therapeutic strategies for multiple sclerosis-related fatigue. *EPMA J.* 2016;7:25. Available from: https://doi.org/10.1186/s13167-016-0073-3.
4. Achiron A, Givon U, Magalashvili D, et al. Effect of Alfacalcidol on multiple sclerosis-related fatigue: a randomized, double-blind placebo-controlled study. *Mult Scler J.* 2015;21(6):767–775. Available from: https://doi.org/10.1177/1352458514554053.
5. Veauthier C, Paul F. [Therapy of fatigue in multiple sclerosis: a treatment algorithm]. *Nervenarzt.* 2016;87 (12):1310–1321. Available from: https://doi.org/10.1007/s00115-016-0128-7.
6. Penner I-K, Paul F. Fatigue as a symptom or comorbidity of neurological diseases. *Nat Rev Neurol.* 2017;13 (11):662–675. Available from: https://doi.org/10.1038/nrneurol.2017.117.
7. Veauthier C, Gaede G, Radbruch H, Gottschalk S, Wernecke K-D, Paul F. Treatment of sleep disorders may improve fatigue in multiple sclerosis. *Clin Neurol Neurosurg.* 2013;115(9):1826–1830. Available from: https://doi.org/10.1016/j.clineuro.2013.05.018.

8. Côté I, Trojan DA, Kaminska M, et al. Impact of sleep disorder treatment on fatigue in multiple sclerosis. *Mult Scler Houndmills Basingstoke Engl.* 2013;19(4):480–489. Available from: https://doi.org/10.1177/1352458512455958.

9. Veauthier C, Gaede G, Radbruch H, Wernecke K-D, Paul F. Sleep disorders reduce health-related quality of life in multiple sclerosis (Nottingham Health Profile Data in patients with multiple sclerosis). *Int J Mol Sci.* 2015;16(7):16514–16528. Available from: https://doi.org/10.3390/ijms160716514.

10. Viana P, Rodrigues E, Fernandes C, et al. InMS: Chronic insomnia disorder in multiple sclerosis—a Portuguese multicentre study on prevalence, subtypes, associated factors and impact on quality of life. *Mult Scler Relat Disord.* 2015;4(5):477–483. Available from: https://doi.org/10.1016/j.msard.2015.07.010.

11. Braley TJ, Segal BM, Chervin RD. Hypnotic use and fatigue in multiple sclerosis. *Sleep Med.* 2015;16(1):131–137. Available from: https://doi.org/10.1016/j.sleep.2014.09.006.

12. Veauthier C. Hypnotic use and multiple sclerosis related fatigue: a forgotten confounder. *Sleep Med.* 2015;16(3):319. Available from: https://doi.org/10.1016/j.sleep.2014.11.012.

13. Italian REMS Study Group, Manconi M, Ferini-Strambi L, et al. Multicenter case-control study on restless legs syndrome in multiple sclerosis: the REMS study. *Sleep.* 2008;31(7):944–952.

14. Li Y, Munger KL, Batool-Anwar S, De Vito K, Ascherio A, Gao X. Association of multiple sclerosis with restless legs syndrome and other sleep disorders in women. *Neurology.* 2012;78(19):1500–1506. Available from: https://doi.org/10.1212/WNL.0b013e3182553c5b.

15. Miri S, Rohani M, Sahraian MA, et al. Restless legs syndrome in Iranian patients with multiple sclerosis. *Neurol Sci.* 2013;34(7):1105–1108. Available from: https://doi.org/10.1007/s10072-012-1186-7.

16. Manconi M, Rocca MA, Ferini-Strambi L, et al. Restless legs syndrome is a common finding in multiple sclerosis and correlates with cervical cord damage. *Mult Scler Houndmills Basingstoke Engl.* 2008;14(1):86–93. Available from: https://doi.org/10.1177/1352458507080734.

17. Minár M, Petrleničová D, Valkovič P. Higher prevalence of restless legs syndrome/Willis–Ekbom disease in multiple sclerosis patients is related to spinal cord lesions. *Mult Scler Relat Disord.* 2017;12:54–58. Available from: https://doi.org/10.1016/j.msard.2016.12.013.

18. Veauthier C, Paul F. Sleep disorders in multiple sclerosis and their relationship to fatigue. *Sleep Med.* 2014;15(1):5–14. Available from: https://doi.org/10.1016/j.sleep.2013.08.791.

19. Braley TJ, Segal BM, Chervin RD. Sleep-disordered breathing in multiple sclerosis. *Neurology.* 2012;79(9):929–936. Available from: https://doi.org/10.1212/WNL.0b013e318266fa9d.

20. Blissitt PA. Sleep-disordered breathing after stroke: nursing implications. *Stroke.* 2017;48(3):e81–e84. Available from: https://doi.org/10.1161/STROKEAHA.116.013087.

21. Devereaux MW, Keane JR, Davis RL. Automatic respiratory failure associated with infarction of the medulla. Report of two cases with pathologic study of one. *Arch Neurol.* 1973;29(1):46–52.

22. Bogousslavsky J, Khurana R, Deruaz JP, et al. Respiratory failure and unilateral caudal brainstem infarction. *Ann Neurol.* 1990;28(5):668–673. Available from: https://doi.org/10.1002/ana.410280511.

23. Levin BE, Margolis G. Acute failure of automatic respirations secondary to a unilateral brainstem infarct. *Ann Neurol.* 1977;1(6):583–586. Available from: https://doi.org/10.1002/ana.410010612.

24. Lyons OD, Ryan CM. Sleep apnea and stroke. *Can J Cardiol.* 2015;31(7):918–927. Available from: https://doi.org/10.1016/j.cjca.2015.03.014.

25. Veauthier C. Sleep disorders in multiple sclerosis. Review. *Curr Neurol Neurosci Rep.* 2015;15(5). Available from: https://doi.org/10.1007/s11910-015-0546-0.

26. Jennum PJ, Kornum BR, Issa NM, et al. Monozygotic twins discordant for narcolepsy type 1 and multiple sclerosis. *Neurol Neuroimmunol Neuroinflammation.* 2016;3(4):e249. Available from: https://doi.org/10.1212/NXI.0000000000000249.

27. Lorenzoni PJ, Werneck LC, Crippa AC, de S, et al. Is there a relationship between narcolepsy, multiple sclerosis and HLA-DQB1*06:02? *Arq Neuropsiquiatr.* 2017;75(6):345–348. Available from: https://doi.org/10.1590/0004-282X20170063.

28. Lysandropoulos AP, Mavroudakis N, Pandolfo M, et al. HLA genotype as a marker of multiple sclerosis prognosis: a pilot study. *J Neurol Sci.* 2017;375:348–354. Available from: https://doi.org/10.1016/j.jns.2017.02.019.

29. Moore P, Harding KE, Clarkson H, Pickersgill TP, Wardle M, Robertson NP. Demographic and clinical factors associated with changes in employment in multiple sclerosis. *Mult Scler J.* 2013;19(12):1647–1654. Available from: https://doi.org/10.1177/1352458513481396.

REFERENCES

30. Hughes AJ, Parmenter BA, Haselkorn JK, et al. Sleep and its associations with perceived and objective cognitive impairment in individuals with multiple sclerosis. *J Sleep Res.* 2017;26(4):428–435. Available from: https://doi.org/10.1111/jsr.12490.

31. Borisow N, Döring A, Pfueller CF, Paul F, Dörr J, Hellwig K. Expert recommendations to personalization of medical approaches in treatment of multiple sclerosis: an overview of family planning and pregnancy. *EPMA J.* 2012;3(1):9. Available from: https://doi.org/10.1186/1878-5085-3-9.

32. Paul F. Pathology and MRI: exploring cognitive impairment in MS. *Acta Neurol Scand.* 2016;134(Suppl 200):24–33. Available from: https://doi.org/10.1111/ane.12649.

33. Goodin DS. The causal cascade to multiple sclerosis: a model for MS pathogenesis. *PLoS One.* 2009;4(2):e4565. Available from: https://doi.org/10.1371/journal.pone.0004565.

34. Pfueller CF, Brandt AU, Schubert F, et al. Metabolic changes in the visual cortex are linked to retinal nerve fiber layer thinning in multiple sclerosis. Kleinschnitz C, ed. PLoS One. 2011;6(4):e18019. <https://doi.org/10.1371/journal.pone.0018019>.

35. Sinnecker T, Mittelstaedt P, Dörr J, et al. Multiple sclerosis lesions and irreversible brain tissue damage: a comparative ultrahigh-field strength magnetic resonance imaging study. *Arch Neurol.* 2012;69(6):739–745. Available from: https://doi.org/10.1001/archneurol.2011.2450.

36. O'Gorman C, Lucas R, Taylor B. Environmental risk factors for multiple sclerosis: a review with a focus on molecular mechanisms. *Int J Mol Sci.* 2012;13(12):11718–11752. Available from: https://doi.org/10.3390/ijms130911718.

37. Zimmermann H, Freing A, Kaufhold F, et al. Optic neuritis interferes with optical coherence tomography and magnetic resonance imaging correlations. *Mult Scler Houndmills Basingstoke Engl.* 2013;19(4):443–450. Available from: https://doi.org/10.1177/1352458512457844.

38. Scheel M, Finke C, Oberwahrenbrock T, et al. Retinal nerve fibre layer thickness correlates with brain white matter damage in multiple sclerosis: a combined optical coherence tomography and diffusion tensor imaging study. *Mult Scler Houndmills Basingstoke Engl.* 2014;20(14):1904–1907. Available from: https://doi.org/10.1177/1352458514535128.

39. Oberwahrenbrock T, Ringelstein M, Jentschke S, et al. Retinal ganglion cell and inner plexiform layer thinning in clinically isolated syndrome. *Mult Scler Houndmills Basingstoke Engl.* 2013;19(14):1887–1895. Available from: https://doi.org/10.1177/1352458513489757.

40. Azevedo CJ, Overton E, Khadka S, et al. Early CNS neurodegeneration in radiologically isolated syndrome. *Neurol Neuroimmunol Neuroinflammation.* 2015;2(3):e102. Available from: https://doi.org/10.1212/NXI.0000000000000102.

41. Solomon AJ, Watts R, Dewey BE, Reich DS. MRI evaluation of thalamic volume differentiates MS from common mimics. *Neurol Neuroimmunol Neuroinflammation.* 2017;4(5):e387. Available from: https://doi.org/10.1212/NXI.0000000000000387.

42. Bakshi R, Yeste A, Patel B, et al. Serum lipid antibodies are associated with cerebral tissue damage in multiple sclerosis. *Neurol Neuroimmunol Neuroinflammation.* 2016;3(2):e200. Available from: https://doi.org/10.1212/NXI.0000000000000200.

43. Bergman J, Dring A, Zetterberg H, et al. Neurofilament light in CSF and serum is a sensitive marker for axonal white matter injury in MS. *Neurol Neuroimmunol Neuroinflammation.* 2016;3(5):e271. Available from: https://doi.org/10.1212/NXI.0000000000000271.

44. Sinnecker T, Kuchling J, Dusek P, et al. Ultrahigh field MRI in clinical neuroimmunology: a potential contribution to improved diagnostics and personalised disease management. *EPMA J.* 2015;6(1):16. Available from: https://doi.org/10.1186/s13167-015-0038-y.

45. Backner Y, Kuchling J, Massarwa S, et al. Anatomical wiring and functional networking changes in the visual system following optic neuritis. *JAMA Neurol.* 2018. Available from: https://doi.org/10.1001/jamaneurol.2017.3880.

46. Pawlitzki M, Neumann J, Kaufmann J, et al. Loss of corticospinal tract integrity in early MS disease stages. *Neurol Neuroimmunol Neuroinflammation.* 2017;4(6):e399. Available from: https://doi.org/10.1212/NXI.0000000000000399.

47. Kuchling J, Brandt AU, Paul F, Scheel M. Diffusion tensor imaging for multilevel assessment of the visual pathway: possibilities for personalized outcome prediction in autoimmune disorders of the central nervous system. *EPMA J.* 2017;8(3):279–294. Available from: https://doi.org/10.1007/s13167-017-0102-x.

48. Srinivasan S, Di Dario M, Russo A, et al. Dysregulation of MS risk genes and pathways at distinct stages of disease. *Neurol Neuroimmunol Neuroinflammation.* 2017;4(3):e337. Available from: https://doi.org/10.1212/NXI.0000000000000337.

49. Villoslada P, Alonso C, Agirrezabal I, et al. Metabolomic signatures associated with disease severity in multiple sclerosis. *Neurol Neuroimmunol Neuroinflammation.* 2017;4(2):e321. Available from: https://doi.org/10.1212/NXI.0000000000000321.

50. Krieger SC, Cook K, De Nino S, Fletcher M. The topographical model of multiple sclerosis: a dynamic visualization of disease course. *Neurol Neuroimmunol Neuroinflammation.* 2016;3(5):e279. Available from: https://doi.org/10.1212/NXI.0000000000000279.

51. Rotstein DL, Healy BC, Malik MT, et al. Effect of vitamin D on MS activity by disease-modifying therapy class. *Neurol Neuroimmunol Neuroinflammation.* 2015;2(6):e167. Available from: https://doi.org/10.1212/NXI.0000000000000167.

52. Zivadinov R, Cerza N, Hagemeier J, et al. Humoral response to EBV is associated with cortical atrophy and lesion burden in patients with MS. *Neurol Neuroimmunol Neuroinflammation.* 2016;3(1):e190. Available from: https://doi.org/10.1212/NXI.0000000000000190.

53. Endriz J, Ho PP, Steinman L. Time correlation between mononucleosis and initial symptoms of MS. *Neurol Neuroimmunol Neuroinflammation.* 2017;4(3):e308. Available from: https://doi.org/10.1212/NXI.0000000000000308.

54. Schlemm L, Giess RM, Rasche L, et al. Fine specificity of the antibody response to Epstein–Barr nuclear antigen-2 and other Epstein–Barr virus proteins in patients with clinically isolated syndrome: a peptide microarray-based case-control study. *J Neuroimmunol.* 2016;297:56–62. Available from: https://doi.org/10.1016/j.jneuroim.2016.05.012.

55. Behrens JR, Rasche L, Gieß RM, et al. Low 25-hydroxyvitamin D, but not the bioavailable fraction of 25-hydroxyvitamin D, is a risk factor for multiple sclerosis. *Eur J Neurol.* 2016;23(1):62–67. Available from: https://doi.org/10.1111/ene.12788.

56. Pfuhl C, Oechtering J, Rasche L, et al. Association of serum Epstein–Barr nuclear antigen-1 antibodies and intrathecal immunoglobulin synthesis in early multiple sclerosis. *J Neuroimmunol.* 2015;285:156–160. Available from: https://doi.org/10.1016/j.jneuroim.2015.06.012.

57. Dörr J, Döring A, Paul F. Can we prevent or treat multiple sclerosis by individualised vitamin D supply? *EPMA J.* 2013;4(1):4. Available from: https://doi.org/10.1186/1878-5085-4-4.

58. Koduah P, Paul F, Dörr J-M. Vitamin D in the prevention, prediction and treatment of neurodegenerative and neuroinflammatory diseases. *EPMA J.* 2017;8(4):313–325. Available from: https://doi.org/10.1007/s13167-017-0120-8.

59. Browne P, Chandraratna D, Angood C, et al. Atlas of Multiple Sclerosis 2013: a growing global problem with widespread inequity. *Neurology.* 2014;83(11):1022–1024. Available from: https://doi.org/10.1212/WNL.0000000000000768.

60. Evans C, Beland S-G, Kulaga S, et al. Incidence and prevalence of multiple sclerosis in the Americas: a systematic review. *Neuroepidemiology.* 2013;40(3):195–210. Available from: https://doi.org/10.1159/000342779.

61. Döring A, Pfueller CF, Paul F, Dörr J. Exercise in multiple sclerosis—an integral component of disease management. *EPMA J.* 2011;3(1):2. Available from: https://doi.org/10.1007/s13167-011-0136-4.

62. Compston A, Coles A. Multiple sclerosis. *Lancet Lond Engl.* 2008;372(9648):1502–1517. Available from: https://doi.org/10.1016/S0140-6736(08)61620-7.

63. Veauthier C, Gaede G, Radbruch H, Wernecke K-D, Paul F. Poor sleep in multiple sclerosis correlates with beck depression inventory values, but not with polysomnographic data. *Sleep Disord.* 2016;2016:8378423. Available from: https://doi.org/10.1155/2016/8378423.

64. Hasselmann H, Bellmann-Strobl J, Ricken R, et al. Characterizing the phenotype of multiple sclerosis-associated depression in comparison with idiopathic major depression. *Mult Scler Houndmills Basingstoke Engl.* 2016. Available from: https://doi.org/10.1177/1352458515622826.

65. Weygandt M, Wakonig K, Behrens J, et al. Brain activity, regional gray matter loss, and decision-making in multiple sclerosis. *Mult Scler Houndmills Basingstoke Engl.* 2017. Available from: https://doi.org/10.1177/1352458517717089. 1352458517717089.

66. Weygandt M, Meyer-Arndt L, Behrens JR, et al. Stress-induced brain activity, brain atrophy, and clinical disability in multiple sclerosis. *Proc Natl Acad Sci USA.* 2016;113(47):13444–13449. Available from: https://doi.org/10.1073/pnas.1605829113.

REFERENCES

67. Weinges-Evers N, Brandt AU, Bock M, et al. Correlation of self-assessed fatigue and alertness in multiple sclerosis. *Mult Scler*. 2010;16(9):1134–1140. Available from: https://doi.org/10.1177/1352458510374202.
68. Urbanek C, Weinges-Evers N, Bellmann-Strobl J, et al. Attention Network Test reveals alerting network dysfunction in multiple sclerosis. *Mult Scler Houndmills Basingstoke Engl*. 2010;16(1):93–99. Available from: https://doi.org/10.1177/1352458509350308.
69. Finke C, Schlichting J, Papazoglou S, et al. Altered basal ganglia functional connectivity in multiple sclerosis patients with fatigue. *Mult Scler J*. 2015;21(7):925–934. Available from: https://doi.org/10.1177/1352458514555784.
70. Kaminska M, Kimoff R, Benedetti A, et al. Obstructive sleep apnea is associated with fatigue in multiple sclerosis. *Mult Scler J*. 2012;18(8):1159–1169. Available from: https://doi.org/10.1177/1352458511432328.
71. Kallweit U, Baumann CR, Harzheim M, et al. Fatigue and sleep-disordered breathing in multiple sclerosis: a clinically relevant association? *Mult Scler Int*. 2013;2013:1–7. Available from: https://doi.org/10.1155/2013/286581.
72. Krupp LB, Alvarez LA, LaRocca NG, Scheinberg LC. Fatigue in multiple sclerosis. *Arch Neurol*. 1988;45(4):435–437. Available from: https://doi.org/10.1001/archneur.1988.00520280085020.
73. Chen J-H, Liu X-Q, Sun H-Y, Huang Y. Sleep disorders in multiple sclerosis in China: clinical, polysomnography study, and review of the literature. *J Clin Neurophysiol*. 2014;31(4):375–381. Available from: https://doi.org/10.1097/WNP.0000000000000067.
74. Penzel T, Zhang X, Fietze I. Inter-scorer reliability between sleep centers can teach us what to improve in the scoring rules. *J Clin Sleep Med*. 2013;9(1):89–91. Available from: https://doi.org/10.5664/jcsm.2352.
75. Danker-Hopfe H, Anderer P, Zeitlhofer J, et al. Interrater reliability for sleep scoring according to the Rechtschaffen & Kales and the new AASM standard. *J Sleep Res*. 2009;18(1):74–84. Available from: https://doi.org/10.1111/j.1365-2869.2008.00700.x.
76. Hossain JL, Ahmad P, Reinish LW, Kayumov L, Hossain NK, Shapiro CM. Subjective fatigue and subjective sleepiness: two independent consequences of sleep disorders? *J Sleep Res*. 2005;14(3):245–253. Available from: https://doi.org/10.1111/j.1365-2869.2005.00466.x.
77. Veauthier C, Paul F. Fatigue in multiple sclerosis: which patient should be referred to a sleep specialist? *Mult Scler Houndmills Basingstoke Engl*. 2012;18(2):248–249. Available from: https://doi.org/10.1177/1352458511411229.
78. Braley TJ, Chervin RD, Segal BM. Fatigue, tiredness, lack of energy, and sleepiness in multiple sclerosis patients referred for clinical polysomnography. *Mult Scler Int*. 2012;2012:1–7. Available from: https://doi.org/10.1155/2012/673936.
79. Brass SD, Li C-S, Auerbach S. The underdiagnosis of sleep disorders in patients with multiple sclerosis. *J Clin Sleep Med*. 2014;10(9):1025–1031. Available from: https://doi.org/10.5664/jcsm.4044.
80. Paralyzed Veterans of America, ed. *Fatigue and multiple sclerosis: evidence-based management strategies for fatigue in multiple sclerosis*. Washington, DC; 1998. <http://mypva.org/images/MS-Fatigue_Management_CPG.pdf>.
81. American Academy of Sleep Medicine, ed. *International Classification of Sleep Disorders*, 3rd Ed. Darien, IL; 2014.
82. Paul F, Veauthier C. Fatigue in multiple sclerosis: a diagnostic and therapeutic challenge. *Expert Opin Pharmacother*. 2012;13(6):791–793. Available from: https://doi.org/10.1517/14656566.2012.667075.
83. Sun X, Zheng B, Lv J, et al. Sleep behavior and depression: findings from the China Kadoorie Biobank of 0.5 million Chinese adults. *J Affect Disord*. 2017;229:120–124. Available from: https://doi.org/10.1016/j.jad.2017.12.058.
84. Schellaert V, Labauge P, Lebrun C, et al. Psychological processes associated with insomnia in patients with multiple sclerosis. *Sleep*. 2018;41(3). Available from: https://doi.org/10.1093/sleep/zsy002.
85. Dolan P. Modeling valuations for EuroQol health states. *Med Care*. 1997;35(11):1095–1108.
86. Ferreira LN, Ferreira PL, Pereira LN, Oppe M. EQ-5D Portuguese population norms. *Qual Life Res*. 2014;23(2):425–430. Available from: https://doi.org/10.1007/s11136-013-0488-4.
87. American Academy of Sleep Medicine. *The International Classification of Sleep Disorders: Diagnostic and Coding Manual*. 2nd ed Westchester, IL: American Academy of Sleep Medicine; 2005.
88. Vitkova M, Rosenberger J, Gdovinova Z, et al. Poor sleep quality in patients with multiple sclerosis: gender differences. *Brain Behav*. 2016;6(11):e00553. Available from: https://doi.org/10.1002/brb3.553.

89. Ohayon MM. Epidemiology of insomnia: what we know and what we still need to learn. *Sleep Med Rev.* 2002;6(2):97−111. Available from: https://doi.org/10.1053/smrv.2002.0186.

90. Ohayon MM. Nocturnal awakenings and comorbid disorders in the American general population. *J Psychiatr Res.* 2008;43(1):48−54. Available from: https://doi.org/10.1016/j.jpsychires.2008.02.001.

91. Morin CM, Stone J, Trinkle D, Mercer J, Remsberg S. Dysfunctional beliefs and attitudes about sleep among older adults with and without insomnia complaints. *Psychol Aging.* 1993;8(3):463−467.

92. Abbasi S, Alimohammadi N, Pahlavanzadeh S. Effectiveness of cognitive behavioral therapy on the quality of sleep in women with multiple sclerosis: a randomized controlled trial study. *Int J Community Based Nurs Midwifery.* 2016;4(4):320−328.

93. Clancy M, Drerup M, Sullivan AB. Outcomes of cognitive-behavioral treatment for insomnia on insomnia, depression, and fatigue for individuals with multiple sclerosis: a case series. *Int J MS Care.* 2015;17 (6):261−267. Available from: https://doi.org/10.7224/1537-2073.2014-071.

94. Baron KG, Corden M, Jin L, Mohr DC. Impact of psychotherapy on insomnia symptoms in patients with depression and multiple sclerosis. *J Behav Med.* 2011;34(2):92−101. Available from: https://doi.org/10.1007/s10865-010-9288-2.

95. Sateia MJ, Buysse DJ, Krystal AD, Neubauer DN, Heald JL. Clinical practice guideline for the pharmacologic treatment of chronic insomnia in adults: an American Academy of Sleep Medicine Clinical Practice Guideline. *J Clin Sleep Med.* 2017;13(2):307−349. Available from: https://doi.org/10.5664/jcsm.6470.

96. Attarian H, Applebee G, Applebee A, et al. Effect of eszopiclone on sleep disturbances and daytime fatigue in multiple sclerosis patients. *Int J MS Care.* 2011;13(2):84−90. Available from: https://doi.org/10.7224/1537-2073-13.2.84.

97. Kolla BP, Mansukhani MP, Bostwick JM. The influence of antidepressants on restless legs syndrome and periodic limb movements: a systematic review. *Sleep Med Rev.* June 2017. Available from: https://doi.org/10.1016/j.smrv.2017.06.002.

98. Farez MF, Mascanfroni ID, Méndez-Huergo SP, et al. Melatonin contributes to the seasonality of multiple sclerosis relapses. *Cell.* 2015;162(6):1338−1352. Available from: https://doi.org/10.1016/j.cell.2015.08.025.

99. Farhadi N, Oryan S, Nabiuni M. Serum levels of melatonin and cytokines in multiple sclerosis. *Biomed J.* 2014;37(2):90−92. Available from: https://doi.org/10.4103/2319-4170.125885.

100. Golan D, Staun-Ram E, Glass-Marmor L, et al. The influence of vitamin D supplementation on melatonin status in patients with multiple sclerosis. *Brain Behav Immun.* 2013;32:180−185. Available from: https://doi.org/10.1016/j.bbi.2013.04.010.

101. Watad A, Azrielant S, Soriano A, Bracco D, Abu Much A, Amital H. Association between seasonal factors and multiple sclerosis. *Eur J Epidemiol.* 2016;31(11):1081−1089. Available from: https://doi.org/10.1007/s10654-016-0165-3.

102. Mehta BK. New hypotheses on sunlight and the geographic variability of multiple sclerosis prevalence. *J Neurol Sci.* 2010;292(1−2):5−10. Available from: https://doi.org/10.1016/j.jns.2010.02.004.

103. Sandyk R, Awerbuch GI. Multiple sclerosis: relationship between seasonal variations of relapse and age of onset. *Int J Neurosci.* 1993;71(1−4):147−157.

104. Ascherio A, Munger KL, Lünemann JD. The initiation and prevention of multiple sclerosis. *Nat Rev Neurol.* 2012;8(11):602−612. Available from: https://doi.org/10.1038/nrneurol.2012.198.

105. Jin Y, de Pedro-Cuesta J, Söderström M, Stawiarz L, Link H. Seasonal patterns in optic neuritis and multiple sclerosis: a meta-analysis. *J Neurol Sci.* 2000;181(1−2):56−64.

106. Brzezinski A. Melatonin in humans. *N Engl J Med.* 1997;336(3):186−195. Available from: https://doi.org/10.1056/NEJM199701163360306.

107. Akpınar Z, Tokgöz S, Gökbel H, Okudan N, Uğuz F, Yılmaz G. The association of nocturnal serum melatonin levels with major depression in patients with acute multiple sclerosis. *Psychiatry Res.* 2008;161 (2):253−257. Available from: https://doi.org/10.1016/j.psychres.2007.11.022.

108. Sandyk R, Awerbuch GI. Nocturnal melatonin secretion in suicidal patients with multiple sclerosis. *Int J Neurosci.* 1993;71(1−4):173−182.

109. Hedström AK, Åkerstedt T, Hillert J, Olsson T, Alfredsson L. Shift work at young age is associated with increased risk for multiple sclerosis. *Ann Neurol.* 2011;70(5):733−741. Available from: https://doi.org/10.1002/ana.22597.

110. Allen RP, Picchietti DL, Garcia-Borreguero D, et al. Restless legs syndrome/Willis−Ekbom disease diagnostic criteria: updated International Restless Legs Syndrome Study Group (IRLSSG) consensus criteria − history, rationale, description, and significance. *Sleep Med*. 2014;15(8):860−873. Available from: https://doi.org/10.1016/j.sleep.2014.03.025.

111. Dauvilliers Y, Winkelmann J. Restless legs syndrome: update on pathogenesis. *Curr Opin Pulm Med*. 2013;19 (6):594−600. Available from: https://doi.org/10.1097/MCP.0b013e328365ab07.

112. Ondo WG. Restless legs syndrome: pathophysiology and treatment. *Curr Treat Options Neurol*. 2014;16(11). Available from: https://doi.org/10.1007/s11940-014-0317-2.

113. Aurora RN, Kristo DA, Bista SR, et al. Update to the AASM Clinical Practice Guideline: "The treatment of restless legs syndrome and periodic limb movement disorder in adults—an update for 2012: practice parameters with an evidence-based systematic review and meta-analyses.". *Sleep*. 2012. Available from: https://doi.org/10.5665/sleep.1986.

114. Garcia-Borreguero D, Patrick J, DuBrava S, et al. Pregabalin versus pramipexole: effects on sleep disturbance in restless legs syndrome. *Sleep*. 2014;37(4):635−643. Available from: https://doi.org/10.5665/sleep.3558.

115. Veauthier C, Gaede G, Radbruch H, Sieb J-P, Wernecke K-D, Paul F. Periodic limb movements during REM sleep in multiple sclerosis: a previously undescribed entity. *Neuropsychiatr Dis Treat*. 2015;2323. Available from: https://doi.org/10.2147/NDT.S83350.

116. Walters AS, Frauscher B, Allen R, et al. Review of diagnostic instruments for the restless legs syndrome/Willis−Ekbom disease (RLS/WED): critique and recommendations. *J Clin Sleep Med*. 2014. Available from: https://doi.org/10.5664/jcsm.4298.

117. Sakurai T, Amemiya A, Ishii M, et al. Orexins and orexin receptors: a family of hypothalamic neuropeptides and G protein-coupled receptors that regulate feeding behavior. *Cell*. 1998;92(4):573−585.

118. de Lecea L, Kilduff TS, Peyron C, et al. The hypocretins: hypothalamus-specific peptides with neuroexcitatory activity. *Proc Natl Acad Sci USA*. 1998;95(1):322−327.

119. Tippmann-Peikert M, Boeve BF, Keegan BM. REM sleep behavior disorder initiated by acute brainstem multiple sclerosis. *Neurology*. 2006;66(8):1277−1279. Available from: https://doi.org/10.1212/01.wnl.0000208518.72660.ff.

120. Plazzi G, Montagna P. Remitting REM sleep behavior disorder as the initial sign of multiple sclerosis. *Sleep Med*. 2002;3(5):437−439.

121. Gómez-Choco MJ, Iranzo A, Blanco Y, Graus F, Santamaria J, Saiz A. Prevalence of restless legs syndrome and REM sleep behavior disorder in multiple sclerosis. *Mult Scler Houndmills Basingstoke Engl*. 2007;13 (6):805−808. Available from: https://doi.org/10.1177/1352458506074644.

122. Marrie RA, Reingold S, Cohen J, et al. The incidence and prevalence of psychiatric disorders in multiple sclerosis: a systematic review. *Mult Scler Houndmills Basingstoke Engl*. 2015;21(3):305−317. Available from: https://doi.org/10.1177/1352458514564487.

123. Greim B, Benecke R, Zettl UK. Qualitative and quantitative assessment of fatigue in multiple sclerosis (MS). *J Neurol*. 2007;254(S2):II58−II64. Available from: https://doi.org/10.1007/s00415-007-2014-5.

124. Bakshi R, Shaikh ZA, Miletich RS, et al. Fatigue in multiple sclerosis and its relationship to depression and neurologic disability. *Mult Scler*. 2000;6(3):181−185. Available from: https://doi.org/10.1177/135245850000600308.

125. Amtmann D, Bamer AM, Kim J, Chung H, Salem R. People with multiple sclerosis report significantly worse symptoms and health related quality of life than the US general population as measured by PROMIS and NeuroQoL outcome measures. *Disabil Health J*. 2017. Available from: https://doi.org/10.1016/j.dhjo.2017.04.008.

126. Hunt SM, McEwen J, McKenna SP. Measuring health status: a new tool for clinicians and epidemiologists. *J R Coll Gen Pract*. 1985;35(273):185−188.

127. Verwimp J, Ameye L, Bruyneel M. Correlation between sleep parameters, physical activity and quality of life in somnolent moderate to severe obstructive sleep apnea adult patients. *Sleep Breath Schlaf Atm*. 2013;17(3):1039−1046. Available from: https://doi.org/10.1007/s11325-012-0796-x.

128. Cadden M, Arnett P. Factors associated with employment status in individuals with multiple sclerosis. *Int J MS Care*. 2015;17(6):284−291. Available from: https://doi.org/10.7224/1537-2073.2014-057.

129. Busche KD, Fisk JD, Murray TJ, Metz LM. Short term predictors of unemployment in multiple sclerosis patients. *Can J Neurol Sci*. 2003;30(2):137−142.

130. Salter A, Thomas N, Tyry T, Cutter G, Marrie RA. Employment and absenteeism in working-age persons with multiple sclerosis. *J Med Econ*. 2017;20(5):493–502. Available from: https://doi.org/10.1080/13696998.2016.1277229.

131. Simpson S, Tan H, Otahal P, et al. Anxiety, depression and fatigue at 5-year review following CNS demyelination. *Acta Neurol Scand*. 2016;134(6):403–413. Available from: https://doi.org/10.1111/ane.12554.

132. Concetta Incerti C, Magistrale G, Argento O, et al. Occupational stress and personality traits in multiple sclerosis: a preliminary study. *Mult Scler Relat Disord*. 2015;4(4):315–319. Available from: https://doi.org/10.1016/j.msard.2015.06.001.

133. Krause I, Kern S, Horntrich A, Ziemssen T. Employment status in multiple sclerosis: impact of disease-specific and non-disease-specific factors. *Mult Scler Houndmills Basingstoke Engl*. 2013;19(13):1792–1799. Available from: https://doi.org/10.1177/1352458513485655.

134. Findling O, Baltisberger M, Jung S, Kamm CP, Mattle HP, Sellner J. Variables related to working capability among Swiss patients with multiple sclerosis—a cohort study. *PLoS One*. 2015;10(4):e0121856. Available from: https://doi.org/10.1371/journal.pone.0121856.

135. Johns MW. Reliability and factor analysis of the Epworth Sleepiness Scale. *Sleep*. 1992;15(4):376–381.

136. Johns MW. A new method for measuring daytime sleepiness: the Epworth sleepiness scale. *Sleep*. 1991;14(6):540–545.

137. Flachenecker P, Kümpfel T, Kallmann B, et al. Fatigue in multiple sclerosis: a comparison of different rating scales and correlation to clinical parameters. *Mult Scler*. 2002;8(6):523–526. Available from: https://doi.org/10.1191/1352458502ms839oa.

138. Paralyzed Veterans of America. *Multiple sclerosis council for clinical practice guidelines. Fatigue and multiple sclerosis: evidence-based management strategies for fatigue in multiple sclerosis*. Washington, DC; 1998.

139. Krupp LB, LaRocca NG, Muir-Nash J, Steinberg AD. The Fatigue Severity Scale. Application to patients with multiple sclerosis and systemic lupus erythematosus. *Arch Neurol*. 1989;46(10):1121–1123.

140. Buysse DJ, Reynolds CF, Monk TH, Berman SR, Kupfer DJ. The Pittsburgh Sleep Quality Index: a new instrument for psychiatric practice and research. *Psychiatry Res*. 1989;28(2):193–213.

141. Bastien C. Validation of the Insomnia Severity Index as an outcome measure for insomnia research. *Sleep Med*. 2001;2(4):297–307. Available from: https://doi.org/10.1016/S1389-9457(00)00065-4.

142. Beck AT, Steer RA. Internal consistencies of the original and revised Beck Depression Inventory. *J Clin Psychol*. 1984;40(6):1365–1367.

143. Branas P. Treatments for Fatigue in Multiple Sclerosis: A Rapid and Systematic Review. *Health Technol Assess*. 2000;4(27):1–61. Available from: https://doi.org/10.3310/hta4270.

144. Gaede G, Tiede M, Lorenz I, et al. Safety and preliminary efficacy of deep transcranial magnetic stimulation in MS-related fatigue. *Neurol Neuroimmunol Neuroinflammation*. 2018;5(1):e423. Available from: https://doi.org/10.1212/NXI.0000000000000423.

145. Valiquette G. Desmopressin in the management of nocturia in patients with multiple sclerosis: a double-blind, crossover trial. *Arch Neurol*. 1996;53(12):1270. Available from: https://doi.org/10.1001/archneur.1996.00550120082020.

146. Mahajan ST, Patel PB, Marrie RA. Under treatment of overactive bladder symptoms in patients with multiple sclerosis: an ancillary analysis of the NARCOMS patient registry. *J Urol*. 2010;183(4):1432–1437. Available from: https://doi.org/10.1016/j.juro.2009.12.029.

147. Riemann D, Baglioni C, Bassetti C, et al. European guideline for the diagnosis and treatment of insomnia. *J Sleep Res*. 2017. Available from: https://doi.org/10.1111/jsr.12594.

148. Garcia-Borreguero D, Silber MH, Winkelman JW, et al. Guidelines for the first-line treatment of restless legs syndrome/Willis–Ekbom disease, prevention and treatment of dopaminergic augmentation: a combined task force of the IRLSSG, EURLSSG, and the RLS-foundation. *Sleep Med*. 2016;21:1–11. Available from: https://doi.org/10.1016/j.sleep.2016.01.017.

149. Atkins D, Eccles M, Flottorp S, et al. Systems for grading the quality of evidence and the strength of recommendations I: critical appraisal of existing approaches The GRADE Working Group. *BMC Health Serv Res*. 2004;4(1):38. Available from: https://doi.org/10.1186/1472-6963-4-38.

150. Barateau L, Lopez R, Dauvilliers Y. Management of narcolepsy. *Curr Treat Options Neurol*. 2016;18(10):43. Available from: https://doi.org/10.1007/s11940-016-0429-y.

REFERENCES

151. Billiard M, Bassetti C, Dauvilliers Y, et al. EFNS guidelines on management of narcolepsy. *Eur J Neurol.* 2006;13(10):1035−1048. Available from: https://doi.org/10.1111/j.1468-1331.2006.01473.x.

152. Mignot EJM. A practical guide to the therapy of narcolepsy and hypersomnia syndromes. *Neurother J Am Soc Exp Neurother.* 2012;9(4):739−752. Available from: https://doi.org/10.1007/s13311-012-0150-9.

153. Kallweit U, Bassetti CL. Pharmacological management of narcolepsy with and without cataplexy. *Expert Opin Pharmacother.* 2017;18(8):809−817. Available from: https://doi.org/10.1080/14656566.2017.1323877.

154. Howell MJ, Schenck CH. Rapid eye movement sleep behavior disorder and neurodegenerative disease. *JAMA Neurol.* 2015;72(6):707−712. Available from: https://doi.org/10.1001/jamaneurol.2014.4563.

155. McGrane IR, Leung JG, St Louis EK, Boeve BF. Melatonin therapy for REM sleep behavior disorder: a critical review of evidence. *Sleep Med.* 2015;16(1):19−26. Available from: https://doi.org/10.1016/j.sleep.2014.09.011.

156. Högl B, Stefani A. REM sleep behavior disorder (RBD): update on diagnosis and treatment. *Somnologie Schlafforschung Schlafmed Somnology Sleep Res Sleep Med.* 2017;21(Suppl 1):1−8. Available from: https://doi.org/10.1007/s11818-016-0048-6.

157. Jung Y, St Louis EK. Treatment of REM sleep behavior disorder. *Curr Treat Options Neurol.* 2016;18(11):50. Available from: https://doi.org/10.1007/s11940-016-0433-2.

158. Sullivan CE, Issa FG, Berthon-Jones M, Eves L. Reversal of obstructive sleep apnoea by continuous positive airway pressure applied through the nares. *Lancet Lond Engl.* 1981;1(8225):862−865.

159. McCall WV. Cognitive behavioral therapy for insomnia (CBT-I): what is known, and advancing the science by avoiding the pitfalls of the placebo effect. *Sleep Med Rev.* 2017. Available from: https://doi.org/10.1016/j.smrv.2017.05.001.

160. van Kessel K, Moss-Morris R, Willoughby E, Chalder T, Johnson MH, Robinson E. A randomized controlled trial of cognitive behavior therapy for multiple sclerosis fatigue. *Psychosom Med.* 2008;70(2):205−213. Available from: https://doi.org/10.1097/PSY.0b013e3181643065.

161. van Kessel K, Wouldes T, Moss-Morris R. A New Zealand pilot randomized controlled trial of a web-based interactive self-management programme (MSInvigor8) with and without email support for the treatment of multiple sclerosis fatigue. *Clin Rehabil.* 2016;30(5):454−462. Available from: https://doi.org/10.1177/0269215515584800.

162. Moss-Morris R, McCrone P, Yardley L, van Kessel K, Wills G, Dennison L. A pilot randomised controlled trial of an Internet-based cognitive behavioural therapy self-management programme (MS Invigor8) for multiple sclerosis fatigue. *Behav Res Ther.* 2012;50(6):415−421. Available from: https://doi.org/10.1016/j.brat.2012.03.001.

163. Jarius S, Wildemann B, Paul F. Neuromyelitis optica: clinical features, immunopathogenesis and treatment. *Clin Exp Immunol.* 2014;176(2):149−164. Available from: https://doi.org/10.1111/cei.12271.

164. Jarius S, Paul F, Franciotta D, et al. Mechanisms of disease: aquaporin-4 antibodies in neuromyelitis optica. *Nat Clin Pract Neurol.* 2008;4(4):202−214. Available from: https://doi.org/10.1038/ncpneuro0764.

165. Paul F, Jarius S, Aktas O, et al. Antibody to aquaporin 4 in the diagnosis of neuromyelitis optica. *PLoS Med.* 2007;4(4):e133. Available from: https://doi.org/10.1371/journal.pmed.0040133.

166. Metz I, Beißbarth T, Ellenberger D, et al. Serum peptide reactivities may distinguish neuromyelitis optica subgroups and multiple sclerosis. *Neurol Neuroimmunol Neuroinflammation.* 2016;3(2):e204. Available from: https://doi.org/10.1212/NXI.0000000000000204.

167. Zekeridou A, Lennon VA. Aquaporin-4 autoimmunity. *Neurol Neuroimmunol Neuroinflammation.* 2015;2(4):e110. Available from: https://doi.org/10.1212/NXI.0000000000000110.

168. Finke C, Heine J, Pache F, et al. Normal volumes and microstructural integrity of deep gray matter structures in AQP4 + NMOSD. *Neurol Neuroimmunol Neuroinflammation.* 2016;3(3):e229. Available from: https://doi.org/10.1212/NXI.0000000000000229.

169. Hertwig L, Pache F, Romero-Suarez S, et al. Distinct functionality of neutrophils in multiple sclerosis and neuromyelitis optica. *Mult Scler Houndmills Basingstoke Engl.* 2016;22(2):160−173. Available from: https://doi.org/10.1177/1352458515586084.

170. Pandit L, Asgari N, Apiwattanakul M, et al. Demographic and clinical features of neuromyelitis optica: a review. *Mult Scler Houndmills Basingstoke Engl.* 2015;21(7):845−853. Available from: https://doi.org/10.1177/1352458515572406.

171. Asgari N, Flanagan EP, Fujihara K, et al. Disruption of the leptomeningeal blood barrier in neuromyelitis optica spectrum disorder. *Neurol Neuroimmunol Neuroinflammation*. 2017;4(4):e343. Available from: https://doi.org/10.1212/NXI.0000000000000343.

172. Takeshita Y, Obermeier B, Cotleur AC, et al. Effects of neuromyelitis optica-IgG at the blood-brain barrier in vitro. *Neurol Neuroimmunol Neuroinflammation*. 2017;4(1):e311. Available from: https://doi.org/10.1212/NXI.0000000000000311.

173. Seok JM, Choi M, Cho EB, et al. Fatigue in patients with neuromyelitis optica spectrum disorder and its impact on quality of life. *PLoS One*. 2017;12(5):e0177230. Available from: https://doi.org/10.1371/journal.pone.0177230.

174. Chavarro VS, Mealy MA, Simpson A, et al. Insufficient treatment of severe depression in neuromyelitis optica spectrum disorder. *Neurol Neuroimmunol Neuroinflammation*. 2016;3(6):e286. Available from: https://doi.org/10.1212/NXI.0000000000000286.

175. Shi Z, Chen H, Lian Z, Liu J, Feng H, Zhou H. Factors that impact health-related quality of life in neuromyelitis optica spectrum disorder: anxiety, disability, fatigue and depression. *J Neuroimmunol*. 2016;293:54−58. Available from: https://doi.org/10.1016/j.jneuroim.2016.02.011.

176. Akaishi T, Nakashima I, Misu T, Fujihara K, Aoki M. Depressive state and chronic fatigue in multiple sclerosis and neuromyelitis optica. *J Neuroimmunol*. 2015;283:70−73. Available from: https://doi.org/10.1016/j.jneuroim.2015.05.007.

177. Muto M, Mori M, Sato Y, et al. Current symptomatology in multiple sclerosis and neuromyelitis optica. *Eur J Neurol*. 2015;22(2):299−304. Available from: https://doi.org/10.1111/ene.12566.

178. Chanson J-B, Zéphir H, Collongues N, et al. Evaluation of health-related quality of life, fatigue and depression in neuromyelitis optica. *Eur J Neurol*. 2011;18(6):836−841. Available from: https://doi.org/10.1111/j.1468-1331.2010.03252.x.

179. He D, Chen X, Zhao D, Zhou H. Cognitive function, depression, fatigue, and activities of daily living in patients with neuromyelitis optica after acute relapse. *Int J Neurosci*. 2011;121(12):677−683. Available from: https://doi.org/10.3109/00207454.2011.608456.

180. Song Y, Pan L, Fu Y, et al. Sleep abnormality in neuromyelitis optica spectrum disorder. *Neurol Neuroimmunol Neuroinflammation*. 2015;2(3):e94. Available from: https://doi.org/10.1212/NXI.0000000000000094.

181. Hyun J-W, Kim S-H, Jeong IH, et al. Increased frequency and severity of restless legs syndrome in patients with neuromyelitis optica spectrum disorder. *Sleep Med*. 2016;17:121−123. Available from: https://doi.org/10.1016/j.sleep.2015.08.023.

182. Pan J, Zhao P, Cai H, et al. Hypoxemia, sleep disturbances, and depression correlated with fatigue in neuromyelitis optica spectrum disorder. *CNS Neurosci Ther*. 2015;21(7):599−606. Available from: https://doi.org/10.1111/cns.12411.

183. Nishino S, Okuro M, Kotorii N, et al. Hypocretin/orexin and narcolepsy: new basic and clinical insights. *Acta Physiol Oxf Engl*. 2010;198(3):209−222. Available from: https://doi.org/10.1111/j.1748-1716.2009.02012.x.

184. Jarius S, Kleiter I, Ruprecht K, et al. MOG-IgG in NMO and related disorders: a multicenter study of 50 patients. Part 3: Brainstem involvement − frequency, presentation and outcome. *J Neuroinflammation*. 2016;13(1):281. Available from: https://doi.org/10.1186/s12974-016-0719-z.

185. Pache F, Zimmermann H, Mikolajczak J, et al. MOG-IgG in NMO and related disorders: a multicenter study of 50 patients. Part 4: Afferent visual system damage after optic neuritis in MOG-IgG-seropositive versus AQP4-IgG-seropositive patients. *J Neuroinflammation*. 2016;13(1):282. Available from: https://doi.org/10.1186/s12974-016-0720-6.

186. Körtvélyessy P, Breu M, Pawlitzki M, et al. ADEM-like presentation, anti-MOG antibodies, and MS pathology: two case reports. *Neurol Neuroimmunol Neuroinflammation*. 2017;4(3):e335. Available from: https://doi.org/10.1212/NXI.0000000000000335.

187. Spadaro M, Gerdes LA, Krumbholz M, et al. Autoantibodies to MOG in a distinct subgroup of adult multiple sclerosis. *Neurol Neuroimmunol Neuroinflammation*. 2016;3(5):e257. Available from: https://doi.org/10.1212/NXI.0000000000000257.

188. Sepúlveda M, Armangué T, Sola-Valls N, et al. Neuromyelitis optica spectrum disorders: comparison according to the phenotype and serostatus. *Neurol Neuroimmunol Neuroinflammation*. 2016;3(3):e225. Available from: https://doi.org/10.1212/NXI.0000000000000225.

REFERENCES

189. Zamvil SS, Slavin AJ. Does MOG Ig-positive AQP4-seronegative opticospinal inflammatory disease justify a diagnosis of NMO spectrum disorder? *Neurol Neuroimmunol Neuroinflammation*. 2015;2(1):e62. Available from: https://doi.org/10.1212/NXI.0000000000000062.
190. Sahraian MA, Rezaali S, Hosseiny M, Doosti R, Tajik A, Naser Moghadasi A. Sleep disorder as a triggering factor for relapse in multiple sclerosis. *Eur Neurol*. 2017;77(5-6):258–261. Available from: https://doi.org/10.1159/000470904.

CHAPTER

17

Gastrointestinal Disorders

Claire J. Han[1,2] *and Margaret M. Heitkemper*[2,3,4]

[1]School of Medicine, Biomedical Informatics and Medical Education, University of Washington, Seattle, WA, United States [2]Department of Biobehavioral Nursing and Health Informatics, School of Nursing, University of Washington, Seattle, WA, United States [3]Department of Nursing, Center for Innovations in Sleep Self-Management, University of Washington, Seattle, WA, United States [4]Division of Gastroenterology, School of Medicine, University of Washington, Seattle, WA, United States

OUTLINE

Introduction	372	Inflammatory Bowel Disease and Sleep	376
		Irritable Bowel Syndrome and Sleep	377
Common Pathophysiological Mechanisms: Sleep and Gastrointestinal Tract	372	Chronic Liver Disease and Sleep	378
Brain—Gut Connection, Gut as a Second Brain	372	Assessment of Sleep Disorders in Gastrointestinal Disorders	379
Gastric, Intestinal, and Colonic Motility and Sleep	373	Management and Prevention of Sleep Disturbances in Persons With Gastrointestinal Disorders	380
Immunity and Inflammation in Gastrointestinal Tract and Sleep	373	Pharmacotherapy	380
Dysbiosis	373	Nonpharmacologic Interventions	383
Epidemiology and Risk Factors of Sleep Disturbances With Gastrointestinal Disorders	374	Clinical Vignette	384
		Areas for Future Research	385
Functional Dyspepsia and Sleep	374	Conclusion	385
Gastroesophageal Reflux Disorder and Sleep	375	References	386
Peptic Ulcer Disease and Sleep	375		

Handbook of Sleep Disorders in Medical Conditions
DOI: https://doi.org/10.1016/B978-0-12-813014-8.00017-2

© 2019 Elsevier Inc. All rights reserved.

INTRODUCTION

Worldwide gastrointestinal (GI) disorders are common and are associated with significant morbidity and health-care resource utilization.[1] Sleep disorders may induce GI disturbances, while GI symptoms also may provoke or worsen sleep quality by creating sleep fragmentation and circadian disruptions.[2,3] In daily clinical practice, clinicians often see persons with comorbid sleep difficulties and GI disorders.

Cross-sectional studies confirm the relationship between sleep and GI symptoms. For example, in a community-based population of 772 men and women, GI symptoms, such as abdominal pain, bloating, constipation, and diarrhea, were more prevalent in persons with chronic insomnia compared with those without insomnia (33.6% vs 9.2%).[2] In a survey ($N = 110{,}752$, Canadian population), Hon[4] noted that those who reported sleep disturbances and insomnia had a threefold increased risk for GI disorders after adjusting for sex, age, and stress. Despite findings such as this, there is limited information on the mechanisms accounting for the interrelationship between sleep and GI disorders. To better understand and manage sleep and GI disorders, the foci of this chapter are twofold; first, to review the literature on what is known about mechanisms and risk factors that link sleep and GI disorders, and second, to examine what is known about the assessment and treatment of sleep disorders in persons with GI disorders. Since sleep and cancer are addressed in Chapter 8, this chapter focuses on the upper and lower functional GI disorders (FGID), inflammatory bowel disease (IBD), and chronic liver disease.

COMMON PATHOPHYSIOLOGICAL MECHANISMS: SLEEP AND GASTROINTESTINAL TRACT

How sleep quality affects GI function, including motility, secretion, intestinal permeability, absorption, and ultimately symptoms, such as abdominal pain, nausea, and bowel alterations, are likely multifactorial. Findings from the literature on mechanisms examining the relationships among sleep and GI disorders are inconsistent due to differences in patient populations, diagnostic criteria, and many lack control for comorbid conditions, body weight, sex, and age.[5] Although no one-consistent mechanism likely accounts for the link between self-report of poor sleep and GI disorders, such as gastroesophageal reflux disorder (GERD) and irritable bowel syndrome (IBS), several mechanisms are being explored.

Brain—Gut Connection, Gut as a Second Brain

The digestive system is controlled in part by both its own enteric nervous system (ENS) within the gut wall and through efferent fibers of the autonomic nervous system (ANS).[5] The ANS contains both sensory afferent neurons from visceral organs to the central nervous system (CNS) and efferent motor neurons that convey impulses from the CNS to GI tract. The ENS consists of involuntary sensory and motor neurons controlling motility and endocrine and exocrine secretions. Although the ENS functions independently, the ANS

COMMON PATHOPHYSIOLOGICAL MECHANISMS: SLEEP AND GASTROINTESTINAL TRACT   373

modulates it. The two-way communication network is referred to as the *gut–brain axis*.[6] An example of this relationship is the response to intense emotional or psychological stimuli. For many individuals, these result in an increase in GI symptoms, such as diarrhea or abdominal pain.[5]

Gastric, Intestinal, and Colonic Motility and Sleep

During sleep, GI motility and intestinal transit decrease, while both increase during nocturnal wake periods.[5,7] The GI migrating motor complex (MMC) has a 90-minute cycle, similar to, but independent from, rapid eye movement (REM) sleep.[7] Although the distribution of MMCs among sleep stages is varied and complex, this periodic motor activity is influenced by sleep. Motility of the GI tract varies depending on sleep stage.[7] A recent study[8] assessed GI motility using 3D-transit during sleep monitored by polysomnography (PSG) in nine healthy individuals. The decline in the mean amplitudes of gastric contractions was related to the depth of sleep. Specifically, gastric, intestinal, and colonic contraction amplitudes in non-REM (NREM) stage 3, "deep sleep" were lower than stage 1 or 2. In contrast, REM sleep was associated with faster gastric emptying and increased colonic pressure.[8] To maintain fecal continence during sleep, the internal anal sphincter retains active pressure independent of external sphincter activity, and anal canal pressure remains above rectal pressure.[8] These pressures prevent stool leakage during sleep.[5,8]

Immunity and Inflammation in Gastrointestinal Tract and Sleep

There are reciprocal relationships among the immune system, inflammation, and sleep. Since many immune factors exhibit a circadian rhythm modulated by the CNS, sleep is an important regulator of immunological homeostasis.[3] Sleep deprivation is associated with an increased susceptibility to infection and a decrease in protective immune response.[3] By combining these result in an increase in the proinflammatory cytokines, potentially leading to or perpetuating chronic inflammatory conditions. Dysregulated immune function accompanying sleep loss may be important in the pathophysiology of GI disorders, such as IBS or postinfectious IBS, IBD, *Clostridium difficile* infection, chronic pancreatitis, colorectal cancer, as well as severity of GI symptoms.[3]

In the gut the immune system plays an important role in the maintenance of the intestinal epithelial barrier. Under normal conditions, this barrier protects against external pathogens. However a compromised intestinal barrier leads to increased permeability "leaky gut" and is linked to the development of GI and systemic disorders.[3] Chronic GI inflammation resulting in peripheral cytokine imbalances may influence sleep through both afferent nerve stimulation as well as the cytokines themselves.

Dysbiosis

The enormous abundance and diversity of microbes in the GI tract perform a range of essential functions. *Dysbiosis* is defined as compositional and functional alterations of the gut microbiome.[9] Alterations in the microbiota diversity may contribute GI disorders, including colorectal cancer, liver cirrhosis, FGIDs, as well as disorders of metabolism and mental illness.[9] Through the brain—gut axis the gut microbiome and its metabolites can potentially influence brain regions controlling sleep. In persons with chronic fatigue syndrome (CFS), Jackson et al.[10] examined the effects of gut microbiota on sleep quality using actigraphy and sleep questionnaires. They[10] found associations between increased colonization of lactic acid—producing fecal Gram-positive bacteria, particularly *Enterococcus* and *Streptococcus* spp. with neurologic dysfunctions (e.g., impaired motor coordination, confusion, excessive irritability, and inability to concentrate) and reduced sleep efficiency. Decreased colonization of lactic acid—producing organisms after antibiotic treatment resulted in both objective and subjective sleep improvement in persons with CFS.[10]

Circadian rhythm disruption also impacts intestinal microbiota composition.[11] In one study,[11] circadian rhythm—disrupted mice were found to have increased *Lachnospiraceae* and *Ruminococcaceae*, and a decrease of *Lactobacillaceae* families, as compared to noncircadian rhythm—disrupted mice. In another study,[12] circadian rhythm—disrupted mice showed evidence of colitis and increased intestinal permeability. In both rodents and humans, chronic short sleep is associated with obesity, diabetes, and cardiometabolic disorders. These conditions are also linked to the composition, that is, diversity of the microbiota, and increased intestinal permeability.[9]

EPIDEMIOLOGY AND RISK FACTORS OF SLEEP DISTURBANCES WITH GASTROINTESTINAL DISORDERS

Functional Dyspepsia and Sleep

Functional dyspepsia (FD) is a common FGID (prevalence range: 5.3%—20.4% in the world).[13] Epigastric pain or discomfort (prevalence in FD of 89%—90%), postprandial fullness (75%—88%), and epigastric pain syndrome (50%—82%) characterize FD (Table 17.1). For many, these symptoms occur following eating, that is, postprandial distress syndrome. A number of factors contribute to FD. These include reduced fundic

TABLE 17.1   Rome IV Functional Gastrointestinal Disorders Criteria[6]

Functional dyspepsia	Irritable bowel syndrome
Postprandial distress syndrome: Bothersome early satiety or postprandial fullness 3 or more days per week in the past 3 months with at least a 6-month history	Recurrent abdominal pain on average at least 1 day a week in the last 3 months associated with two or more of the following:
Epigastric pain syndrome: Bothersome epigastric pain or epigastric burning 1 or more days per week in the past 3 months with at least a 6-month history	1. Related to defecation 2. Associated with a change in a frequency of stool 3. Associated with a change in form (consistency) of stool

Note: Criteria fulfilled for the last 3 months with symptom onset at least 6 months prior to diagnosis.

accommodation, gastric hypersensitivity, abnormal duodenojejunal motility, duodenal motor, and sensory dysfunction, infections, such as *Helicobacter pylori*, and psychological distress.[13] Sleep disturbances are associated with FD, specifically FD symptom severity and higher reports of anxiety.[14–17] In a study with 137 persons with FD, 46% self-reported sleep awakenings due to reflux symptoms (heartburn and/or regurgitation) during the night.[14] In another study, 70% of the 131 persons with FD-reported problems falling asleep, or staying asleep 1–3 days per month on the Pittsburgh Sleep Quality Index (PSQI).[15]

Gastroesophageal Reflux Disorder and Sleep

GERD is a common and chronic condition with symptoms of heartburn and regurgitation. At least 20% of the adults in the United States have GERD.[18] There is a well-established bidirectional relationship between sleep disturbances and GERD. The lower esophageal sphincter (LES) normally acts, in conjunction with the diaphragm, as a barrier to prevent reflux of acidic stomach contents into the esophagus. During sleep, the LES relaxes, thus facilitating the reflux of gastric acid. When awake, acidification of the distal esophagus produces a marked increase in the secretion of saliva, swallowing, and peristalsis.[18] During sleep, these protective mechanisms along with swallowing and salivary secretions decrease. The sensation of "heartburn," as a warning sign, is also absent during sleep, due to limited protective mechanisms against gastric acid reflux. As a result, recumbent sleep facilitates gastric acid migration into the esophagus.[18,19] Subsequently, there is increased risk of mucosal irritation and inflammation and ultimately esophagitis and GERD.[18]

A population-based, cross-sectional, case-control study of 65,333 Norwegian participants[20] found a positive bidirectional link between GERD with self-reported insomnia (OR = 3.2; 95% CI = 2.7–3.7), sleeplessness (OR = 3.3; 95% CI = 2.9–3.8), and problems of falling asleep (OR = 3.1; 95% CI = 2.5–3.8), even after controlling for age, sex, tobacco smoking, obesity, and socioeconomic status. Persons with GERD report heartburn and acid reflux events, awakenings from sleep, and reduced sleep quality.[14,21–24] In four studies in persons with GERD,[25–28] arousals and awakenings measured by PSG were significantly associated with subsequent GERD events, measured by esophageal pH monitoring.

Obstructive sleep apnea (OSA) is also independently associated with GERD.[29] In one population-level analysis,[29] multivariate analysis demonstrated a significant positive association between GERD and OSA (OR = 2.13; 95% CI = 1.2–3.9), when adjusted for covariate factors, such as age, sex, race/ethnicity, sinonasal and laryngopharyngeal disorders, and obesity. In persons with OSA, there is a decrease in intrathoracic pressure and an increase in diaphragmatic pressure, which, in turn, increases nighttime acid reflux.[19] This nighttime gastric reflux can further disrupt sleep.[18]

Peptic Ulcer Disease and Sleep

Peptic ulcer disease (PUD), including gastric and duodenal ulcers, remains a relatively common disorder worldwide (prevalence = 4.1%, incidence = 10% of people develop PUD during their lifetime).[30] In the Western countries, the main etiological factor of the PUD is H. pylori, followed by use of nonsteroidal antiinflammatory drugs and acetylsalicylic acid.[30] It is suggested that poor sleep may play a role in the development of PUD with impaired mucosal healing and gastric acid secretion throughout the night.[30]

In healthy adults, gastric acid secretion decreases in deeper stages of sleep, especially during REM sleep. Normally continued secretion of bicarbonate and gastric mucosal blood flow buffers the acid. When these defensive mechanisms are impaired, for example, reduced gastric mucosal blood flow and gastric bicarbonate efflux, the gastric mucosa is left undefended against hydrochloric acid and the proteolytic enzyme pepsin. If prolonged and/or repeated, lack of buffer results in gastric mucosal damage, gastritis, and ulceration.[5,31] The risk of PUD is greater in those with sleep apnea. In a nationwide population-based study in Taiwan with 35,480 persons with sleep apnea, those with sleep apnea experienced a 2.4-fold higher risk for peptic ulcer bleeding after adjusting for other variables.[32] On the other hand, longer sleep duration is associated with reduced risk of PUD. In one epidemiological study ($N = 14,290$), after adjusting for covariates, for example, smoking, drinking, exercise, education, income, and spouse, longer sleep duration (≥ 9 hours) was associated with reduced risk of PUD.[31]

Inflammatory Bowel Disease and Sleep

IBD is an idiopathic GI disorder associated with a dysregulated immune response and chronic or recurring inflammation of GI tract. Ulcerative colitis and Crohn's disease occur in an estimated 1.5 million Americans, 2.2 million people in Europe, and several hundred thousands more, worldwide. Of those affected, 750,000 to 1 million report sleep disturbances.[33] The association between sleep and IBD is likely bidirectional. Increased inflammatory disease activity disrupts sleep through afferent fiber stimulation of neurochemical mediators, such as cytokines. Conversely, sleep disturbances can exacerbate inflammation and delay healing of the GI mucosa.[3,33,34] Moreover, symptoms of IBD (e.g., abdominal cramping pain, bloody diarrhea, fever, malnutrition) affect sleep quality.[33]

Cytokines play an important role in the pathophysiology of IBD.[33] In a review of 12 studies focused on IBD and sleep,[33] altered immune function [e.g., increased inflammatory cytokines such as interleukin (IL)-1β, tumor necrosis factor (TNF)-α, leukocytosis, and natural killer cells] were positively associated with NREM suppression and increased brief arousal events during sleep in persons with IBD. Increased levels of proinflammatory cytokines, such as IL-1α, IL-1β, IL-6, and TNF-α, are positively associated with disease activity and risk of relapse of IBD.[33]

When compared to healthy individuals, persons with IBD self-report more sleep disturbances, such as prolonged sleep latency, frequent awakenings, a higher rate of sleeping pill use, decreased daytime functioning, increased fatigue, and overall poor self-reported sleep quality.[35–39] Keefer et al.[38] examined subjective (PSQI) and objective (PSG) sleep indices in persons with IBD Researchers found increased subjective sleep complaints and decreased objective total sleep time in both the groups. One small study[40] assessed objective sleep using a wrist-worn actigraphy in four persons with IBD. In this study,[40] increased sleep-onset latency and decreased sleep efficiency were observed in persons with IBD when compared to healthy controls.

Irritable Bowel Syndrome and Sleep

IBS is a common chronic FGID, characterized by abdominal pain associated with alterations in defecation or stool frequency and consistency. Globally, the prevalence of IBS is higher in women. Based on the Rome IV diagnostic criteria (Table 17.1), IBS is subcategorized as diarrhea-predominant IBS (IBS-D), constipation-predominant IBS (IBS-C), IBS with mixed bowel habits (IBS-M), and IBS unclassified.[6] The prevalence of IBS is 3%–20% in the Western countries and 11% of the population globally, and the overall prevalence of IBS in women is 67% higher than in men.[41] The etiology of IBS is unknown, but alterations in motility, visceral sensitivity, intestinal barrier function, genetics, intestinal infection, bile acid malabsorption, dysbiosis, sleep disturbances, imbalanced autonomic dysfunction (sympathetic dominance), and psychological disorders, such as global anxiety disorders and major depressive disorders, are implicated in its pathogenesis.[41]

The relevance of sleep in IBS has recently received more attention.[42] In a population-based study ($N = 2269$),[43] 33% of the participants with sleep disturbances were diagnosed with IBS.

In addition the risk of having IBS is 60% greater in persons with sleep disturbances. Persons with IBS commonly self-reported sleep disturbances, such as difficulty in falling asleep, shortened sleep time, frequent arousals and awakenings, and/or nonrestorative sleep.[42] The sleep abnormalities in persons with IBS are associated with IBS-related pain, greater severity and frequency of GI symptoms, psychological distress, and poorer quality of life.[42,44] Using rectal manometry, Chen et al.[45] found increased rectal urge sensitivity in IBS patients with a high PSQI score (poor sleepers) as compared to healthy controls. In one study in persons with IBS,[46] both self-report and objective (i.e., actigraphy) sleep disturbance predicted worsening of next-day abdominal pain symptoms.

IBS symptoms often interrupt sleep.[42] In a systematic review of data-based studies of sleep in IBS,[42] 71% of the studies used objective measures, such as PSG or actigraphy. In this review,[42] the majority of studies found that women with IBS had prolonged sleep-onset latency, increased percentage of stage-2 sleep, increased latency of first REM onset, increased arousals and awakenings and wake time after sleep onset, and decreased total sleep time and reduced sleep efficiency, compared to women in healthy control groups. There is insufficient data available about men with IBS and their sleep quality. However the majority of studies used only one night of recording and did not account for "first- night effect."[42] Heitkemper et al.[47] reported a "first-night effect" of delayed sleep onset and greater sleep disruption on Night 1 (adaptation) when compared with Night 2 (baseline) in both healthy control women as well as those with IBS. However, this first-night effect was greater in the women with IBS. This highlights the importance in mechanistic sleep studies to assess at least two nights of recording to minimize the first-night effect.

Several investigators describe measures of ANS function during sleep in persons with IBS. Using heart rate variability captured by electrocardiography to determine low-to-high frequency and balance indicators, Jarrett et al.[48] found that women with IBS compared to healthy controls, have elevated sympathovagal balance. In another study, those with IBS compared to healthy controls had greater sympathetic dominance during REM sleep.[49] It can be hypothesized that this imbalance is associated with vagal withdrawal, in particular a lower parasympathetic nervous system activity during REM sleep. Lower vagal is noted particularly in those with IBS-C.[5,49] Women with IBS-D demonstrated significantly increased parasympathetic and lower sympathetic/parasympathetic nervous system balance across sequential NREM periods and REM cycles compared to both IBS-C and IBS-M subtypes.[50] Jarrett et al.[50] also described differences in noradrenaline, adrenaline, and cortisol levels during sleep among bowel pattern subtypes of persons with IBS. In a subsequent study, the introduction of a stressor, that is, public speaking anticipation, resulted in higher levels of plasma cortisol during the night in women with IBS relative to healthy control women.[47] Together these results suggest that ANS imbalance and hypothalamic—pituitary—adrenal activation during sleep may be present in a subgroup of persons with IBS.

Chronic Liver Disease and Sleep

Chronic liver disease is prevalent throughout the world. Liver cirrhosis, chronic infection with hepatitis C virus (HCV) and hepatitis B virus (HBV), alcoholic liver disease, and nonalcoholic fatty liver disease (NAFLD) occur in varying rates across the world.[51,52] According to the World Health Organization, about 46% of the global diseases and 59% of the mortality is because of chronic liver disease and almost 35 million people in the world die from this disease.[51] Persons with chronic liver disease with or without hepatic encephalopathy exhibit sleep changes. In those with hepatic encephalopathy, elevated ammonia levels lead to cerebral dysfunction as determined by the electroencephalography characteristics and neurological symptoms. The latter includes sleep—wake inversions, resulting in a combination of sleep fragmentation and excessive daytime sleepiness.[9]

Common self-reported sleep complaints in persons with chronic liver disease include prolonged time to fall asleep (longer sleep latency), daytime sleepiness (somnolence), shortened sleep duration, frequent nocturnal awakenings, and poor sleep quality.[52–56] In a recent case-control study using PSG,[57] persons with cirrhosis showed shorter sleep time, reduced sleep efficiency, increased sleep latency, increased REM latency, and reduced REM sleep.

In a systematic review[58] of 15 studies focused on sleep disturbances in persons with HCV, the overall prevalence of sleep disturbances was 60%—65%, independent of antiviral therapy. The presence of insomnia is more prevalent among individuals with cirrhosis due to HCV compared to HBV (e.g., 57% HCV vs 36% HBV, $N = 200$).[59] Other sleep disturbances associated with chronic liver disease include hypersomnia, restless leg syndrome (RLS), and OSA. RLS is a sleep-related movement disorder, which causes an urge to move the legs and uncomfortable sensations in the legs.[60] In one study, 62% of the 141 persons with chronic liver disease had RLS.[60] RLS and subsequent sleep disruptions in persons with liver disease contributes to reductions in the quality of life.

NAFLD is one of the types of fatty liver disease, which occurs when fat is deposited (steatosis) in the liver due to causes other than excessive alcohol use. These include metabolic syndrome, obesity, insulin resistance, and hypoxemia.[53] Persons with NAFLD also present with complaints of daytime sleepiness, fatigue, short sleep duration, insomnia, poor sleep quality, and OSA.[52] OSA is associated with the risk of NAFLD, indicating that hypoxia secondary to OSA may lead to an increase in fat accumulation in the liver, contributing to NAFLD and/or nonalcoholic steatohepatitis. Hepatic inflammation associated with NAFLD may also contribute to sleep disturbances, such as insomnia and daytime sleepiness, through a liver—brain-signaling mechanism that involves cytokine activation.[52]

ASSESSMENT OF SLEEP DISORDERS IN GASTROINTESTINAL DISORDERS

The assessment of sleep disorders in persons with GI disorders is the first step toward effective management. Several examples of objective and subjective sleep measures are described in Table 17.2. The most frequently used subjective sleep measure in GI disorders is PSQI, followed by other self-reported sleep questionnaire, and Epworth Sleepiness

TABLE 17.2 Subjective and Objective Sleep Measures Used in Studies of Gastrointestinal (GI) Disorders

GI Disorders	Subjective Sleep Measures	Objective Sleep Measures
Functional dyspepsia	Pittsburgh Sleep Quality Index[13,15,16]	None
	Insomnia Severity Index[15]	
	Other self-reported sleep questionnaires[14,17]	
Gastroesophageal reflex diseases	Pittsburgh Sleep Quality Index[23–25] Epworth Sleepiness Scale[24,25]	Actigraphy[19,24]
	Insomnia Severity Index[24,25]	Polysomnography[26–29]
	Other self-reported sleep questionnaires[14,20,22]	
Peptic ulcer disease	Self-reported sleep questionnaire[31]	None
Inflammatory bowel disease	Pittsburgh Sleep Quality Index[35–39] Epworth Sleepiness Scale[37,38]	Actigraphy[40]
	Insomnia Severity Index[39]	Polysomnography[38]
	Patient-Reported Outcomes Measurement Information System online survey[34]	
Irritable bowel syndrome	Systematic review with 21 articles:[42]	Systematic review with 21 articles:[42]
	Pittsburgh Sleep Quality Index (study $n = 12$)	Actigraphy (study $n = 20$)
	Epworth Sleepiness Scale (study $n = 2$)	Polysomnography (study $n = 15$)
	Sleep diary (study $n = 9$)	
Chronic liver disease	Pittsburgh Sleep Quality Index[54] Epworth Sleepiness Scale[53,54]	Actigraphy[55]
	Other self-reported sleep questionnaires[55,56]	Polysomnography[57,59]

Scale. Only a few studies used the Insomnia Severity Index with persons with FD, GERD, and IBD (Table 17.2). The Patient-Reported Outcomes Measurement Information System was used to measure sleep quality in one study of persons with IBD.[34] In IBS, self-reported sleep diaries are commonly used. Objective sleep measures used in GI disorders are actigraphy and PSG. Interestingly, similarly to the general population, several studies have found a discrepancy between subjective and objective measures of sleep in persons with IBS as well as other GI disorders.[5,42]

In assessment, special considerations are given to the use of medications. Clinicians should monitor GI side effects of sleep agents with anticholinergic properties or excessive daytime sleepiness, and drowsiness of GI motility agents with opioid receptor agonist, such as loperamide (Imodium). Nocturnal acid reflex events in upper GI disorders frequently occur during sleep, beginning around 10 p.m., or between 1 and 4 a.m.[23] Thus appropriate timing of taking proton-pump inhibitors (PPIs) (e.g., twice daily, before breakfast and before dinner) and dosage (e.g., omeprazole 20 mg) should be assessed to maximize its acid suppressive effect during the nocturnal period in persons with upper GI disorders.[23] Since the risk for OSA is more prevalent in persons with GERD,[29] patients need to be assessed for the presence of OSA. The risk factors and manifestations of OSA, including high blood pressure, overweight/obesity, use of alcohol or sedatives, smoking, narrowed airway, diabetes, asthma, and chronic nasal congestion, should be also assessed and managed.[29]

MANAGEMENT AND PREVENTION OF SLEEP DISTURBANCES IN PERSONS WITH GASTROINTESTINAL DISORDERS

GI disorders are frequently comorbid with psychiatric conditions (e.g., anxiety, major depression), migraine, and other somatic disorders, such as fibromyalgia and CFS.[1] Symptom overlap and comorbidity between GI disorders (e.g., IBS and GERD; FD and GERD; IBS and IBD) are also frequent.[1] Thus the strategies to improve sleep should be implemented by considering multiple comorbid symptoms, including GI symptoms and psychological distress, and with systematic approaches in persons with GI disorders.

Pharmacotherapy

Effects of Sleep-Enhancing Agents on Gastrointestinal Tract

Although there are no unique sleep medications, specific for sleep problems associated with GI disorders, multiple drug treatments exist for sleep disturbances, such as insomnia.[5] Managing poor sleep reduces overall symptom distress in persons with GI disorders.[5]

Sleep medications prescribed for poor sleep in persons with GI include benzodiazepines (e.g., alprazolam), antidepressants [e.g., tricyclic antidepressants (TCAs), selective serotonin reuptake inhibitor (SSRI)], imidazopyridines (e.g., zolpidem, clonazepam), pyrazolopyrimidines (e.g., zaleplon), cyclopyrrolones (e.g., zopiclone), anticonvulsants (pregabalin/gabapentin), barbiturates (e.g., amobarbital), α-2-adrenergic agonists (e.g., clonidine), antipsychotics (e.g., haloperidol), and melatonin or melatonin receptor agonists (e.g., ramelteon).[6] Because many of these have anticholinergic properties, many produce GI side effects (e.g., dry mouth, nausea, vomiting, stomach pain, and constipation), which can aggravate symptoms of GI disorders.[6] The majority of sleep medications do not result in acute liver injury or transient serum liver enzyme elevations. However, there are case reports of alprazolam, clonazepam, doxepin, trazodone, zolpidem, and gamma hydroxybutyrate producing elevations in liver enzymes.[61]

MELATONIN

The melatonin may be helpful for both sleep and visceral sensitivity.[62] Melatonin from the pineal gland is associated with increased duration of REM and deep sleep.[63] Anecdotal evidence shows that the melatonin is useful for improving symptoms and reducing sleep disturbances in persons with GERD and FD,[58] IBD,[64] and IBS.[65] Melatonin has antiinflammatory benefits, such as a reduction in mucosal inflammation severity in animal models of autoimmune diseases.[61] The enterochromaffin cells of the GI tract mucosa also produce melatonin. Levels of melatonin production in the gut greatly exceed peripheral levels of melatonin. In the gut, melatonin acting through autocrine and paracrine routes affects motility, secretion, and afferent fiber stimulation.[61]

Two small sample reports suggest that exogenous melatonin reduces GI symptom distress in persons with IBS. Lu et al.[66] reported that 3 mg melatonin nightly for 8 weeks produced an improvement in IBS symptom scores as compared to placebo. Similarly, in a sample of 18 IBS patients, Saha et al.[65] found that 3 mg of melatonin for 8 weeks produced a decrease in both individual IBS symptoms as well as overall IBS symptom severity. There are several potential mechanisms (e.g., blockage of nicotinic channels or calcium-activated channels) via which melatonin may modulate visceral sensitivity in persons with IBS. When bowel distension protocol (barostat) is used, melatonin administration reduces visceral pain perception.[62] More studies are needed to decipher whether the benefits of melatonin administration to individuals with GI complaints are due to central versus peripheral mechanisms. Although melatonin has sleep-promoting effects, it can also produce diarrhea, abnormal feces, stomach pain, vomiting, nausea, cramping, increased appetite or altered taste perception. As a result, melatonin should be used cautiously.[61]

PSYCHOTROPIC AGENTS

Psychological distress, for example, anxiety and/or depression, often accompanies sleep-related symptoms in persons with GI disorders.[67] Thus psychotropic agents, such as TCAs and SSRIs, are beneficial and often used for the treatment of sleep disturbances in persons with comorbid psychological distress.[61] Mirtazapine (Remeron), as an atypical antidepressant, is efficacious and safe for treating interferon (IFN)-α-induced insomnia in persons with HCV, with promoting sleep through its 5-HT$_2$ and H$_1$ blockade.[68] Benzodiazepines (e.g., lorazepam, oxazepam, and temazepam) are also used for acute treatment of insomnia in persons with HCV.[69] However, many psychotropic drugs require dosage adjustment in persons with hepatic impairment.[68]

Effects of Gastrointestinal Symptom Management Medications on Sleep

ACID NEUTRALIZATION

To reduce the likelihood of upper GI disorder—induced irritation and inflammation during sleep, PPIs are the most widely used class of drugs to block acid secretions for persons with upper GI disorders, such as FD, PUD, and GERD. Three randomized controlled trials (RCTs) in persons with GERD plus sleep disturbances demonstrate the efficacy of PPIs to reduce self-reported sleep disturbances (e.g., esomeprazole 20 or 40 mg, rabeprazole 20 mg).[23] H_2-receptor antagonists are also effective in reducing nocturnal irritation caused by gastric reflux (measured by pH monitoring).[70]

GASTROINTESTINAL MOTILITY AGENTS

Antispasmodic agents act by inhibiting smooth muscle contractions in the GI tract. The most common antispasmodics contain anticholinergic properties [e.g., dicyclomine (Bentyl), hyoscyamine (Levsin, NuLev, Levbid), octylonium bromide]. These are helpful in relieving symptoms, such as abdominal pain and diarrhea.[52] The antidiarrheal medication, loperamide, is an opioid μ-receptor agonist. It is effective in reducing diarrhea and can improve sleep quality in persons, unable to sleep secondary to diarrhea and abdominal pain.[1] However, agents, such as loperamide, induce CNS side effects, including insomnia, drowsiness, or sedation.[1] Thus side effects of daytime sleepiness or insomnia can be the limiting factor in the use of CNS-active agents, specifically in older adults.[1]

ANTIINFLAMMATORY, ANTIBIOTICS, AND ANTIVIRAL AGENTS

Antiinflammatory drugs (e.g., aminosalicylates, corticosteroids) and immunosuppressants (e.g., cyclosporine) reduce inflammation and improve GI symptoms in persons with IBD.[71] Antibiotics (e.g., metronidazole, ciprofloxacin) eradicate infections, such as *H. pylori* in PUD. Metronidazole and vancomycin are also used to treat *C. difficile* infection.[72] Although the use of antiinflammatory agents or antibiotics alleviates GI symptoms by treating or suppressing inflammation, these agents may disrupt the ecology of the gut microbiome. This, in turn, may contribute to sleep disturbances in humans and animals. Rifaximin (Xifaxan), as an antibiotic, is effective to improve objective sleep parameters in GI disorders, such as diarrhea,[73] IBS,[73] and hepatic encephalopathy.[74] Rifaximin was effective in improving sleep disturbances measured by PSG, relieving RLS symptoms and GI symptoms in IBS patients with coexisting RLS ($N = 90$).[73] In one pilot study in 15 persons with hepatic encephalopathy,[74] rifaximin treatment (550 mg twice daily for 28 days) improved an objective sleep parameter (increased REM sleep on PSG).

HCV treatment with IFN-α confers an additional risk of sleep disturbance. In one study, 24% of persons receiving immunotherapy reported insomnia.[69] One side effect of IFN-α is depression, which in turn increases the risk of poor sleep quality and excessive daytime sleepiness.[69] Side effects of newer drugs for HCV, such as a combination of antiviral drugs like ombitasvir/paritaprevir/ritonavir (Technivie) and sofosbuvir/velpatasvir (Epclusa), also include insomnia.[73]

In sum, pharmacologic interventions used to alleviate GI and psychological symptoms are often used to manage sleep disturbances. Some GI medications have adverse effects on sleep,[69] or some sleep aids worsen GI symptoms.[75] Thus GI medications and sleep aids should be used cautiously. Different dosages and durations and interactions with other drugs must be considered before using these medications for persons with comorbid GI complaints and sleep problems.

Nonpharmacologic Interventions

Lifestyle Modifications

Sleep hygiene practices combined with a healthy lifestyle are the first priority for managing sleep disturbances in persons with GI disorders. Detailed recommendations of healthy sleep hygiene strategies (e.g., maintain regular sleep–wake schedule even on weekends) for sleep disorders have been outlined by the American Academy of Sleep Medicine.[76] For example, the persons with upper GI disorders are instructed to avoid of foods and beverages especially near bedtime (e.g., chocolate, coffee, spicy foods, alcohol) associated with an increase in gastric acid secretions, reflex, and heartburn. Other behavioral factors are avoidance of smoking, maintenance of normal body weight, refraining from eating 2–3 hours prior to going to bed, avoidance of tight-fitting clothing, and elevate the head of bed $\geq 30°$. These behavioral modifications can be applied for managing comorbid upper GI symptoms and sleep disturbances.

Elimination Diet

For some groups of patients a special emphasis on diet should be considered. For example, in some IBS patients, specific foods triggers account for abdominal pain, bloating, and diarrhea, which can interfere with sleep. Consultation with a dietitian may be helpful in identifying those food triggers. Examples of dietary interventions include instituting a low-fat diet, limiting dairy products, increasing low-residue fibers are suggested. An elimination diet in which short chain carbohydrates, also known as FODMAPs (fermentable oligosaccharides, disaccharides, monosaccharides, and polyols), restrict several dietary substances, such as lactose, alcohol, gas-producing foods (e.g., hot spicy foods, rich, fatty foods, caffeine, cabbage, broccoli), and foods containing sorbitol (e.g., artificial sweeteners, ice cream).[77] A metaanalysis of 22 studies supports the efficacy of a low-FODMAP diet in the treatment of functional GI symptoms in persons with IBS or IBD. However the long-term efficacy of low FODMAPs remains to be determined.[77] To date, no studies have specifically looked at sleep and whether sleep is affected by a low-FODMAP diet.[77]

Psychotherapy

There is mounting evidence that supports the positive relationships among sleep disturbances, GI symptoms, and psychological distress (e.g., stress, traumatic event, depression, and anxiety).[67] The central emotional motor system is subject to the influence of various physiological abnormalities (e.g., visceral hypersensitivity, hyperalgesia, abnormalities in autonomic, and neuroendocrine and immune functions). But the mechanisms underlying the impact of psychotherapy in GI disorders are still unclear.[78] With psychotherapy, the goal is to modulate GI functioning through changes in the how sensory signals from the GI tract are perceived.[78] Since psychological distress is associated with poor sleep, stress-management strategies, such as mind–body exercise, cognitive-behavioral therapy (CBT), and hypnotherapy, may be effective as adjuvant therapies to improve sleep.[79] In a RCT of mindfulness-based intervention combined with CBT for IBS symptoms, bowel symptom severity decreased, psychological distress improved, health-related quality of life improved, and stress decreased in persons with IBS.[80] Hypnotherapy is useful to relieve overall GI symptoms for the persons with FD and IBS. Currently, the evidence of the efficacy of mindfulness-based intervention, CBT or hypnotherapy in persons with GI disorders, are limited due to small sample size studies, and the lack of rigorous study designs and methodologies.[81] However the benefits of behavioral therapies on GI and psychological symptoms[78] could indirectly influence sleep.[5]

Complementary and Alternative Medicine

Some forms of complementary and alternative medicine, such as probiotics, herbals, and acupuncture, appear to have promise in GI disorders, such as FD, IBD, IBS, and chronic liver disease as well as sleep problems. But there are few well-designed studies of their safety and effectiveness.[82] This is particular of concern regarding the use of probiotics and herbals in patients with chronic liver disease.

Continuous Positive Airway Pressure

The continuous positive airway pressure (CPAP) is commonly used in clinical practice to treat OSA in patients with GI disorders.[83] However, there is a lack of evidence supporting its efficacy and safety in this specific population.[83] One side effect of CPAP is aerophagia that is air entering the esophagus and stomach rather than the lungs. Gastric distension can increase GERD and indigestion.[83] Therefore CPAP should be used with caution in patients with GI disorders, and GI symptoms should be regularly monitored in individuals on CPAP.[83]

CLINICAL VIGNETTE

A 48-year-old White woman with a Body Mass Index (BMI) 32 kg/m² (obese) arrives at her primary care provider office. She reports a 15-year history of occasional heartburn, reports having had worsening heartburn for the past 6 months, with daily symptoms of difficulty getting to sleep, frequent awakenings, and awakening feeling unrefreshed. She denies GI bleeding, dysphasia, or weight loss but reports episodes of depression and moderate fatigue. What would you advise regarding her evaluation and treatment?

CONCLUSION

385

Treatment plans:

- *Behavioral management: Incorporate sleep hygiene practices to improve sleep quality; behavioral modifications (e.g., avoid foods and beverages associated with an increase in heartburn symptoms, reduce smoking, reduce weight, refrain from eating <2–3 hours prior to going to bed, wear loose-fitting clothing) to reduce hydrochloric acid reflux.*
- *PPIs: PPIs suppress hydrochloric acid secretion via the inhibition of the hydrogen—potassium-activated adenosine triphosphatase proton pump.*

Rationales:

- *PPIs are the most widely used class of drugs for the treatment of GERD.*
- *PPIs and nonpharmacologic approaches provide rapid GI symptom relief and reduce sleep disturbances in many patients.[23]*

AREAS FOR FUTURE RESEARCH

Further research is needed to develop better patient-centered and validated sleep measurement instruments or health information techniques to examine the directionality of the association between sleep disturbances and GI disorders. Persons with GI disorders who report poor sleep quality should be further evaluated for comorbid sleep conditions, such as OSA, insomnia, and RLS. In addition, they may need to be further evaluated for underlying mechanisms of GI disorders, such as inflammation, and dysbiosis with population-based, larger sample studies. Further study of possible links between gut microorganisms and sleep is warranted. Lastly, future studies are also suggested to evaluate the effects of sleep improvements interventions on the clinical outcomes of GI disorders, and vice versa. Manipulation of gut microbiota is a promising treatment for persons with sleep and GI issues together that also warrants to be assessed.

CONCLUSION

There are complex bidirectional associations between sleep disturbances and GI disorders. Dietary modifications, good sleep hygiene, and drug therapies targeted to the specific disease conditions may be warranted to reduce disease progression, reduce sleep disruptions, and enhance quality of life. However, our knowledge and understanding of the associations between sleep and the digestive system is still limited. Further studies using well-vetted patient group should use validated subjective and objective sleep measures. A better understanding of the relationships between sleep and GI disorders may allow the health-care professionals to provide patient-centered management for persons with both GI disorders and sleep disturbances more effectively.

17. GASTROINTESTINAL DISORDERS

KEY PRACTICE POINTS

Assessment of Sleep Disturbances With GI Disorders

- Determine the presence of OSA in patients with GERD.
- Consider the risk for RLS in patients with chronic liver disease.
- Screen patients with chronic liver disease for hepatic encephalopathy.
- Ask about use of antisecretory agents and their timing.
- Pain associated with FD, GERD, and PUD can disrupt sleep.
- Query patients about their use of sleep hygiene practices and lifestyle factors.
- For all patients with a GI disorder, conduct a sleep medication assessment with attention to both efficacy and GI side effects. Sleep disorders in patients with chronic liver disease occur due to metabolic changes secondary to impaired clearance of medications.

Treatment and Prevention of Sleep Disturbances with GI Disorders

- The important management and prevention strategy of sleep disturbances with GI disorders is focus on either sleep management or symptom management of GI disorders, or both sleep and symptom management together.
- Psychotropic agents and sleep aids are cautiously used because they may worsen GI symptoms.
- Benzodiazepines are associated with heartburn during sleep and gastric acid reflux events. Benzodiazepines decrease the LES pressure and increase the probability of GERD.
- Zolpidem inhibits the arousal response to nighttime reflux and increases the duration of acid reflux incidents with heartburn in persons with and without GERD.
- Trazodone is often used as a hypnotic at lower doses for sleep disturbances in patients with HCV. Hepatotoxicity with trazodone has been reported, thus further studies are required to confirm the safety of trazodone in HCV.

References

1. Wu JCY. Psychological co-morbidity in functional gastrointestinal disorders: epidemiology, mechanisms and management. *Neurogastroenterol Motil.* 2012;18(1):13–18.
2. Ali T, Choe J, Awab A, Wagener TL, Orr WC. Sleep, immunity and inflammation in gastrointestinal disorders. *World J Gastroenterol.* 2013;19(48):9231–9239.
3. Taylor DJ, Mallory LJ, Lichstein KL, Durrence HH, Riedel BW, Bush AJ. Comorbidity of chronic insomnia with medical problems. *Sleep.* 2007;30(2):213–218.
4. Hon C-Y. Examining the association between insomnia and bowel disorders in Canada: is there a trend? *UBC Med J.* 2010;2(1):11–15.
5. Orr WC, Chen CL. Sleep and the gastrointestinal tract. *Neurol Clin.* 2005;23(4):1007–1024.
6. Drossman DA, Hasler WL. Rome IV—functional GI disorders: disorders of gut–brain interaction. *Gastroenterology.* 2016;150(6):1257–1261.
7. Kumar D. Sleep as a modulator of human gastrointestinal motility. *Gastroenterology.* 1994;107(5):1548–1550.
8. Haase A-M, Fallet S, Otto M, Scott SM, Schlageter V, Krogh K. Gastrointestinal motility during sleep assessed by tracking of telemetric capsules combined with polysomnography – a pilot study. *Clin Exp Gastroenterol.* 2015;8:327–332.

REFERENCES

387

9. Robles Alonso V, Guarner F. Linking the gut microbiota to human health. *Br J Nutr*. 2013;109(S2):S21–S26.
10. Jackson ML, Butt H, Ball M, Lewis DP, Bruck D. Sleep quality and the treatment of intestinal microbiota imbalance in chronic fatigue syndrome: a pilot study. *Sleep Sci*. 2015;8(3):124–133.
11. Voigt RM, Forsyth CB, Green SJ, et al. Circadian disorganization alters intestinal microbiota. *PLoS One*. 2014;9 (5):e97500.
12. Summa KC, Voigt RM, Forsyth CB, et al. Disruption of the circadian clock in mice increases intestinal permeability and promotes alcohol-induced hepatic pathology and inflammation. *PLoS One*. 2013;8(6):e67102.
13. Futagami S, Shimpuku M, Yamawaki H, et al. Sleep disorders in functional dyspepsia and future therapy. *J Nippon Med Sch*. 2013;80(2):104–109.
14. Vakil N, Wernersson B, Wissmar J, Dent J. Sleep disturbance due to heartburn and regurgitation is common in patients with functional dyspepsia. *United Eur Gastroenterol J*. 2016;4(2):191–198.
15. Lacy BE, Everhart K, Crowell MD. Functional dyspepsia is associated with sleep disorders. *Clin Gastroenterol Hepatol*. 2011;9(5):410–414.
16. Chen Y, Wang C, Wang J, et al. Association of psychological characteristics and functional dyspepsia treatment outcome: a case-control study. *Gastroenterol Res Pract*. 2016;5984273.
17. Welén K, Faresjö Å, Faresjö T. Functional dyspepsia affects women more than men in daily life: a case-control study in primary care. *Gender Med*. 2008;5(1):62–73.
18. Dent J, Holloway RH, Eastwood PR. Systematic review: relationships between sleep and gastro-oesophageal reflux. *Aliment Pharmacol Ther*. 2013;38(7):657–673.
19. Poh CH, Gasiorowska A, Allen L, et al. Reassessment of the principal characteristics of gastroesophageal reflux during the recumbent period using integrated actigraphy-acquired information. *Am J Gastroenterol*. 2010;105(5):1024–1031.
20. Jansson C, Nordenstedt H, Wallander MA, et al. A population-based study showing an association between gastroesophageal reflux disease and sleep problems. *Clin Gastroenterol Hepatol*. 2009;7(9):960–965.
21. Fass R. The relationship between gastroesophageal reflux disease and sleep. *Curr Gastroenterol Rep*. 2009;11 (3):202–208.
22. Johnson DA, Orr WC, Crawley JA, et al. Effect of esomeprazole on nighttime heartburn and sleep quality in patients with GERD: a randomized, placebo-controlled trial. *Am J Gastroenterol*. 2005;100(9):1914–1922.
23. Hiramoto K, Fujiwara Y, Ochi M, et al. Effects of esomeprazole on sleep in patients with gastroesophageal reflux disease as assessed on actigraphy. *Intern Med*. 2015;54(6):559–565.
24. Iwakura N, Fujiwara Y, Shiba M, et al. Characteristics of sleep disturbances in patients with gastroesophageal reflux disease. *Intern Med*. 2016;55(12):1511–1517.
25. Yang YX, Spencer G, Schutte-Rodin S, Brensinger C, Metz DC. Gastroesophageal reflux and sleep events in obstructive sleep apnea. *Eur J Gastroenterol Hepatol*. 2013;25(9):1017–1023.
26. Penzel T, Becker H, Brandenburg U, Labunski T, Pankow W, Peter J. Arousal in patients with gastro-oesophageal reflux and sleep apnoea. *Eur Respir J*. 1999;14(6):1266–1270.
27. Mello-Fujita L, Roizenblat S, Frison CR, et al. Gastroesophageal reflux episodes in asthmatic patients and their temporal relation with sleep architecture. *Braz J Med Biol Res*. 2008;41:152–158.
28. Jaimchariyatam N, Tantipornsinchai W, Desudchit T, Gonlachanvit S. Association between respiratory events and nocturnal gastroesophageal reflux events in patients with coexisting obstructive sleep apnea and gastroesophageal reflux disease. *Sleep Med*. 2016;22:33–38.
29. Gilani S, Quan SF, Pynnonen MA, Shin JJ. Obstructive sleep apnea and gastroesophageal reflux. *Otolaryngology—Head Neck Surgery*. 2015;154(2):390–395.
30. Sung J, Kuipers E, El-Serag H. Systematic review: the global incidence and prevalence of peptic ulcer disease. *Aliment Pharmacol Ther*. 2009;29(9):938–946.
31. Ko S-H, Baeg MK, Ko SY, Han K-D. Women who sleep more have reduced risk of peptic ulcer disease; Korean national health and nutrition examination survey (2008–2009). *Sci Rep*. 2016;6:36925.
32. Shiao T-H, Liu C-J, Luo J-C, et al. Sleep apnea and risk of peptic ulcer bleeding: a nationwide population-based study. *Am J Med*. 2013;126(3):249–255. e241.
33. Kinnucan JA, Rubin DT, Ali T. Sleep and inflammatory bowel disease: exploring the relationship between sleep disturbances and inflammation. *Gastroenterol Hepatol*. 2013;9(11):718–727.
34. Ananthakrishnan AN, Long MD, Martin CF, Sandler RS, Kappelman MD. Sleep disturbance and risk of active disease in patients with Crohn's disease and ulcerative colitis. *Clin Gastroenterol Hepatol*. 2013;11(8):965–971.

II. SLEEP DISORDERS IN SPECIFIC MEDICAL CONDITIONS

35. Ranjbaran Z, Keefer L, Farhadi A, Stepanski E, Sedghi S, Keshavarzian A. Impact of sleep disturbances in inflammatory bowel disease. *J Gastroenterol Hepatol*. 2007;22(11):1748−1753.
36. Ali T, Madhoun MF, Orr WC, Rubin DT. Assessment of the relationship between quality of sleep and disease activity in inflammatory bowel disease patients. *Inflamm Bowel Dis*. 2013;19(11):2440−2443.
37. Graff LA, Vincent N, Walker JR, et al. A population-based study of fatigue and sleep difficulties in inflammatory bowel disease. *Inflamm Bowel Dis*. 2011;17(9):1882−1889.
38. Keefer L, Stepanski EJ, Ranjbaran Z, Benson LM, Keshavarzian A. An initial report of sleep disturbance in inactive inflammatory bowel disease. *J Clin Sleep Med*. 2006;2(4):409−416.
39. Banovic I, Gilibert D, Cosnes J. Crohn's disease and fatigue: constancy and co-variations of activity of the disease, depression, anxiety and subjective quality of life. *Psychol Health Med*. 2010;15(4):394−405.
40. Burgess HJ, Swanson GR, Keshavarzian A. Endogenous melatonin profiles in asymptomatic inflammatory bowel disease. *Scand J Gastroenterol*. 2010;45(6):759−761.
41. Canavan C, West J, Card T. The epidemiology of irritable bowel syndrome. *Clin Epidemiol*. 2014;6:71.
42. Tu Q, Heitkemper M, Jarrett M, Buchanan D. Sleep disturbances in irritable bowel syndrome: a systematic review. *Neurogastroenterol Motil*. 2017;29(3):e12946.
43. Vege SS, Locke GR, Weaver AL, Farmer SA, Melton LJ, Talley NJ. Functional gastrointestinal disorders among people with sleep disturbances: a population-based study. *Mayo Clin Proc*. 2004;79(12):1501−1506.
44. Patel A, Hasak S, Cassell B, et al. Effects of disturbed sleep on gastrointestinal and somatic pain symptoms in irritable bowel syndrome. *Aliment Pharm Ther*. 2016;44(3):246−258.
45. Chen CL, Liu TT, Yi CH, Orr WC. Evidence for altered anorectal function in irritable bowel syndrome patients with sleep disturbance. *Digestion*. 2011;84(3):247−251.
46. Buchanan DT, Cain K, Heitkemper M, et al. Sleep measures predict next-day symptoms in women with irritable bowel syndrome. *J Clin Sleep Med*. 2014;10(9):1003−1009.
47. Heitkemper MM, Cain KC, Deechakawan W, et al. Anticipation of public speaking and sleep and the hypothalamic-pituitary-adrenal axis in women with irritable bowel syndrome. *Neurogastroenterol Motil*. 2012;24(7):626−631. e270-621.
48. Jarrett M, Heitkemper M, Cain KC, Burr RL, Hertig V. Sleep disturbance influences gastrointestinal symptoms in women with irritable bowel syndrome. *Dig Dis Sci*. 2000;45(5):952−959.
49. Thompson JJ, Elsenbruch S, Harnish MJ, Orr WC. Autonomic functioning during REM sleep differentiates IBS symptom subgroups. *Am J Gastroenterol*. 2002;97(12):3147−3153.
50. Jarrett ME, Burr RL, Cain KC, Rothermel JD, Landis CA, Heitkemper MM. Autonomic nervous system function during sleep among women with irritable bowel syndrome. *Dig Dis Sci*. 2008;53(3):694−703.
51. Murray CJ, Lopez AD. Evidence-based health policy—lessons from the global burden of disease study. *Science (New York, NY)*. 1996;274(5288):740−743.
52. Khanijow V, Prakash P, Emsellem HA, Borum ML, Doman DB. Sleep dysfunction and gastrointestinal diseases. *Gastroenterol Hepatol*. 2015;11(12):817−825.
53. Mostacci B, Ferlisi M, Baldi Antognini A, et al. Sleep disturbance and daytime sleepiness in patients with cirrhosis: a case control study. *Neurol Sci*. 2008;29(4):237−240.
54. Newton JL, Gibson GJ, Tomlinson M, Wilton K, Jones D. Fatigue in primary biliary cirrhosis is associated with excessive daytime somnolence. *Hepatology*. 2006;44(1):91−98.
55. Cordoba J, Cabrera J, Lataif L, Penev P, Zee P, Blei AT. High prevalence of sleep disturbance in cirrhosis. *Hepatology*. 1998;27(2):339−345.
56. Mir HM, Stepanova M, Afendy H, Cable R, Younossi ZM. Association of sleep disorders with nonalcoholic fatty liver disease (NAFLD): a population-based study. *J Clin Exp Hepatol*. 2013;3(3):181−185.
57. Teodoro VV, Júnior MAB, Lucchesi LM, et al. Polysomnographic sleep aspects in liver cirrhosis: a case control study. *World J Gastroenterol*. 2013;19(22):3433−3438.
58. Carlson MD, Hilsabeck RC, Barakat F, Perry W. Role of sleep disturbance in chronic hepatitis C infection. *Curr Hepat Rep*. 2010;9(1):25−29.
59. Mabrouk AA, Nooh MA, Azab NY, Elmahallawy II, Elshenawy RHM. Sleep pattern changes in patients with liver cirrhosis. *Egypt J Chest Dis Tuberc*. 2012;61(4):447−451.
60. Franco RA, Ashwathnarayan R, Deshpandee A, et al. The high prevalence of restless legs syndrome symptoms in liver disease in an academic-based hepatology practice. *J Clin Sleep Med*. 2008;4(1):45−49.
61. Proctor A, Bianchi MT. Clinical pharmacology in sleep medicine. *Pharmacology*. 2012;914168.

REFERENCES

62. Song GH, Leng PH, Gwee KA, Moochhala SM, Ho KY. Melatonin improves abdominal pain in irritable bowel syndrome patients who have sleep disturbances: a randomised, double blind, placebo controlled study. *Gut.* 2005;54(10):1402−1407.

63. Uchiyama M, Hamamura M, Kuwano T, Nishiyama H, Nagata H, Uchimura N. Evaluation of subjective efficacy and safety of ramelteon in Japanese subjects with chronic insomnia. *Sleep Med.* 2011;12(2):119−126.

64. Mozaffari S, Abdollahi M. Melatonin, a promising supplement in inflammatory bowel disease: a comprehensive review of evidences. *Curr Pharm Des.* 2011;17(38):4372−4378.

65. Saha L, Malhotra S, Rana S, Bhasin D, Pandhi P. A preliminary study of melatonin in irritable bowel syndrome. *J Clin Gastroenterol.* 2007;41(1):29−32.

66. Lu WZ, Gwee KA, Moochhalla S, Ho KY. Melatonin improves bowel symptoms in female patients with irritable bowel syndrome: a double-blind placebo-controlled study. *Aliment Pharm Ther.* 2005;22(10):927−934.

67. Koloski NA, Talley NJ, Boyce PM. Does psychological distress modulate functional gastrointestinal symptoms and health care seeking? A prospective, community cohort study. *Am J Gastroenterol.* 2003;98(4):789−797.

68. Schaefer M, Schwaiger M, Garkisch AS, et al. Prevention of interferon-alpha associated depression in psychiatric risk patients with chronic hepatitis C. *J Hepatol.* 2005;42(6):793−798.

69. Sockalingam S, Abbey SE, Alosaimi F, Novak M. A review of sleep disturbance in hepatitis C. *J Clin Gastroenterol.* 2010;44(1):38−45.

70. Fackler WK, Ours TM, Vaezi MF, Richter JE. Long-term effect of H2RA therapy on nocturnal gastric acid breakthrough. *Gastroenterology.* 2002;122(3):625−632.

71. Leitner GC, Vogelsang H. Pharmacological- and non-pharmacological therapeutic approaches in inflammatory bowel disease in adults. *World J Gastrointest Pharmacol Ther.* 2016;7(1):5−20.

72. Wannmacher L. Review of the evidence for *H. pylori* treatment regimens. In: *18th Expert Committee on the Selection and Use of Essential Medicines.* 2011;17(1):1−33.

73. Basu PP, Shah NJ, Krishnaswamy N, Pacana T. Prevalence of restless legs syndrome in patients with irritable bowel syndrome. *World J Gastroenterol.* 2011;17(39):4404−4407.

74. Bruyneel M, Serste T, Libert W, et al. Improvement of sleep architecture parameters in cirrhotic patients with recurrent hepatic encephalopathy with the use of rifaximin. *Eur J Gastroenterol Hepatol.* 2017;29(3):302−308.

75. Fass R, Quan SF, O'Connor GT, Ervin A, Iber C. Predictors of heartburn during sleep in a large prospective cohort study. *Chest.* 2005;127(5):1658−1666.

76. Morgenthaler T, Kramer M, Alessi C, et al. Practice parameters for the psychological and behavioral treatment of insomnia: an update. An American Academy of Sleep Medicine report. *Sleep.* 2006;29(11):1415−1419.

77. Marsh A, Eslick EM, Eslick GD. Does a diet low in FODMAPs reduce symptoms associated with functional gastrointestinal disorders? A comprehensive systematic review and meta-analysis. *Eur J Nutr.* 2016;55 (3):897−906.

78. Justin CYW. Psychological co-morbidity in functional gastrointestinal disorders: epidemiology, mechanisms and management. *Neurogastroenterol Motil.* 2012;18(1):13−18.

79. Hofmann SG, Asnaani A, Vonk IJJ, Sawyer AT, Fang A. The efficacy of cognitive behavioral therapy: a review of meta-analyses. *Cognit Ther Res.* 2012;36(5):427−440.

80. Gaylord SA, Palsson OS, Garland EL, et al. Mindfulness training reduces the severity of irritable bowel syndrome in women: results of a randomized controlled trial. *Am J Gastroenterol.* 2011;106(9):1678−1688.

81. Palsson OS. Hypnosis treatment of gastrointestinal disorders: a comprehensive review of the empirical evidence. *Am J Clin Hypn.* 2015;58(2):134−158.

82. Tillisch K. Complementary and alternative medicine for functional gastrointestinal disorders. *Gut.* 2006;55 (5):593−596.

83. Shepherd K, Hillman D, Eastwood P. Symptoms of aerophagia are common in patients on continuous positive airway pressure therapy and are related to the presence of nighttime gastroesophageal reflux. *J Clin Sleep Med.* 2013;9(1):13−17.

CHAPTER

18

Sleep in Pediatric Patients

Teryn Bruni[1], Emma Gill[2] and Dawn Dore-Stites[1]

[1]Department of Pediatrics, Michigan Medicine, Ann Arbor, MI, United States
[2]University of Michigan, Ann Arbor, MI, United States

OUTLINE

Introduction	391	Assessment	399
Pediatric Sleep Patterns	392	Interventions	401
		Managing the Sleep of Children with	
Consequences of Insufficient Sleep in		*Chronic Health Condition: A*	
Children and Adolescents	392	*Framework*	404
Pediatric Sleep Problems	393	**Areas for Future Research**	405
Sleep in Chronic Health Condition:			
Trends	399	Conclusion	405
Sleep in Chronic Health Condition:		References	406
Assessment and Intervention	399		

INTRODUCTION

The number of pediatric patients diagnosed with a chronic health condition (CHC) [e.g., chronic illness (CI); neurodevelopmental condition; and behavioral health diagnosis] has increased over the past several decades.[1] Medical advances that have increased life expectancy and the accompanying focus on quality of life allow a broader focus on optimizing health across all domains for these populations. As a result, promoting optimal sleep is prioritized given its critical relationship with health.

Handbook of Sleep Disorders in Medical Conditions
DOI: https://doi.org/10.1016/B978-0-12-813014-8.00018-4

© 2019 Elsevier Inc. All rights reserved.

PEDIATRIC SLEEP PATTERNS

To understand sleep in special populations, it is first crucial to understand sleep among typically developing children. Sleep changes dramatically in the first two decades of life. In one of the few large-scale studies measuring sleep patterns across childhood, average sleep duration in infants was longer (16.1 hours at 6 months) relative to older children and adolescents (8.1 hours at 16 years).[2] There was also greater variability in average sleep duration in groups of infants and toddlers relative to adolescents.[2]

Average sleep time across populations only tell one part of the sleep story in childhood. More recently, an expert panel conducted an exhaustive review of the current literature with the goal of identifying recommended sleep duration in different age-groups based upon outcomes across several domains (e.g., cognitive functioning, physical health).[3] Current recommendations for sleep duration associated with optimal health across ages are listed in Table 18.1.

Knowing where an individual child lies within the range of recommendations for sleep duration requires recognition of symptoms of sleepiness. For prepubescent children, sleepiness often is associated with *increased* physical activity and behavioral dysregulation.[4] Adolescents tend to demonstrate sleepiness in similar ways to adults (e.g., verbal complaints of fatigue; falling asleep at inappropriate times).[5] Other signs of sleep deprivation include requiring parental intervention to wake in the morning, falling asleep in classes, or sleeping significantly longer on weekends or holidays relative to the school year.[6]

TABLE 18.1 Sleep Duration by Age Associated With Optimal Outcomes as Suggested by Paruthi et al.[3]

Age	Recommended Duration of Sleep (per 24 hours)
Infants: 4 months to 12 months	12—16 hours[a]
Children: 1—2 years of age	11—14 hours[a]
Children: 3—5 years of age	10—13 hours[a]
Children: 6—12 years of age	9—12 hours
Adolescents: 13—18 years of age	8—10 hours

[a]Includes naps.

CONSEQUENCES OF INSUFFICIENT SLEEP IN CHILDREN AND ADOLESCENTS

Insufficient or poor quality sleep is associated with negative outcomes in physical, mental, and cognitive health for children and adolescents. Sleep problems have been associated with increased rates of depression,[7] suicidal thoughts and behaviors,[8] increased rates of obesity,[9] decreased attention, [10] and increased risk behaviors.[11] Sleep problems in children and adolescents also have impacts on the larger family, including parental sleep patterns, mood, and stress.[12]

Pediatric Sleep Problems

Typically developing populations: Pediatric sleep problems are common with estimates across childhood and adolescence ranging from 20% to 50%.[13] The symptoms of sleep problems can vary across ages. As an example, difficulties with sleep initiation look different for a toddler (frequent exits from the bedroom) relative to an adolescent (lying awake feeling anxiety or frustration). Common pediatric sleep conditions are listed in Table 18.2.

Chronic health conditions: CHCs, including neurodevelopmental diagnoses (NDs), psychiatric conditions, or CI, are at higher risk for sleep problems.[22] While rates vary largely by condition, prevalence of sleep problems among children with ND can range from 30% to 80%.[13] Among those with CI, prevalence of sleep problems demonstrates similar variability. It is suggested that sleep problems among pediatric patients with CHC also tend to be more enduring relative to those without chronic conditions.[22]

CHCs present with a host of unique risk factors associated with poor sleep, including symptoms associated with the condition (e.g., obesity, pain)[14,22]; regimen-related factors, including medication side effects,[14,22] hospitalizations[14,22,23], and family stress related to parenting a child with CHC.[14] As an example a pediatric patient with type I diabetes may have sleep disruptions related to 3 a.m. blood glucose checks (regimen-related factor) or frequent hypoglycemic episodes (condition-related factor). They may also experience anxiety at sleep onset leading to delayed sleep-onset latency (SOL) due to the stress of a new diagnosis or past experiences of hypoglycemia (family stress related to condition). Hypersomnia can occur due to elevated blood glucose levels secondary to inadequate insulin dosages (medication side effect) or nonadherence. Each CHC yields its own unique factors that should be considered when forming diagnostic impressions regarding sleep.

Within this literature, there are considerations to be made when interpreting the data. First, overall research is limited with more data available on ND and psychiatric conditions relative to CI. The limitations in number of studies lead to challenges identifying overall trends in sleep patterns among children with CHC. In addition, most studies measure sleep using subjective report rather than more objective means, such as polysomnography (PSG).

Table 18.3 presents data on prevalence rates of sleep problems and disease- and treatment-related impacts on sleep across several common pediatric CHC. Data populating the table include review articles, book chapters, and individual studies. Emphasis was placed on more comprehensive reviews in order to capture more general trends rather than isolated findings from an individual study. In addition, many NDs and CIs are associated with comorbid symptoms of behavioral disorders (e.g., depression, anxiety). In summarizing the trends, focus was placed when possible on sleep disorders unique to the CHC and not the comorbidities.

TABLE 18.2 Common Medical and Behavioral Sleep Conditions

Sleep Condition	Symptoms	Prevalence: Typically Developing Pediatric Population
MEDICAL CONDITIONS		
Sleep-disordered breathing including Obstructive sleep apnea[6,14,15]	• Mouth breathing; snoring • Gasping; apneic pauses • Restlessness; odd sleep positions • Diaphoresis • Frequent arousals • Waking with headache, dry mouth, and/or sore throat • Excessive daytime sleepiness	Sleep-disordered breathing: 10%−20%[16] Obstructive sleep apnea: 1%−4%[14]
Restless legs syndrome/Periodic limb movement disorder[6,14]	• Urge to move legs • Growing pains or leg pains • Restless sleep • Kicking and jerking during sleep • Excessive daytime sleepiness	Restless legs syndrome: 1%−6%[14] Periodic limb movement disorder: 5%−27%[14]
Narcolepsy[6,14]	• Excessive daytime sleepiness • Sleep attacks • Cataplexy • Sleep paralysis • Hallucinations at sleep onset or waking	3−16 per 10,000[14]
Delayed sleep−wake phase disorder[6,14]	• Long sleep-onset latency at earlier bedtimes • Minimal difficulty with sleep maintenance • Difficulty waking • Excessive daytime sleepiness generally due to decreased sleep duration	3.3%−16% depending on source[17−19]
BEHAVIORAL CONDITIONS		
Behavioral insomnia of childhood: sleep-onset association subtype[14,20,21]	• Short sleep-onset latency if in presence of sleep-onset association (e.g., parent) • Frequent and lengthy night waking if sleep-onset association removed • Insufficient sleep duration • Excessive daytime sleepiness or hyperactivity in younger children	In one paper, 30% of children fell asleep with parent present, and this was correlated with significant increase in night waking[14]
Behavioral insomnia of childhood: Limit-setting subtype[6,14]	• Long sleep-onset latency • Insufficient sleep duration • Appropriate sleep maintenance • Excessive daytime sleepiness or hyperactivity in younger children	10%−30% of toddlers and preschoolers demonstrate resistance to bedtime[14]

TABLE 18.3 Common Pediatric Conditions and Associated Sleep-Related Problems (SRP)

	Condition	Rates of SRP	Common Sleep Patterns and Conditions	Condition-Related Considerations	Regimen-Related Considerations
Neurodevelopmental diagnosis and psychiatric disorders	Autism	53%–78%[24]	↑Sleep-onset latency[25-27] ↑Nocturnal arousals[28] Waking early[29] ↑Parasomnias[29]	Difficulties with limit setting at bedtime due to behavioral sequelae of autism spectrum disorder[25,26] Medical comorbidities (e.g., epilepsy)[30]	None noted in current review of literature
	Attention deficit/ hyperactivity disorder	50%[31,32]	↑Sleep-onset latency[31,33] Restless legs syndrome[33] periodic limb movement disorder[14,32] obstructive sleep apnea[31-33]	Difficulties with limit setting at bedtime due to behavioral sequelae ofAttention deficit/ hyperactivity disorder[31,34]	Medication effects[35]: use or timing of stimulants may lead to difficulties with sleep onset
	Anxiety	1 SRP: 88%[36] ≥ 3 SRP: 55%[36]	↑Sleep-onset latency[36] ↑Night wakings[37,38]	If anxiety related to school, improvements on nights and weekends[37]	Medication effects[36]: some anxiety medications can lead to excessive sleepiness
Atopic conditions	Allergic rhinitis	Not found in current review of literature	Sleep-disordered breathing/obstructive sleep apnea[16]	Congestion can lead to upper airway obstructions[14]	Medication effects: use of antihistamines may lead to drowsiness, daytime lethargy, and exacerbate restless legs syndrome[22] Inhaled corticosteroids may improve both allergic rhinitis and sleep-disordered breathing[39]
	Eczema	60%[40] 83% during exacerbations[40]	↓Sleep efficiency secondary to ↑ sleep-onset latency and ↑arousals[40]; sleep disturbances may be increased earlier at night[40]	Pruritus	None noted in current review of literature

(Continued)

TABLE 18.3 (Continued)

	Condition	Rates of SRP	Common Sleep Patterns and Conditions	Condition-Related Considerations	Regimen-Related Considerations
Pulmonary disorders	Asthma	60%[13]	↓Deep sleep ↑Arousals Early morning wakings[41] Sleep-disordered breathing/obstructive sleep apnea[42]	Increased rates of sleep-disordered breathing in children with poorly controlled asthma symptoms[42] Asthma severity partially defined by presence of nocturnal symptoms (e.g., coughing)[43]	Medication effects (albuterol; steroids): steroids are associated with increased sleep-onset latency, increased nocturnal arousals, and decreased rapid eye movement sleep[22]
	Cystic fibrosis	Overall: 50% with 10% of patients reporting moderate/severe sleep concerns that persist[44] Pediatric rates: 32% with small-to-moderate sleep problems[44] Adolescents: 39%–44% with complaints of ↑Sleep-onset latency and ↑nocturnal arousals and 74% with excessive daytime sleepiness[44]	Changes in sleep architecture[22] ↑Arousals[22] Nocturnal hypoxemia[22] ↓Sleep efficiency[22,44] ↑Time torapid eye movement[22] ↓Rapid eye movement percentage[22] ↓Sleep duration[44] Risk of obstructive sleep apnea[22] excessive daytime sleepiness[44]	Exacerbations of symptoms associated with wake after sleep onset and ↓rapid eye movement[44]	Percutaneous endoscopic gastrostomy feedings associated with ↓sleep duration[45] Medication side effects: steroids can impact sleep-onset latency, sleep maintenance, and decrease rapid eye movement sleep[22]
Endocrinological disorders	Type I diabetes	Not found during current literature review	↑Frequency and duration of night wakings[46]	Even mild sleep-disordered breathing correlated with hyperglycemia[46]	Timing of insulin: if sleep duration varies widely night to night, considerations should be made about the timing of insulin dosages[47] Treatment of hypoglycemia during the night can disrupt sleep
	Prader–Willi syndrome	Overall prevalence: between 1:12,000 and 1:25,000 births[48] Prevalence of excessive daytime sleepiness: data limited but some estimates of 40%[48] of those with Prader–Willi syndrome Prevalence of obstructive sleep apnea: data demonstrate mixed results[48]	↓Latency to rapid eye movement[49] Sleep architecture changes[49] Central and obstructive sleep apnea[49] excessive daytime sleepiness with narcolepsy-like symptoms (e.g., quick onset of rapid eye movement)[49]	Weight gain and hyperphagia associated with Prader–Willi places patients at higher risk of obstructive sleep apnea due to weight gain[49]	Treatment with growth hormone is contraindicated in patients with untreated obstructive sleep apnea[50] and recommendations made to have baseline polysomnography prior to starting growth hormone[49]

Neurological diagnoses	Headaches (including migraines)	Restless sleep: 30%−35%[51] (controls: ∼20%) Sleep-disordered breathing: 21.9%−49%[51,52]	Parasomnias[22] Bruxism[22] Sleep-disordered breathing[22] ↓Slow wave sleep[22] ↓Rapid eye movement sleep[22] ↓Sleep duration[22] ↑Sleep-onset latency[22] ↑Bedtime problems[51] ↑Restless sleep[51] ↑Rates of restless legs syndrome[53]	Bidirectional relationship between headaches and sleep: sleep deprivation (and in small group, sleep excess) can cause headache, and headache pain can negatively impact sleep[51] Morning headache is one of the symptoms of sleep-disordered breathing/obstructive sleep apnea[51] Sleep sometimes used to manage headaches[51]	Medication effects: several classes of prophylactic agents used in migraine associated with sedative effects and somnolence[54]
	Epilepsy	Sleep-related epilepsy: ∼30% of seizure disorders in pediatrics[14] obstructive sleep apnea: prevalence rates variable but some findings suggest higher rates[55]	↓Total sleep time[56] ↓Rapid eye movement sleep[56] ↓Sleep efficiency[56] ↑Sleep-disordered breathing[22] and obstructive sleep apnea[56] ↑Restless legs syndrome/periodic limb movement disorder[56]	Sleep deprivation can trigger seizures, and electroencephalogram activity can lead to sleep disruptions[22,56] The clinical presentation of seizure activity can mimic other sleep disorders (e.g., parasomnias)[14]	Antiepileptic medications have variable impacts on sleep but can be associated with sedative effects and changes in sleep architecture.[14] Changes in sleep patterns can also occur secondary to withdrawal of medications[14]
Hematology/Oncology conditions	Cancer	Overall prevalence unknown[57] Hypothalamic tumors: excessive daytime sleepiness most common complaint[58]	↑Sleep-onset latency[59] ↑Nocturnal arousals[59] excessive daytime sleepiness[58]	Site of cancer: ↑Risk of SRP if site is close to neurological controls of sleep[58] Brain tumors associated with ↑risks of central apneas and obstructive sleep apnea[58] and are most common cause of secondary narcolepsy[58]	Chemotherapy associated with ↑sleep duration, excessive daytime sleepiness, and ↑nocturnal arousals[58] Cranial radiation: ↑Hypersomnia[14] Steroids: ↑Risk for SRP[59]
	Sickle cell disease	40% with nocturnal oxygen desaturation[14] Sleep-disordered breathing: 36%[52] periodic limb movement disorder: 23%[52]	Nocturnal desats associated with ↓sleep duration, ↑latency to rapid eye movement; however, some data suggest that children with sickle cell disease and without sleep-disordered breathing are relatively similar in sleep patterns relative to those with sleep-disordered breathing[52]	Hypoxemia elevates risk of pain crises, and pain can affect sleep patterns[14]	Medication effects—Opioids can disrupt sleep staging and daytime fatigue[22]

(Continued)

TABLE 18.3 (Continued)

Condition		Rates of SRP	Common Sleep Patterns and Conditions	Condition-Related Considerations	Regimen-Related Considerations
Other conditions	Craniofacial anomalies	Estimates of rates of obstructive sleep apnea range from 7% to 67%[60]; however, more precise rates elusive due to lack of data and likely variance based upon presentation of craniofacial anomoly[61]	Sleep-disordered breathing including obstructive sleep apnea and central apneas primarily studied	Structural factors associated with increased obstructive sleep apnea rates: retrognathia; midface hypoplasia; micrognathia[61] Increased cranial pressure observed in patients with craniosynostosis resulting in higher risk of central apneas[60]	Surgical procedures common in this population[61] leading to elevated risk for SRP due to pain, hospitalizations, and medications
	Trisomy 21	Obstructive sleep apnea: 30%–80%[14] Sleep-related complaints: 70%[14]	↑Rates of obstructive sleep apnea[14] ↑Arousals[14] ↑Fragmented sleep[14] ↑Periodic limb movements[14]	Structural factors: midface hypoplasia; mandibular hypoplasia; hypotonia; tonsillar hypertrophy[61] Other factors: obesity[62]	None noted in the current literature search

Sleep in Chronic Health Condition: Trends

The data consistently demonstrate that sleep problems follow similar presentations as those in typically developing populations, including lengthy SOL, frequent nocturnal arousals, and short sleep duration. In addition, bidirectional relationships exist between sleep and many symptoms of the CHC (e.g., pain leading to difficulties with sleep onset and the associated decreased sleep duration leading to exacerbation of headache pain). Among children with psychiatric conditions, in addition to the sleep patterns noted above, there is also higher prevalence of parasomnias, restless sleep, and sleep-related anxiety.[13]

As noted above, much of the research in sleep problems among CHC populations rely on parent and self-report to the exclusion of more objective indices of sleep (e.g., actigraphy, PSG). Therefore it is important to note that our understanding of the *perception* of sleep problems in this population is greater than knowing *how* the CHC impacts sleep. Despite this limitation of the data, higher rates of sleep problems exist among pediatric patients with CHC across much of the literature. In addition, sleep problems appear more chronic. However, it remains unclear whether the chronicity of the sleep problems is related to the CHC course directly or is secondary to behavioral considerations in managing a CI.

Overall, sleep problems among pediatric patients with CHC tend to be more frequent and more chronic in nature. This may be secondary to several factors unique to a population with CHC. Sleep concerns often look similar to those in typically developing populations; however, careful assessment is needed to understand factors driving the concerns.

SLEEP IN CHRONIC HEALTH CONDITION: ASSESSMENT AND INTERVENTION

Assessment

Conceptual underpinnings: Sleep problems and daytime behaviors or symptoms often exist in a bidirectional relationship[7] leading to the need for nuanced assessment in any population. In essence, it is critical to differentiate sleep as a causal factor versus a symptom. As an example, in an adolescent with depression, poor sleep stemming from other factors (e.g., poor sleep hygiene) can worsen mood; however, disrupted sleep patterns are also a common symptom in depression itself. Understanding which symptoms (e.g., sleep disruptions or depressive symptoms) occurred first can help in disentangling the causal factors. This becomes more critical for children with CHC given the array of risk factors at play.

Strong assessment focuses not only on the description of symptoms but also the function they may possess. Teens who nap during the day may be suffering from suboptimal sleep duration at night due to heavy homework loads or extracurricular activities. They may also be opting to escape from family interactions during the day by sleeping and then remaining up at night when parents are in bed. The function of the behavior is critical to selecting an effective intervention.

Assessment strategies: Given the complexities of pediatric sleep in general, and those with CHC specifically, thorough assessment should ideally use multiple methods to understand sleep patterns, the variables amenable to intervention, and the associated impacts on daytime functioning.[37] Some common strategies used in pediatric sleep are described below. Descriptions of PSG and multiple sleep latency tests are not included as the details are beyond the brief scope of this chapter.

Screening: Screening for sleep-related concerns is critical given the high prevalence and risk for negative outcomes among those with CHC. One of the most researched screening tools is the BEARS interview.[63] BEARS is a mnemonic representing five areas of pediatric sleep-related concerns, including *B*edtime issues, *E*xcessive daytime sleepiness (EDS), night *A*wakenings, *R*egularity and duration of sleep, and *S*noring. Use of the BEARS screening led to increased discussion between primary-care physicians and patients about sleep concerns as well as increased identification of sleep problems.[63]

Clinical interview: A clinical interview exploring details around sleep and daytime functioning can help direct further assessment. Clinical assessment of pediatric sleep should include at a minimum: sleep patterns (SOL; wake after sleep onset or WASO; napping) across school days, weekends, and holidays; sleep hygiene (routines, caffeine use) and environment; and behavioral difficulties circumscribed to bedtime and associated parental responses [e.g., laying down with child to keep them in bed leading to problematic sleep-onset association (SOA)]. In addition, details around the consequences of the sleep problems on academic functioning (e.g., absences to school), mood, or behavior can provide insight into whether sleep is a primary concern or secondary to other factors. Finally, assessment of medical symptoms associated with sleep, developmental history, and overall behavioral functioning is needed.[14,64] For children and adolescents with CHC, further assessment of their current treatment regimens, symptoms, and associated disruptions in routines due to hospitalizations, medical visits, school absences, and/or procedures is also warranted.

Brief questions related to parental sleep difficulties as a result of the child's sleep patterns and/or regimen can be helpful in later formulation of treatment plans. This is related to two factors. First, both parents of typically developing children and of those with CHC are at risk for disrupted sleep patterns if their children have sleep concerns.[12] Second, for parents of children with CHC, their child's symptoms or regimen-related demands (e.g., blood glucose checks in the middle of the night) may lead to parental sleep disruption.[65] Given that parents would likely be the primary interventionists for any behavioral treatments related to sleep, understanding their sleep patterns may lead to more individualized behavioral protocols that accommodate parental functioning.

Sleep diaries/logs: In concert with a clinical interview, sleep logs or diaries serve as the basis for a good clinical assessment. Parents often maintain the logs for children who are younger or developmentally delayed. For older children and adolescents, they may be able to fill out the logs independently as parents often have more limited opportunities to observe their sleep schedules relative to younger children. Sleep logs capture data on bedtimes, SOL, WASO, wake times, and napping. In addition, space can be made for capturing aspects of a child's illness, such as pain ratings or medication changes.

Recording data over a 2-week period can give a better picture of day-to-day variability relative to general trends often noted in clinical interviews.[14]

Standardized assessment: Several surveys (self-, parent, and/or teacher report) of varying aspects of pediatric sleep exist. Measures vary in domains assessed (e.g., sleep-related cognitions; EDS), ages captured, psychometric properties, respondents, and research basis.[66] Preferred surveys are multidimensional (i.e., include assessment of varying aspects of sleep, including behavioral and medical features).[66] No surveys specific to sleep in a CHC population were found in this review with the exception of the Family Inventory of Sleep used in children with autism.[67] There is also work in modifying existing surveys to be more relevant to children with CHC (as an example, see Katz et al.[68]). Please refer to Lewandowski et al.[66] for an exhaustive review of measures.

As mentioned previously the features of sleep for a patient with a CHC may appear similar to a typically developing child; however, the factors driving the symptom may be different. Therefore, given that, there are few surveys specifically designed to capture all elements that may come into play for a child with CHC, use of the questionnaires merit, use of other assessment strategies, including a comprehensive clinical interview, before deciding upon intervention plans.

Actigraphy: An actigraph is a small portable device worn by the patient for a 1–2 weeks period in their normal sleep environment. The device captures movement in order to determine sleep–wake patterns. Some devices also capture light exposure. An actigraph must be worn both during the day and night to accurately capture sleep–wake patterns. Sleep diaries should be kept while actigraphy is used.[14]

Actigraphy was previously only used in research settings; however, data guiding clinical practice has been increasing over the past 5 years. The literature within pediatrics, and specifically validation studies, remain scarce.[69] In addition, few data are available to understand the potential impact of the limited mobility of children with disability or medical conditions on the algorithm used to track sleep–wake patterns.[70]

In typically developing children and those with no limitations on mobility during the day, actigraphy can capture sleep patterns (e.g., bedtimes and wake times).[71] In addition to assessment, they can also capture responses to treatment interventions.[71] Caution should be used when interpreting actigraphy data on total sleep time, SOL, or WASO due to its limited ability to discriminate wakefulness from restless sleep.[69] Commercially available accelerometers (e.g., FitBit©, Jawbone©) are not as reliable in determining sleep patterns relative to PSG or actigraphy.[72]

Interventions

Most sleep problems persist without intervention, especially in children with CHC.[6,9] Improving the sleep patterns in children and adolescents also often leads to improved daytime functioning—without directly targeting the daytime symptoms.[73] Thorough discussion of all aspects of interventions for pediatric sleep conditions is beyond the scope of this chapter; however, several excellent and empirically driven summaries exist.[14,74–76]

Behavioral/Nonpharmacological interventions: Behavioral interventions are considered the first-line treatments for sleep disorders once any symptoms suggestive of medical etiologies [e.g., obstructive sleep apnea (OSA)] are either ruled out or more thoroughly evaluated and/or managed.[73] As an example, if a child presents with concerns about nocturnal arousals and has features warranting a sleep study (e.g., risk factors, such as obesity; symptoms, including mouth breathing or sweaty sleep), completion of the overnight sleep study is prioritized to ensure that the arousals are not related to OSA. Should behavioral interventions be used aggressively before understanding whether OSA is present, there is a higher risk of treatment failure since the symptoms are not exclusively related to challenging sleep habits. This can lead to erosion of family and patient motivation to address sleep concerns.

In typically developing pediatric populations, sleep-related concern with behavioral etiologies are more common than those with mainly medical etiologies.[77] Behavioral interventions include an array of strategies targeting reinforcement of sleep-promoting behaviors. For younger children, most strategies necessarily involve modification of parental response to a child's behavior at bedtime making the need for thorough assessment of parental stress and sleep patterns critical. Table 18.4 outlines common behavioral treatments of pediatric sleep problems.

Compared to the use of medication in pediatric sleep, behavioral interventions enjoy a larger research base[64] although data are more prevalent for interventions targeting younger children relative to older children and adolescents.[78] Within typically developing younger children, efficacy of behavioral interventions is strong. Morgenthaler et al.'s[74] 2006 review found that 94% of studies reviewed were efficacious with over 80% of the participants showing improvements classified as clinically significant. A more recent review by Meltzer and Mindell[78] described moderate level of support for the use of behavioral treatments in pediatric insomnia among younger children. For older children, adolescents, and those with CHC, evidence was lacking due to a paucity of research.[78]

As noted above and consistent with much of the data involving pediatric sleep and CHC, study of behavioral strategies targeting sleep in these populations is limited leading to difficulties drawing strong conclusions based upon evidence.[78,79] Within this subset of the literature, research on sleep interventions in ND and psychiatric conditions is more prevalent relative to data in pediatric patients with CI. Clinically derived guidelines typically support the use of behavioral interventions in CHC.[24,64,80] Implementation of behavioral interventions for those with special needs should account for the associated challenges, including regimen- and disease-related factors, disruptions in routines secondary to school absences, medical visits and/or hospitalizations, and parental stress. The limited research available detailing behavioral interventions in CHC relies on strategies, such as those reviewed in Table 18.4 with more individualized attention given to the condition-related factors that may present as barriers without accommodation (e.g., see Allen et al.[81] or Malow et al.[82]).

As an example, if a patient presents with a SOA (requires parental presence to fall asleep), the preferred strategy would be a graduated extinction approach in which the parent stays out of the room at sleep onset for progressively longer periods of time. In typically developing children, this often leads to decreased nocturnal arousals as they become more able to fall asleep independently and maintain sleep through the night.

For a child with a CHC, this intervention may require modification. For example, if the child has a condition leading to failure to thrive and they vomit when crying (as they may

SLEEP IN CHRONIC HEALTH CONDITION: ASSESSMENT AND INTERVENTION 403

TABLE 18.4 Common Behavioral Interventions in Pediatric Sleep

Intervention	Description	Associated Behavioral Sleep Conditions
Bedtime routines	• Increasing appropriate before bedtime behavior by directing consistent routines and providing positive feedback for compliance[64,74,77]	Behavioral insomnia of childhood: Limit-setting subtype Delayed sleep–wake phase disorder
Faded bedtime	• Often paired with establishing bedtime routines[77] • Temporary delay of bedtime to decrease sleep-onset latency, followed by shift of bedtime earlier until reaching the optimal bedtime[64,74,77]	Behavioral insomnia of childhood: Limit-setting subtype Delayed sleep–wake phase disorder
Sleep hygiene	• Establishment of positive sleep habits early in life as means of preventing sleep problems[64] • Includes: consistent routines, eliminating screens before bedtime and creating a comfortable sleeping environment[77]	Behavioral insomnia of childhood: Limit-setting subtype Behavioral insomnia of childhood: sleep-onset association subtype Delayed sleep–wake phase disorder
Scheduled awakenings	• Parent wakes child 15–30 min prior to parasomnia event or arousal[74,77] • Timing of event/arousal must be somewhat predictable[74]	Behavioral insomnia of childhood: sleep-onset association subtype Parasomnia (e.g., sleep terrors)
Extinction	• All inappropriate behaviors are ignored from bedtime through scheduled wake time[77] • Significant escalation in behavior in response to removal of reinforcement (i.e., extinction bursts) is likely and may be difficult for parents to tolerate[77]	Behavioral insomnia of childhood: Limit-setting subtype Behavioral insomnia of childhood: sleep-onset association subtype
Graduated extinction	• Parents ignore all inappropriate bedtime behavior for gradually longer periods of time[74,77] • Duration between checks is determined by time, not child's behavior[77] • Relative to extinction, can reduce extinction bursts, and be more tolerable to parents[77]	Behavioral insomnia of childhood: Limit-setting subtype Behavioral insomnia of childhood: sleep-onset association subtype
Extinction with parental presence	• Parent sleeps in child's bedroom in separate bed for 1 week, feigning sleep while ignoring all behavior of the child including attempts to interact with parent directly. Parental presence faded over time[77]	Behavioral insomnia of childhood: Limit-setting subtype
Phase advance	• Gradual shift of bedtime and wake time earlier[74,77] • Maintenance of new sleep schedule over several weeks critical[77]	Delayed sleep–wake phase disorder
Phase delay (chronotherapy)	• Gradually and systematically delaying a child's bedtime by 2–3 h each day until the desired bedtime is reached[74,77] • Maintenance of new sleep schedule over several weeks critical[77]	Delayed sleep–wake phase disorder

II. SLEEP DISORDERS IN SPECIFIC MEDICAL CONDITIONS

be prone to do so during a graduated extinction protocol), changes in treatment planning may be needed to limit risk of vomiting. As an another example, for the child with asthma, they may be prone to late-night coughing and require parental presence for use of medications. For the tired parent who is also not sleeping, assessing whether they struggle to remain awake with the child during nebulizer treatments may also be necessary in order to inform a pragmatic treatment plan.

Pharmacological interventions: Both prescription and nonprescription agents (e.g., supplements, herbal remedies) are often used in pediatric sleep; however, relative to the more robust data on behavioral interventions, there are few data guiding use of medication in treatment.[64] Pharmacological interventions generally target insomnia or EDS with the former having more research relative to the latter. The Food and Drug Administration also has no medications approved for the treatment of pediatric insomnia.[64] Therefore treatment is largely directed by consensus statements and clinical guidelines.

Owens et al.[64] published one of the initial guidelines outlining sound practices for use of pharmacological interventions in pediatric insomnia, defined as difficulties initiating or maintaining sleep. The review is notable for detailing the increasing rates of prescriptions, especially among patients with CHC. Key recommendations from the panel included promotion of good sleep hygiene as an initial strategy, avoiding use of medication as a first-line or exclusive intervention, and the absence of need for medication management in the youngest of patients.[64] The authors outlined indications for use of pharmacological interventions with a notable contraindication being the presence of sleep-disordered breathing (SDB).[64]

Among pediatric patients with CHC, reviews acknowledge the unique risk factors (e.g., biological predisposition to changes in sleep/wake cycles; multiple demands due to medical regimen) in the population and the challenges associated with implementation of behavioral interventions[64,80] leading to modified practices with pharmacological interventions relative to the typically developing population. The modifications may include longer duration of use of pharmacological agents, more aggressive treatment, and/or need for closer care coordination with other providers.[64] The use of pharmacotherapy in these populations requires significant knowledge related to the disease process (e.g., knowing which medications may impact respiratory parameters) and very close follow-up to track outcomes. Excellent reviews in this area highlight the priorities placed on behavioral interventions prior to use of medications as well as the subtleties to consider when prescribing.[24,25,64,80]

Managing the Sleep of Children with Chronic Health Condition: A Framework

Pediatric sleep problems are common among typically developing populations and exist at equivalent frequencies or higher among those with CHC. For primary-care providers or specialists involved in the care of those with special needs the below framework for managing sleep-related concerns is proposed.

1. *Screen:* Pediatric patients with CHC should be screened due to the high prevalence of sleep-related concerns.[24,64]
2. *Assess:* If screening yields symptoms suggestive of sleep problems, further medical evaluation and/or behavioral assessment should occur. Symptoms associated with

medical sleep disorders (e.g., SDB/OSA) should be evaluated by an accredited pediatric sleep center (lists available through the Pediatric Sleep Council or American Academy of Sleep Medicine websites). Brief behavioral assessments (e.g., clinical interview, sleep logs) can be directed by a variety of providers. It would be critical to identify factors related to the patient's condition that could be impacting sleep.

3. *Initial intervention:* Sleep-inhibiting behaviors can often be identified. These are often first addressed by helping the family establish good sleep hygiene and bedtime routines. Owens et al.[64] or Mindell and Owens[14] have provided outlines of elements of good sleep hygiene. Helping caregivers or patients identify one or two behaviors to work on can be helpful in moving toward improved sleep practices.

4. *Further consultation:* More thorough assessment is indicated for pediatric patients with special needs as management strategies may be more challenging.[64] Due to the burden on the family given the likely presence of several other providers, as well as the difficulty identifying behavioral sleep specialists, it may be possible to only refer patients who fail to respond to basic behavioral interventions as long as close monitoring is in place.

AREAS FOR FUTURE RESEARCH

As noted above, much of the data on sleep and CHC are the result of parent- and self-report measures. Future research should prioritize multimethod assessment of sleep patterns with use of both self-report measures and objective tracking of sleep through actigraphy or PSG. In addition, as we understand more about the role sleep plays in physiological processes, tracking changes related to the disease (e.g., blood glucose values in type I diabetes) concurrent with measuring sleep may provide insight into how sleep and CHC interrelate.

From a clinical standpoint, adaptation of assessment and intervention strategies with known efficacy to CHC populations is critical. As these populations are small, the adaptation would likely require increased collaboration among behavioral sleep medicine professionals as well as multicenter trials.

CONCLUSION

Children and adolescents with CHC are at higher risk for developing sleep problems that persist over time.[22] Given the multiple needs patients with CHC may have, sleep may be overlooked; however, prioritizing screening, assessment, and eventual treatment of sleep-related concerns may yield positive impacts across several domains. Current research, while limited, suggests that sleep problems in pediatric patients with CHC are largely similar to those seen in a typically developing population. This opens up the larger literature on behavioral strategies commonly used in pediatric sleep medicine although modifications must be made to accommodate the complex care needs of patients with CHC.

KEY PRACTICE POINTS

- Pediatric sleep problems are common and may be more prevalent in pediatric patients with CHC.
- Assessment should attempt to discern whether sleep problems are primary or secondary to condition or treatments.
- Clinical guidelines support the use of behavioral interventions as a first-line treatment. If medications or supplements are recommended, behavioral strategies should be incorporated as an adjunctive treatment.
- Current data suggest that sleep problems in pediatric patients with CHC are largely similar to that seen in the typically developing population affording the ability to draw on the larger literature in pediatric sleep medicine.
- Further research, especially incorporating use of more objective indices of sleep, is warranted given the prevalence of sleep problems and the deleterious consequences of poor sleep in pediatric patients with CHC.
- Assessment and treatment of sleep-related concerns in pediatric patients with CHC employs similar strategies as often used in general pediatric sleep medicine; however, accommodations (e.g., incorporating medical regimen tasks into the bedtime routine; accommodating for middle-of-the-night care needs) are needed to address the unique risk factors inherent in the various conditions.

References

1. Van Der Lee J, Mokkink L, Grootenhuis M, Heymans H, Offringa M. Definitions and measurement of a systematic review. *JAMA*. 2007;297(24):2741–2751. Available from: https://doi.org/10.1001/jama.297.24.2741.
2. Iglowstein I, Jenni OG, Molinari L, Largo RH. Sleep duration from infancy to adolescence: reference values and generational trends. *Pediatrics*. 2003;111(2):302–307. Available from: https://doi.org/10.1542/peds.111.2.302.
3. Paruthi S, Brooks LJ, D'Ambrosio CD, et al. Recommended amount of sleep for pediatric populations: a consensus statement of the American Academy of Sleep Medicine. *J Clin Sleep Med*. 2016;12(6):785–786. Available from: https://doi.org/10.5664/jcsm.5866.
4. Owens J. Classification and epidemiology of childhood sleep disorders. *Prim Care*. 2008;35:533–546.
5. Hoban TF, Chervin RD. Sleepiness in children. *Semin Pediatr Neurol*. 2001;8(4):216–228. Available from: https://doi.org/10.1016/j.jsmc.2005.11.006.
6. Meltzer LJ, Mindell JA. Sleep and sleep disorders in children and adolescents. *Psychiatr Clin North Am*. 2006;29(4):1059–1076. Available from: https://doi.org/10.1016/j.psc.2006.08.004.
7. Owens JA. Insufficient sleep in adolescents and young adults: an update on causes and consequences. *Pediatrics*. 2014;134(3):e921–e932. Available from: https://doi.org/10.1542/peds.2014-1696.
8. Pigeon WR, Pinquart M, Conner K. Meta-analysis of sleep disturbance and suicidal thoughts and behaviors. *J Clin Psychiatry*. 2012;73(9):1160–1167. Available from: https://doi.org/10.4088/JCP.11r07586.
9. Owens JA. Classification and epidemiology of sleep disorders. *Prim Care*. 2008;35:533–546. Available from: https://doi.org/10.1016/j.chc.2009.04.005.
10. Fallone G, Owens JA, Deane J. Sleepiness in children and adolescents: clinical implications. *Sleep Med Rev*. 2002;6(4):287–306. Available from: https://doi.org/10.1053/smrv.2001.0192.
11. Wheaton AG, Chapman DP, Croft JB. School start times, sleep, behavioral, health, and academic outcomes: a review of the literature. *J Sch Health*. 2016;86(5):363–381. Available from: https://doi.org/10.1111/josh.12388.

REFERENCES

12. Meltzer LJ, Montgomery-Downs HE. Sleep in the family. *Pediatr Clin North Am*. 2011;58(3):765−774. Available from: https://doi.org/10.1016/j.pcl.2011.03.010.
13. Owens J. Epidemiology of sleep disorders during childhood. In: Sheldon S, Ferber R, Kryger MH, eds. *Principles and Practice of Pediatric Sleep Medicine*. Philadelphia, PA: Elsevier; 2005:27−33.
14. Mindell JA, Owens JA. *A Clinical Guide to Pediatric Sleep: Diagnosis and Management of Sleep Problems*. 2nd ed Philadelphia, PA: Lippincott, Williams & Wilkins; 2010.
15. Schwengel DA, Dalesio NM, Stierer TL. Pediatric obstructive sleep apnea. *Anesthesiol Clin*. 2014;32(1):237−261. Available from: https://doi.org/10.1016/j.anclin.2013.10.012.
16. Kimple A, Ishman S. Allergy and sleep-disordered breathing. *Curr Opin Otolaryngol Head Neck Surg*. 2013;21:277−281. Available from: https://doi.org/10.1097/MOO.0b013e32835ff132.
17. Sivertsen B, Pallesen S, Stormark K, Bøe T, Lundervold AJ, Hysing M. Delayed sleep phase syndrome in adolescents: prevalence and correlates in a large population based study. *BMC Public Health*. 2013;13(1):1163. Available from: https://doi.org/10.1186/1471-2458-13-1163.
18. Saxvig I, Pallesen S, Wilhelmsen-Langeland A, Molde H, Bjorvatn B. Prevalence and correlates of delayed sleep phase in high school students. *Sleep Med*. 2012;13(2):193−199. Available from: https://doi.org/10.1016/j.sleep.2011.10.024.
19. Bartlett D, Biggs S, Armstrong S. Circadian rhythm disorders among adolescents: assessment and treatment options. *Med J Aust*. 2013;199(October):16−20. Available from: https://doi.org/10.5694/mja13.10912.
20. Mindell JA, Kuhn B, Lewin DS, Meltzer LJ, Sadeh A. Behavioral treatment of bedtime problems and night wakings in infants and young children − an American Academy of Sleep Medicine review. *Sleep*. 2006;29 (10):1263−1276.
21. Meltzer LJ. Clinical management of behavioral insomnia of childhood: treatment of bedtime problems and night wakings in young children. *Behav Sleep Med*. 2010;8(3):172−189. Available from: https://doi.org/10.1080/15402002.2010.487464.
22. Lewandowski AS, Ward TM, Palermo TM. Sleep problems in children and adolescents with common medical conditions. *Pediatr Clin North Am*. 2011;58(3):699−713. Available from: https://doi.org/10.1016/j.pcl.2011.03.012.
23. Meltzer LJ, Davis KF, Mindell JA. Patient and parent sleep in a children's hospital. *Pediatr Nurs*. 2012;38(2).
24. Malow BA, Byars K, Johnson K, et al. A practice pathway for the identification, evaluation, and management of insomnia in children and adolescents with autism spectrum disorders. *Pediatrics*. 2012;130(Supplement): S106−S124. Available from: https://doi.org/10.1542/peds.2012-0900I.
25. Grigg-Damberger M, Ralls F. Treatment strategies for complex behavioral insomnia in children with neurodevelopmental disorders. *Curr Opin Pulm Med*. 2013;19(6):616−625. Available from: https://doi.org/10.1097/MCP.0b013e328365ab89.
26. Liu X, Hubbard JA, Fabes RA, Adam JB. Sleep disturbances and correlates of children with autism spectrum disorders. *Child Psychiatry Hum Dev*. 2006;37(2):179−191. Available from: https://doi.org/10.1007/s10578-006-0028-3.
27. Malow BA, Marzec ML, McGrew SG, Wang L, Henderson LM, Stone WL. Characterizing sleep in children with autism spectrum disorders: a multidimensional approach. *Sleep*. 2006;29(12):1563−1571.
28. Meltzer LJ. Brief report: sleep in parents of children with autism spectrum disorders. *J Pediatr Psychol*. 2008;33 (4):380−386. Available from: https://doi.org/10.1093/jpepsy/jsn005.
29. Cuomo BM, Vaz S, Lee EAL, Thompson C, Rogerson JM, Falkmer T. Effectiveness of sleep-based interventions for children with autism spectrum disorder: a meta-synthesis. *Pharmacotherapy*. 2017;37(5):555−578. Available from: https://doi.org/10.1002/phar.1920.
30. Melke J, Goubran Botros H, Chaste P, et al. Abnormal melatonin synthesis in autism spectrum disorders. *Mol Psychiatry*. 2008;13(1):90−98. Available from: https://doi.org/10.1038/sj.mp.4002016.
31. Cortese S, Faraone SV, Konofal E, Lecendreux M. Sleep in children with attention-deficit/hyperactivity disorder: meta-analysis of subjective and objective studies. *J Am Acad Child Adolesc Psychiatry*. 2009;48(9):894−908. Available from: https://doi.org/10.1097/CHI.0b013e3181ac09c9.
32. Golan N, Shahar E, Ravid S, Pillar G. Sleep disorders and daytime sleepiness in children with attention-deficit/hyperactive disorder. *Sleep*. 2004;27(2):261−266.
33. Walters AS, Silvestri R, Zucconi M, Chandrashekariah R, Konofal E. Review of the possible relationship and hypothetical links between Attention Deficit Hyperactivity Disorder (ADHD) and the simple sleep related

movement disorders, parasomnias, hypersomnias, and circadian rhythm disorders. *J Clin Sleep Med.* 2008;4 (6):591–600.

34. O'Brien LM, Ivanenko A, Crabtree VM, et al. Sleep disturbances in children with Attention Deficit Hyperactivity Disorder. *Pediatr Res.* 2003;54(2):237–243. Available from: https://doi.org/10.1203/01. PDR.0000072333.11711.9A.

35. Huang Y-S, Tsai M-H, Guilleminault C. Pharmacological treatment of ADHD and the short and long term effects on sleep. *Curr Pharm Des.* 2011;17(15):1450–1458. Available from: https://doi.org/10.2174/138161211796197179.

36. Alfano CA, Ginsburg GS, Kingery JN. Sleep-related problems among children and adolescents with anxiety disorders. *J Am Acad Child Adolesc Psychiatry.* 2007;46(2):224–232. Available from: https://doi.org/10.1097/01. chi.0000242233.06011.8e.

37. Gregory AM, Sadeh A. Sleep, emotional and behavioral difficulties in children and adolescents. *Sleep Med Rev.* 2012;16(2):129–136. Available from: https://doi.org/10.1016/j.smrv.2011.03.007.

38. Hudson JL, Gradisar M, Gamble A, Schniering CA, Rebelo I. The sleep patterns and problems of clinically anxious children. *Behav Res Ther.* 2009;47(4):339–344. Available from: https://doi.org/10.1016/j. brat.2009.01.006.

39. Lin SY, Melvin T-AN, Boss EF, Ishman SL. The association between allergic rhinitis and sleep-disordered breathing in children: a systematic review. *Int Forum Allergy Rhinol.* 2013;3(6):504–509. Available from: https://doi.org/10.1002/alr.21123.

40. Fishbein AB, Vitaterna O, Haugh IM, et al. Nocturnal eczema: review of sleep and circadian rhythms in children with atopic dermatitis and future research directions. *J Allergy Clin Immunol.* 2015;136(5):1170–1177. Available from: https://doi.org/10.1016/j.jaci.2015.08.028.

41. Banasiak NC. Understanding the relationship between asthma and sleep in the pediatric population. *J Pediatr Heal Care.* 2016;30(6):546–550. Available from: https://doi.org/10.1016/j.pedhc.2015.11.012.

42. Ross K. Sleep-disordered breathing and childhood asthma: clinical implications. *Curr Opin Pulm Med.* 2013;19 (1):79–83. Available from: https://doi.org/10.1097/MCP.0b013e32835b11a1.

43. NHLBI. National asthma education and prevention program. *Children.* 2007;120(5 Suppl):S94–S138. Available from: https://doi.org/10.1016/j.jaci.2007.09.029.

44. Katz ES. Cystic fibrosis and sleep. *Clin Chest Med.* 2014;35(3):495–504. Available from: https://doi.org/10.1016/j.ccm.2014.06.005.

45. Vandeleur M, Walter LM, Armstrong DS, Robinson P, Nixon GM, Horne RSC. What keeps children with cystic fibrosis awake at night? *J Cyst Fibros.* 2017. Available from: https://doi.org/10.1016/j.jcf.2017.04.012.

46. Perfect MM, Patel PG, Scott RE, et al. Sleep, glucose, and daytime functioning in youth with type 1 diabetes. *Sleep.* 2012;35(1). Available from: https://doi.org/10.5665/sleep.1590.

47. Yeshayahu Y, Mahmud FH. Altered sleep patterns in adolescents with type 1 diabetes: implications for insulin regimen. *Diabetes Care.* 2010;33(11):2010. Available from: https://doi.org/10.2337/dc10-1536.

48. Camfferman D, Doug McEvoy R, O'Donoghue F, Lushington K. Prader–Willi Syndrome and excessive daytime sleepiness. *Sleep Med Rev.* 2008;12(1):65–75. Available from: https://doi.org/10.1016/j. smrv.2007.08.005.

49. Cassidy SB, Schwartz S, Miller JL, Driscoll DJ. Prader–Willi syndrome. *Genet Med.* 2012;14(1):10–26. Available from: https://doi.org/10.1097/GIM.0b013e31822bead0.

50. Deal CL, Tony M, Hoybye C, et al. Growth hormone research society workshop summary: consensus guidelines for recombinant human growth hormone therapy in Prader–Willi syndrome. *J Clin Endocrinol Metab.* 2013;98(6):E1072–E1087. Available from: https://doi.org/10.1210/jc.2012-3888.

51. Bellini B, Panunzi S, Bruni O, Guidetti V. Headache and sleep in children. *Curr Pain Headache Rep.* 2013;17 (6):335. Available from: https://doi.org/10.1007/s11916-013-0335-x.

52. Valrie CR, Bromberg MH, Palermo T, Schanberg LE. A systematic review of sleep in pediatric pain populations. *J Dev Behav Pediatr.* 2013;34(2):120–128. Available from: https://doi.org/10.1097/DBP.0b013e31827d5848.

53. Dosi C, Figura M, Ferri R, Bruni O. Sleep and headache. *Semin Pediatr Neurol.* 2015;22(2):105–112. Available from: https://doi.org/10.1016/j.spen.2015.04.005.

54. Tajti J, Szok D, Csáti A, Vécsei L. Prophylactic drug treatment of migraine in children and adolescents: an update. *Curr Pain Headache Rep.* 2016;20(1):1–9. Available from: https://doi.org/10.1007/s11916-015-0536-6.

REFERENCES

409

55. Cielo CM, Konstantinopoulou S, Hoque R. OSAS in specific pediatric populations. *Curr Probl Pediatr Adolesc Health Care*. 2016;46(1):11−18. Available from: https://doi.org/10.1016/j.cppeds.2015.10.008.

56. Parisi P, Bruni O, Pia Villa M, et al. The relationship between sleep and epilepsy: the effect on cognitive functioning in children. *Dev Med Child Neurol*. 2010;52(9):805−810. Available from: https://doi.org/10.1111/j.1469-8749.2010.03662.x.

57. Rosen GM. Sleep in children who have cancer. *Sleep Med Clin*. 2007;2(3):491−500. Available from: https://doi.org/10.1016/j.jsmc.2007.05.011.

58. Rosen GM, Shor AC, Geller TJ. Sleep in children with cancer. *Curr Opin Pediatr*. 2008;20(6):676−681. Available from: https://doi.org/10.1097/MOP.0b013e328312c7ad.

59. Olson K. Sleep-related disturbances among adolescents with cancer: a systematic review. *Sleep Med*. 2014;15 (5):496−501. Available from: https://doi.org/10.1016/j.sleep.2014.01.006.

60. Tan HL, Kheirandish-Gozal L, Abel F, Gozal D. Craniofacial syndromes and sleep-related breathing disorders. *Sleep Med Rev*. 2016;27:74−88. Available from: https://doi.org/10.1016/j.smrv.2015.05.010.

61. Cielo CM, Marcus CL. Obstructive sleep apnoea in children with craniofacial syndromes. *Paediatr Respir Rev*. 2015;16(3):189−196. Available from: https://doi.org/10.1016/j.prrv.2014.11.003.

62. Lal C, White DR, Joseph JE, Van Bakergem K, LaRosa A. Sleep-disordered breathing in down syndrome. *Chest*. 2015;147(2):570−579. Available from: https://doi.org/10.1378/chest.14-0266.

63. Owens JA, Dalzell V. Use of the "BEARS" sleep screening tool in a pediatric residents' continuity clinic: a pilot study. *Sleep Med*. 2005;6(1):63−69. Available from: https://doi.org/10.1016/j.sleep.2004.07.015.

64. Owens JA, Babcock D, Blumer J, et al. The use of pharmacotherapy in the treatment of pediatric insomnia in primary care: rational approaches. A consensus meeting summary. *J Clin Sleep Med*. 2005;1(1):49−59.

65. Meltzer LJ, Moore M. Sleep disruptions in parents of children and adolescents with chronic illnesses: prevalence, causes, and consequences. *J Pediatr Psychol* 2008;33(3):279−291. <https://doi.org/10.1093/jpepsy/jsm118>.

66. Lewandowski AS, Toliver-Sokol M, Palermo TM. Evidence-based review of subjective pediatric sleep measures. *J Pediatr Psychol*. 2011;36(7):780−793. Available from: https://doi.org/10.1093/jpepsy/jsq119.

67. Malow BA, Crowe C, Henderson L, et al. A sleep habits questionnaire for children with autism spectrum disorders. *J Child Neurol*. 2009;24(1):19−24. Available from: https://doi.org/10.1177/0883073808321044.

68. Katz T, Shui AM, Johnson CR, et al. Modification of the children's sleep habits questionnaire for children with autism spectrum disorder. *J Autism Dev Disord*. 2018;. Available from: https://doi.org/10.1007/s10803-018-3520-2.

69. Meltzer LJ, Wong P, Biggs SN, et al. Validation of actigraphy in middle childhood. *Sleep*. 2016;39 (6):1219−1224. Available from: https://doi.org/10.5665/sleep.5836.

70. Laakso ML, Leinonen L, Lindblom N, Joutsiniemi SL, Kaski M. Wrist actigraphy in estimation of sleep and wake in intellectually disabled subjects with motor handicaps. *Sleep Med*. 2004;5(6):541−550. Available from: https://doi.org/10.1016/j.sleep.2004.05.002.

71. Morgenthaler T, Alessi C, Friedman L, et al. Practice parameters for the use of actigraphy in the assessment of sleep and sleep disorders: an update for 2007. *Sleep*. 2007;30(4):519−529.

72. Meltzer LJ, Hiruma LS, Avis K, Montgomery-Downs H, Valentin J. Comparison of a commercial accelerometer with polysomnography and actigraphy in children and adolescents. *Sleep*. 2015;38(8):1323−1330. Available from: https://doi.org/10.5665/sleep.4918.

73. Sadeh A. Cognitive-behavioral treatment for childhood sleep disorders. *Clin Psychol Rev*. 2005;25(5):612−628. Available from: https://doi.org/10.1016/j.cpr.2005.04.006.

74. Morgenthaler TI, Owens J, Alessi C, et al. Practice parameters for behavioral treatment of bedtime problems and night wakings in infants and young children. *Sleep*. 2006;29(10):1277−1281.

75. Moore M, Meltzer LJ, Mindell JA. Bedtime problems and night wakings in children. *Sleep Med Clin*. 2007;2 (3):377−385. Available from: https://doi.org/10.1016/j.jsmc.2007.05.008.

76. Blake MJ, Sheeber LB, Youssef GJ, Raniti MB, Allen NB. Systematic review and meta-analysis of adolescent cognitive−behavioral sleep interventions. *Clin Child Fam Psychol Rev*. 2017;20(3):227−249. Available from: https://doi.org/10.1007/s10567-017-0234-5.

77. Kuhn BR, Elliott AJ. Treatment efficacy in behavioral pediatric sleep medicine. *J Psychosom Res*. 2003;54 (6):587−597. Available from: https://doi.org/10.1016/S0022-3999(03)00061-8.

II. SLEEP DISORDERS IN SPECIFIC MEDICAL CONDITIONS

410   18. SLEEP IN PEDIATRIC PATIENTS

78. Meltzer LJ, Mindell JA. Systematic review and meta-analysis of behavioral interventions for pediatric insomnia. *J Pediatr Psychol*. 2014;39(8). Available from: https://doi.org/10.1093/jpepsy/jsu041.
79. Brown CA, Kuo M, Phillips L, Berry R, Tan M. Non-pharmacological sleep interventions for youth with chronic health conditions: a critical review of the methodological quality of the evidence. *Disabil Rehabil*. 2013;35(15):1221−1255. Available from: https://doi.org/10.3109/09638288.2012.723788.
80. Bruni O, Angriman M, Calisti F, et al. Practitioner review: treatment of chronic insomnia in children and adolescents with neurodevelopmental disabilities. *J Child Psychol Psychiatry*. 2017. Available from: https://doi.org/10.1111/jcpp.12812.
81. Allen KD, Kuhn BR, DeHaai KA, Wallace DP. Evaluation of a behavioral treatment package to reduce sleep problems in children with Angelman Syndrome. *Res Dev Disabil*. 2013;34(1). Available from: https://doi.org/10.1016/j.ridd.2012.10.001.
82. Malow BA, Adkins KW, Reynolds A, et al. Parent-based sleep education for children with autism spectrum disorders. *J Autism Dev Disord*. 2014;44(1):216−228. Available from: https://doi.org/10.1007/s10803-013-1866-z.

CHAPTER 19

Sleep in Hospitalized Patients

Melissa P. Knauert and Margaret A. Pisani
Internal Medicine, Pulmonary, Critical Care, and Sleep Medicine, Yale University School of Medicine, New Haven, CT, United States

OUTLINE

Introduction	411	Mechanical Ventilation	419
Sleep Disruption in Hospitalized Patients	412	Medications	419
		Assessment of Sleep in Hospitalized Patients	421
Environmental Factors Affecting Sleep in Hospitalized Patients	415	Protocols to Improve Sleep in Hospitalized Patients	423
Elevated Sound Levels	415		
Noncircadian Light Patterns	416	Medications to Treat Sleep Disruption in the Hospital	425
Patient Care Interactions	416		
Preexisting Sleep Disorders and Poor Hospital Sleep	417	Clinical Vignette	428
		Areas for Future Research	429
Risk Factors for Poor Sleep in Hospital Settings	418	Conclusion	429
Pain	418	References	429
Severity of Illness and Sepsis	418		
Postsurgical	418		

INTRODUCTION

Sleep disturbance in hospitalized patients has been reported in the medical literature for over 30 years. The impact of sleep deprivation on hospital safety and outcomes is

412   19. SLEEP IN HOSPITALIZED PATIENTS

difficult to measure because it is dynamic and varies with illness severity. Studies that report on sleep disturbances in hospitalized patients often focus on specific subgroups of patients such as critically ill or postsurgical or those with a specific disease process such as chronic obstructive pulmonary disease (COPD) or obstructive sleep apnea (OSA). Regardless of this, sleep disturbance that occurs during a hospital stay has been posited to contribute to "post hospital syndrome," which is and acquired, transient period of increased vulnerability after discharge, and may play a role in rehospitalizations.[1]

This chapter will report on what is known regarding sleep disturbances in a variety of hospital settings and in certain disease states. We will review the data regarding the importance of noise, light, and in-room patient care activity contributing to disrupted sleep in hospitalized patients. Patient-related risk factors for poor sleep in the hospital, including severity of illness, comorbidities, pain, anxiety, and delirium, will be discussed. Both objective and subjective assessment tools for assessing sleep in hospitalized patients will be reviewed. Finally, we will review the literature regarding sleep enhancing protocols to improve sleep in hospitalized patients and the role of medications to treat insomnia in the hospital.

SLEEP DISRUPTION IN HOSPITALIZED PATIENTS

Sleep disruption is common among hospitalized patients, and acute illness can be a precipitant for insomnia.[2] Table 19.1 lists common comorbidities and their impact on sleep. Hospitalization, especially in an intensive care unit (ICU), can increase a patient's risk for fragmented sleep and altered sleep architecture. Older patients are particularly vulnerable to sleep deprivation and may have more difficulty falling asleep, have more frequent awakenings, and may be more easily aroused from sleep by noise or other environmental stimuli.[3] About half of patients on a general medical ward of a Veterans Affair Hospital reported significant insomnia, excessive daytime somnolence, or both.[4] Sleep disturbance is even greater among critically ill patients. Surveys of ICU patients demonstrate a high frequency of self-reported sleep disturbance up to 100% in some studies.[5,6] Studies utilizing polysomnography (PSG) in critically ill patients have consistently shown that these patients experience decreased total sleep time, increased sleep during daytime hours consistent with circadian misalignment, frequent arousals, and abnormal sleep architecture with increased N1 and N2 stages and reduced or absent N3 and rapid eye movement (REM) stages.[7-10]

Sleep disruption can occur due to the reduced quantity of sleep, reduced quality, or misalignment of the circadian rhythm. Both acute and chronic sleep disruption can precipitate physiologic perturbations and worsen patient outcomes.[11] Sleep impairment has important implications for hospitalized patients. Lack of sleep can potentiate alterations in cellular immune function and changes in the immune system with sleep deprivation are well described in healthy individuals.[12-17] Important for hospitalized patients is that sleep disruption has been demonstrated to increase stress responses, delay wound healing, and increase susceptibility to viral and bacterial infection.[18-20] In addition, there are detrimental cardiovascular, metabolic, and endocrine effects with acute sleep deprivation.[12,21]

II. SLEEP DISORDERS IN SPECIFIC MEDICAL CONDITIONS

SLEEP DISRUPTION IN HOSPITALIZED PATIENTS 413

TABLE 19.1 Impact of Comorbidities on Sleep

Underlying Disease	Sleep Problem	Treatments to Improve Sleep
Obstructive sleep apnea	Snoring, arousals, hypoxemia	Minimize Central nervous system depressants, avoid supine position, continuous positive airway pressure, encourage patient to use home continuous positive airway pressure during admission
Asthma	Arousal due to nocturnal exacerbation	Inhaled corticosteroids and/or long-acting beta-adrenergic medications
Chronic obstructive pulmonary disease	Nocturnal hypoxemia, cor pulmonale, decreased respiratory muscle strength in rapid eye movement, decreased supine Functional Reserve Capacity	Treat hypoxemia to SaO_2 above 88%, avoid sedative-hypnotics that cause respiratory depression, head of bed elevated above 30 degrees
Congestive Heart Failure	Orthopnea, paroxysmal nocturnal dyspnea, nocturnal diuresis, Cheyne–Stokes respiration	Head of bed elevated above 30 degrees, treat hypoxemia to SaO_2 above 88%, diuresis during daytime, continuous positive airway pressure for congestive heart failure
Chronic Kidney Disease	Increased risk of restless leg syndrome and periodic limb movement disorder	Ambulation for restless leg syndrome, correction of hyperphosphatemia and uremia
Gastroesophageal reflux disease	Arousals, symptoms increased when supine	Head of bed elevated >30 degrees, no food for 4 or more hours before bed
Stroke	Hypersomnia, insomnia, increased risk of obstructive sleep apnea	Head of bed elevated >30 degrees, frequent suctioning of secretions, continuous positive airway pressure

Research also demonstrates that sleep deprivation impacts postural control and this may contribute to falls in hospitalized patients.[22] Acute sleep deprivation impairs psychomotor performance, short-term memory, and executive functioning.[23] Mood disturbances such as fatigue, irritability, difficulty concentrating, disorientation, anxiety, depression, and paranoia also occur.[24] Clinicians should be particularly alert to sleep disruption and subsequent alterations in patient mood or cognition when having end-of-life and goals-of-care discussions with patients.

Similarities between sleep deprivation symptomatology and delirium have prompted many experts to draw links between the two and question the role of sleep deprivation in causing ICU delirium (Fig. 19.1).

This is important since delirium has been demonstrated to be associated with adverse consequences including prolonged length of stay, long-term cognitive impairment, functional decline, and increased 1-year mortality.[25,26] Poor sleep is linked to delirium through several possible mechanisms including impairing mobilization efforts in the hospital,[27] prolonging ventilator weaning, and accentuating all the consequences of bed rest including nosocomial pneumonia, deep vein thrombosis, and skin breakdown.[28,29] The literature has suggested that poor sleep may be an important modifiable risk factor for delirium in

II. SLEEP DISORDERS IN SPECIFIC MEDICAL CONDITIONS

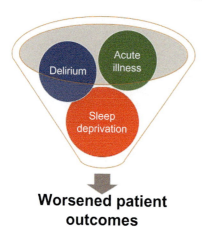

FIGURE 19.1 Schematic of this interrelationship between illness, sleep, delirium, and outcomes.

hospitalized patients.[30,31] A multicomponent intervention prevention trial in hospitalized patients aged ≥70 years successfully reduced the rate of delirium from 15% to 9%; in addition, the total number of days of delirium and the number of instances of delirium were reduced in the intervention group.[31] The intervention consisted of standardized protocols for the management of six risk factors for delirium: cognitive impairment, sleep deprivation, immobility, visual impairment, hearing impairment, and dehydration. Nonpharmacological sleep promotion was one of the successful components of this intervention, which reduced the use of pharmacological sleep aids from 46% to 35% of patients.

Sleep deprivation has a detrimental impact on the respiratory system and can diminish the respiratory response to hypoxemia and hypercapnea.[32] Important for patients with asthma and COPD is data demonstrating both forced expiratory volume in 1 second and forced vital capacity decrease in sleep-deprived patients.[33] Poor sleep quality or atypical sleep patterns may also predict late failure of noninvasive ventilation in cases of acute hypercarbic respiratory failure.[34]

Normal endocrine functions are also influenced by sleep. Growth hormone and prolactin, hormones necessary for cell differentiation and proliferation, follow the sleep–wake cycle and are suppressed during sleep restriction.[35] Similarly, cortisol, which rises in the early morning and peaks in the late morning, loses its periodicity during sleep loss and reentrains during sleep recovery.[36,37]

The clinical significance of circadian disruption in hospitalized patients is difficult to assess in isolation. Studies that have examined circadian rhythm in critically ill patients have found severe circadian misalignment. The molecular clocks in various peripheral organs share the same basic genetic makeup, but they differ in terms of their sensitivity to particular synchronizers (e.g., light, mobility, and feeding), and as a result can be disassociated from one another.[38,39] The misalignment of the suprachiasmatic nucleus (SCN) and peripheral organ clocks is often referred to as "internal desynchronization" and the body appears to interpret this as a form of stress that is additive to, but independent from sleep deprivation.[40] Circadian rhythms on a molecular level are likely to have a bearing on hospitalized patient outcomes, especially those who are critically ill or with prolonged length of stay.

ENVIRONMENTAL FACTORS AFFECTING SLEEP IN HOSPITALIZED PATIENTS

Environmental factors are a major source of sleep disruption in hospitalized patients. High and/or variable sound levels can directly cause arousal from sleep.[9,41] In addition, depending on the source, sound can cause increased patient stress and anxiety.[42] Hospital light levels also contribute to sleep disruption in hospitalized patients; brighter than normal overnight light levels and dimmer than normal daytime light levels have the potential to disrupt normal sleep and circadian function.[10,43] Finally, overnight in-room patient care activities frequently disrupt sleep in hospitalized patients;[44,45] these activities cause both direct disturbance and cause local increases in sound and light which disrupt both the patient receiving care and nearby patients. We will discuss each type of disturbance in detail below.

Elevated Sound Levels

Excessive hospital sound levels have been extensively reported. The World Health Organization recommends an average sound level (L_{eq}) of 35 A-weighted decibels (dBA) and maximum sound levels (L_{max}) no higher than 40 dBA.[46] Several studies have examined sound levels at various locations within the hospital, and there is evidence that hospital sound levels have increased over time (from the 1960s to 2005).[47] Recent work by Shield et al. included an extensive survey of 31 different hospital locations; sound measurements were completed in a variety of patient rooms including single rooms and multibed rooms, and common areas such as nursing stations. Consistent with prior literature, daytime sound levels varied between 50 and 51 dBA and night sound levels varied between 41 and 51 dBA. The frequency of sound peaks >70 dBA varied between 4 and 9 peaks per hour during the overnight period. ICU-based investigations have also demonstrated average sound levels between 43 and 66 dBA[10,48−50] and peak sound levels between 80 and 90 dBA.[48−55] Studies have been mixed regarding the association between severity of illness and sound levels.[50,56]

Regarding the effect of sound on sleep, PSG studies in critically ill patients have demonstrated a correlation between sound peaks >80 dBA and approximately 20% of arousals from sleep.[9,41] In the general wards, postoperative orthopedic patients reported that pain (45%) and noise (23%) were the most common factors affecting sleep.[57] Healthy volunteers exposed to simulated ICU sounds experienced decreased REM sleep and reported prolonged sleep latency, increased arousals, and poorer sleep quality.[58,59]

Some experts suggest that compliance with overnight sound guidelines may not be possible in modern hospitals due to unavoidable machine noise inside patient rooms. Concern regarding sound from air-handlers and structural building sound has been raised.[47,50] One study demonstrated that recommended noise levels could only be achieved in unoccupied side rooms with all equipment switched off.[60] Though we may need to reconsider our ability to eliminate background machine noise, a reasonable first step toward patient sleep improvement may be to address sound peaks and avoidable disturbing sounds such as staff conversations and unnecessary equipment alarms.

Noncircadian Light Patterns

Data on light levels and sleep in the hospital is more limited, but there are important biologic reasons to consider abnormal light exposure in hospitalized patients as a source of sleep disruption.[61] Day–night light patterns are the most important cue (zeitgeber) for the entrainment of circadian rhythms. Circadian rhythms are key determinants in the timing of sleep and, more indirectly, determinants in the quality of sleep.[62] The normal decrease in light exposure prior to habitual bedtime allows melatonin secretion and thus promotes sleep. Conversely, the normal morning increase in light suppresses melatonin and allows the promotion of wake. Studies of light measurements in hospital settings reveal a common pattern: (1) dim overnight light punctuated with multiple, brief exposures to bright light and (2) low daytime light levels that are insufficient to promote normal circadian entrainment.[63,64] In a study of light, sound, and sleep in elderly hospitalized patients, overnight light levels were low with an average of 3 periods of elevated light levels (mean, 64 lx) lasting an average of 1.75 hours. Patients did have poor sleep and there was a nonsignificant trend in association between worse sleep and higher overnight light (and sound) levels.[65] One ICU study demonstrated low average light intensity during the day (mean 80 lx) and night (4 lx) in the rooms of sedated mechanically ventilated patients.[43] Another ICU study demonstrated a low nighttime light levels (median <2 lx) and low daytime light levels (median 74 lx).[10] Though sleep promotion interventions aimed at controlling overnight light levels have shown some promise, there is a risk of these protocols lowering overall light levels, but creating either more light peaks or more light variability. This could occur if care providers turn lights on and off frequently in an attempt to comply with both the protocol and provide bedside care.[66] At this time, it is unknown what the impact of light variability is on sleep quality.

Daytime bright light interventions have demonstrated improved sleep in elderly hospitalized patients,[67,68] including delirious elderly patients,[69] and in cardiology ward patients.[70] On the other hand, bright light did not improve sleep for hospitalized cirrhotic patients.[71] In related studies, bright light therapy or exposure to natural light has been linked to reduced rates of delirium (postoperative patients)[72] and mortality (postmyocardial infarction patients).[73]

Patient Care Interactions

Sleep disturbances in the general wards and ICUs may also occur during nocturnal care in the patient room. This in-room patient care includes medication administration, vital sign measurement, bathing or other hygiene care, and phlebotomy. These nocturnal care events cause primary disturbance and increase local sound and light levels. An examination of 160 patient-nights in a hematopoietic stem cell transplant unit reported an average of 41 nocturnal care interactions per patient per night.[74] Retrospective chart reviews of 147 patient-nights across 4 ICUs and 180 patient-nights in a single surgical ICU demonstrated averages of 43 and 51 nocturnal care interactions per patient-night, respectively.[44,75] An observational study of 1831 nighttime patient interactions across 200 patients in 5 ICUs estimated that 14% of nocturnal interactions were not time critical.[45] In a study of cancer ward patients, sleep disturbance determined by the Verran/Snyder-Halpern Scale was worse in patients with more overnight disturbances.[76] Given that nocturnal care activities occur frequently and are often not time-critical, efforts to reschedule and cluster nocturnal patient care activities have been recommended by expert guidelines for sleep promotion as a means of delirium prevention.[77]

PREEXISTING SLEEP DISORDERS AND POOR HOSPITAL SLEEP

Chronic sleep disorders such as sleep disordered breathing (SDB) or insomnia can significantly contribute to sleep disruption in hospitalized patients. Insomnia is one of the most common sleep disorders in the world, and, though data is very limited, patients with preexisting insomnia are likely to experience an exacerbation of their sleep complaints during and following hospitalization.[2,78] Similarly, patients with undiagnosed or untreated SDB are at risk for chronic sleep deprivation that is exacerbated by multiple facets of acute illness and hospitalization. SDB is prevalent in the general population and patients with coronary artery disease, diabetes, and stroke have increased likelihood of having SDB.[79–84] SDB is often underdiagnosed in patients presenting to the hospital for admission and unrecognized by the medical staff caring for them.

In addition to potentiating sleep disruption, the presence of SDB carries and increased risk of complications and poor outcomes during hospitalization.[85] Patients with SDB who are undergoing surgery are at increased risk of respiratory, cardiac, and neurologic complications. In those patients who require general anesthesia, there is an increased risk of airway complications compared to those without SDB (17% vs 4%).[86] In addition to risk of airway complications, patients with SDB have higher rates of adverse outcomes including increase length of hospital stay in patients undergoing orthopedic surgery for hip or knee replacement.[87]

While there is a paucity of data on SDB prevalence in general medical patients, several hospital admission diagnoses carry increased risk for having coexisting SDB. These include congestive heart failure (CHF), COPD, and stroke. OSA affects up to 55% of patient with cardiovascular disease and untreated it triples the mortality risk compared to those without OSA.[88] In patients with heart failure, with either preserved or reduced ejection fraction, rates of SDB have been reported between 50% and 80%.[89–93] Patients with CHF also present with Cheyne–Stokes respiration with risk factors that include advanced age, male sex, hypocapnia, atrial fibrillation, and treatment with diuretics.[94]

A percentage of 10–12 patients with COPD have concomitant OSA,[95] and patients with COPD are at risk for sleep deprivation while hospitalized. Studies have demonstrated that COPD patients experience increased sleep latency, decreased total sleep time, and increased arousals.[96] Patients with diagnosis of both COPD and OSA have decreased PaO_2, decreased central respiratory drive, and higher $PaCO_2$.[97,98]

Several studies have reported high prevalence of SDB in patients hospitalized with cerebrovascular accidents (CVA) with rates between 60% and 96%.[83,84] Patients with CVA and concomitant OSA have an increased risk for recurrent stroke, increased length of stay, and 6-month mortality.[99–101] Location and severity of the CVA, poststroke neurologic function, and medications influence the severity of SDB post-CVA.[101]

418   19. SLEEP IN HOSPITALIZED PATIENTS

All hospitalized patients with suspected SDB should be referred for a formal sleep evaluation. Patients who have known SDB should be maintained on their home positive airway pressure settings whenever medically feasible. There is data to support the use of positive airway pressure treatment in improving sleep parameters and other outcomes in all the above subgroups of patients with CHF, COPD, and stroke.

RISK FACTORS FOR POOR SLEEP IN HOSPITAL SETTINGS

Pain

There are associations between poor sleep and pain that are assumed to carry over to hospitalized patients. Patients who experience chronic pain also report disrupted sleep with a prevalence of between 50% and 70%.[102] Pain causes sleep fragmentation by increasing cortical arousals. There is also data to suggest that sleep deprivation increases pain sensitivity by inhibiting opioid protein synthesis or reducing affinity for the opioid receptor.[103] In a study of burn patients, pain was associated with increased arousals and prolonged periods of nocturnal wakefulness.[104] In addition, on subsequent days, these patients had poorer pain tolerance and greater pain intensity. It can be assumed from these data that patients with acute or chronic pain are at risk for sleep disruption during hospitalization.

Severity of Illness and Sepsis

Severity of illness will influence the clinical interventions required by the medical staff and can limit availability of rest periods. Increased illness severity is also likely associated with increased pain, anxiety, and delirium, which have all been linked to poor patient sleep.[105,106] Patients admitted with sepsis are also at risk for disrupted sleep patterns including reductions in REM sleep with increases in N1 and N2 along with low-voltage, mixed frequency electroencephalogram (EEG) with variable theta and delta waves which has been described as septic encephalopathy.[107] In addition, a loss of normal circadian rhythm has been reported in septic patients. There is lack of periodic excretion of urinary 6-SMT, a melatonin metabolic, in septic awake patients compared to nonseptic ICU patients.[108] It is possible that some of these sleep changes may be adaptive in patients with sepsis. REM sleep is associated with oxygen desaturation and cardiovascular variability and hence reduced REM sleep may be protective in septic patients with hemodynamic instability.[109] In animal models of sepsis, sleep deprivation has been shown to increase mortality, but there is no comparable human data.[110]

Postsurgical

Studies of postsurgical patients have reported pain (discussed above) as the most common cause of sleep disruption.[57,111,112] In the immediate postoperative period, patients who have undergone abdominal surgery, including gynecologic surgery, have absent or markedly depressed REM sleep along with increases in stages N1 and N2.[112–114] Several

II. SLEEP DISORDERS IN SPECIFIC MEDICAL CONDITIONS

factors may be playing a role in REM suppression postoperatively including increases in catecholamine and cortisol levels. Much of the REM suppression seen may be secondary to the receipt of opioids for pain control.[112] Studies that have examined patients after abdominal surgery, over several nights, have reported REM rebound occurring between days 3 and 6.[114] REM rebound may have an impact on respiratory function as a study examining patients after major abdominal surgery reported significant nocturnal desaturations on the second and third postoperative nights.[115,116]

Mechanical Ventilation

Patients who are critically ill and mechanically ventilated have poor sleep quality. The causes of sleep disturbance in these patients include patient-ventilator asynchrony, central apneas caused by overventilation, and increased respiratory effort secondary to improper settings or air leaks, pain, anxiety, and patient care activities.[117,118] Mechanically ventilated patients have increased sleep fragmentation, increased stage N1 and N2 sleep, and loss of circadian rhythm.[8,9,41,118–125] PSG studies of sleep in mechanically ventilated patients have found variable total sleep times, but all found that sleep was distributed over the 24-hour period. Small studies have attempted to examine the effect of ventilator mode on sleep. Patients placed on pressure support ventilation experienced greater sleep fragmentation and more central apneas than those on assist control. When dead space was added to the pressure support mode to increase the $PaCO_2$, most of the central apneas and sleep disruption were eliminated.[118] In addition, a study of patients receiving noninvasive ventilation for hypercapnic respiratory failure was found to have circadian sleep cycle disruption, decreased REM time, as well as subsequent delirium.[34]

Medications

Many medications prescribed to hospitalized patients have been demonstrated to alter sleep architecture. Table 19.2 lists commonly prescribed medications in hospitalized patients and their impact on sleep. Often times medications such as antihistamines, antipsychotics, and benzodiazepines are given off-label for their sedative properties to promote sleep but may actually inhibit deep sleep stages (REM and N3) and are associated with side effects such as delirium.[126,127] Drugs that most commonly impair sleep include antiepileptic agents, selective serotonin reuptake inhibitors, monoamine oxidase inhibitors, tricyclic antidepressants (TCAs), antihypertensives, antihistamines, antibiotics, vasopressors, beta-agonists, and corticosteroids. While sedatives and opioids may depress the level of consciousness and create the subjective appearance of sleep, they have a negative impact on sleep quality and architecture.

Benzodiazepines are gamma-aminobutyric acid A ($GABA_A$) receptor agonists and disrupt sleep architecture via decreases in N3 and REM sleep stages.[128,129] Studies in chronic insomnia patients demonstrate improved sleep latency and total sleep time with benzodiazepines, but these benefits are weighed against high abuse potential, risk of falls and delirium particularly in the elderly, rebound insomnia following drug discontinuation when used on a daily basis, and daytime cognitive impairment.[130,131] Narcotics are frequently prescribed in the inpatient setting and are known to provoke nocturnal awakenings, suppress N3 and

TABLE 19.2 Effects of Medications on Sleep

Medication	Mechanism of Action	Effects on Sleep Architecture
Midazolam	Modulator of GABAA receptor	↓rapid eye movement (sleep), ↓N3, ↑percent of total sleep time spent in N2
Lorazepam	Modulator of GABAA receptor	Enhances β waves
Diazepam	Modulator of GABAA receptor	May ↑sleep efficiency (% of time "in bed" spent asleep) and ↓sleep latency
Propofol	Bind to β_3 of Gamma-aminobutyric acid A; additional effects on glutamate and cannabinoid receptors	↓rapid eye movement (sleep) Enhances γ waves and dose-dependent burst suppression May not affect sleep efficiency (% of time "in bed" spent asleep) or NREM
Dexmedetomidine	α_2 Agonist	↓rapid eye movement (sleep), ↓N3 Enhances N2 spindle activity

OPIOIDS

Medication	Mechanism of Action	Effects on Sleep Architecture
Morphine	μ (κ and δ to lesser extent) opioid receptor agonists	↓N3, ↑percent of total sleep time spent in N2
Fentanyl	μ (κ and δ to lesser extent) opioid receptor agonists	Likely ↓rapid eye movement (sleep)
Hydromorphone	μ (κ and δ to lesser extent) opioid receptor agonists	May ↑rapid eye movement (sleep) latency and total sleep time

ANTIPSYCHOTICS

Medication	Mechanism of Action	Effects on Sleep Architecture
Haloperidol[a]	D_2 receptor antagonist	↑N2, ↑sleep efficiency (% of time "in bed" spent asleep) Limited effects on rapid eye movement (sleep) or N3
Risperidone[a,**]	D_2 and 5-HT$_2$ receptor antagonists	↓rapid eye movement (sleep), ↑percent of total sleep time spent in N2
Olanzapine[a,†,‡]	D_2 and 5-HT$_2$ receptor antagonists	↑N3, ±rapid eye movement (sleep), May ↑total sleep time and ↑sleep efficiency (% of time "in bed" spent asleep)
Quetiapine[a]	D_2 and 5-HT$_2$ receptor antagonists	↑percent of total sleep time spent in N2, ↑sleep efficiency (% of time "in bed" spent asleep), ↑total sleep time

CARDIOVASCULAR

Medication	Mechanism of Action	Effects on Sleep Architecture
β-Blockers	CNS β-receptor agonist	↓rapid eye movement (sleep), ↑nightmares
Dopamine	D_2, β_1, α_1 receptor agonist	↓rapid eye movement (sleep), ↓N3
Norepinephrine	α and β receptor agonist	↓rapid eye movement (sleep), ↓N3
Epinephrine	α and β receptor agonist	↓rapid eye movement (sleep), ↓N3
Phenylephrine	α_1 receptor agonist	↓rapid eye movement (sleep), ↓N3

(Continued)

ASSESSMENT OF SLEEP IN HOSPITALIZED PATIENTS

TABLE 19.2 (Continued)

Medication	Mechanism of Action	Effects on Sleep Architecture
ANTIDEPRESSANTS		
Selective serotonin reuptake inhibitor	5-HT$_2$ receptor antagonists	↓rapid eye movement (sleep), ↓total sleep time
Tricyclic antidepressant	5-HT$_2$ receptor antagonists	↓rapid eye movement (sleep), ↓total sleep time, ↑periodic limb movements
Trazodone	5-HT$_2$ receptor antagonists, H$_1$ receptor antagonist	↓sleep latency, ↑N3, ±sleep efficiency (% of time "in bed" spent asleep), ±rapid eye movement (sleep)
OTHER MEDICATIONS		
Antihistamines	H$_1$ receptor antagonist	↓sleep latency, ↓rapid eye movement (sleep), ±sleep efficiency (% of time "in bed" spent asleep), ±N3
Corticosteroids	Decrease melatonin levels	↓rapid eye movement (sleep), ↓N3

[a]All neuroleptics carry a black box warning for "increased mortality in elderly patients with dementia-related psychosis." Also, the use of neuroleptics for the management of general agitation, delirium, or sleep disturbances in the ICU is off-label.
[**]All neuroleptics carry a black box warning for "increased mortality in elderly patients with dementia-related psychosis." Also, the use of neuroleptics for the management of general agitation, delirium, or sleep disturbances in the ICU is off-label.
[†]Olanzapine is available as an orally disintegrating tablet (ODT) under the brand name Zyprexa Zydis. The traditional 5 mg tablet can be broken in half to administer 2.5 mg, but the 5 mg ODT is too fragile for this purpose.
[‡]Intramuscular olanzapine causes significant respiratory suppression and in general should not be administered with parenteral benzodiazepines unless a patient is on mechanical ventilation.
±, equivocal effect on indicated sleep feature; ↓, decrease in indicated sleep feature; ↑, increase in indicated sleep feature.

REM sleep, and cause central apneas.[112,132–134] Propofol has also been shown to suppress REM and worsen sleep quality.[135] Antihistamines, such as diphenhydramine, reversibly antagonize histamine H$_1$ receptors and inhibit histamine-induced wakefulness, but PSG studies demonstrate that sleep quality is not improved.[136] Trazodone, an antidepressant that inhibits serotonin reuptake and antagonizes the H$_1$ histamine receptor, is very sedating and associated with arrhythmias and drug–drug interactions. Trazodone is prescribed frequently in hospitalized patients with sleep complaints. Unfortunately, there is limited PSG data on the effects on sleep stages.[137] Beta blockers have been demonstrated to cause insomnia and nightmares secondary to suppressed REM sleep.[127] Quinolone antibiotics can disrupt sleep by inhibition of GABA receptors in the brain.[138] Fig. 19.2 is a schematic demonstrating the multitude of factors that impact sleep and circadian rhythm.

ASSESSMENT OF SLEEP IN HOSPITALIZED PATIENTS

Sleep measurement has been a challenge for the clinicians and investigators of hospital sleep.[139] Sleep measurement devices cannot conflict with medical devices or procedures and be comfortable enough to not affect sleep. Also, because sleep is highly fragmented

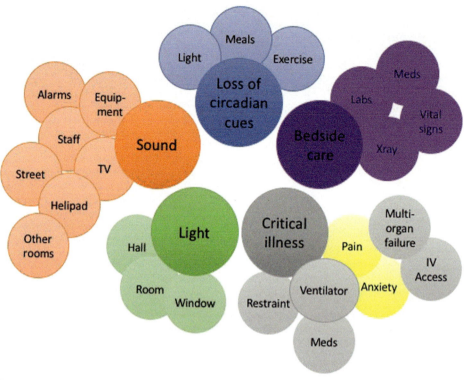

FIGURE 19.2 Patient, environmental, and care factors that impact sleep and circadian disruption.

and occurs during the day and night, sleep should be assessed around the clock in hospitalized patients.

PSG is the gold standard for sleep measurement and offers complete physiologic recording of sleep state, arousals, limb movements, and respiratory events. The EEG, electrooculogram, and electromyogram leads which allow sleep staging and the measurement of wake, sleep, and arousals are particularly important.[140] Unfortunately, even with the advent of portable systems, PSG is costly, requires specialized equipment and staff, and is poorly tolerated by patients particularly during daytime hours.[7] In addition, acutely ill patients do not always demonstrate typical sleep stages, which makes traditional scoring of sleep stages a challenge.[141,142] Given the interest in EEG data, sleep time, and sleep stages, portable devices which record cardiorespiratory data are not helpful in overcoming these challenges. Because of these limitations, alternatives to PSG have been sought.

Actigraphy is an objective measure of sleep that correlates reasonably with PSG and offers better feasibility.[143–145] Actigraphy uses accelerometer technology mounted on a wristwatch-like device to characterizes rest versus activity of the wearer. This data device allows time series monitoring with a high sampling frequency and is easy to use many days in a row. Total sleep time, sleep efficiency, and wake after sleep onset are the most commonly reported parameters.[146] Total sleep time has the best agreement with PSG though there is modest overestimation.[147,148] There is more significant overestimation of sleep efficiency and underestimation of wake after sleep onset due to actigraphy's inability to differentiate between quiet rest and sleep.[145,146] This is a particular concern in bedbound ill patients in the hospital. Actigraphy has nonetheless been successfully used in general ward and ICU settings.[149–151]

Direct observation of patients by clinical or research staff has been utilized. However, this method is staff intensive and results in significant overestimation of sleep.[145,152,150] There are a wide variety of other devices intended to measure sleep. These include commercially available sleep trackers,[147,148] and contact-free sleep-monitoring devices.[145] Though promising, these novel devices await further validation before clinical or research use can be recommended.

Validated sleep questionnaires are a useful, low-cost alternative which can examine both quality of sleep and functional outcomes of sleep. They can be combined with the above discussed objective measures to assess sleep in hospitalized patients or used alone. Sleep questionnaires are more informative when used for several days in a row for trend analysis in a single patient. One of the most commonly used assessment tools, the Richards Campbell Sleep Questionnaire,[153] was developed and validated to assess sleep in critically ill patients. It was originally a five-item visual analog scale evaluating the perception of depth of sleep, sleep onset latency, number of awakenings, time spent awake, and overall sleep quality (some investigators have added an additional question regarding perceived sound levels).[154] It is estimated that the scale takes 2–5 minutes to complete.[155] The Verran and Snyder-Halpern Sleep Scale[156] was designed to assess sleep quality in hospitalized patients without preexisting sleep disorders. Sleep is examined in two domains, sleep disturbance and sleep effectiveness. Questions are answered via 15 visual analog scale questions, and it is estimated that the scale takes 10–15 minutes to complete.[155] Other brief assessment scales such as the St. Mary's Hospital Sleep Questionnaire[157] are available but have not been as commonly or effectively used in hospitalized patient populations.[158]

Sleep diaries are especially limited by perception, delirium, and acute illness and are not validated or likely to provide a valid assessment in hospitalized patients. Tools that assess current sleepiness such as the Stanford Sleepiness Scale[159,160] or the Karolinska Sleepiness Scale[70,161] can be used and are valid measures of sleepiness but are very focused on sleepiness at the moment of completing the scale and can be expected to vary widely even during a single day. This provides limited information about overall sleep quality and function during a particular night or a series of nights. Questionnaires that assess daytime sleepiness and sleep such as the Epworth Sleepiness Scale, Insomnia Severity Index, Pittsburgh Sleep Quality Index, or the Functional Outcomes of Sleep Questionnaire are not appropriate for assessment during hospitalization because they require the patient to assess their sleepiness and sleep over weeks to months. Table 19.3 lists common assessment tools used to measure sleep in hospitalized patients.

PROTOCOLS TO IMPROVE SLEEP IN HOSPITALIZED PATIENTS

Patients, clinical staff, and hospital administrators share a growing awareness that sleep is severely disrupted in hospitalized patients. Studies have demonstrated that providers

424
19. SLEEP IN HOSPITALIZED PATIENTS

TABLE 19.3 Selected Common Assessment Tools for Measurement of Sleep in Hospitalized Patients

Assessment Tool	Strengths	Weaknesses
OBJECTIVE TOOLS		
Polysomnography (portable)	• Gold standard for staging and arousals	• Cost and feasibility limitations • Poor patient tolerance • Prolonged (>24 h) recording difficult • Limitations to scoring in acute illness
Actigraphy	• Acceptable correlation with polysomnography • Low cost, high feasibility • Good patient tolerance • Prolonged recording facile	• Overestimation of sleep; sedated or inactive patients a concern • No sleep staging data
SUBJECTIVE TOOLS		
Richards-Campbell Sleep Questionnaire	• Validated in Intensive care unit patients • Low cost, high feasibility • Good patient tolerance • Prolonged measurement facile	• No sleep staging data • Limited in delirious patients • Recall bias
Verran and Snyder-Halpern Sleep Scale	• Validated in hospitalized patients	• No sleep staging data • Limited in delirious patients • Recall bias

and patients value sleep and believe that sleep is important to recovery.[162–164] Despite this interest, there remains a large gap between the belief that sleep is important and implementation of sleep promotion protocols.[162] A study of patients, physicians, and nurses from a general medicine floor demonstrated agreement among all three groups that pain, vital signs, and tests were the top three disrupters of patient sleep.[164] In the ICU, clinicians appear more focused on environmental causes of sleep disruption while patients are more focused on emotional causes of sleep disruption.[163] Because sleep disruption is believed to contribute significantly to delirium, and because delirium is prevalent in the ICU,[165] there has been a significant research effort toward established best practice for ICU sleep promotion.[166] Current expert critical care guidelines for delirium prevention in hospitalized critically ill patients recommend sleep promotion via a multidisciplinary, bundled approach aimed at addressing modifiable environmental disruptors of sleep in the ICU setting.[77]

Nonpharmacologic hospital-based sleep promotion protocols variably combine one or more elements of the following: sound control, light control, rescheduling of routine patient care to provide a rest period, provision of eye masks, provision of ear plugs, sleep education, treatment of anxiety, promotion of relaxation, and (when applicable) adjustment of ventilator settings.[166,167] Interventions which provided daytime bright light to patients are discussed above under "noncircadian light patterns." A pilot study of a nurse-delivered sleep-promoting intervention that included sleep hygiene education and environmental control for patients on a general medical floor showed an improvement in

II. SLEEP DISORDERS IN SPECIFIC MEDICAL CONDITIONS

total sleep time as recorded in sleep diaries; validated questionnaire and actigraphic sleep measures had trends toward improved sleep as well.[168] ICU sleep promotion interventions have commonly included environmental noise and light reduction via "quiet time" protocols and clustering of patient care activities.[66,166,169,170] One bundled intervention did show a reduction in the incidence of delirium/coma (odds ratio: 0.46; 95% confidence interval, 0.23−0.89; $P = .02$) and an increase in delirium/coma-free days (odds ratio: 1.64; 95% confidence interval, 1.04−2.58; $P = .03$). This is despite not showing a change in subjective sleep scores.[170] Ear plugs and eye masks can be helpful to patients but are not universally tolerated.[170,171] Sleep education in the use and implementation of eye mask, ear plugs, and a white noise machine showed improvement in fatigue scores in hospitalized patients on a non-ICU cardiac-monitored floor.[172] Provision of a signal to sleep and patient education about the importance of sleep may also be helpful and promote sleep self-efficacy.[173] Small studies support the use of complementary therapies that promote relaxation and anxiolysis. Therapies include music therapy,[174] massage,[175] acupressure,[176] and aromatherapy[177] as safe options for sleep promotion. A systematic review of nonpharmacologic strategies does suggest benefit and modest improvement of sleep, though the quality of evidence is low to very low.[167]

In sum, these are low risk, but staff intensive and high complexity protocols which are demanding to implement. Limitations in sleep measurement have made it difficult to prove impact on patient outcomes. Table 19.4 presents key stakeholders in the development of a sleep-promoting protocol.

MEDICATIONS TO TREAT SLEEP DISRUPTION IN THE HOSPITAL

While sleep disruption is common in hospitalized patients, there is a paucity of randomized controlled trials of pharmacologic agents for treatment in this patient population. Data on the scope of pharmacological sleep aids in the inpatient setting is limited and consist of older studies. Historically rates of sleep aid use in hospitalized patients have ranged from 29% to 42%, mostly consisting of benzodiazepines.[178,179] A more recent single-center study demonstrated 26% of all patients admitted to the general adult medical and surgical units of a tertiary care center received a medication for sleep.[180] Median days of sleep aid use, while admitted, was 2 with an interquartile range of 1−5 days. Of these patients, 69% had no known history of prior insomnia or sleep medication use. Drugs most frequently prescribed were trazodone (30%), lorazepam (24%), and zolpidem tartrate (18%). Most concerning was that more than a third of patients who had never been on a sleep medication previously were discharged with a prescription for one.

Research on medications to treat sleep disorders have focused on healthy outpatients, and while data can be extrapolated to some extent to hospitalized patients, one needs to be especially cautious regarding side effects and drug interactions. Many of the medications prescribed for sleep in hospitalized patients are non-Food and Drug Administration (FDA) approved for insomnia.

The most common "off-label" medications prescribed for sleep complaints include antidepressants such as mirtazapine, trazodone, and TCAs. Both mirtazapine and trazodone require lower doses for the sedative effects compared to dosing when used for depression.

426

19. SLEEP IN HOSPITALIZED PATIENTS

TABLE 19.4 Hospital Stakeholders That Need to be Involved in Sleep Promotion

Stakeholder	Role	Intervention
Physician	Overall care plan	Avoidance of diagnostic testing, procedures and medications during sleep period, reassurance, communication
Nurse[a]	1:1 or 1:2 bedside care	Avoidance of nonurgent bedside care during sleep period
Respiratory therapist	Ventilator management	Avoidance of routine ventilator checks and suctioning during sleep period
Pharmacist	Drug choice, delivery mode and timing	Pharmacy ordering protocols that prevent nonurgent medication administration during sleep period
Nutrition	Tube feeding formulation and timing	Intermittent daytime tube feeding schedule
Physical and occupational therapist	Mobilization	Provision of daytime exercise
Hospital administration	Hospital policies, staffing and workflow	Increased staffing during day shifts
Facilities	Maintenance and cleaning	No trash and laundry pickup overnight, no maintenance overnight
Lab/Diagnostic radiology	Processing of patient samples	Increased staffing during day shifts

[a]Gatekeeper.

Both may be beneficial in patients with comorbid depression and insomnia. A recent meta-analysis was conducted on polysomnographic studies using TCAs for the treatment of insomnia in the general population.[181] This analysis demonstrated that the use of TCAs was associated with increased total sleep time and improvements in sleep efficiency. These findings were not modulated by participant age, sample size, or TCA type. Of note, 82% of patients receiving TCA reported daytime somnolence.

While the antihistamines diphenhydramine and hydroxyzine are the most common over-the-counter outpatient medications used for sleep, they should not be used in hospitalized patients for this indication. They can cause residual daytime sedation, delirium, orthostatic hypotension, prolonged QT syndrome, blurred vision, and urinary retention. All of these side effects can lead to adverse outcomes including falls, decreased mobility, urinary infections, and arrhythmias in already vulnerable hospitalized patients.

Atypical antipsychotics including quetiapine and olanzapine are also frequently prescribed "off-label" for sleep in hospitalized patients. Quetiapine is the most sedating of the atypical antipsychotics but is not recommended as a sleep aid unless there is a comorbid psychiatric disorder. It is sometimes used in critically ill patients with delirium and nocturnal agitation, but there are no randomized controlled trials to support its use for this indication and the Society of Critical Care Medicine guidelines do not recommend it.[77] These medications can cause hyperglycemia and hyperlipidemia.

II. SLEEP DISORDERS IN SPECIFIC MEDICAL CONDITIONS

There are three general classes of medications approved for the treatment of insomnia. These include benzodiazepine $GABA_A$ receptor agonists, nonbenzodiazepine $GABA_A$ receptor agonists, and melatonin receptor agonists. While benzodiazepine receptor agonists decrease sleep latency and increase total sleep time, they decrease N3 and REM sleep and can cause daytime sedation, delirium, dependence, and tolerance. Medications in this class include estazolam, flurazepam, temazepam, and triazolam. Flurazepam and quazepam have long half-lives and should not be used in hospitalized patients. The use of these medications should be minimized in all patients and should not be prescribed to older patients or those at high risk for delirium.[182–184]

The nonbenzodiazepine receptor agonists have fewer side effects compared to the benzodiazepine receptor agonists but should also be used with caution especially in older patients. Medications in this class include eszopiclone, zaleplon, and zolpidem. These medications have less next-day sedation and psychomotor dysfunction, less REM rebound, and lower abuse potential.[185] Despite their reduced side-effect profile, risks still exist. Zolpidem tartrate use has been shown to be a risk factor for falls (odds ratio of 4.37) in hospitalized patients.[186]

Melatonin is a hormone secreted primarily by the pineal gland with increased secretion occurring during darkness.[187] Its synthesis is controlled by the SCN in the hypothalamus.[188] Ramelteon is a selective melatonin-receptor agonist which is FDA approved for insomnia. It reduces sleep latency and extends the duration of sleep by diminishing arousal signals generated by the SCN.[189] In addition, melatonin does not produce the next-day psychomotor and memory impairments that are common with other medications to treat insomnia and has no abuse potential. A recent metaanalysis of 12 randomized controlled studies examined the efficacy of melatonin in the treatment of primary sleep disorders.[190] Data to support the use of melatonin was found in reducing sleep onset latency in primary insomnia ($P = .002$), delayed sleep phase syndrome ($P < .0001$), and in regulating sleep–wake patterns in blind patients compared with placebo.

There are several small prospective studies that have examined the role of melatonin on sleep and delirium in ICU patients, which obtained mixed results. These studies examined different doses of melatonin and a variety of outcome measures making it hard to compare or combine them to draw conclusions.[191] In one study of eight hemodynamically stable ICU patients and six patients on a general medical floor with COPD or heart failure, the authors reported that melatonin induced sleep in the participants and that no side effects were observed. Study methods and end points were not well described.[192] Another study of 32 patients who had tracheostomies was randomized to 3 mg of melatonin or placebo.[193] Melatonin levels were measured to confirm absorption. Sleep duration was similar in the two groups and agitation was higher in the melatonin group but this difference did not reach statistical significance. These investigators concluded that nocturnal melatonin administration did not improve observed nocturnal sleep or decrease agitation in this specific patient population. Another randomized, double-blind, placebo-controlled trial of 24 critically ill patients examined the effects of 10 mg of melatonin given for 4 consecutive days.[194] A mean of 1-hour increased sleep time was seen in the patients receiving melatonin compared to placebo, but the difference in sleep efficiency was not statistically significant. This study had marked differences in age and delirium between the melatonin and placebo groups, with increased age in the melatonin group (mean 69.9 years melatonin vs 58.7 years placebo) and increased delirium in the melatonin group (33.3% melatonin vs

8.3% placebo). Both age and delirium are risk factors for sleep disturbance. Large randomized controlled trials examining melatonin and sleep with well-defined, measurable outcomes are needed. There is currently an ongoing trial of melatonin and sleep in the ICU which may provide more data to guide clinical decision-making.[195]

There is one clinical trial which examined ramelteon in the prevention of delirium in a study of 67 older ICU patients.[196] Patients received ramelteon or placebo nightly until the development of delirium or up to 7 days. This study also allowed the use of hydroxyzine as a sleep aid. There was significantly less delirium in the ramelteon group compared to placebo (3% ramelteon vs 32% placebo, $P = .003$) with no significant difference in the use of as-needed hydroxyzine. There were no adverse events reported in patients receiving ramelteon. This study also contributes data to the link between delirium and sleep deprivation and circadian misalignment by implicating a melatonin pathway.

The decision regarding whether to prescribe a medication for sleep to hospitalized patients will depend on local hospital formularies, patient age, and medical comorbidities. In general, agents with a short half-life and minimal drug−drug interactions should be used at the lowest effective dose and for a limited duration. Nightly scheduled medications should be avoided to reduce risks for tolerance and dependence and other nonpharmacologic sleep interventions should be pursued along with prescribed medication. Sleep aids that are prescribed for inpatient insomnia should not be carried over as outpatient prescriptions upon hospital discharge.

CLINICAL VIGNETTE

An 83-year-old man with a past medical history significant for colon cancer was admitted to the medical ICU (MICU) for urosepsis; he was delirious and had a moderate severity of illness as indicated by an APACHE II score of 21. His admission time was 10:00 a.m. He was mechanically ventilated and initiated on broad antibiotic coverage and vasopressors. The mean (L$_{eq}$) overnight sound level between 8:00 p.m. and 8:00 a.m. was 54 dBA. The mean sound level between 8:00 p.m. and midnight was 55 dBA and the mean sound level between midnight and 4:00 a.m. was 52 dBA, which exceed the World Health Organization's recommendations. There were 21, 24, and 17 peaks per hour during the full overnight (8:00 p.m. to 8:00 a.m.), beginning of the night (8:00 p.m. to midnight) and middle of the night (midnight to 4:00 a.m.) time periods, respectively. Average light levels were 47, 29, and 49 lx during the full overnight, beginning of the night, and middle of the night periods, respectively. The patient's room was entered an average of 2.0 times per hour overnight. Of the 36 hours surveyed (three overnight periods from 8:00 p.m. to 8:00 a.m.), there were 17 rest periods of 30 consecutive minutes; there were 6 rest periods of 60 consecutive minutes. On the first night of admission, the patient had a routine bath, skin care, and sheet change from 2:17 a.m. to 3:26 a.m. The patient stayed in the MICU for 7 days and in the hospital for 16 days. Ultimately, he died of further infectious complications.

Based on these and similar observations about the MICU environment, we developed an overnight, nonpharmacologic sleep promotion protocol at our institution. The protocol, called Naptime, was designed to redirect routine care away from the middle of the night period and to promote sleep between midnight and 04:00 a.m. Provision of this rest period requires changes in workflow for physicians, nurses, respiratory therapists, pharmacists, laboratory medicine, diagnostic imaging, and facilities staff. Routine medications, bathing and skin care, and routine phlebotomy were the most commonly rescheduled care tasks.[197] *Nonpharmacologic sleep promotion protocols such as this are in alignment with expert guidelines for the prevention and treatment of ICU delirium, but very difficult to implement and maintain.*

AREAS FOR FUTURE RESEARCH

There are numerous investigative questions that remain in the area of hospital sleep. Under the domain of the hospital environment, there are questions regarding how to implement and maintain sleep promotion protocols that provide patients with an overnight sleep opportunity. This includes significant changes in existing workflow and staff culture. Closely related to such investigations are studies of devices or techniques that further mitigate the hospital environment; this includes such things as white noise, ear plugs, noise canceling headphones, music therapy, and alternative relaxation techniques. To motivate these complex and expensive changes in hospital care, we need greater biologic evidence regarding the short- and long-term impact of the acute sleep and circadian disruption that occurs during hospitalization. Outcomes such as delirium, length of stay, mortality, and quality of life metrics are likely to be impacted by acute sleep and circadian disruption, but this has not yet been proven. Finally, there is a lack of high-quality evidence regarding safe and effective medications that can be used to promote sleep in this population. Though nonpharmacologic sleep promotion is critically important, there is also a need for pharmacologic aids in some cases. Medications which do not have high abuse potential or dangerous daytime side effects, such as melatonin and its agonists, are currently the most promising agents; however, studies supporting their efficacy and safety are very limited at this time.

CONCLUSION

Sleep disruption is common in hospitalized patients. There is emerging evidence of the importance of sleep in maintaining immune, cardiovascular, and cognitive function, but despite this, there is limited data regarding the benefits of sleep during acute illness.[16,110,198,199] Multicomponent interventions have been shown to be successful in minimizing sedative–hypnotic use and improving sleep.[31,178] In addition, sleep promotion and minimization of pharmacological sleep aids are part of the strategy to reduce delirium, length of stay, and hospital costs.[200]

References

1. Krumholz HM. Post-hospital syndrome—an acquired, transient condition of generalized risk. *N Engl J Med.* 368(2), 2013, 100–102.
2. Parsons EC, Kross EK, Caldwell ES, et al. Post-discharge insomnia symptoms are associated with quality of life impairment among survivors of acute lung injury. *Sleep Med.* 2012;13(8):1106–1109.

3. Bliwise DL. Sleep in normal aging and dementia. *Sleep*. 1993;16(1):40–81.
4. Meissner HH, Riemer A, Santiago SM, Stein M, Goldman MD, Williams AJ. Failure of physician documentation of sleep complaints in hospitalized patients. *West J Med*. 1998;169(3):146–149.
5. Elliott R, Rai T, McKinley S. Factors affecting sleep in the critically ill: an observational study. *J Crit Care*. 2014;29(5):859–863.
6. Freedman NS, Kotzer N, Schwab RJ. Patient perception of sleep quality and etiology of sleep disruption in the intensive care unit. *Am J Respir Crit Care Med*. 1999;159(4 Pt 1):1155–1162.
7. Knauert MP, Yaggi HK, Redeker NS, Murphy TE, Araujo KL, Pisani MA. Feasibility study of unattended polysomnography in medical intensive care unit patients. *Heart Lung*. 2014;43(5):445–452.
8. Cooper AB, Thornley KS, Young GB, Slutsky AS, Stewart TE, Hanly PJ. Sleep in critically ill patients requiring mechanical ventilation. *Chest*. 2000;117(3):809–818.
9. Freedman NS, Gazendam J, Levan L, Pack AI, Schwab RJ. Abnormal sleep/wake cycles and the effect of environmental noise on sleep disruption in the intensive care unit. *Am J Respir Crit Care Med*. 2001;163 (2):451–457.
10. Elliott R, McKinley S, Cistulli P, Fien M. Characterisation of sleep in intensive care using 24-hour polysomnography: an observational study. *Crit Care*. 2013;17(2):R46.
11. Bonnet M. Acute sleep deprivation. In: Kryger M, Roth T, Dement W, eds. *Principles and Practice of Sleep Medicine*. 5th ed Philadelphia, PA: Saunders/Elsevier; 2011:54–66.
12. Faraut B, Boudjeltia KZ, Vanhamme L, Kerkhofs M. Immune, inflammatory and cardiovascular consequences of sleep restriction and recovery. *Sleep Med Rev*. 2012;16(2):137–149.
13. Spiegel K, Leproult R, Van Cauter E. Impact of sleep debt on metabolic and endocrine function. *Lancet*. 1999;354(9188):1435–1439.
14. Spiegel K, Sheridan JF, Van Cauter E. Effect of sleep deprivation on response to immunization. *JAMA*. 2002;288(12):1471–1472.
15. Benedict C, Dimitrov S, Marshall L, Born J. Sleep enhances serum interleukin-7 concentrations in humans. *Brain Behav Immun*. 2007;21(8):1058–1062.
16. Irwin M, McClintick J, Costlow C, Fortner M, White J, Gillin JC. Partial night sleep deprivation reduces natural killer and cellular immune responses in humans. *FASEB J*. 1996;10(5):643–653.
17. Irwin M. Effects of sleep and sleep loss on immunity and cytokines. *Brain Behav Immun*. 2002;16(5):503–512.
18. Gouin JP, Kiecolt-Glaser JK. The impact of psychological stress on wound healing: methods and mechanisms. *Immunol Allergy Clin North Am*. 2011;31(1):81–93.
19. Patel SR, Malhotra A, Gao X, Hu FB, Neuman MI, Fawzi WW. A prospective study of sleep duration and pneumonia risk in women. *Sleep*. 2012;35(1):97–101.
20. Prather AA, Janicki-Deverts D, Hall MH, Cohen S. Behaviorally assessed sleep and susceptibility to the common cold. *Sleep*. 2015;38(9):1353–1359.
21. Knutson KL, Spiegel K, Penev P, Van Cauter E. The metabolic consequences of sleep deprivation. *Sleep Med Rev*. 2007;11(3):163–178.
22. Patel M, Gomez S, Berg S, et al. Effects of 24-h and 36-h sleep deprivation on human postural control and adaptation. *Exp Brain Res*. 2008;185(2):165–173.
23. Pilcher JJ, Huffcutt AI. Effects of sleep deprivation on performance: a meta-analysis. *Sleep*. 1996;19(4):318–326.
24. Kahn-Greene ET, Killgore DB, Kamimori GH, Balkin TJ, Killgore WD. The effects of sleep deprivation on symptoms of psychopathology in healthy adults. *Sleep Med*. 2007;8(3):215–221.
25. Pisani MA, Kong SY, Kasl SV, Murphy TE, Araujo KL, Van Ness PH. Days of delirium are associated with 1-year mortality in an older intensive care unit population. *Am J Respir Crit Care Med*. 2009;180(11):1092–1097.
26. Thomason JW, Shintani A, Peterson JF, Pun BT, Jackson JC, Ely EW. Intensive care unit delirium is an independent predictor of longer hospital stay: a prospective analysis of 261 non-ventilated patients. *Crit Care*. 2005;9(4):R375–R381.
27. Hopkins RO, Spuhler VJ, Thomsen GE. Transforming ICU culture to facilitate early mobility. *Crit Care Clin*. 2007;23(1):81–96.
28. Schweickert WD, Pohlman MC, Pohlman AS, et al. Early physical and occupational therapy in mechanically ventilated, critically ill patients: a randomised controlled trial. *Lancet*. 2009;373(9678):1874–1882.
29. Engel HJ, Needham DM, Morris PE, Gropper MA. ICU early mobilization: from recommendation to implementation at three medical centers. *Crit Care Med*. 2013;41(9Suppl 1):S69–S80.

REFERENCES

30. Weinhouse GL, Schwab RJ, Watson PL, et al. Bench-to-bedside review: delirium in ICU patients—importance of sleep deprivation. *Crit Care*. 2009;13(6):234.
31. Inouye SK, Bogardus ST, Charpentier PA, et al. A multicomponent intervention to prevent delirium in hospitalized older patients. *N Engl J Med*. 1999;340(9):669–676.
32. White DP, Douglas NJ, Pickett CK, Zwillich CW, Weil JV. Sleep deprivation and the control of ventilation. *Am Rev Respir Dis*. 1983;128(6):984–986.
33. Phillips BA, Cooper KR, Burke TV. The effect of sleep loss on breathing in chronic obstructive pulmonary disease. *Chest*. 1987;91(1):29–32.
34. Roche Campo F, Drouot X, Thille AW, et al. Poor sleep quality is associated with late noninvasive ventilation failure in patients with acute hypercapnic respiratory failure. *Crit Care Med*. 2010;38(2):477–485.
35. Griffin Sr JO. *Textbook of endocrine physiology*. 5th ed New York: Oxford University Press; 2004.
36. Schussler P, Uhr M, Ising M, et al. Nocturnal ghrelin, ACTH, GH and cortisol secretion after sleep deprivation in humans. *Psychoneuroendocrinology*. 2006;31(8):915–923.
37. Lee-Chiong T. *Sleep medicine: essentials and review*. New York: Oxford University Press; 2008.
38. Sujino M, Furukawa K, Koinuma S, et al. Differential entrainment of peripheral clocks in the rat by glucocorticoid and feeding. *Endocrinology*. 2012;153(5):2277–2286.
39. Yoo SH, Yamazaki S, Lowrey PL, et al. PERIOD2::LUCIFERASE real-time reporting of circadian dynamics reveals persistent circadian oscillations in mouse peripheral tissues. *Proc Natl Acad Sci USA*. 2004;101(15):5339–5346.
40. Archer SN, Laing EE, Möller-Levet CS, et al. Mistimed sleep disrupts circadian regulation of the human transcriptome. *Proc Natl Acad Sci USA*. 2014;111(6):E682–E691.
41. Gabor JY, Cooper AB, Crombach SA, et al. Contribution of the intensive care unit environment to sleep disruption in mechanically ventilated patients and healthy subjects. *Am J Respir Crit Care Med*. 2003;167(5):708–715.
42. Topf M. Hospital noise pollution: an environmental stress model to guide research and clinical interventions. *J Adv Nurs*. 2000;31(3):520–528.
43. Gehlbach BK, Chapotot F, Leproult R, et al. Temporal disorganization of circadian rhythmicity and sleep-wake regulation in mechanically ventilated patients receiving continuous intravenous sedation. *Sleep*. 2012;35(8):1105–1114.
44. Tamburri LM, DiBrienza R, Zozula R, Redeker NS. Nocturnal care interactions with patients in critical care units. *Am J Crit Care*. 2004;13(2):102–112. quiz114-105.
45. Le A, Friese RS, Hsu CH, Wynne JL, Rhee P, O'Keeffe T. Sleep disruptions and nocturnal nursing interactions in the intensive care unit. *J Surg Res*. 2012;177(2):310–314.
46. Schwela DH. The new World Health Organization guidelines for community noise. *Noise Control Eng J*. 2001;49(4):193–198.
47. Busch-Vishniac IJ, West JE, Barnhill C, Hunter T, Orellana D, Chivukula R. Noise levels in Johns Hopkins Hospital. *J Acoust Soc Am*. 2005;118(6):3629–3645.
48. Falk SA, Woods NF. Hospital noise—levels and potential health hazards. *N Engl J Med*. 1973;289(15):774–781.
49. Snyder-Halpern R. The effect of critical care unit noise on patient sleep cycles. *CCQ*. 1985;7(4):41–51.
50. Knauert M, Jeon S, Murphy TE, Yaggi HK, Pisani MA, Redeker NS. Comparing average levels and peak occurrence of overnight sound in the medical intensive care unit on A-weighted and C-weighted decibel scales. *J Crit Care*. 2016;36:1–7.
51. Meyer TJ, Eveloff SE, Bauer MS, Schwartz WA, Hill NS, Millman RP. Adverse environmental conditions in the respiratory and medical ICU settings. *Chest*. 1994;105(4):1211–1216.
52. Akansel N, Kaymakci S. Effects of intensive care unit noise on patients: a study on coronary artery bypass graft surgery patients. *J Clin Nurs*. 2008;17(12):1581–1590.
53. Lawson N, Thompson K, Saunders G, et al. Sound intensity and noise evaluation in a critical care unit. *Am J Crit Care*. 2010;19(6):e88–e98. quize99.
54. Cordova AC, Logishetty K, Fauerbach J, Price LA, Gibson BR, Milner SM. Noise levels in a burn intensive care unit. *Burns*. 2013;39(1):44–48.
55. Xie H, Kang J. The acoustic environment of intensive care wards based on long period nocturnal measurements. *Noise Health*. 2012;14(60):230–236.

II. SLEEP DISORDERS IN SPECIFIC MEDICAL CONDITIONS

56. Park M, Vos P, Vlaskamp BN, Kohlrausch A, Oldenbeuving AW. The influence of APACHE II score on the average noise level in an intensive care unit: an observational study. *BMC Anesthesiol.* 2015;15:42.
57. Buyukyilmaz FE, Sendir M, Acaroglu R. Evaluation of night-time pain characteristics and quality of sleep in postoperative Turkish orthopedic patients. *Clin Nurs Res.* 2011;20(3):326–342.
58. Topf M, Bookman M, Arand D. Effects of critical care unit noise on the subjective quality of sleep. *J Adv Nurs.* 1996;24(3):545–551.
59. Topf M, Davis JE. Critical care unit noise and rapid eye movement (REM) sleep. *Heart Lung.* 1993;22 (3):252–258.
60. Darbyshire JL, Young JD. An investigation of sound levels on intensive care units with reference to the WHO guidelines. *Crit Care.* 2013;17(5):R187.
61. Oldham MA, Lee HB, Desan PH. Circadian rhythm disruption in the critically ill: an opportunity for improving outcomes. *Crit Care Med.* 2016;44(1):207–217.
62. Kryger MH, Roth T, Dement WC. *Principles and Practice of Sleep Medicine.* 6th ed. Philadelphia, PA: Elsevier; 2017.
63. Fan EP, Abbott SM, Reid KJ, Zee PC, Maas MB. Abnormal environmental light exposure in the intensive care environment. *J Crit Care.* 2017;40:11–14.
64. Duffy JF, Wright Jr. KP. Entrainment of the human circadian system by light. *J Biol Rhythms.* 2005;20 (4):326–338.
65. Missildine K, Bergstrom N, Meininger J, Richards K, Foreman MD. Sleep in hospitalized elders: a pilot study. *Geriatr Nurs.* 2010;31(4):263–271.
66. Walder B, Francioli D, Meyer JJ, Lancon M, Romand JA. Effects of guidelines implementation in a surgical intensive care unit to control nighttime light and noise levels. *Crit Care Med.* 2000;28(7):2242–2247.
67. Fukuda N, Kobayashi R, Kohsaka M, et al. Effects of bright light at lunchtime on sleep in patients in a geriatric hospital II. *Psychiatry Clin Neurosci.* 2001;55(3):291–293.
68. Kobayashi R, Fukuda N, Kohsaka M, et al. Effects of bright light at lunchtime on sleep of patients in a geriatric hospital I. *Psychiatry Clin Neurosci.* 2001;55(3):287–289.
69. Chong MS, Tan KT, Tay L, Wong YM, Ancoli-Israel S. Bright light therapy as part of a multicomponent management program improves sleep and functional outcomes in delirious older hospitalized adults. *Clin Interv Aging.* 2013;8:565–572.
70. Gimenez MC, Geerdinck LM, Versteylen M, et al. Patient room lighting influences on sleep, appraisal and mood in hospitalized people. *J Sleep Res.* 2017;26(2):236–246.
71. De Rui M, Middleton B, Sticca A, et al. Sleep and circadian rhythms in hospitalized patients with decompensated cirrhosis: effect of light therapy. *Neurochem Res.* 2015;40(2):284–292.
72. Ono H, Taguchi T, Kido Y, Fujino Y, Doki Y. The usefulness of bright light therapy for patients after oesophagectomy. *Intensive Crit Care Nurs.* 2011;27(3):158–166.
73. Beauchemin KM, Hays P. Dying in the dark: sunshine, gender and outcomes in myocardial infarction. *J R Soc Med.* 1998;91(7):352–354.
74. Hacker ED, Patel P, Stainthorpe M. Sleep interrupted: nocturnal care disturbances following hematopoietic stem cell transplantation. *Clin J Oncol Nurs.* 2013;17(5):517–523.
75. Celik S, Oztekin D, Akyolcu N, Issever H. Sleep disturbance: the patient care activities applied at the night shift in the intensive care unit. *J Clin Nurs.* 2005;14(1):102–106.
76. Sheely LC. Sleep disturbances in hospitalized patients with cancer. *Oncol Nurs Forum.* 1996;23(1):109–111.
77. Barr J, Fraser GL, Puntillo K, et al. Clinical practice guidelines for the management of pain, agitation, and delirium in adult patients in the intensive care unit. *Crit Care Med.* 2013;41(1):263–306.
78. Altman MT, Knauert MP, Pisani MA. Sleep disturbance after hospitalization and critical illness: a systematic review. *Ann Am Thorac Soc.* 2017;14(9):1457–1468.
79. Peppard PE, Young T, Barnet JH, Palta M, Hagen EW, Hla KM. Increased prevalence of sleep-disordered breathing in adults. *Am J Epidemiol.* 2013;177(9):1006–1014.
80. Sanner BM, Konermann M, Doberauer C, Weiss T, Zidek W. Sleep-disordered breathing in patients referred for angina evaluation—association with left ventricular dysfunction. *Clin Cardiol.* 2001;24(2):146–150.
81. Hetzenecker A, Buchner S, Greimel T, et al. Cardiac workload in patients with sleep-disordered breathing early after acute myocardial infarction. *Chest.* 2013;143(5):1294–1301.

REFERENCES

82. Schober AK, Neurath MF, Harsch IA. Prevalence of sleep apnoea in diabetic patients. *Clin Respir J.* 2011;5(3):165–172.
83. Dyken ME, Somers VK, Yamada T, Ren ZY, Zimmerman MB. Investigating the relationship between stroke and obstructive sleep apnea. *Stroke.* 1996;27(3):401–407.
84. Shahar E, Whitney CW, Redline S, et al. Sleep-disordered breathing and cardiovascular disease: cross-sectional results of the Sleep Heart Health Study. *Am J Respir Crit Care Med.* 2001;163(1):19–25.
85. Lindenauer PK, Stefan MS, Johnson KG, Priya A, Pekow PS, Rothberg MB. Prevalence, treatment, and outcomes associated with OSA among patients hospitalized with pneumonia. *Chest.* 2014;145(5):1032–1038.
86. Loube DI, Erman MK, Reed W. Perioperative complications in obstructive sleep apnea patients. *Sleep Breath.* 1997;2(1):3–10.
87. Gupta RM, Parvizi J, Hanssen AD, Gay PC. Postoperative complications in patients with obstructive sleep apnea syndrome undergoing hip or knee replacement: a case-control study. *Mayo Clin Proc.* 2001;76(9):897–905.
88. Marin JM, Carrizo SJ, Vicente E, Agusti AG. Long-term cardiovascular outcomes in men with obstructive sleep apnoea–hypopnoea with or without treatment with continuous positive airway pressure: an observational study. *Lancet.* 2005;365(9464):1046–1053.
89. Somers VK, White DP, Amin R, et al. Sleep apnea and cardiovascular disease: an American Heart Association/American College of Cardiology Foundation Scientific Statement from the American Heart Association Council for High Blood Pressure Research Professional Education Committee, Council on Clinical Cardiology, Stroke Council, and Council on Cardiovascular Nursing. *J Am Coll Cardiol.* 2008;52(8):686–717.
90. Bitter T, Faber L, Hering D, Langer C, Horstkotte D, Oldenburg O. Sleep-disordered breathing in heart failure with normal left ventricular ejection fraction. *Eur J Heart Fail.* 2009;11(6):602–608.
91. Peker Y, Kraiczi H, Hedner J, Loth S, Johansson A, Bende M. An independent association between obstructive sleep apnoea and coronary artery disease. *Eur Respir J.* 1999;14(1):179–184.
92. Bradley TD, Floras JS. Obstructive sleep apnoea and its cardiovascular consequences. *Lancet.* 2009;373(9657):82–93.
93. Kasai T, Floras JS, Bradley TD. Sleep apnea and cardiovascular disease: a bidirectional relationship. *Circulation.* 2012;126(12):1495–1510.
94. Yumino D, Wang H, Floras JS, et al. Prevalence and physiological predictors of sleep apnea in patients with heart failure and systolic dysfunction. *J Card Fail.* 2009;15(4):279–285.
95. Lopez-Acevedo MN, Torres-Palacios A, Elena Ocasio-Tascon M, Campos-Santiago Z, Rodriguez-Cintron W. Overlap syndrome: an indication for sleep studies?: A pilot study. *Sleep Breath.* 2009;13(4):409–413.
96. Cormick W, Olson LG, Hensley MJ, Saunders NA. Nocturnal hypoxaemia and quality of sleep in patients with chronic obstructive lung disease. *Thorax.* 1986;41(11):846–854.
97. Chaouat A, Weitzenblum E, Krieger J, Oswald M, Kessler R. Pulmonary hemodynamics in the obstructive sleep apnea syndrome. Results in 220 consecutive patients. *Chest.* 1996;109(2):380–386.
98. Chaouat A, Weitzenblum E, Krieger J, et al. Prognostic value of lung function and pulmonary haemodynamics in OSA patients treated with CPAP. *Eur Respir J.* 1999;13(5):1091–1096.
99. Dziewas R, Humpert M, Hopmann B, et al. Increased prevalence of sleep apnea in patients with recurring ischemic stroke compared with first stroke victims. *J Neurol.* 2005;252(11):1394–1398.
100. Cherkassky T, Oksenberg A, Froom P, Ring H. Sleep-related breathing disorders and rehabilitation outcome of stroke patients: a prospective study. *Am J Phys Med Rehabil.* 2003;82(6):452–455.
101. Bassetti CL, Milanova M, Gugger M. Sleep-disordered breathing and acute ischemic stroke: diagnosis, risk factors, treatment, evolution, and long-term clinical outcome. *Stroke.* 2006;37(4):967–972.
102. Barczi S, Juergens T. Comorbidities: psychiatric, medical, medications, and substances. *Sleep Med Clinic.* 2006;231–245.
103. Lautenbacher S, Kundermann B, Krieg JC. Sleep deprivation and pain perception. *Sleep Med Rev.* 2006;10(5):357–369.
104. Raymond I, Ancoli-Israel S, Choinière M. Sleep disturbances, pain and analgesia in adults hospitalized for burn injuries. *Sleep Med.* 2004;5(6):551–559.
105. Novaes MA, Aronovich A, Ferraz MB, Knobel E. Stressors in ICU: patients' evaluation. *Intensive Care Med.* 1997;23(12):1282–1285.

II. SLEEP DISORDERS IN SPECIFIC MEDICAL CONDITIONS

106. Little A, Ethier C, Ayas N, Thanachayanont T, Jiang D, Mehta S. A patient survey of sleep quality in the Intensive Care Unit. *Minerva Anestesiol.* 2012;78(4):406−414.
107. Weinhouse GL, Schwab RJ. Sleep in the critically ill patient. *Sleep.* 2006;29(5):707−716.
108. Mundigler G, Delle-Karth G, Koreny M, et al. Impaired circadian rhythm of melatonin secretion in sedated critically ill patients with severe sepsis. *Crit Care Med.* 2002;30(3):536−540.
109. Parthasarathy S, Tobin MJ. Sleep in the intensive care unit. *Intensive Care Med.* 2004;30(2):197−206.
110. Friese RS, Bruns B, Sinton CM. Sleep deprivation after septic insult increases mortality independent of age. *J Trauma.* 2009;66(1):50−54.
111. Closs SJ. Patients' night-time pain, analgesic provision and sleep after surgery. *Int J Nurs Stud.* 1992;29(4):381−392.
112. Cronin AJ, Keifer JC, Davies MF, King TS, Bixler EO. Postoperative sleep disturbance: influences of opioids and pain in humans. *Sleep.* 2001;24(1):39−44.
113. Gay PC. Sleep and sleep-disordered breathing in the hospitalized patient. *Respir Care.* 2010;55(9):1240−1254.
114. Knill RL, Moote CA, Skinner MI, Rose EA. Anesthesia with abdominal surgery leads to intense REM sleep during the first postoperative week. *Anesthesiology.* 1990;73(1):52−61.
115. Rosenberg J, Wildschiødtz G, Pedersen MH, von Jessen F, Kehlet H. Late postoperative nocturnal episodic hypoxaemia and associated sleep pattern. *Br J Anaesth.* 1994;72(2):145−150.
116. Rosenberg J, Rasmussen GI, Wøjdemann KR, Kirkeby LT, Jørgensen LN, Kehlet H. Ventilatory pattern and associated episodic hypoxaemia in the late postoperative period in the general surgical ward. *Anaesthesia.* 1999;54(4):323−328.
117. Roussos M, Parthasarathy S, Ayas NT. Can we improve sleep quality by changing the way we ventilate patients? *Lung.* 2009;188(1):1−3.
118. Parthasarathy S, Tobin MJ. Effect of ventilator mode on sleep quality in critically ill patients. *Am J Respir Crit Care Med.* 2002;166(11):1423−1429.
119. Valente M, Placidi F, Oliveira AJ, et al. Sleep organization pattern as a prognostic marker at the subacute stage of post-traumatic coma. *Clin Neurophysiol.* 2002;113(11):1798−1805.
120. Edwards GB, Schuring LM. Pilot study: validating staff nurses' observations of sleep and wake states among critically ill patients, using polysomnography. *Am J Crit Care.* 1993;2(2):125−131.
121. Bosma K, Ferreyra G, Ambrogio C, et al. Patient-ventilator interaction and sleep in mechanically ventilated patients: pressure support versus proportional assist ventilation. *Crit Care Med.* 2007;35(4):1048−1054.
122. Toublanc B, Rose D, Glerant JC, et al. Assist-control ventilation vs. low levels of pressure support ventilation on sleep quality in intubated ICU patients. *Intensive Care Med.* 2007;33(7):1148−1154.
123. Aurell J, Elmqvist D. Sleep in the surgical intensive care unit: continuous polygraphic recording of sleep in nine patients receiving postoperative care. *Br Med J (Clin Res Ed).* 1985;290(6474):1029−1032.
124. Ozsancak A, D'Ambrosio C, Garpestad E, Schumaker G, Hill NS. Sleep and mechanical ventilation. *Crit Care Clin.* 2008;24(3):517−531. vi-vii.
125. Alexopoulou C, Kondili E, Vakouti E, Klimathianaki M, Prinianakis G, Georgopoulos D. Sleep during proportional-assist ventilation with load-adjustable gain factors in critically ill patients. *Intensive Care Med.* 2007;33(7):1139−1147.
126. Schweitzer PK. Drugs that disturb sleep and wakefulness. In: Kryger MH, Roth T, Dement WC, eds. *Principles and Practice of Sleep Medicine.* Philadelphia, PA: Elsevier/Saunders; 2005:499−515.
127. Bourne RS, Mills GH. Sleep disruption in critically ill patients—pharmacological considerations. *Anaesthesia.* 2004;59(4):374−384.
128. Achermann P, Borbely AA. Dynamics of EEG slow wave activity during physiological sleep and after administration of benzodiazepine hypnotics. *Hum Neurobiol.* 1987;6(3):203−210.
129. Borbely AA, Mattmann P, Loepfe M, Strauch I, Lehmann D. Effect of benzodiazepine hypnotics on all-night sleep EEG spectra. *Hum Neurobiol.* 1985;4(3):189−194.
130. Holbrook AM, Crowther R, Lotter A, Cheng C, King D. Meta-analysis of benzodiazepine use in the treatment of insomnia. *CMAJ.* 2000;162(2):225−233.
131. Buscemi N, Vandermeer B, Friesen C, et al. The efficacy and safety of drug treatments for chronic insomnia in adults: a meta-analysis of RCTs. *J Gen Intern Med.* 2007;22(9):1335−1350.
132. Kay DC, Eisenstein RB, Jasinski DR. Morphine effects on human REM state, waking state and NREM sleep. *Psychopharmacologia.* 1969;14(5):404−416.

REFERENCES

435

133. Dimsdale JE, Norman D, DeJardin D, Wallace MS. The effect of opioids on sleep architecture. *J Clin Sleep Med.* 2007;3(1):33–36.

134. Wang D, Teichtahl H. Opioids, sleep architecture and sleep-disordered breathing. *Sleep Med Rev.* 2007;11 (1):35–46.

135. Kondili E, Alexopoulou C, Xirouchaki N, Georgopoulos D. Effects of propofol on sleep quality in mechanically ventilated critically ill patients: a physiological study. *Intensive Care Med.* 2012;38(10):1640–1646.

136. Morin CM, Koetter U, Bastien C, Ware JC, Wooten V. Valerian-hops combination and diphenhydramine for treating insomnia: a randomized placebo-controlled clinical trial. *Sleep.* 2005;28(11):1465–1471.

137. Buysse D. Clinical pharmacology of other drugs used as hypnotics. In: Kryger M, Roth T, Dement W, eds. *Principles and Practice of Sleep Medicine.* 5th ed Philadelphia, PA: Saunders/Elsevier; 2011:492–509.

138. Unseld E, Ziegler G, Gemeinhardt A, Janssen U, Klotz U. Possible interaction of fluoroquinolones with the benzodiazepine-GABAA-receptor complex. *Br J Clin Pharmacol.* 1990;30(1):63–70.

139. Watson PL. Measuring sleep in critically ill patients: beware the pitfalls. *Crit Care.* 2007;11(4):159.

140. Iber C, Ancoli-Israel S, Chesson A, Quan SF. *The AASM Manual for the Scoring of Sleep and Associated Events: Rules, Terminology, and Technical Specification.* 1st ed Westchester, IL: American Academy of Sleep Medicine; 2007.

141. Watson PL, Pandharipande P, Gehlbach BK, et al. Atypical sleep in ventilated patients: empirical electroencephalography findings and the path toward revised ICU sleep scoring criteria. *Crit Care Med.* 2013;41 (8):1958–1967.

142. Bridoux A, Thille AW, Quentin S, et al. Sleep in ICU: atypical sleep or atypical electroencephalography? *Crit Care Med.* 2014;42(4):e312–e313.

143. Ancoli-Israel S, Clopton P, Klauber MR, Fell R, Mason W. Use of wrist activity for monitoring sleep/wake in demented nursing-home patients. *Sleep.* 1997;20(1):24–27.

144. Kushida CA, Chang A, Gadkary C, Guilleminault C, Carrillo O, Dement WC. Comparison of actigraphic, polysomnographic, and subjective assessment of sleep parameters in sleep-disordered patients. *Sleep Med.* 2001;2(5):389–396.

145. Van de Water AT, Holmes A, Hurley DA. Objective measurements of sleep for non-laboratory settings as alternatives to polysomnography—a systematic review. *J Sleep Res.* 2011;20(1 Pt 2):183–200.

146. Marino M, Li Y, Rueschman MN, et al. Measuring sleep: accuracy, sensitivity, and specificity of wrist actigraphy compared to polysomnography. *Sleep.* 2013;36(11):1747–1755.

147. Mantua J, Gravel N, Spencer RM. Reliability of sleep measures from four personal health monitoring devices compared to research-based actigraphy and polysomnography. *Sensors (Basel).* 2016;16(5):1–11.

148. Montgomery-Downs HE, Insana SP, Bond JA. Movement toward a novel activity monitoring device. *Sleep Breath.* 2012;16(3):913–917.

149. Redeker NS, Wykpisz E. Effects of age on activity patterns after coronary artery bypass surgery. *Heart Lung.* 1999;28(1):5–14.

150. Beecroft JM, Ward M, Younes M, Crombach S, Smith O, Hanly PJ. Sleep monitoring in the intensive care unit: comparison of nurse assessment, actigraphy and polysomnography. *Intensive Care Med.* 2008;34 (11):2076–2083.

151. Kamdar BB, Kadden DJ, Vangala S, et al. Feasibility of continuous actigraphy in medical intensive care unit patients. *Am J Crit Care.* 2017;26(4):329–335.

152. Nicolas A, Aizpitarte E, Iruarrizaga A, Vazquez M, Margall A, Asiain C. Perception of night-time sleep by surgical patients in an intensive care unit. *Nurs Crit Care.* 2008;13(1):25–33.

153. Richards KC, O'Sullivan PS, Phillips RL. Measurement of sleep in critically ill patients. *J Nurs Meas.* 2000;8 (2):131–144.

154. Kamdar BB, King LM, Collop NA, et al. The effect of a quality improvement intervention on perceived sleep quality and cognition in a medical ICU. *Crit Care Med.* 41(3), 2013, 800–809.

155. Shahid A. *STOP, THAT and One Hundred Other Sleep Scales.* New York: Springer; 2012.

156. Snyderhalpern R, Verran JA. Instrumentation to describe subjective sleep characteristics in healthy-subjects. *Res Nurs Health.* 1987;10(3):155–163.

157. Ellis BW, Johns MW, Lancaster R, Raptopoulos P, Angelopoulos N, Priest RG. The St. Mary's Hospital sleep questionnaire: a study of reliability. *Sleep.* 1981;4(1):93–97.

II. SLEEP DISORDERS IN SPECIFIC MEDICAL CONDITIONS

436 19. SLEEP IN HOSPITALIZED PATIENTS

158. Hoey LM, Fulbrook P, Douglas JA. Sleep assessment of hospitalised patients: a literature review. *Int J Nurs Stud*. 2014;51(9):1281−1288.
159. Hoddes E, Zarcone V, Smythe H, Phillips R, Dement WC. Quantification of sleepiness: a new approach. *Psychophysiology*. 1973;10(4):431−436.
160. Herscovitch J, Broughton R. Sensitivity of the Stanford sleepiness scale to the effects of cumulative partial sleep deprivation and recovery oversleeping. *Sleep*. 1981;4(1):83−91.
161. Kaida K, Takahashi M, Akerstedt T, et al. Validation of the Karolinska sleepiness scale against performance and EEG variables. *Clin Neurophysiol*. 2006;117(7):1574−1581.
162. Kamdar BB, Knauert MP, Jones SF, et al. Perceptions and practices regarding sleep in the intensive care unit. A survey of 1,223 critical care providers. *Ann Am Thorac Soc*. 2016;13(8):1370−1377.
163. Ding Q, Redeker NS, Pisani MA, Yaggi HK, Knauert MP. Factors influencing patients' sleep in the intensive care unit: perceptions of patients and clinical staff. *Am J Crit Care*. 2017;26(4):278−286.
164. Grossman MN, Anderson SL, Worku A, et al. Awakenings? Patient and hospital staff perceptions of night-time disruptions and their effect on patient sleep. *J Clin Sleep Med*. 2017;13(2):301−306.
165. Jones SF, Pisani MA. ICU delirium: an update. *Curr Opin Crit Care*. 18(2), 2012, 146−151.
166. Kamdar BB, Kamdar BB, Needham DM. Bundling sleep promotion with delirium prevention: ready for prime time? *Anaesthesia*. 2014;69(6):527−531.
167. Hu RF, Jiang XY, Chen J, et al. Non-pharmacological interventions for sleep promotion in the intensive care unit. *Cochrane Database Syst Rev*. 2015;(10)CD008808.
168. Gathecha E, Rios R, Buenaver LF, Landis R, Howell E, Wright S. Pilot study aiming to support sleep quality and duration during hospitalizations. *J Hosp Med*. 2016;11(7):467−472.
169. Olson DM, Borel CO, Laskowitz DT, Moore DT, McConnell ES. Quiet time: a nursing intervention to promote sleep in neurocritical care units. *Am J Crit Care*. 2001;10(2):74−78.
170. Kamdar BB, King LM, Collop NA, et al. The effect of a quality improvement intervention on perceived sleep quality and cognition in a medical ICU. *Crit Care Med*. 2013;41(3):800−809.
171. Richardson A, Allsop M, Coghill E, Turnock C. Earplugs and eye masks: do they improve critical care patients' sleep? *Nurs Crit Care*. 2007;12(6):278−286.
172. Farrehi PM, Clore KR, Scott JR, Vanini G, Clauw DJ. Efficacy of sleep tool education during hospitalization: a randomized controlled trial. *Am J Med*. 2016;129(12):1329.e9−1329.e17.
173. Adachi M, Staisiunas PG, Knutson KL, Beveridge C, Meltzer DO, Arora VM. Perceived control and sleep in hospitalized older adults: a sound hypothesis? *J Hosp Med*. 2013;8(4):184−190.
174. Zimmerman L, Nieveen J, Barnason S, Schmaderer M. The effects of music interventions on postoperative pain and sleep in coronary artery bypass graft (CABG) patients. *Sch Inq Nurs Pract*. 1996;10(2):153−170. discussion171-154.
175. Richards KC. Effect of a back massage and relaxation intervention on sleep in critically ill patients. *Am J Crit Care*. 1998;7(4):288−299.
176. Chen JH, Chao YH, Lu SF, Shiung TF, Chao YF. The effectiveness of valerian acupressure on the sleep of ICU patients: a randomized clinical trial. *Int J Nurs Stud*. 2012;49(8):913−920.
177. Moeini M, Khadibi M, Bekhradi R, Mahmoudian SA, Nazari F. Effect of aromatherapy on the quality of sleep in ischemic heart disease patients hospitalized in intensive care units of heart hospitals of the Isfahan University of Medical Sciences. *Iran J Nurs Midwifery Res*. 2010;15(4):234−239.
178. Bartick MC, Thai X, Schmidt T, Altaye A, Solet JM. Decrease in as-needed sedative use by limiting nighttime sleep disruptions from hospital staff. *J Hosp Med*. 2010;5(3):E20−E24.
179. Frighetto L, Marra C, Bandali S, Wilbur K, Naumann T, Jewesson P. An assessment of quality of sleep and the use of drugs with sedating properties in hospitalized adult patients. *Health Qual Life Outcomes*. 2004;2:17.
180. Gillis CM, Poyant JO, Degrado JR, Ye L, Anger KE, Owens RL. Inpatient pharmacological sleep aid utilization is common at a tertiary medical center. *J Hosp Med*. 2014;9(10):652−657.
181. Liu Y, Xu X, Dong M, Jia S, Wei Y. Treatment of insomnia with tricyclic antidepressants: a meta-analysis of polysomnographic randomized controlled trials. *Sleep Med*. 2017;34:126−133.
182. Agostini JV, Zhang Y, Inouye SK. Use of a computer-based reminder to improve sedative−hypnotic prescribing in older hospitalized patients. *J Am Geriatr Soc*. 2007;55(1):43−48.

II. SLEEP DISORDERS IN SPECIFIC MEDICAL CONDITIONS

REFERENCES

183. Tamblyn R, Abrahamowicz M, du Berger R, McLeod P, Bartlett G. A 5-year prospective assessment of the risk associated with individual benzodiazepines and doses in new elderly users. *J Am Geriatr Soc.* 2005;53 (2):233−241.
184. Glass J, Lanctot KL, Herrmann N, Sproule BA, Busto UE. Sedative hypnotics in older people with insomnia: meta-analysis of risks and benefits. *BMJ.* 2005;331(7526):1169.
185. Zammit GK, Corser B, Doghramji K, et al. Sleep and residual sedation after administration of zaleplon, zolpidem, and placebo during experimental middle-of-the-night awakening. *J Clin Sleep Med.* 2006;2(4):417−423.
186. Kolla BP, Lovely JK, Mansukhani MP, Morgenthaler TI. Zolpidem is independently associated with increased risk of inpatient falls. *J Hosp Med.* 2013;8(1):1−6.
187. Claustrat B, Brun J, Chazot G. The basic physiology and pathophysiology of melatonin. *Sleep Med Rev.* 2005;9 (1):11−24.
188. Reiter RJ, Tan DX, Fuentes-Broto L. Melatonin: a multitasking molecule. *Prog Brain Res.* 2010;181:127−151.
189. Neubauer DN. A review of ramelteon in the treatment of sleep disorders. *Neuropsychiatr Dis Treat.* 2008;4 (1):69−79.
190. Auld F, Maschauer EL, Morrison I, Skene DJ, Riha RL. Evidence for the efficacy of melatonin in the treatment of primary adult sleep disorders. *Sleep Med Rev.* 2017;34:10−22.
191. Bellapart J, Boots R. Potential use of melatonin in sleep and delirium in the critically ill. *Br J Anaesth.* 108(4), 2012, 572−580.
192. Shilo L, Dagan Y, Smorjik Y, et al. Effect of melatonin on sleep quality of COPD intensive care patients: a pilot study. *Chronobiol Int.* 2000;17(1):71−76.
193. Ibrahim MG, Bellomo R, Hart GK, et al. A double-blind placebo-controlled randomised pilot study of nocturnal melatonin in tracheostomised patients. *Crit Care Resusc.* 2006;8(3):187−191.
194. Bourne RS, Mills GH, Minelli C. Melatonin therapy to improve nocturnal sleep in critically ill patients: encouraging results from a small randomised controlled trial. *Crit Care.* 2008;12(2):R52.
195. Huang H, Jiang L, Shen L, et al. Impact of oral melatonin on critically ill adult patients with ICU sleep deprivation: study protocol for a randomized controlled trial. *Trials.* 2014;15:327.
196. Hatta K, Kishi Y, Wada K, et al. Preventive effects of ramelteon on delirium: a randomized placebo-controlled trial. *JAMA Psychiatry.* 2014;71(4):397−403.
197. Knauert M, Redeker NS, Yaggi HK, Bennick M, Pisani M. Creating naptime: an overnight, nonpharmacologic intensive care unit sleep promotion protocol. *J Patient Exp.* 2018;5(3):180−187.
198. Arora VM, Chang KL, Fazal AZ, et al. Objective sleep duration and quality in hospitalized older adults: associations with blood pressure and mood. *J Am Geriatr Soc.* 2011;59(11):2185−2186.
199. Pisani MA, Friese RS, Gehlbach BK, Schwab RJ, Weinhouse GL, Jones SF. Sleep in the intensive care unit. *Am J Respir Crit Care Med.* 2015;191(7):731−738.
200. Zaubler TS, Murphy K, Rizzuto L, et al. Quality improvement and cost savings with multicomponent delirium interventions: replication of the Hospital Elder Life Program in a community hospital. *Psychosomatics.* 2013;54(3):219−226.

Index

Note: Page numbers followed by "*f*", "*t*", and "*b*" refer to figures, tables, and boxes, respectively.

A

AASM. *See* American Academy of Sleep Medicine (AASM)
Abatacept, 318
Aβ. *See* Amyloid beta (Aβ)
Abrupt awakening, 11, 226
Acetaminophen, 36
Acetylcholin, 260–261
Acid neutralization, 382
ACR. *See* American College of Rheumatology (ACR)
Actigraphy, 17–19, 165–166, 187, 209–210, 242, 401, 422–423
   actigraphic technologically derived data, 312
   actigraphy-based studies, 280–281
Acupuncture for insomnia, 37–38
Acute HIV infection, 296
Acute phase after TBI, sleep–wake alterations in, 224–225
Acute RBD, 90
AD. *See* Alzheimer's disease (AD)
Adaptive servo-ventilation (ASV), 55*t*
   for CSA due to opioid use, 65
   for CSA treatment, 64
   for OSA treatment, 57
   for treatment-emergent central sleep apnea, 66
Adenotonsillar hypertrophy, 297
Adherence, poor, 295
Adherence monitoring, 58
Adipose tissue, 183–184
Adolescent Sleep Habits Scale (ASHS), 209
Adolescent Sleep–Wake Scale (ASWS), 209
Adolescents, 202
   insufficient sleep consequences in, 392–399
Adults, 202, 207
   sleep disorders and sleep deficiency, 105–107
   survivors of childhood cancer, 181–182
Advanced sleep phase type disorder, 14
Advanced sleep–wake phase disorder (ASWPD), 85–86
Aggravated sleep disturbances in older adults, 261–262
Aging, 176

   cognitive, 254–258
   normal cognitive aging process, 254
   sleep and circadian rhythm changes with, 258–262
      aggravated sleep disturbances in older adults, 261–262
      mechanisms underlying sleep and cognition, 260–261
      structure of sleep predict cognitive function in older people, 260
Agonists, 34
   receptor agonists, 380–381
AHI. *See* Apnea–hypopnea index (AHI)
AHI_{flow}, 58, 60
AIDS, 297
Alcohol ingestion, acute, 62
Alcoholic liver disease, 378
Allodynia, 327–328
α-2-adrenergic agonists, 380–381
Alpha-2δ ligand, 81, 356
Alpha-EEG, 330
Alpha–delta sleep, 330
Alprazolam, 380–381
ALS. *See* Amyotrophic lateral sclerosis (ALS)
Alveolar hypoventilation, 51–52, 139–140
Alzheimer's disease (AD), 11, 254, 257
American Academy of Sleep Medicine (AASM), 82–83, 356
   Clinical Practice Guideline, 349–350, 355–356
   publication, 5
American College of Rheumatology (ACR), 327
American Psychiatric Association, 5
Aminosalicylates, 382
Amitriptyline, 35, 336–337
Amobarbital, 380–381
Amphetamine, 79
Amyloid beta (Aβ), 263
Amyotrophic lateral sclerosis (ALS), 69, 138
Androgen-deprivation therapy, 177
Anecdotal evidence, 381
Anemia, 105
Ankylosing spondylitis (AS), 311
Anosognosia, 231

ANS. *See* Autonomic nervous system (ANS)
Anti-IL-6 antibodies, 310–311
Anti-TNF
  anti-TNF-α
    antibodies, 310–311
    therapy, 317–318
  therapies, 314
Antibiotics agents, 382–383
Anticoagulants, 283
Anticonvulsants, 36, 380–381
Anticyclic citrullinated peptide autoantibody, 310
Antidepressants, 35, 352
Antihistamines, 79, 353, 419–421
  diphenhydramine, 426
Antihypertensives, 283, 419
Antiinflammatory agents, 382–383
Antinausea medications, 79
Antiplatelets, 283
Antipsychotics, 35–36, 233, 380–381, 419
Antispasmodic agents, 382
Antiviral agents, 382–383
Anxiety, 228, 333. *See also* Depression
APAP mode. *See* Autotitrating continuous positive
    airway pressure mode (APAP mode)
Apigenin, 36–37
Apnea–hypopnea index (AHI), 19, 58, 60, 127–128,
    156, 179–180, 239
Apolipoprotein E (ApoE), 257
Appendicular skeleton joints, 310
Appetite, 157–158
Aquaporin-4 (AQP4), 357–358
ARIRANG study, 163
Armodafinil, 82–83
Arousal, 13
  confusional, 88
Arthritis, 211, 309–310
AS. *See* Ankylosing spondylitis (AS)
ASHS. *See* Adolescent Sleep Habits Scale (ASHS)
Association of Sleep Medicine, 301
Asthma, 122
  and sleep disorders, 130–132
    asthma–OSA overlap syndrome, 132
    nocturnal asthma, 130–131, 131t
    sleep quality, 132
ASV. *See* Adaptive servo-ventilation (ASV)
ASWPD. *See* Advanced sleep–wake phase disorder
    (ASWPD)
ASWS. *See* Adolescent Sleep–Wake Scale (ASWS)
Asymptomatic HIV infection. *See* Chronic HIV
    infection
Atlantoaxial involvement in region, 310
Atrial fibrillation, 101
Augmentation, 81, 356

Autogenic training, 301
Autonomic nervous system (ANS), 372–373
  activity during sleep, 331–332
Autosomal dominant cerebellar ataxia, deafness, and
    narcolepsy, 12
Autosomal dominant narcolepsy, obesity, and type 2
    diabetes, 12
"Autosomal reduced penetrance" pattern, 15
Autosuggestion, 301
Autotitrating continuous positive airway pressure
    mode (APAP mode), 52–53
  for OSA treatment, 60
Average VAPS (AVAPS), 54–57
Axial skeleton, 310
Axial spondyloarthropathy (AxSpA), 310–311

B
Barbiturates, 380–381
BDI-II. *See* Beck Depression Inventory (BDI-II)
BEARS sleep screening tool, 400
Beck Depression Inventory (BDI-II), 354
Bedtime routines, 403t
Behavioral
  assessment for sleep–wake disorder, 16–20
  behavioral/nonpharmacological interventions,
    401–402
  factors, 102, 222, 232
  interventions, 237
    for disturbed sleep in dementia, 264–265
    in pediatric sleep, 403t
  management, 385
  modification for OSA, 61–62
  risk factor, 257–258
  sleep conditions, 394t
Behavioral Risk Factor Surveillance System (BRFSS),
    156
Benzodiazepine receptor agonists (BzRAs), 33–34, 233,
    427
Benzodiazepines, 33, 129, 233, 266, 380–381, 386b,
    419–421
Berlin Questionnaire, 18t, 312, 314–315
Berlin treatment algorithm, 357
Bilevel PAP Timed mode (BPAP-T mode), 53–54
Bilevel PAP-Spontaneous mode (BPAP-S mode), 53–54
  for obesity hypoventilation syndrome, 67
  for OSA treatment, 60–61
Bilevel PAP-Spontaneous/Timed mode (BPAP-ST
    mode), 53–54
  for CCHS, 68
  for CSA due to opioid use, 65
  for CSA treatment, 64
  for obesity hypoventilation syndrome, 67
  for treatment-emergent central sleep apnea, 66

INDEX

441

Bilevel positive airway pressure mode (BPAP mode), 51–52, 55*t*
  advanced, 54–57
  basic, 53–54
Biologic therapies and sleep in inflammatory arthritis, 317–318
BMI. *See* Body mass index (BMI)
Body mass index (BMI), 133, 139–140, 154
BPAP mode. *See* Bilevel positive airway pressure mode (BPAP mode)
BPAP-S mode. *See* Bilevel PAP-Spontaneous mode (BPAP-S mode)
BPAP-ST mode. *See* Bilevel PAP-Spontaneous/Timed mode (BPAP-ST mode)
BPAP-T mode. *See* Bilevel PAP Timed mode (BPAP-T mode)
Brain plasticity, 281–282
Brain–gut axis, 373–374
Brain–gut connection, 372–373
Brainstem lesions, 352
Breast cancer, 176
Breathing devices, 210
Breathing-related disorders. *See also* Sleep-disordered breathing (SDB)
  CSA treatment, 63–66
  OSA treatment, 59–63
  PAP devices, 52–58
  sleep disorders, 5, 6*t*, 13
  sleep-related hypoventilation disorder treatment, 66–70
    CCHS, 68–69
    obesity hypoventilation syndrome, 67–68
BRFSS. *See* Behavioral Risk Factor Surveillance System (BRFSS)
Bright light therapy
  for cancer, 191
  for insomnia, 37
  for TBI patients, 241
BZD receptor agonists. *See* Benzodiazepine receptor agonists (BzRAs)
BzRAs. *See* Benzodiazepine receptor agonists (BzRAs)

C

C-reactive protein (CRP), 111, 310–311
C-related peptides, 161–162
Caffeine reduction, 301
Canadian Haida populations, 311
Canadian Problem Checklist (CPC), 186
Cancer, 176, 189, 203–204
  areas for future research, 193
  cancer-related insomnia, 189
  clinical vignette, 192–193
  medical condition, 176–177

sleep difficulties
  assessment, 186–187
  in special populations, 181–182
  treatment, 188–191
sleep disorders
  consequences, 184–186
  epidemiology, 177–181
  etiology, 182–184
sleep disturbance assessment in patients with, 194*b*
Carbidopa, 80
Cardiometabolic
  disease risk factors, 166
  health, 154
Cardiovascular disease (CVD), 100, 129–130, 263, 277–278
  areas for future research, 112–113
  case study/vignette, 113
  clinical approaches targeting sleep to reducing CVD risk, 110–111
  consequences of sleep deficiency and sleep disorders, 107–108
  contributions of sleep deficiency to development comorbid sleep disorders, 105
    insomnia and sleep quality, 102–104
    OSA, 104
    RLS, 104–105
    sleep duration, 100–102
  interventions to improving sleep impairments in people with, 109–112
  outcomes in people with chronic, 110–111
  sleep disorders and sleep deficiency among adults, 105–107
  sleep disturbance assessment in people with, 108–109
  treatment of sleep-disordered breathing, 111–112
Caregiver-reported sleep characteristics, 263–264, 267
Cataplexy, 83
Caucasians, 257
CBT. *See* Cognitive-behavioral therapy (CBT)
CCHS. *See* Congenital central alveolar hypoventilation syndrome (CCHS)
CD4 + lymphocytes, 295
CD8 + lymphocytes, 295
CDC. *See* Centers for Disease Control and Prevention (CDC)
Cellular restorative processes, 263
Centers for Disease Control and Prevention (CDC), 328
Central emotional motor system, 383–384
Central nervous system (CNS), 83, 372–373
Central respiratory depression, 140–141
Central sleep apnea (CSA), 6*t*, 13, 51–52, 105–107, 346
  comorbid with opioid use. *See* Arousal
  due to opioid use, 64–65

442 INDEX

Central sleep apnea (CSA) (*Continued*)
  primary or idiopathic, 65
  treatment, 63–66
    approach to patient care, 64
    CSA–CSB, 63–64
    indications, 63
    type of devices, 64
  treatment-emergent, 66
Cerebral
  atrophy, 223
  health, 257
  recovery, 224
Cerebrospinal fluid (CSF), 12, 206, 257, 351
Cerebrovascular accidents (CVA), 417
CF. *See* Cystic fibrosis (CF)
CFS. *See* Chronic fatigue syndrome (CFS)
Chamomile, 36–37
CHANGE-HIV System, 301
CHC. *See* Chronic health condition (CHC)
CHD. *See* Coronary heart disease (CHD)
Chemo-brain, 177
Chemotherapy, 176–177, 183. *See also* Cancer
Chest wall disorders, 69–70
Cheyne–Stokes Breathing (CSB), 5–10, 13, 63, 106, 417
  CSA with, 63–64
CHF. *See* Congestive heart failure (CHF)
Children
  evaluation of sleep, 207, 230
  insufficient sleep consequences in, 392–399
  psychosocial problem effect, 230
  sleep deficiency in, 202
  sleep management with chronic health condition
    (CHC), 404–405
Children's Sleep Habits Questionnaire (CSHQ), 209
Chronic fatigue syndrome (CFS), 326–328, 373–374
  areas for future research, 337
  assessment and treatment of sleep
    disorders/problems in medical condition, 333
  consequences of sleep disorders and abnormalities,
    332–333
  diagnosis, 326–327
  etiology, 327–328
  prevalence, 327
  sleep abnormalities in patients without sleep
    disorders, 328–332
  sleep disorders in, 328
Chronic health condition (CHC), 391, 393
  sleep in, 399–405
Chronic HIV infection, 296–297
Chronic obstructive pulmonary disease (COPD), 122,
  411–412
  sleep disorders in, 123–130
    insomnia and, 128–129

restless legs syndrome, 129–130
sleep-related hypoxemia and hypoventilation in,
  124–125, 125f
Chronic obstructive pulmonary disease–obstructive
  sleep apnea overlap syndrome (COPD–OSA
  overlap syndrome), 125–128
  treatment of patients with COPD and, 127–128
Chronic pain, 201–202, 206–207
  sleep deficiency in chronic pain conditions, 202–204
Chronic phase
  functioning in, 229
  insomnia in chronic phase of stroke, 284–285
  sleep architecture in chronic phase after TBI,
    225–226
Chronobiotic agents, 34
Chronotherapy
  in advanced sleep–wake phase disorder (ASWPD),
    86
  in delayed sleep–wake phase disorder (DSWPD), 84
Chronotherapy, 403t
Cigarette smoking, 310
Ciprofloxacin, 382
Circadian disruptions, 191, 414
Circadian pacemaker, 87
Circadian rhythm sleep–wake disorders (CRSWD), 5,
  6t, 14, 223, 242–243
  areas for future research, 243
  assessment of circadian rhythm disorders, 242
  pharmacological and nonpharmacological treatment,
    242–243
Circadian rhythms, 285, 414, 416
  changes with normal aging, 258–262
  disorders, 242
    delayed sleep–wake phase disorder (DSWPD),
      84–85
  disruption, 374
Clarithromycin, 83–84
Classic "rheumatoid factor", 310
Clinical latency period of HIV infection. *See* Chronic
  HIV infection
Clinical particularities, 356
Clinical vignette
  cancer, 192–193
  insomnia disorder, 231
  multiple sclerosis, 357–358
  sleep and gastrointestinal disorders, 384–385
  sleep in hospitalized patients, 428–429
Clonazepam, 91, 380–381
Clonidine, 91, 380–381
Cluster analysis, 299–300
CNS. *See* Central nervous system (CNS)
Cognition, 260–261
Cognitive aging, 254–258

normal cognitive aging process, 254
prevalence and symptoms of MCI and dementia, 254–257
Cognitive factors, 206
Cognitive functioning, 228
Cognitive processes, 277–278
Cognitive therapy, 32, 212
Cognitive-behavioral therapy (CBT), 29, 92, 188–190, 210, 234, 235t, 265, 267, 300, 313–314, 335–336, 349, 383–384
  for fatigue and physical exercise, 355
Cognitive-behavioral therapy for insomnia (CBT-I), 30–33, 110, 129, 188, 210–211, 282–283, 285–286
  clinical practice recommendations, 39t
  effect sizes from metaanalyses for insomnia, 42f
  indications and rationale, 33
  multifaceted cognitive-behavioral therapy, 33
  outcome evidence, 38–45
    areas for future research, 45
    clinical and practical considerations, 44–45
    combined cognitive-behavioral therapy and medication, 43
    short-term and long-term outcomes, 43–44
    treatment delivery models, 44
    impact of treatment on nighttime and daytime symptoms, 38–43
    relaxation-based interventions and mindfulness, 31–32
    sleep hygiene education, 32
    sleep restriction, 30–31
    stimulus control therapy, 31
Colonic motility and sleep, 373
Combined cognitive-behavioral therapy and medication, 43
Comorbid, 10–11, 21
  psychiatric disorders, 128–129
  sleep disorders, 105
  in ILD, 136
Complex sleep apnea syndrome. See Treatment-emergent central sleep apnea
Confidence interval (CI), 102–103
Confusion, 83
Confusional arousals, 88
Congenital central alveolar hypoventilation syndrome (CCHS), 66–69
Congestive heart failure (CHF), 417
Consensus sleep diary, 187
Constipation-predominant IBS (IBSC), 377
Constraint-induced movement therapy, 281
Continuous positive airway pressure (CPAP), 52–53, 55t, 111–112, 191, 241, 264, 352–353, 357, 384
  for CCHS, 68

for CSA due to opioid use, 65
for CSA treatment, 64
effect on inflammation and pain in inflammatory arthritis, 318–319
for obesity hypoventilation syndrome, 67
for OSA treatment, 59–60, 63
for sleep apnea, 241
therapy, 127–128, 141, 347–348
for treatment-emergent central sleep apnea, 66
COPD. See Chronic obstructive pulmonary disease (COPD)
Core body temperature nadir, 84–85
Coronary heart disease (CHD), 100, 102
Corticosteroids, 314, 382
Cortisol, 414
CPAP. See Continuous positive airway pressure (CPAP)
CPC. See Canadian Problem Checklist (CPC)
Crohn's disease, 376
CRP. See C-reactive protein (CRP)
CRSWD. See Circadian rhythm sleep–wake disorders (CRSWD)
CSA. See Central sleep apnea (CSA)
CSB. See Cheyne–Stokes Breathing (CSB)
CSF. See Cerebrospinal fluid (CSF)
CSHQ. See Children's Sleep Habits Questionnaire (CSHQ)
CVA. See Cerebrovascular accidents (CVA)
CVD. See Cardiovascular disease (CVD)
Cyclopyrrolones, 380–381
Cyclosporine, 382
Cystic fibrosis (CF), 122, 134
  sleep in, 132–134
    prevalence, predictors, and consequences, 132–134
    treatment, 134
Cytokines, 317, 376

D

Daily self-monitoring, 17
Date rape drug, 83
Daytime sleepiness, 181, 281, 299. See also Excessive daytime sleepiness (EDS)
DBAS-16. See Dysfunctional Beliefs About Sleep Questionnaire-16 (DBAS-16)
Decongestant medications, 79
Delayed sleep phase type disorder, 6t, 14
Delayed sleep–wake phase disorder (DSWPD), 84–85
Delirium, 413–414
Delta pressure (ΔP), 53
Dementia, 254, 258–262
  areas for future research, 267–268
  genetic, physiological, and behavioral risk factors, 257–258

Dementia (*Continued*)
  with Lewy bodies, 15–16
  prevalence and symptoms, 254–257
  sleep disordered breathing in, 262–264
  treatment of sleep disorders, 264–267
Depression, 102, 123, 163, 206–207, 228, 258, 281–282, 298–299, 333, 352–353
  sleep disorders relationship with, 352–353
Depressive disorders, 11
Detrended fluctuation analysis method, 332
Dextroamphetamine, 79, 83
Diabetes, 104, 166. *See also* Obesity
  sleep and, 158–160
    diagnosis, 158, 159*t*
    etiology, 159
    future directions, 166–167
    insomnia, 160
    insufficient sleep, 160
    OSA, 160
    public health relevance, 159
Diagnostic and Statistical Manual of Mental Disorders (DSM), 10–11
Diagnostic and Statistical Manual of Mental Disorders, Fifth Edition (DSM-5), 3–5, 6*t*, 12, 254
  DSM-5/ICSD-3 systems, 21
Diaphragm pacing, 53–54
Diarrhea-predominant IBS (IBS-D), 377
Dicyclomine, 382
Diffusion tensor imaging fractional anisotropy, 351
Digestive system, 372–373
Diphenhydramine, 79, 349–350
Disability, sleep disorders relationship with, 352–353
Disease course, sleep disorders relationship with, 352–353
Disrupted/disturbed sleep, 313–314
  architecture, 358
  sleep in CFS patients, 327
Distress thermometer, 186
Dizziness, 83
DM. *See* Myotonic dystrophy (DM)
Dopamine, 80, 206, 260–261
Dopaminergic
  drugs, 356
  medications, 80
Doxepin, 35, 233, 336–337
Dream-enacting episodes, 352
Dry mouth, 82–83
DSM. *See* Diagnostic and Statistical Manual of Mental Disorders (DSM)
DSWPD. *See* Delayed sleep–wake phase disorder (DSWPD)
Duke Structured Interview for Sleep Disorders, 10–11, 187
Dysbiosis, 373–374

Dysfunctional Beliefs About Sleep Questionnaire-16 (DBAS-16), 18*t*
Dyslipidemia, 163, 165
Dysrhythmias, 100, 104
Dyssomnias, 5–10

E

Early birds. *See* Morning lark
ED. *See* Emergency department (ED)
Edmonton Symptom Assessment System (ESAS), 186
EDS. *See* Excessive daytime sleepiness (EDS)
Effectiveness monitoring, 58
Electroencephalography (EEG), 3–5, 226, 418
Electronic medical record systems (EMR systems), 108
Elevated sound levels, 415
Elevated triglycerides, 161
Elimination diet, 383
Emergency department (ED), 223
EMR systems. *See* Electronic medical record systems (EMR systems)
ENS. *See* Enteric nervous system (ENS)
Enteric nervous system (ENS), 372–373
Enterococcus spp., 373–374
Enuresis, 83
Environmental factors, 222, 225
  affecting sleep in hospitalized patients, 415–417
    elevated sound levels, 415
    noncircadian light patterns, 416
    patient care interactions, 416–417
EPAP. *See* Expiratory positive airway pressure (EPAP)
Epworth Sleepiness Scale (ESS), 18*t*, 165–166, 233, 241, 312–313, 333, 354
ESAS. *See* Edmonton Symptom Assessment System (ESAS)
ESS. *See* Epworth Sleepiness Scale (ESS)
Estazolam, 266, 427
Eszopiclone, 336–337, 427
Etanercept, 314, 317–318
EURLSSG. *See* European RLS Study Group (EURLSSG)
European guideline for the diagnosis and treatment of insomnia, 355–356
European RLS Study Group (EURLSSG), 355–356
Excessive daytime sleepiness (EDS), 82, 132–133, 222–223, 238–242, 351, 400
  in context of TBI, 240
  nonpharmacological treatment for, 241–242
  pharmacological treatment for, 240–241
Exercise, 61, 190–191, 334–335
  CBT for, 355
Exertion, 334
Expiratory positive airway pressure (EPAP), 52
Expiratory volume in 1 second (FEV1), 123
Extinction, 403*t*
Extra therapy mode features for comfort, 58

INDEX

445

F

Factor analysis, 299–300
Faded bedtime, 403*t*
Fatigue, 134, 139, 184, 203–204, 222, 281, 296, 298–299
  CBT for, 355
  sleep disorders relationship with, 352–353
  sleep disturbance impacts on, 332–333
  treatment, 357
Fatigue Severity Scale (FSS), 347–348
FD. *See* Functional dyspepsia (FD)
FDA. *See* Food and Drug Administration (FDA)
Fermentable oligosaccharides, disaccharides,
    monosaccharides, and polyols (FODMAPs), 383
FEV1. *See* Expiratory volume in 1 second (FEV1)
FGID. *See* Functional GI disorders (FGID)
Fibromyalgia (FM), 204, 211, 326–328
  areas for future research, 337
  assessment and treatment of sleep disorders/
    problems, 333
  consequences of sleep disorders and abnormalities,
    332–333
  diagnosis, 326–327
  etiology, 327–328
  prevalence, 327
  sleep abnormalities in patients without sleep
    disorders, 328–332
  sleep disorders in, 328
Fitbit, 17–19
  wearable fitness trackers, 17–19
Flu-like infection, 296
Fluid intelligence, 254
Flumazenil, 83–84
Flurazepam, 266, 427
FM. *See* Fibromyalgia (FM)
FODMAPs. *See* Fermentable oligosaccharides,
    disaccharides, monosaccharides, and polyols
    (FODMAPs)
Food and Drug Administration (FDA), 404, 425
Food intake, 157–158
Fractal scaling exponent, 332
Framingham Risk Scores, 101
FSS. *See* Fatigue Severity Scale (FSS)
Fukuda Case Definition, 326–327
Functional dyspepsia (FD), 374–375
  and sleep, 374–375
Functional GI disorders (FGID), 372, 374*t*

G

GABA. *See* Gamma-aminobutyric acid (GABA)
GABAA receptor. *See* Gamma-aminobutyric acid type
    A receptor (GABAA receptor)
Gabapentin, 80, 233, 356, 380–381
Gabapentin enacarbil, 80

Gamma-aminobutyric acid (GABA), 34, 266
  receptor, 83–84
    GABA$_A$ receptor, 83–84, 419–421
Gamma-aminobutyric acid type A receptor (GABAA
    receptor), 34
Garmin, 17–19
Gastric motility and sleep, 373
Gastroesophageal reflux disorder (GERD), 123, 126,
    132, 372, 413*t*
  comorbid, 131
  in ILD, 136
  and sleep, 375
Gastrointestinal (GI)
  disorders, 372
    epidemiology and risk factors of sleep
      disturbances, 374–379
    management and prevention of sleep
      disturbances, 380–384
    sleep disorders assessment, 379–380
  motility agents, 382
  symptom management medications effects on sleep,
    382–383
    acid neutralization, 382
    antiinflammatory, antibiotics, and antiviral agents,
      382–383
    gastrointestinal motility agents, 382
  symptoms, 372
  tract, 372–374
    chronic GI inflammation, 373
    immunity and inflammation in, 373
    sleep-enhancing agents effects on, 380–381
GCS. *See* Glasgow Coma Scale (GCS)
Genetic factors, 257–258, 267
GERD. *See* Gastroesophageal reflux disorder (GERD)
Ghrelin, 158
Glasgow Coma Scale (GCS), 223
Glasses, 85
Glymphatic system, 261
Grading of Recommendations Assessment,
    Development, and Evaluation approach
    (GRADE approach), 356
Graduated extinction, 403*t*
Guided imagery technique, 301
Gut as second brain, 372–373
Gut microbiota
  effects, 373–374
  manipulation, 385
Gut–brain axis, 372–373

H

H$_2$-receptor antagonists, 382
HAART. *See* Highly active antiretroviral therapy
    (HAART)

INDEX

Habitual short sleep duration, 163
Haloperidol, 380–381
Hamilton Depression Inventory, 178–179
Hazard ratio (HR), 102–103
HbA1C. *See* Hemoglobin A1c (HbA1C)
HBV. *See* Hepatitis B virus (HBV)
HCRT. *See* Hypocretin (HCRT)
HCV. *See* Hepatitis C virus (HCV)
HDL. *See* High-density lipoprotein (HDL)
Head cancer, 179
Headache, 82–83, 204
Health assessment questionnaire disability index, 316
Health-related QoL (HRQoL), 353
Healthy sleep habits, 109–110
Heart failure (HF), 59, 100, 102
Heart rate variability (HRV), 331
Hemiparesis or constraint-induced aphasia therapy, 281
Hemoglobin A1c (HbA1C), 158
Hemorrhagic stroke, 278–279
Hepatic inflammation, 378–379
Hepatitis B virus (HBV), 378
Hepatitis C virus (HCV), 378
Hering–Breuer inflation reflex, 59–60
HF. *See* Heart failure (HF)
HF with preserved ejection fraction (HFpEF), 63
HF with reduced ejection fraction (HFrEF), 63
High vitamin D serum levels, 350
High-density lipoprotein (HDL), 161
High-sensitivity C-reactive protein (hsCRP), 101, 103
Highly active antiretroviral therapy (HAART), 294
HIV infection, 294
    correlates of sleep quality in, 298–299
    sleep disturbances and, 294–295
    stages, 296–297
    symptom clusters in, 299–300
HLA. *See* Human leucocyte antigen (HLA)
HLA DRB1*15:01, 351–352
Hormone therapy, 176–177, 183
Hormone-dependent cancers, 176–177
Hospital(ization), 222, 412
    hospitalized patients
        environmental factors affecting sleep
            in, 415–417
        interrelationship between illness, sleep, delirium,
            414*f*
        protocols to improving sleep in hospitalized
            patients, 423–425
        sleep disruption in, 412–414
        sleep measurement in, 421–423
    medications to treat sleep disruption in, 425–428
    poor hospital sleep, 417–418
    risk factors for poor sleep in hospital settings,
        418–421
        mechanical ventilation, 419

medications, 419–421
pain, 418
postsurgical, 418–419
severity of illness and sepsis, 418
stakeholders, 426*t*
HR. *See* Hazard ratio (HR)
HRQoL. *See* Health-related QoL (HRQoL)
HRV. *See* Heart rate variability (HRV)
hsCRP. *See* High-sensitivity C-reactive protein (hsCRP)
Human leucocyte antigen (HLA), 346
Hydroxyzine, 426
Hyoscyamine, 382
Hyperalgesia, 327–328
Hypercapnia, 122–123, 136
Hypercholesterolemia, 163
Hypersomnia, 5, 6*t*, 11, 82–84, 243, 297, 393. *See also*
    Insomnia
    central disorders of, 5
    IH, 82
    MS, 351–352
    nonorganic, 5–10
    posttraumatic, 238–242
    treatment, 82–84
Hypersomnolence disorder. *See* Hypersomnia
Hypertension, 100–102, 104–105, 166, 263
    insomnia and MetS components, 163–164
    OSA and MetS components, 165
    sleep duration and MetS components, 162
Hypertension, chronic, 263
Hypnotherapy, 383–384
Hypnotic(s), 188
    hypnotic–mortality association, 267
    use due to insomnia and relationship with fatigue,
        353–354
Hypocretin (HCRT), 351
    deficiency syndrome. *See* Narcolepsy type 1 (NT1)
    HCRT-1 and 2, 351
Hypoglossal nerve stimulation, 51–52, 62
Hypothalamic–pituitary–adrenal axis functions,
    260–261
Hypothalamus, 351
Hypoxemia, 122–123, 136
    acute, 140–141
Hypoxemic disorders, 51–52

I

IBD. *See* Inflammatory bowel disease (IBD)
IBS. *See* Irritable bowel syndrome (IBS)
IBS-D. *See* Diarrhea-predominant IBS (IBS-D)
IBS-M. *See* Irritable bowel syndrome with mixed bowel
    habits (IBS-M)
IBSC. *See* Constipation-predominant IBS (IBSC)
ICD-10. *See* International Classification of Diseases,
    Tenth Edition (ICD-10)

INDEX

ICSD. *See* International Classification of Sleep
    Disorders (ICSD)
ICU. *See* Intensive care unit (ICU)
Idiopathic CSA, 13
Idiopathic hypersomnia (IH), 82
Idiopathic pulmonary fibrosis (IPF), 136
Idiopathic RBD, 352
IFN. *See* Interferon (IFN)
IH. *See* Idiopathic hypersomnia (IH)
IL. *See* Interleukin (IL)
ILD. *See* Interstitial lung disease (ILD)
Illness severity, 418
Image rehearsal therapy (IRT), 91
Imidazopyridines, 380–381
Immune system, 373
Immunity in gastrointestinal tract and sleep, 373
In-home titration, 60
Inflammatory arthritis
    biologic therapies and sleep in, 317–318
    effect of CPAP on inflammation and pain in,
        318–319
    nature and prevalence of sleep disturbance/
        disorders in, 311–315
        insomnia and rheumatic diseases, 313–314
        OSA, 314–315
        RLS, 315
        sleep quality, 312–313
    polysomnographic studies in, 316
Inflammatory bowel disease (IBD), 372, 376
    and sleep, 376
Inflammatory/inflammation, 206
    cells, 310–311
    cytokines, 103
        molecules, 310–311
    in gastrointestinal tract and sleep, 373
    hepatic, 378–379
    markers, 161–162, 206
Injury severity, 231
Insomnia, 5, 6t, 105, 154
    acute, 284–285
    cancer, 177–179, 182
        consequences of sleep disorders, 184–185
        etiology of sleep disorders, 182–183
    and cardiovascular disease, 100, 107, 109
    chronic, 102, 128, 346
        stroke, 284–285
    chronic obstructive pulmonary disease and
        prevalence, predictors, and assessment, 128–129
        treatment, 129
    and components of metabolic syndrome, 163–164
        hypertension, 163–164
        insomnia and lipid disorders, 164
        insulin resistance, 164

conceptual model, 29
and diabetes, 160
diagnostic considerations, 28–29
disorder, 5, 10
epidemiology, 28
interview schedule, 178, 187
mood disturbance, 206–207
multiple sclerosis, 348–350
nonpharmacological treatment for, 300–302
and obesity, 155–156
pharmacological treatment for, 302
posttraumatic brain injury insomnia
    assessment, 232–233
    nonpharmacological interventions for, 234–237
    pharmacological interventions for, 233
preexisting, 417
prevalence, 296
and rheumatic diseases, 313–314
risk factors
    behavioral factors, 232
    injury severity, 231
    medications, 231–232
and sickle cell disease, 204
and sleep quality, 102–104
and stage of HIV infection, 296–297
in stroke, 279
syndrome, 178–179
in traumatic brain injury, 231–237
treatment, 29–38, 92, 188–191
    acupuncture, 37–38
    bright light therapy, 37, 191
    cognitive-behavioral therapy, 30–33, 188–190
    exercise, 190–191
    mindfulness-based stress reduction, 190
    natural products, 36–37
    pharmacotherapy, 33–36, 188
Insomnia Severity Index (ISI), 18t, 108, 186–187, 209,
    232, 333, 354
Inspiratory positive airway pressure (IPAP), 52
Institute of Medicine, 326–327
Insufficient sleep
    consequences in children and adolescents, 392–399
        pediatric sleep problems, 393–398
        sleep in CHC, 399
    and diabetes, 160
    and obesity, 155
Insulin resistance, 162–164
    syndrome, 161
Intensive care unit (ICU), 223–224, 412, 423–424
    sleep promotion interventions, 424–425
Interferon (IFN), 381
Interleukin (IL), 376
    IL-1, 103

448   INDEX

Interleukin (IL) (*Continued*)
   IL-6, 103, 161−162, 310−311, 317−318
Internal desynchronization, 414
International Association for Study of Pain, 201−202
International Classification of Diseases, Tenth Edition
   (ICD-10), 3−10, 6*t*, 21
International Classification of Sleep Disorders (ICSD),
   10−11
   ICSD-2, 51−52
   ICSD-3, 5, 6*t*, 348, 350
International Restless Legs Syndrome Study Group
   (IRLSSG), 312, 350
Internet-delivered CBT-I, 44
Interstitial lung disease (ILD), 134
   sleep in, 134−136
      comorbid sleep disorders, 136
      sleep architectural disturbances, 135
      sleep-related respiratory changes in, 134−135
Intestinal motility and sleep, 373
IPAP. *See* Inspiratory positive airway pressure (IPAP)
IPF. *See* Idiopathic pulmonary fibrosis (IPF)
IRLSSG. *See* International Restless Legs Syndrome
   Study Group (IRLSSG)
Irregular sleep−wake rhythm disorder (ISWRD), 87
Irregular sleep−wake type disorder, 6*t*, 14
Irritable bowel syndrome (IBS), 372, 377
   and sleep, 377−378
Irritable bowel syndrome with mixed bowel habits
   (IBS-M), 377
IRT. *See* Image rehearsal therapy (IRT)
Ischemic strokes, 278−279
ISI. *See* Insomnia Severity Index (ISI)
ISWRD. *See* Irregular sleep−wake rhythm disorder
   (ISWRD)

J

J receptors, 59−60
Jawbone, 17−19
Jenkins Sleep Scale, 335
Juvenile idiopathic arthritis (JIA), 203

K

Karolinska Sleepiness Scale, 423
Kleine−Levin syndrome, 84

L

Laboratory assessment for sleep−wake disorder,
   16−20
Lachnospiraceae, 374
LDL. *See* Low-density lipoprotein (LDL)
Leptin, 158
LES. *See* Lower esophageal sphincter (LES)

Levodopa, 80
Lifestyle-based sleep interventions, 166
Light, 37, 412, 416
   sleep, 258−259
   therapy, 84−85, 285
Lipid disorders, insomnia and, 164
Liver cirrhosis, 378
Liver disease and sleep, chronic, 378−379
Locus coeruleus, 227
Long sleep duration, 102
Long-term outcomes in CBT-I, 43−44
Longitudinal care of patients on PAP, 57−58
Loperamide, 382
Lorazepam, 425
Low-density lipoprotein (LDL), 161
Lower esophageal sphincter (LES), 375
Lung cancer, 176
Lung diseases
   asthma and sleep disorders, 130−132
   OHS, 139−141
   respiratory physiology during normal sleep,
      122−123
   sleep
      in CF, 132−134
      in chronic respiratory failure due to
         neuromuscular disease, 137−139
      disorders in COPD, 123−130
      in ILD, 134−136
      after lung transplantation, 143
      in musculoskeletal disorders, 136−137
      in PH, 141−143

M

Macrostructure of sleep, 328−330
Magnetic resonance imaging (MRI), 257, 348, 351
Magnetoencephalography, 228
Maintenance of wakefulness test (MWT), 16, 20
Major depression, 352−353
Maladaptive sleep behaviors, 183
Mask fitting, 57−58
MBCR intervention. *See* Mindfulness-based cancer
   recovery intervention (MBCR intervention)
MBSR. *See* Mindfulness-based stress reduction (MBSR)
MCI. *See* Mild cognitive impairment (MCI)
Medical ICU (MICU), 428
Medical Outcome Study Sleep Scale (MOS-SS),
   312−313, 318
Medical sleep conditions, 394*t*
Medications, 210, 231−232, 259−260, 282−283, 301,
   419−421
   effects on sleep, 420*t*
   patient, environmental, and care factors, 422*f*
   to treat sleep disruption in hospital, 425−428

INDEX

449

Melatonin, 79, 85–86, 91, 265, 302, 336–337, 350, 380–381, 427
  deficiency, 350
  exogenous, 265, 381
  levels, 184–185
Memory consolidation, 283
Mesopontine tegmentum, 227
Metabolic syndrome (MetS), 154–155
  future directions, 166–167
  insomnia and components, 163–164
  OSA and components, 164–165
  sleep and, 161–165
    diagnosis, 161
    etiology, 161–162
    public health relevance, 162
  sleep duration and components
    hypertension, 162
    insulin resistance, 162–163
    sleep duration and other unhealthy behaviors, 163
Methamphetamine, 83
Methotrexate, 314
Methylphenidate, 79, 83
Metoclopramide, 79
Metronidazole, 382
MetS. *See* Metabolic syndrome (MetS)
Micro-longitudinal studies, 205
Microstructure of sleep, 330
MICU. *See* Medical ICU (MICU)
Migrating motor complex (MMC), 373
Mild cognitive impairment (MCI), 254, 258–262
  areas for future research, 267–268
  genetic, physiological, and behavioral risk factors, 257–258
  prevalence and symptoms, 254–257
  sleep disordered breathing in, 262–264
  treatment of sleep disorders, 264–267
Mindfulness, 31–32
Mindfulness-based cancer recovery intervention (MBCR intervention), 190
Mindfulness-based stress reduction (MBSR), 190
Mini–mental state examination (MMSE), 260
Mirtazapine, 35, 79, 349–350, 381, 425–426
MMC. *See* Migrating motor complex (MMC)
MMSE. *See* Mini–mental state examination (MMSE)
Modafinil, 82–83
MOG. *See* Myelin oligodendrocyte glycoprotein (MOG)
Montreal Cognitive Assessment, 228
Mood, 206–207
  disturbances, 412–413
Morning headache, 210
Morning lark, 14
Morningness–Eveningness Questionnaire, 18*t*, 242

MOS-SS. *See* Medical Outcome Study Sleep Scale (MOS-SS)
Movement-related sleep disorders, 16
MRI. *See* Magnetic resonance imaging (MRI)
MS. *See* Multiple sclerosis (MS)
MSA. *See* Multiple system atrophy (MSA)
MSLT. *See* Multiple sleep latency test (MSLT)
MT1 receptor, 34
MT2 receptor, 34
Multifaceted cognitive-behavioral therapy, 33
Multiple sclerosis (MS), 346–347
  clinical vignette, 358
  differential diagnosis, 355
  fatigue and employment status, 354
  future research directions, 358–359
  hypnotic use due to insomnia and relationship with fatigue, 353–354
  impact and quality of life, 353
  insomnia, 347–348
  MS-related fatigue, 347
  narcolepsy and hypersomnia, 351–352
  periodic limb movement disorder, 350–351
  polygraphic recordings and polysomnography, 354–355
  questionnaires, 354
  RBDs, 352
  RLS, 350–351
  sleep disorders in, 347
    assessment, 354–355
    and fatigue in NMOSDs, 357–358
    relationship, 352–353
    treatment, 355–357
  sleep-related breathing disorders, 347–348
Multiple sleep latency test (MSLT), 11, 20, 133, 238, 351
Multiple system atrophy (MSA), 11, 15–16, 138
Muscarinic receptors, 36
Musculoskeletal conditions, 202
Musculoskeletal disorders, sleep in
  description, predictors, and consequences, 136–137
  treatment, 136–137
MWT. *See* Maintenance of wakefulness test (MWT)
Myelin oligodendrocyte glycoprotein (MOG), 358
Myotonic dystrophy (DM), 137–138
  DM1, 137–138
  DM2, 138

N

N1 sleep. *See* "Stage 1" sleep
N2 sleep. *See* "Stage 2" sleep
N24SWD. *See* Non-24-hour sleep–wake disorder (N24SWD)
N3 sleep. *See* "Stage 3" sleep
NAFLD. *See* Nonalcoholic fatty liver disease (NAFLD)

Naps, 241–242
Naptime, 428–429
Narcolepsy, 5–12, 6t, 20–21, 82–83, 227, 332
  without cataplexy but with hypocretin deficiency, 12
  with cataplexy but without hypocretin deficiency, 12
  MS and, 351–352
  narcoleptic features, 239
  narcoleptic triad, 91–92
  secondary to another medical condition, 12
Narcolepsy type 1 (NT1), 346, 351
Narcotics, 419–421
Nasal EPAP devices, 63
Nasal masks, 57–58
Nasal pillows/cushions, 57–58
National Health and Nutrition Examination Survey (NHANES), 103–104, 155
National Health Interview Survey (NHIS), 159
National Healthy Sleep Awareness Program, 109
Natural products, 36–37
Nausea, 82–83
NCD. *See* Neurocognitive disorders (NCD)
NDs. *See* Neurodevelopmental diagnoses (NDs)
Near-infrared light, 79
Neck cancer, 179
Negative mood, 333
Neoplasm, 257
Neurobiological factor, 222
Neurocognitive disorders (NCD), 254, 255t, 262–263
Neurodegeneration processes, 223
Neurodegenerative diseases, 138–139
Neurodevelopmental diagnoses (NDs), 393
Neurogenesis, 260–261
Neuroimaging techniques, 257
Neuromodulation, 62
Neuromuscular diseases, 69–70
  sleep in chronic respiratory failure due to, 137–139
  myotonic dystrophy, 137–138
  neurodegenerative diseases, 138–139
Neuromyelitis optica spectrum disorder (NMOSD), 346–347
  sleep disorders and fatigue in, 357–358
Neuroplasticity, 283
Neuropsychological tests, 228
Neurorehabilitation, 283
Neurotransmitter, 351
NHANES. *See* National Health and Nutrition Examination Survey (NHANES)
NHIS. *See* National Health Interview Survey (NHIS)
NHP. *See* Nottingham Health Profile (NHP)
Night owls, 14
Nightmare disorder, 5, 6t, 15–16, 91
Nighttime sleep, poor, 205

NIPPV. *See* Noninvasive positive-pressure ventilation (NIPPV)
NMOSD. *See* Neuromyelitis optica spectrum disorder (NMOSD)
Nocturnal
  asthma, 130–131, 131t
  dyspnea, 128–129
  hypercapnia, 133
  hypoxemia, 133, 135
  PSG, 11
Noise, 412, 415
Non-24-hour sleep–wake disorder (N24SWD), 6t, 14, 86–87
Non-BzRA. *See* Nonbenzodiazepines (Non-BzRA)
Non-REM. *See* Nonrapid eye movement (NREM)—sleep
Nonalcoholic fatty liver disease (NAFLD), 378–379
Nonamphetamine stimulants, 79
Nonbenzodiazepines (Non-BzRA), 267
  medications for insomnia, 233
  receptor agonists, 427
Noncircadian light patterns, 416
Noninvasive positive-pressure ventilation (NIPPV), 128, 134, 137, 139
Noninvasive ventilation, 135, 137, 210
Nonorganic
  hypersomnia, 5–10
  insomnia, 5–10
  sleep disorders, 5–10
Nonpharmacologic(al)
  hospital-based sleep promotion protocols, 424–425
  interventions, 222, 334–336, 357, 383–384
    CBT, 335–336
    complementary and alternative medicine, 384
    CPAP, 384
    current treatment, 336–337
    elimination diet, 383
    exercise, 334–335
    lifestyle modifications, 383
    for posttraumatic brain injury insomnia, 234–237
    psychotherapy, 383–384
  sleep promotion protocols, 428–429
  treatment for excessive daytime sleepiness
    bright light therapy, 241
    continuous positive airway pressure for sleep apnea, 241
    naps, 241–242
  treatment for insomnia, 300–302
  treatment of circadian disturbances, 242–243
Nonrapid eye movement (NREM), 122–123, 203, 225–226, 330, 332, 373
  parasomnias, 88–90
    behavior and pharmacological treatment, 90t

INDEX   **451**

confusional arousals, 88
  sleep terrors, 89
  somnambulism, 88–89
  SREDs, 89–90
sleep, 3–5, 16
  arousal disorders, 5, 6*t*
Nonvolitional activity, 102
Noradrenergic neurons, 227
Norepinephrine, 260–261
Normal sleep, respiratory physiology during, 122–123
Nottingham Health Profile (NHP), 353
NREM. *See* Nonrapid eye movement (NREM)
NT1. *See* Narcolepsy type 1 (NT1)
Nurse practitioners, 57

O

OA. *See* Osteoarthritis (OA)
OAT. *See* Oral appliance therapy (OAT)
Obese patients with COPD/OSA overlap syndrome, 126
Obesity, 104, 112, 183–184, 203. *See also* Diabetes
Obesity hypoventilation syndrome (OHS), 67–68, 122, 139–141, 140*t*, 157
Obstructive sleep apnea (OSA), 5, 13, 51–52, 104–105, 136, 154, 166, 177, 179–180, 183–186, 228, 262, 279, 295, 297, 312, 314–315, 375, 378–379, 401–402, 411–412
  and components of MetS, 164–165
    dyslipidemia, 165
    hypertension, 165
  and diabetes, 160
  hypopnea, 6*t*
  and obesity, 156–157
  treatment
    approach to patient care, 63
    behavioral modification, 61–62
    device types, 59
    indications, 59–63
    OAT, 61
    other therapies, 62–63
Obstructive sleep apnea syndrome (OSAS), 346, 348, 353
Octylonium bromide, 382
Off-label use of pharmacotherapy for insomnia, 35–36
OHS. *See* Obesity hypoventilation syndrome (OHS)
Olanzapine, 426
Older adults, aggravated sleep disturbances in, 261–262
Opioids, 51–52, 64, 181
  opiate medications, 80
Oral appliance therapy (OAT), 61, 264
Oral pressure therapy (OPT), 63

Orexin, 351
  receptor antagonist, 35
OSA. *See* Obstructive sleep apnea (OSA)
OSAS. *See* Obstructive sleep apnea syndrome (OSAS)
Osteoarthritis (OA), 203
Over-the-counter
  hypnotic use, 353
  medications, 79
  sleep aids, 36
Overlap syndrome, 124
Overweight patients with COPD/OSA overlap syndrome, 126
Overweight–obesity, sleep and, 154–158
  diagnosis, 154
  etiology, 154
  future directions, 166–167
  insomnia, 155–156
  insufficient sleep, 155
  obesity classification among adults, 154*t*
  OSA, 156–157
  public health relevance, 155
  sleep, appetite, and food intake, 157–158
Oxygen saturation (SpO$_2$), 122–123

P

p-tau. *See* Phosphorylated-tau (p-tau)
PaCO$_2$. *See* Partial pressure of carbon dioxide (PaCO$_2$)
Pain, 203, 309–310, 418
  impacts on, 332–333
  intensity, 205
  sleep and, 205–207
Palliative care, 181
Pannus formation, 310–311
PaO$_2$. *See* Partial pressure of oxygen (PaO$_2$)
PAP devices. *See* Positive airway pressure devices (PAP devices)
Parasomnias, 5–10, 6*t*, 14–15, 88–92, 223
  NREM, 88–90
  rapid eye movement, 90–92
Paresis, 240
Parkinson's disease, 11, 15–16, 138, 285
Partial pressure of carbon dioxide (PaCO$_2$), 122–123, 135
Partial pressure of oxygen (PaO$_2$), 122–123
Pathophysiological mechanisms, 372–374
  brain–gut connection, 372–373
  dysbiosis, 373–374
  gastric, intestinal, and colonic motility and sleep, 373
  immunity and inflammation in GI tract and sleep, 373
Patient care interactions, 416–417
Patient-Reported Outcomes Measurement Information System, 379–380

Peak expiratory flow rate (PEFR), 130–131, 131f
Pediatric
  conditions and associated SRP, 395t
  sleep patterns, 392
    sleep duration by age, 392t
  sleep problems, 393–398
PEFR. *See* Peak expiratory flow rate (PEFR)
People living with HIV/AIDS (PLWHA), 294, 298–299
Peptic ulcer disease (PUD), 375–376
  and sleep, 375–376
Periodic limb movement (PLM), 351
Periodic limb movement disorder (PLMD), 16, 81, 353
  exacerbating factors, 78–79
  MS and, 350–351
  prevalence, 78
  treatment, 79–81, 81t
Periodic limb movements during sleep (PLMS), 81,
    104–107, 136, 180–181, 184, 186, 232
Perpetuating factors, 29, 183, 282
Persistent hypoventilation during sleep, 136
Persistent insomnia, 28
PFTs. *See* Pulmonary function tests (PFTs)
PG. *See* Polygraphy (PG)
PH. *See* Pulmonary hypertension (PH)
Pharmacological interventions, 222, 265–267, 313,
    333–334, 355–356, 404
  for posttraumatic brain injury insomnia, 233
Pharmacological treatment
  of circadian disturbances, 242–243
  for excessive daytime sleepiness, 240–241
  for insomnia, 302
Pharmacotherapy, 188
  GI symptom management medications effects on
      sleep, 382–383
    acid neutralization, 382
    antiinflammatory, antibiotics, and antiviral agents,
      382–383
    gastrointestinal motility agents, 382
  for insomnia, 33–36
    BZD receptor agonists, 34
    chronobiotic agents, 34
    clinical practice recommendations, 39t
    off-label use of pharmacotherapy, 35–36
    orexin receptor antagonist, 35
    over-the-counter sleep aids, 36
  sleep-enhancing agents effects on GI tract, 380–381
Phase advance, 403t
Phase delay, 403t
Phosphorylated-tau (p-tau), 263
Phrenic nerve stimulation, 64
"Physical abilities" category, 353
Physical therapy, 210
Physician assistants, 57

Physiological factor, 257–258
Pittsburgh Sleep Quality Index (PSQI), 18t, 165–166,
    177, 180–182, 190–191, 209, 232, 296, 312,
    333–334, 354, 374–375
PLM. *See* Periodic limb movement (PLM)
PLMD. *See* Periodic limb movement disorder (PLMD)
PLMS. *See* Periodic limb movements during sleep
    (PLMS)
PLWHA. *See* People living with HIV/AIDS (PLWHA)
Pneumatic compression, 79
Polydipsia, 158
Polygraphic recordings, MS, 354–355
Polygraphy (PG), 347–348
Polyphagia, 158
Polysomnography (PSG), 16, 19, 127, 165–166, 177,
    187, 202, 210, 225–226, 260, 347, 373, 393, 412,
    422
  MS and, 354–355
  Polysomnogram studies, 314
  sleep test, 3–5
  studies, 328–329
    in inflammatory arthritis, 316
  technologically derived data, 312
Polyuria, 158
Population-based approaches, 109–110
Positive airway pressure devices (PAP devices), 52–58
  advanced BPAP modes, 54–57
  basic BPAP modes, 53–54
  CPAP mode, 52–53
  download interpretation, 58
  longitudinal care of patients, 57–58
  mode waveforms, 53f
  PAP therapy, 127–128
  profiles, 55t
Post-traumatic stress disorder (PTSD), 91, 229–230
Poststroke insomnia, 280
Postsurgical patients, 418–419
Posttraumatic amnesia, 223–224, 228
Posttraumatic brain injury
  fatigue, 230–231
  insomnia
    assessment, 232–233
    pharmacological interventions for, 233
  psychopathology, 229–230
Posttraumatic hypersomnia, 238–242
Posttraumatic pleiosomnia, 239–240
PPIs. *See* Proton-pump inhibitors (PPIs)
Pramipexole, 80, 89–90, 350
Prazosin, 91
Precipitating factors, 29, 183, 282
Predisposing factors, 29, 182, 282
Preexisting sleep disorders, 417–418
Pregabalin, 80, 356, 380–381

INDEX

453

Pressure release during exhalation, 58
Pressure support (PS), 53
Prochlorperazine, 79
Progressive muscle relaxation, 301
Proinflammatory cytokines, 206, 376
Prolonged sleep duration, 102
Prolonged-release melatonin for insomnia, 34
PROMIS-Sleep Disturbance, 209
Propofol, 419–421
Prostaglandins, 206
Prostate cancer, 176
Proton-pump inhibitors (PPIs), 380, 385
Proximal myotonic myopathy. *See* Myotonic dystrophy (DM)—DM2
PS. *See* Pressure support (PS)
PSG. *See* Polysomnography (PSG)
PSQI. *See* Pittsburgh Sleep Quality Index (PSQI)
Psychiatric conditions, children with, 399
Psychoeducation, 92
Psychological
    assessment for sleep–wake disorder, 16–20
    changes, 284
    distress, 381
    factor, 222
    interventions, 300, 357
Psychotherapy, 383–384
Psychotropic agents, 381
PTSD. *See* Post-traumatic stress disorder (PTSD)
PUD. *See* Peptic ulcer disease (PUD)
Pulmonary disorders, 122
Pulmonary function tests (PFTs), 123
Pulmonary hypertension (PH), 122, 141–142
    sleep in, 141–143
Pyrazolopyrimidines, 380–381

Q

Quality of life (QoL), 346
Quantitative sleep testing, 316
Quazepam, 266, 427
Questionnaire
    instruments, 316
    -based studies, 312
    MS, 354
    self-report, 17
Quetiapine, 426
Quinolone, 419–421

R

"R" codes, 5–10
RA. *See* Rheumatoid arthritis (RA)
Radiation therapy, 176–177, 179–180, 183
Ramelteon, 34, 302, 380–381, 427
Ramp feature, 58
Randomized controlled trials (RCTs), 190, 303, 346, 382

Raphe nucleus, 227
Rapid eye movement (REM), 83, 179, 224, 258, 280, 412, 418–419
    parasomnias, 90–92
        behavior and pharmacological treatment, 92*t*
        nightmare disorder, 91
        RBD, 90–91
        RISP, 91–92
    REM-related hypoxemia, 135
    sleep, 3–5, 4*t*, 6*t*, 122–123, 135, 373, 418
RBD. *See* REM sleep behavior disorder (RBD)
RCTs. *See* Randomized controlled trials (RCTs)
RDI. *See* Respiratory disturbance index (RDI)
Reassurance, 92
Receptor strategies, 310–311
"Recovery" behaviors, 232
Recurrent isolated sleep paralysis (RISP), 90–92
Rehabilitation, 237
Relapsing-remitting MS (RR-MS), 349
Relaxation
    relaxation-based interventions, 31–32
    techniques, 265
    training, 301
REM. *See* Rapid eye movement (REM)
REM sleep behavior disorder (RBD), 5, 15–16, 90–91, 138, 346
    MS, 352
Remeron, 381
Respiratory disturbance index (RDI), 19, 113
Respiratory physiology during normal sleep, 122–123
Respiratory suppressants, 62
Restless legs syndrome (RLS), 5, 6*t*, 16, 100, 104–105, 128–130, 136, 180–181, 184, 186, 223, 280, 312, 315, 346, 378
    in COPD, 129–130
        treatment, 130
    exacerbating factors, 78–79
    MS, 350–351
    prevalence, 78
    scale, 18*t*
    treatment, 79–81, 81*t*
Restrictive lung disease, 122
Retrograde amnesia, 223
Rheumatic diseases, 313–314
Rheumatoid arthritis (RA), 310
Rheumatoid synovium, 310–311
Rifaximin, 382
RISP. *See* Recurrent isolated sleep paralysis (RISP)
Ritanserin, 83–84
RLS. *See* Restless legs syndrome (RLS)
RLS Foundation (RLSF), 355–356
Ropinirole, 80, 350
Rotigotine, 80, 350
RR-MS. *See* Relapsing-remitting MS (RR-MS)
Ruminococcaceae, 374

454

INDEX

S

Sacroiliac joints, 311
Sarcoidosis, 12
SCD. *See* Sickle cell disease (SCD)
Scheduled awakenings, 403*t*
Scheduled naps, 82–83
SCN. *See* Suprachiasmatic nucleus (SCN)
Screening, 186
SDB. *See* Sleep-disordered breathing (SDB)
SDRB, 241–242
Sedentary behavior, 163
Selective serotonin reuptake inhibitors (SSRIs), 79,
    380–381
Selegiline, 83–84
Self-rating questionnaires, 354–355
Self-reported sleep characteristics, 263–264
Sepsis severity, 418
Sero-negative spondyloarthropathies, 311
Sero-positivity, 310
    sero-positive for disease-associated autoantibodies,
        310
Serotonergic neurons, 227
Serotonin, 260–261
Shift work type disorder, 6*t*, 14
Short sleep, 101, 155
Short-onset REM period (SOREMP), 12
Short-term HRV, 332
Short-term outcomes in CBT-I, 43–44
Sickle cell disease (SCD), 201–202, 204
Sleep, 3–5, 122–123, 154, 157–158, 166–167, 207, 222,
        238–242, 277–278, 317, 372–374
    abnormalities in patients without sleep disorders,
        328–332
        autonomic nervous system activity, 331–332
        macrostructure of sleep, 328–330
        microstructure of sleep, 330
        sleep stage dynamics, 331
    in America poll, 309–310
    architecture
        in chronic phase after TBI, 225–226
        disturbances in ILD, 135
    assessment, 207–210
    assessment tools for sleep measurement in
        hospitalized patients, 424*t*
    in CF, 132–134
    in CHC, 399–405
        assessment, 399–401
    in chronic respiratory failure due to neuromuscular
        disease, 137–139
    and circadian rhythm changes with normal aging,
        258–262, 259*f*
    comorbidities impact, 413*t*
    compression, 286

deprivation, 137, 158, 373, 414
diaries, 165–166, 187, 209, 233, 400–401, 423
difficulties assessment in cancer
    actigraphy, 187
    clinical interview, 187
    PSG, 187
    questionnaires, 187
    screening, 186
difficulties in special populations
    adult survivors of childhood cancer, 181–182
    advanced cancer and palliative care, 181
difficulties treatment in cancer
    insomnia treatment, 188–191
    sleep disorders treatment, 191
disruptions, 206–207
drunkenness. *See* Abrupt awakening
duration, 154
dysfunction, 332
effects on breathing, 123*t*
efficiency, 43–44
in hospitalized patients
    areas for future research, 429
    clinical vignette, 428–429
    environmental factors affecting, 415–417
    medications to treat sleep disruption in hospital,
        425–428
    preexisting sleep disorders and poor hospital
        sleep, 417–418
    protocols to improving sleep in hospitalized
        patients, 423–425
    risk factors for poor sleep in hospital settings,
        418–421
    sleep disruption in hospitalized patients, 412–414
    sleep measurement in hospitalized patients,
        421–423
hygiene, 92, 300–301, 403*t*
    education, 32, 301
in ILD, 134–136
immunity and inflammation in, 373
impairment, 295
    interventions to improving, 109–112
    population-based approaches, 109–110
inertia. *See* Abrupt awakening
interventions, 210–211, 285–286
    CBT-I, 285–286
    light therapy, 285
loss, 155
after lung transplantation, 143
maintenance, 296
in musculoskeletal disorders, 136–137
and pain, 205–207
    interconnection between, 205
    shared mechanisms, 205–207

paralysis, 91–92
predicting cognitive function in older people, 260
problems, 309–311
in pulmonary hypertension, 141–143
quality, 102–104, 258, 295, 312–313
  asthma and, 132
  correlates in HIV/AIDS, 298–299
restriction, 30–31, 189, 302
stage dynamics, 331
stages and general characteristics, 3–5, 4*t*
terrors, 15, 89
treatment, 210–212
window, 30–31
Sleep apnea, 5–10, 100, 241, 332. *See also* Obstructive
    sleep apnea (OSA)
  interventions for, 264
Sleep deficiency, 100, 105, 155, 202, 213
  among adults with chronic CVD, 105–107
  in chronic pain conditions, 202–204
  consequences in people with chronic CVD, 107–108
  sleep deficiency contributions to CVD development,
    100–105
Sleep disorders, 100, 105, 122, 138, 165–166, 244, 254,
    280–281, 372
  among adults with chronic CVD, 105–107
  areas for future research, 20–21
  assessment and treatment of in medical condition,
    333
  assessment in GI disorders, 379–380
  asthma and, 130–132
  in cancer, 177–181
    insomnia, 178–179
    OSA, 179–180
    PLMS/RLS, 180–181
  in CFS and FM, 328
  chronic, 417
  chronic primary, 284–285
  consequences in cancer
    insomnia, 184–185
    OSA, 185–186
    PLMS/RLS, 186
  consequences in people with chronic CVD, 107–108
  consequences of sleep disorders and abnormalities,
    332–333
  in COPD, 123–130
  diagnostic classification systems, 5–10, 6*t*
  etiology in cancer, 182–184
    insomnia, 182–183
    OSA, 183–184
    PLMS/RLS, 184
  in MS, 347
  relationship with fatigue, depression, disease course,
    and disability, 352–353

in stroke, 279–283
  SDB, 279–280
  sleep–stroke interaction, 280–281
treatment, 191
  in multiple sclerosis, 355–356
treatment in MCI and dementia
  behavioral interventions for disturbed sleep in
    dementia, 264–265
  interventions for sleep apnea, 264
  pharmacological interventions, 265–267
Sleep disturbance scale, 190–191
Sleep disturbances, 112–113, 123, 177, 186, 191, 222,
    411–412, 416–417
  assessment
    with GI disorders, 386*b*
    in patients with cancer, 194*b*
    in people with CVD, 108–109
  assessments, 165–166
  epidemiology and risk factors with GI disorders
    chronic liver disease and sleep, 378–379
    FD and sleep, 374–375
    GERD and sleep, 375
    IBD and sleep, 376
    IBS and sleep, 377–378
    PUD and sleep, 375–376
  with GI disorders, 374–379
  and HIV disease, 294–295, 298
  management and prevention in persons, 380–384
    nonpharmacologic interventions, 383–384
    pharmacotherapy, 380–383
  treatment, 166
Sleep Heart Health Study, 104
Sleep Timing Questionnaire, 165–166, 242
Sleep-disordered breathing (SDB), 106, 109, 122, 124,
    143, 202, 238–242, 258, 279–280, 328, 404, 417.
    See also Breathing-related disorders
  treatment, 111–112
  in MCI/dementia
    diagnostic considerations, 263–264
    epidemiology, 262–263
    mechanisms, 263
Sleep-enhancing agent effects on GI tract, 380–381
  melatonin, 381
  psychotropic agents, 381
Sleep-onset association (SOA), 400
Sleep-onset latency (SOL), 43–44, 393
Sleep-related breathing disorders, 5, 347–348
Sleep-related bruxism, 204
Sleep-related eating disorders (SREDs), 88–90
Sleep-related hypoventilation, 6*t*, 13–14, 51–52
  in COPD, 124–125
  disorder treatment, 66–70
    CCHS, 68–69

Sleep-related hypoventilation (*Continued*)
  obesity hypoventilation syndrome, 67–68
  due to medical disorder, neuromuscular diseases, or chest wall disorders, 69–70
Sleep-related hypoxemia, 135
  in COPD, 124–125, 125f
Sleep-related movement disorders, 5, 6t
Sleep-related respiratory changes in ILD, 134–135
Sleep-related symptoms, 332
Sleep-related ventilatory depression, 122–123
Sleep–recovery interaction model, 283–285
  insomnia in chronic phase of stroke, 284–285
  physiological changes, 283
  psychological changes, 284
Sleep–stroke interaction, 280–281
Sleep–stroke recovery model, 281–282, 282f
Sleep–wake alterations in acute phase after TBI, 224–225
Sleep–wake cycle, 84, 260
Sleep–wake disorder
  clinical features and diagnostic considerations, 10–16
  nonorganic disorder of, 5–10
  psychological, behavioral, and laboratory assessments to aid diagnosis, 16–20
  questionnaires used to assess sleep-related concerns, 18t
Sleep–wake disturbances, 222
  on evolution of condition after TBI
    cognitive functioning, 228
    functioning in chronic phase, 229
    interaction with pain, 229
    interaction with posttraumatic brain injury psychopathology, 229–230
    posttraumatic brain injury fatigue, 230–231
  after TBI, 226–227
Sleep–wake phase disorder, 242
Sleepwalking. *See* Somnambulism
Slow-wave sleep (SWS). *See* "Stage 3" sleep
Snoring, 5–10
  risk factor of stroke, 279
SOA. *See* Sleep-onset association (SOA)
Sodium oxybate, 83
SOL. *See* Sleep-onset latency (SOL)
Somnambulism, 15, 83, 88–89
SOREMP. *See* Short-onset REM period (SOREMP)
Spasticity, 240
Spectral analysis, 330
Spielman's 3P Model of insomnia adapted for Stroke, 282
Spielman's model, 29, 182
SpO₂. *See* Oxygen saturation (SpO₂)
SREDs. *See* Sleep-related eating disorders (SREDs)

SSRIs. *See* Selective serotonin reuptake inhibitors (SSRIs)
"Stage 1" sleep, 3–5, 4t
"Stage 2" sleep, 3–5, 4t
"Stage 3" sleep, 3–5, 4t, 225–226, 258–259
Stanford Sleepiness Scale, 423
Stevens–Johnson syndrome, 82–83
Stimulus control therapy, 31, 302
STOP-Bang Questionnaire, 18t, 108, 165–166, 354
Streptococcus spp., 373–374
Stroke, 100, 277–278
  clinical features, 278–279
  clinical practice considerations, 286
  incidence and prevalence, 278
  risk factors, 279
  sleep disorders in stroke, 279–283
Subjective and objective sleep measures, 379t
Substance/medication-induced sleep disorder, 5, 6t
Sundowning state, 261–262
Supplemental oxygen, 134
Suprachiasmatic nucleus (SCN), 260, 414
Surgery, 176–177
  therapy, 179–180
Sustained hypoxemia, 140–141
Symptom worsening, complaints of, 334–335
"Syndrome X", 161
Synucleinopathy diagnosis, 15–16

T

t-tau. *See* Total-tau (t-tau)
T2DM. *See* Type 2 diabetes mellitus (T2DM)
Tai Chi Chih (TCC), 111
Taiwan national database, 314
TBI. *See* Traumatic brain injury (TBI)
TCAs. *See* Tricyclic antidepressants (TCAs)
TCC. *See* Tai Chi Chih (TCC)
tcPCO₂. *See* Transcutaneous CO₂ (tcPCO₂)
Temazepam, 266, 427
Temporomandibular joint abnormality severity, 316
Tension-type headache, 204
Tentorium, 226–227
3P model. *See* Spielman's model
Tiagabine, 233
TNF-α. *See* Tumor necrosis factor-alpha (TNF-α)
Total sleep time, 43–44
Total-tau (t-tau), 263
Traditional Chinese Medicine, 37–38
Transcutaneous CO₂ (tcPCO₂), 127
Transitional sleep, 258–259
Trauma, 257
Traumatic brain injury (TBI), 222
  CRSWD, 242–243
  excessive daytime sleepiness, 238–242

INDEX

457

insomnia, 231–237
  pathophysiology of sleep–wake disturbances, 226–227
  prevalence, etiology, treatment, 223–224
  sleep architecture in chronic phase after, 225–226
  sleep–wake
    alterations in acute phase after, 224–225
    disturbances on evolution of condition, 228–231
Trazodone, 35, 233, 386b, 419–421, 425–426
Treatment Delivery Models, 44
Treatment-emergent central sleep apnea, 66
Triazolam, 266, 427
Tricyclic antidepressants (TCAs), 79, 336–337, 380–381, 419, 425–426
Tryptophan, 37
Tumor necrosis factor-alpha (TNF-α), 103, 310–311, 317, 376
Tumor-related pain, 203–204
Type 2 diabetes mellitus (T2DM), 155, 158–159

U

Ulcerative colitis, 376
Unhealthy behaviors, sleep duration and, 163
Unrefreshing sleep, 327, 332
  symptoms of, 330
Unspecified sleep–wake disorder, 6t
Unspecified type disorder, 6t
Upper airway surgery, 51–52

V

Valerian, 36
  root, 336–337

Vancomycin, 382
VAPS. *See* Volume assured pressure support (VAPS)
Vascular dementia, 254–257
Vascular insufficiency, 263
Verran and Snyder-Halpern Sleep Scale, 416–417, 423
Vibrating devices, 79
Volitional activity, 102
Volume assured pressure support (VAPS), 54–57, 55t
  for obesity hypoventilation syndrome, 67

W

Wake after sleep onset (WASO), 43–44, 296, 400
Weight loss, 61
Whipple's disease, 12
Widespread pain index (WPI), 327
Willis–Ekbom disease. *See* Restless legs syndrome (RLS)
World Health Organization, 5–10, 277–278, 378, 415

X

Xifaxan, 382

Z

Z-drugs. *See* Benzodiazepine receptor agonists (BzRAs)
"Z" codes, 5–10
Zaleplon, 33, 233, 267, 380–381, 427
Zeitgeber, 260
Zolpidem, 33, 233, 267, 336–337, 380–381, 386b, 425, 427
Zopiclone, 33, 233, 267, 380–381